Molecular Luminescence Spectroscopy

Methods and Applications: Part 2

STEPHEN G. SCHULMAN

College of Pharmacy
University of Florida
Gainesville, Florida

A WILEY-INTERSCIENCE PUBLICATION

JOHN WILEY & SONS

New York / Chichester / Brisbane / Toronto / Singapore

Library of Congress Cataloging in Publication Data:

(Revised for vol. 2)

Molecular luminescence spectroscopy.
(Chemical analysis, ISSN 0069-2883 ; v. 77)
"A Wiley-Interscience publication."
Includes bibliographies and indexes.
1. Luminescence spectroscopy. I. Schulman, Stephen G. (Stephen Gregory), 1940–

QD96.L85M65 1985 543′.085 84-21880
ISBN 0-471-86848-5 (v. 1)
ISBN 0-471-63684-3 (v. 2)

Printed in the United States of America

10 9 8 7 6 5 4 3 2 1

To the memories of
Abram Hulshoff
1943–1986
and
Jeffrey S. O'Neal
1957–1985
Scientists
Colleagues
and
Friends

CONTRIBUTORS

Harry G. Brittain
Squibb Institute for Medical Research
New Brunswick, NJ

J. N. Demas
Chemistry Department
University of Virginia
Charlottesville, VA

C. Gooijer
Department of General and Analytical Chemistry
Free University
Amsterdam, The Netherlands

J. W. Hofstraat
Department of General and Analytical Chemistry
Free University
Amsterdam, The Netherlands

Robert J. Hurtubise
Chemistry Department
University of Wyoming
Laramie, WY

Richard N. Kelly
College of Pharmacy
University of Florida
Gainesville, FL

Stephen G. Schulman
College of Pharmacy
University of Florida
Gainesville, FL

N. H. Velthorst
Department of General and Analytical Chemistry
Free University
Amsterdam, The Netherlands

Otto S. Wolfbeis
Institute of Organic Chemistry
Division of Analytical Chemistry
Karl Franzens University
Graz, Austria

PREFACE

If there was a pervasive aspect of the expositions of analytical luminescence spectroscopy in Part 1 of *Molecular Luminescence Spectroscopy*, it was that of the currency of the methodologies covered. For example, the luminescence spectroscopies of pharmaceuticals, natural products, and inorganics are practiced daily in thousands of laboratories around the world. Fluorescence detection in chromatography and luminescence immunoassay have emerged at the forefront of the analysis of real samples, especially those of biological origin.

Where, then, can analytical luminescence spectroscopy be expected to go from here? In Part 2, we attempt to answer or at least give some insight into possible answers to this question.

In recent years, analytical luminescence spectroscopists have turned their attention from the simple spectral analysis of homogeneous solutions to more elegant methodologies and more complex media. Herein we consider various aspects of the analysis of solids and solid solutions. In addition, we describe analysis based on the temporal and phase characteristics of luminescence spectra and the application of these forms of analysis to the evaluation of very fast chemical reactions. Finally, we consider the theory and applications of fluorescence optical sensors.

Some of these subjects have already found their way into analytical practice; others have yet to do so. In any event, it is the editor's opinion that these will constitute the "hot" areas of analytical luminescence spectroscopy, at least until the turn of the century. As in Part 1, the various chapters in Part 2 have been written by scientists who are at the "cutting edges" of their respective areas.

The editor would like to express his appreciation to Ms. Virginia Tomat and Ms. Vada Taylor for technical assistance with preparation of the manuscript.

STEPHEN G. SCHULMAN

Gainesville, Florida
October 1987

CONTENTS

Molecular Luminescence Spectroscopy

Methods and Applications: Part 2

CHAPTER

1

LUMINESCENCE FROM SOLID SURFACES

ROBERT J. HURTUBISE

Chemistry Department
University of Wyoming
Laramie, WY 82071

1.1. INTRODUCTION

The analytical measurement of the fluorescence or phosphorescence of components adsorbed on solid materials constitutes the general area of solid-surface luminescence analysis. A considerable number of solid materials have been used in chemical analysis. These include filter paper, silica gel, aluminum oxide, silicone rubber, sodium acetate, potassium bromide, and cellulose. Several of the arguments presented for sensitivity and selectivity in solution luminescence analysis can be applied to solid-surface luminescence analysis (1–4).

One important difference between solid-surface luminescence and solution luminescence is that in solid-surface luminescence the luminescent molecules are usually adsorbed on small particles or a solid-surface like filter paper. However, in solution luminescence, the molecules are dissolved in a solvent. The adsorbed molecules and the solid matrix will cause both the source and luminescent radiation to be scattered. The scattered source radiation and scattered luminescent radiation are reflected from the surface of the solid matrix and can also be transmitted through the solid material. Of course, for transmitted radiation, the experimental conditions have to allow for the transmission of the radiation. Wendlandt and Hecht (5) discussed the difference between specular reflection and diffuse reflection. These phenomena are important in solid-surface luminescence analysis. Specular reflection or mirror reflection is defined by Fresnel's equations and occurs from a very smooth surface. Diffuse reflection of exciting radiation results from penetration of the incident radiation into the interior of the solid substrate, and multiple scattering occurs at the boundaries of individual particles. Ideal diffuse reflection takes place when the angular distribution of the reflected radiation is independent of the angle of incidence of source radiation (6). Körtum (6) has pointed out that specular reflection and diffuse reflection are two important limiting cases, and all possible variations are found, in practice, between these two extremes. Normally, with solid-surface luminescence analysis, diffuse luminescence is measured. A fraction of the sample of interest penetrates into the solid matrix, and the sample luminescence is excited at the surface and within the solid matrix at a given depth depending on the characteristics of the solid matrix. Usually, the excited luminescence is scattered diffusely. In this chapter, reflected luminescence refers to diffusely reflected luminescence, and it appears at the same side as the excitation radiation. Generally, any specular reflection is considered insignificant. Transmitted luminescence refers to diffusely transmitted luminescence, and the luminescence appears at the unexcited surface.

Several commercial and laboratory-constructed instruments are used to measure solid-surface luminescence. Commercial instruments for solid-surface fluorescence measurements became available about 1968. Laboratory-constructed instruments and modifications to commercial instruments have appeared in the last 10 years for measuring both fluorescence and phosphorescence from solid surfaces. Hundreds of applications have appeared in the literature in areas such as environmental research, forensic science, pesticide analysis, food analysis, pharmaceutical analysis, biochemistry, medicine, and clinical chemistry. Most of the applications have been with solid-surface fluorescence. However, more and more applications with room-temperature phosphorescence (RTP) are appearing in the literature. In this chapter, emphasis will be given to recent developments in the theoretical and practical aspects of solid-surface RTP. A recent monograph details the theory, instrumentation, and applications in solid-surface fluorescence and phosphorescence (7). In addition, another recent monograph provides a basic introduction to the analytical principles and practice of RTP (8).

1.2. PRACTICAL CONSIDERATIONS

In this section, the application of the sample to the solid surface, choice of substrate, and other experimental conditions will be discussed. Several procedural details for thin-layer chromatography (TLC) with subsequent quantitation of separated components have been considered elsewhere (9,10). Many of these procedural details are applicable to solid-surface luminescence analysis. Hurtubise (7) has considered several procedural aspects in solid-surface luminescence analysis, and Vo-Dinh (8) has discussed in detail the practical aspects of RTP.

1.2.1. Application of the Sample to the Solid Surface

Two general methods have been used to deposit luminescent components on solid surfaces. Syringes or micropipets have been employed which can deliver microliter amounts of solution to a solid material such as filter paper. Also, luminescent components have been adsorbed onto powders by evaporating the solution in which the components are dissolved. For solutions deposited on flat surfaces, the size of the initial spot should be as uniform as possible. If a syringe is used, it is possible that some of the solution can "creep back" on the outside of the stem. Part of the drop can curl back around the tip of the syringe and remain after the sample is placed on the surface. This source of error can be minimized with a very fine tip or by coating the outside of the stem with silicone (11). However, generally, adequate precision can be obtained without taking special precautions. Folded nichrome wire loops were used by Samuels and Fisher (12) to apply nanoliter volumes to flat solid surfaces. Normally, 1–5 μL volumes are employed to deposit the sample on the surface in solid-surface luminescence analysis.

The technique used by von Wandruszka and Hurtubise (13) for the deposition of various samples on sodium acetate for RTP studies is typical of the approaches for dealing with powdered substrates. A 25-μL volume of ethanol was introduced into a 4 × 0.4 cm test tube from a micropipet, and then 1–6 μL volumes of standard or sample solution were added from a 10-μL Hamilton syringe. A constant amount of sodium acetate was added to each tube with a measuring spoon which had the same volume as the depressions in a special sample plate (13). The tube was placed in an oven at 80°C until the ethanol was evaporated. The dry solid was transferred quantitatively to a small mortar and pestle, which was used to gently break up conglomerate particles. The powder was then transferred to a special sample plate.

Spotting a sample on a flat surface with a microsyringe or a micropipet is more rapid than the solvent evaporation technique used for powders. However, particularly in RTP work, a powdered substrate may offer greater selectivity or have a lower luminescence background than flat surface material.

1.2.2. Choice of Substrate

If solid-surface fluorescence analysis is used in conjunction with thin-layer or paper chromatography, one is normally limited by the choice of adsorbent because the chromatographic separation is usually the overriding factor. However, if two chromatographic adsorbents give identical or similar separation of components, but one adsorbent yields greater fluorescence from the components than the other adsorbent, then the former should be chosen for the experiment. Sawicki (14) and Sawicki and Sawicki (15) have discussed several aspects of the use of thin-layer chromatography adsorbents with room-temperature fluorescence (RTF) and low-temperature phosphorescence in air pollution research. Also, Hurtubise (7) has summarized numerous examples of the use of fluorescence with thin-layer chromatography and some examples for paper chromatography. Solid-surface RTP has not yet been used extensively in thin-layer and paper chromatography. This is due to the special conditions needed to induce RTP from a solid surface. As one application, Ford and Hurtubise (16) have shown how the phthalic acid isomers could be separated by thin-layer chromatography and then detected by their RTP signals from the chromatoplate.

Only certain substrates are useful for inducing RTP from adsorbed organic compounds. Filter paper is the most widely used solid surface to date

Table 1.1. Several Solid Surfaces for Inducing RTP

Solid Surface	References
Several brands of filter paper	8, 17–19
Polyacrylic acid–treated filter paper	20
Ion-exchange filter papers	21,22
Sodium acetate	
Powder	8, 23
Peliets	8, 24
Impregnated paper	8, 24, 25
Silica gel chromatoplates with a salt of polyacrylic acid as a binder	8, 26
Polyacrylic acid–sodium chloride or sodium bromide mixtures	8, 27–30
Chalk, H_3BO_3/T-7 clay/NaOH, $CaHPO_4$/T-7 clay/cornstarch/NaOH	8, 31

for inducing RTP. Because several aspects of solid surface RTP are not understood, solid surfaces for inducing RTP are still chosen somewhat empirically. However, several solid substrates are available for RTP, which allows for the analysis of numerous organic compounds.

Vo-Dinh (8) has considered several of the solid substrates available for inducing RTP and various criteria for the selection of solid substrates. No detailed guidelines have been developed despite the importance of selecting the proper solid surface for good sensitivity and selectivity. This points out the need for more research in this particular area. It is important to consult the literature to determine if compounds of a similar nature have been investigated by RTP. Table 1.1 lists several of the solid surfaces that have been used in RTP work. Generally, a good starting surface is filter paper.

1.2.3. Other Experimental Considerations

The control of experimental conditions for inducing RTF from organic compounds adsorbed on solid surfaces is not as important as controlling the experimental conditions for solid surface RTP. This is partially due to the short lifetime of fluorescence and the relatively long lifetime for phosphorescence, although several other factors are involved. Sawicki (14) has considered several examples of earlier work for enhancing and quenching of solid-surface RTF. Hurtubise (7) has discussed a variety of the more recent experimental aspects of solid-surface RTF. It is beyond the scope of this chapter to consider in detail the experimental conditions for solid-surface RTF, and refs. 7 and 14 can be consulted by the reader. Because of the rapid growth of solid-surface RTP, a general outline of important experimental conditions for RTP will be given.

Vo-Dinh (8) has discussed in detail the variety of experimental needs for solid-surface RTP. Solvents, drying time, moisture, background luminescence, and variation of the properties of similar solid substrates are some of the more important experimental aspects. Solvents can have an important effect on RTP. For example, von Wandruszka and Hurtubise (23) found that common alcohol solvents were useful solvents for the RTP of *p*-aminobenzoic acid adsorbed on sodium acetate. However, aprotic solvents such as ether, acetone, dimethylformamide, and cyclohexane permitted no RTP from *p*-aminobenzoic acid adsorbed on sodium acetate. Ramasamy and Hurtubise (28) showed that a 0.1 *M* HBr–methanol solution of benzo[*f*]quinoline adsorbed onto 0.5% polyacrylic acid–sodium chloride yielded an RTP signal 3.5 times greater than a comparable sample adsorbed from a 0.1 *M* HBr ethanol solution. Dalterio and Hurtubise (29) used ethanol–water solutions of 4-phenylphenol to enhance the RTP of the hydroxyl aromatic adsorbed on 0.8% polyacrylic acid–sodium bromide. Some general considerations for solvents are the solubility of the luminescent component, the chemical compatibility between the solvent and the solid surface, and the purity of the solvent. Ethanol has been the most widely used solvent for RTP work (8).

Water can have a deleterious effect on the RTP of many compounds; thus it is important to dry the sample and substrate prior to the RTP measurement step. Various techniques can be used to dry the samples. For example, blowing hot air onto the sample, keeping the sample inside a desiccator, heating the sample in an oven, and drying the sample under an infrared heating lamp can be employed. Vo-Dinh (8) has given a detailed discussion of drying samples for RTP measurements. McAleese et al. (32) used sodium citrate–treated filter paper to minimize moisture an oxygen quenching of the RTP of several polar compounds.

Background luminescence from solid surfaces can cause problems in both solid-surface fluorescence and phosphorescence. Filter paper and chromatoplates can be developed in ethanol several times to move impurities to one end before spotting the sample. However, this approach is not always satisfactory. In RTP work, Lue-Yen Bower et al. (18) reported several pretreatment methods for substrates. These included baking for various periods at different oven temperatures, eluting with polar and nonpolar solvents, and irradiation with several light sources. They found no combination of substrates and pretreatments that would enhance the signal-to-noise ratio. Bateh and Winefordner (33) conducted further studies on the treatment of cellulose materials as substrates in RTP. The treatments involved soaking techniques using dioxane–water, diethylenetriamine–pentaacetic acid: water, ether, boiling water, sodium hydroxide, and periodic acid. These authors concluded that pretreatment would improve the adsorption characteristics of filter paper but that phosphorescence background probably would not be reduced substantially. Vo-Dinh (8) has discussed general background correction techniques for use in quantitative RTP, and Su et al. (34) have described a computer approach for background correction in RTP measurements. Recently, McAleese and Dunlap (35) showed that the phosphorescence background of filter paper could be reduced considerably by exposing the filter paper to 285-nm radiation or white light from a xenon lamp for several hours. Of all the approaches investigated so far to reduce background phosphorescence from filter paper, the irradiation technique by McAleese and Dunlap seems to be the most promising.

Variation among different lots of paper and chromatoplates of the same type can contribute to errors in the final results. Bateh and Winefordner (33) compared various lots of one type of filter paper. They found that the various lots were consistent in quality for use in RTP work. The relative standard deviation for the blank signal varied from 3.5 to 4.7%, and the relative standard deviation for the sample (*p*-aminobenzoic acid) changed from 2.2% to 3.5%. Ford and Hurtubise (36) showed that with 100 ng of benzo[*f*]quinoline adsorbed on three different chromatoplates, the relative RTP intensities of 62.0, 54.8, and 57.6 were obtained. To minimize the problem of variations from the same chromatoplate, reference standards and the unknown sample can be applied to the same chromatoplate. In comparing results from chromatoplate to chromatoplate, the same reference standard can be applied to the different chromatoplates and the RTP intensities normalized to the RTP intensity of a given chromatoplate. This approach assumes that instrumental conditions are the same in all measurements. As Vo-Dinh (8) has pointed out, lot-to-lot variation with filter paper is not a major problem because one RTP assay requires little material, and it takes a large number of measurements to consume one lot.

Other factors that the analyst should consider are optimal sample volume, choice of drying gas, optimal excitation and emission wavelengths, sample decomposition, and contamination related to sample holders (8).

1.3. INSTRUMENTATION

1.3.1. Commercial Instruments

Commercial instruments for quantitative thin-layer chromatography have been available since about 1968. These instruments can be used for quantitative and qualitative analysis of a variety of luminescent components adsorbed on surfaces, such as silica gel, alumina, filter paper, gels, potassium bromide, silicone rubber, and sodium acetate. Several reviews have appeared which describe self-contained units and motorized thin-film scanners as attachments to spectrofluorometers (37–40). Commercial instruments have been modified for use in solid-surface luminescence analysis, and some of these modifications will be discussed in the next section. Some of the companies that supply spectrodensitometers or scanning attachments are Carl Zeiss, Inc.; Kontes Glass Co.; Camag; Perkin-Elmer; and Shimadzu Scientific Instruments, Inc. The yearly Labguide issue of *Analytical Chemistry* and yearly Buyer's Guide edition of *American Laboratory* list several companies that manufacture luminescence instrumentation.

1.3.2. Research Instruments, Modified Instruments, and Accessories

Researchers have designed and built their own instruments and accessories. For the measurement of luminescence from solid surfaces, Hurtubise (7) has discussed instruments for both fluorescence and phosphorescence measurements, while recently Vo-Dinh (8) has considered instruments for phosphorescence measurements. This section will summarize some of the instruments designed, accessories, and innovative experimental techniques. First, equipment for fluorescence measurements will be considered.

1.3.2.1. Fluorescence Equipment

Goldman and Goodall (41) modified a Chromoscan densitometer (manufactured by Joyce Loebl & Co. Ltd., Gateshead, England) to use for absorption measurements for components separated on thin-layer chromatoplates. Even though this instrument was not designed for fluorescence measurements, it is of historical interest and the design principles could be used in the construction of an instrument for fluorescence measurements. The most important modification involved a scanning apparatus of their design that fit into the sample compartment of the Chromoscan densitometer. The scanning apparatus resulted in a sawtooth motion to the chromatoplate. This approach is called the flying-spot technique, and the sawtooth motion of the scanning device allowed for more accurate absorbance measurements. Later Goldman and Goodall (42) built a very sophisticated instrument that permitted the measurement of transmittance of components on silica gel chromatoplates in the ultraviolet region down to 240 nm. The ultraviolet transmittance at 0.5 × 0.5 mm intervals in the two dimensions of the chromatoplate was calculated and recorded on computer tape. Silica gel absorbs radiation in the region 200–280 nm, which is an important consideration in absorption analysis and luminescence analysis.

Goldman and Goodall (42) showed that the absorption by silica gel decreased the transmittance of a silica gel chromatoplate to 0.001 of that normally observed above 280 nm. The computer system employed with the Goldman and Goodall instrument allowed for two iterative procedures at each data point, aligned the data, determined the length of each spot zone, and interpolated the background transmittance in the zone. Goodall (43) modified the instrument just discussed so that transmission could be presented continuously as a negative logarithm on a flatbed recorder in the ultraviolet or visible regions. With the first version of the instrument, the signals were digitized and recorded on paper tape, and sometimes there were delays in processing the tapes. The modified instrument permitted both encoded digital tape and instantaneous graphical recordings. This feature allowed rapid review of results to decide if more accurate computer processing was required.

Instrument theory and design for quantitative absorption analysis of components on solid surfaces such as thin-layer chromatoplates have been considered extensively by Pollak and Boulton (44–51), Boulton and Pollak (52,53), and Pollak (54–59). Pollak and Boulton (50) and Pollak (57,60–62) discussed theoretically the performance of photometric methods for the quantitative evaluation of components on thin-layer chromatoplates by fluorescence. An instrument designed by Pollak and Boulton (63) and Boulton et al. (64) that can be used in the reflectance and transmittance absorption modes and the fluorescence mode will be discussed briefly. This illustrates the similarity of instrument design for absorption and luminescence measurements. The main thrust of Pollak and Boulton's research was in designing an instrument that could be used in all types of photometric determinations which are used in thin-layer chromatography and related techniques. They discussed several aspects for the optimum performance of the instrument. One aspect was a double-beam design that avoided any spatial or time separation of the beams before the beams interacted with the chromatoplate (Fig. 1.1). They adopted the principle of flying-spot scanning to compensate for nonhomogeneous distribution of the sample. The two beam signals were linearized in terms of concentration by logarithm forming for transmittance and simple inversion for reflectance measurements. The reasons for logarithm forming were twofold. First, optical noise was decreased considerably, and the signal was almost independent of the output of the light source. Second, a signal was obtained which was almost a linear function of the concentration. The previous aspect is very important for a flying-spot system to function efficiently. For solid-surface luminescence analysis, linearization is not required (50,57). Pollak (60) mentioned that the double-beam method is advantageous for fluorescence measurements, but the combination of the two beams to minimize optical noise is less

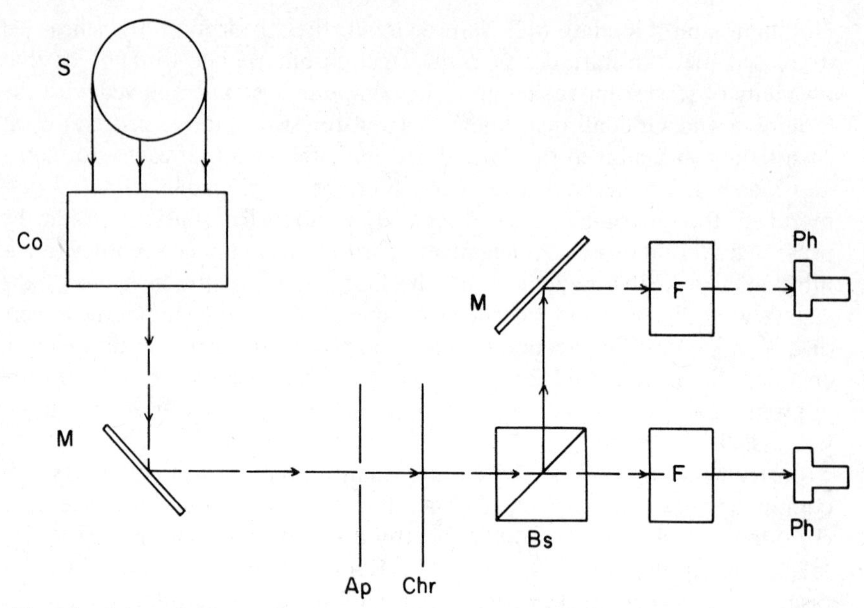

Fig. 1.1. Schematic diagram of the optical path of a double-beam scanning device. S, light source; Co, collimator; M, mirror; Ap, aperture; Chr, chromatogram; Bs, beam splitter; F, optical filter or monochromator; Ph, photodetector. (From ref. 63.)

straightforward compared to transmission or reflection absorption measurements. In addition, the theoretical principles for luminescence measurements from solid surfaces need to be further clarified for instrument design. Fluorescence was measured in the single-beam mode with Pollak and Boulton's instrument. Boulton et al. (64) reported data on the reproducibility and on the stability of their new instrument. The fluorophor, dansyl ethylamine, was separated by paper chromatography and used to evaluate the performance of the instrument. In addition, the instrument was used only in the fluorescence transmission mode. A straight-line relationship was obtained for fluorescence when log area versus log sample (nanograms) was plotted. With 500 ng the deviation from the mean was $\pm 0.07\%$, while with 10 ng the deviation from the mean was $\pm 12\%$. The instrument was capable of handling thin-layer absorbents on supports such as glass, polyethylene, and aluminum, provided that their width was 5 cm or less.

Guilbault and co-worker (65–69) designed equipment and developed methods for the assays of enzymes, substrates, activators, and inhibitors by solid-surface fluorescence analysis. The approach developed by these workers was to place an Aminco filter instrument on its side, and a cell holder was adapted to accept a metal slide on which a silicone rubber pad was

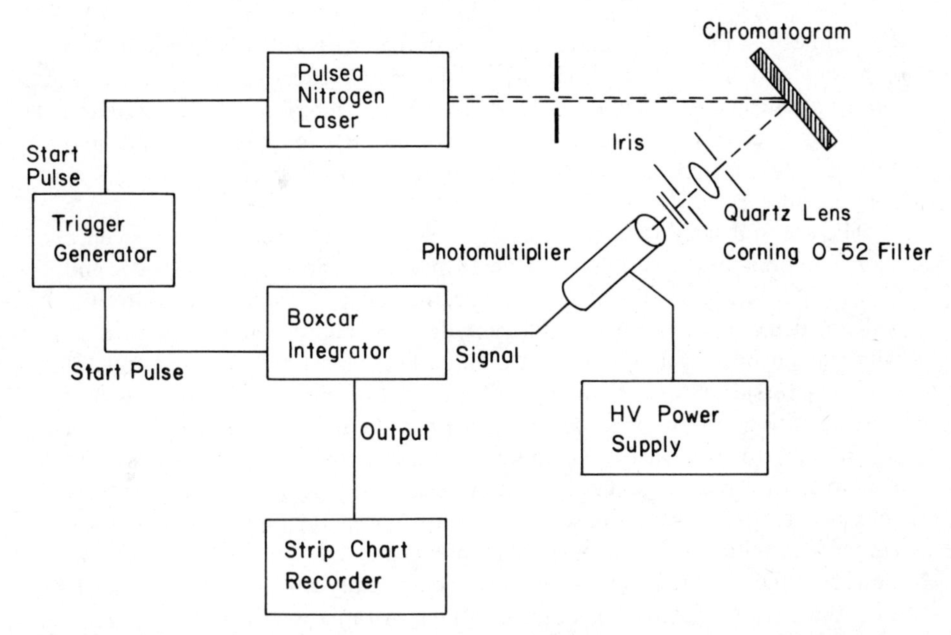

Fig. 1.2. Experimental system for laser-induced fluorescence of aflatoxins on thin-layer chromatoplates. Reprinted with permission from M. R. Berman and R. N. Zare, "Laser fluorescence analysis of chromatograms: sub-nanogram detection of aflatoxins," *Analytical Chemistry*, **47**, 1201 (1975). Copyright 1975 American Chemical Society.

placed that contained the reagents for an assay. The position of the filter fluorometer prevented any reagents from falling off the surface of the pad. The concentration of the substance to be determined was related to the change in fluorescence with respect to time. The silicone rubber pads were made by pressing uncured rubber between a glass plate and a stainless steel mold. Several steps were used in the final preparation of the pads. However, the reagent pads were simple to prepare and hundreds could be easily manufactured at one time (69).

Lasers have been used sparingly in solid-surface fluorescence analysis and certainly much potential exists for the application of lasers in this area. Berman and Zare (70) used time-resolved and wavelength-resolved laser-induced fluorescence for the analysis of aflatoxins on thin-layer chromatoplates. The experimental system is shown in Fig. 1.2. A pulsed nitrogen laser (337.1 nm) was used as the source and an RCA 7265 photomultiplier with appropriate aperatures and wavelength filters was employed for the detection of fluorescence. Berman and Zare stated that the technique of laser fluorescence analysis would have several advantages over fluorescence detection by standard techniques. For example, the laser radiation source is coherent and may be focused on the spot more effectively, the effect of phosphorescence and scattered laser radiation is minimized because gated detection electronics are employed, and the use of time-resolved and wavelength-resolved detection should allow for the analysis of mixtures of fluorescent species.

The use of a silicon vidicon detector system to detect chemiluminescence from components on thin-layer chromatoplates was developed by Curtis and Seitz (71). The emission from the chromatoplate was focused onto the vidicon detector surface, and different detector channels corresponded to different positions on the chromatoplate. The cathode-ray-tube output of intensity versus channel number gave a record of emission intensity versus plate position along one axis. A silicon vidicon detector gives several advantages for measuring chemiluminescence from chromatoplates. For example, sensitivity is generally greater because the detector simultaneously measures emission from all areas of the chromatoplate along one dimension rather than focusing at one particular spot at a given time. With chemiluminescence, the luminescence is generated chemically on the chromatoplate; thus there is no need for an external source of radiation.

Shanfield et al. (72) discovered that a wide variety of organic compounds adsorbed on silica gel chromatoplates gave intense fluorescence after a special electrical treatment. For the treatment, a piece of the plate with adsorbed compound was placed into an electrical discharge tube which was evacuated to about 0.2 torr and the system was discharged for 5 s to 5 min. After treatment in the tube, the plate was heated on a hot plate at about 130°C. The electrical discharge was produced by a Tesla coil-type high-frequency generator which was rated at a maximum of 20,000 V, 0.5 MHz. Segura and Gotto (73) used the technique described above and induced fluorescence in several organic compounds by exposing them to vapors of ammonium hydrogen sulfate.

1.3.2.2. Phosphorescence Equipment

In the area of RTP, several sample holders have been designed. Paynter et al. (17) discussed a finger-type sample holder which was placed into a standard phosphoroscope accessory of an Aminco-Bowman spectrophotometer. A modification of the previous sample holder was developed by Lue-Yen Bower and Winefordner (74). With this sample holder handling time was reduced by using individual holder tips. A similar sample holder was used by Vo-Dinh and Martinez (75), and its main advantage was that it could be inserted directly on top of the phosphoroscope assembly without any mounting or positioning procedure (8). A bar-type sample holder for multiple sampling was designed by Ward et al. (76). This holder could be inserted into the sample compartment of an Aminco–Bowman spectrofluorimeter. The bar-type device could hold four 0.6-cm-diameter filter paper disks at one time. De Lima and de M. Nicola (77) constructed a multiple-sample paper chromatography accessory for measurement of RTP. A dried chromatography Whatman No. 1 paper strip with sample on it was attached to the accessory, and a Helipot dial was used to manually locate the spots. Warren et al. (78) employed a rotating cylinder for a sample holder. The filter paper with sample was attached to the sample holder, and the sample holder was then rotated with a small dc motor at several thousand revolutions per minute. Schulman and Parker (19) used an Aminco-Keirs spectrophosphorimeter with a rotating can phosphoroscope and a sample holder similar to the one described by Paynter et al. (17). Jakovljevic (79) constructed a sample holder in which filter paper strips were inserted and used in an Aminco–Bowman spectrophotofluorometer with a cold finger phosphoroscope to obtain phosphorescence excitation and emission spectra. Fattah et al. (80) designed a sample holder for use with samples on paper at room temperature. The sample holder replaced the standard quartz Dewar assembly normally used at 77 K. Lloyd (81) described a unique packed flow-through cell for RTP measurements. A flow-through cell (i.d. 0.8 mm) was dry-packed with a mixture of lint scraped from Whatman No. 541 paper and crushed spectrosil quartz (<170 μm). The packed cell was mounted within the rotating can of a phosphoroscope assembly, and the cell was prepared for use by a special washing procedure. Using the technique reported, over 100 determinations could be obtained with a cell before replacement of the packing was necessary. Von Wandruszka and Hurtubise (13) described a blackened brass plate with holes drilled into it that was used with powdered samples and a modified spectrodensitometer. Dalterio and Hurtubise (30) obtained RTP data from powdered surfaces and filter paper with a metal sample holder that had a support bar held by two screws in the top of the holder. The height of the holder in a cell compartment could be adjusted by moving the two screws. Two circular depressions of different depths were cut into the front of the holders. One depression was for holding powders and the other for holding filter paper circles. McCall and Winefordner (82) cooled samples for low-temperature phosphorescence measurements with a conduction cooling bar. The sample holder was constructed of a copper hemispherical rod with a small hole for alignment and a large hold or well for the filter paper with sample. The temperature obtained at the sample holder was 90 K.

Commercial instruments can readily be used for RTP measurements; however, special sample holders have to be used. Many types of sample holders were discussed above. Modifications to commercial instruments can be relatively minor or very sophisticated depending on individual needs.

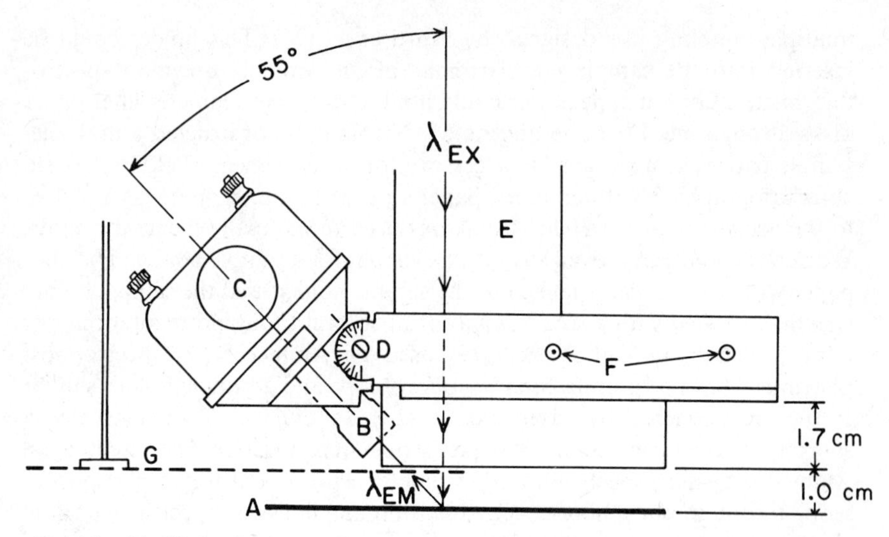

Fig. 1.3. Modified reflection mode assembly. A, RTP solid support; B, detector lens; C, photomultiplier tube housing; D, photomultiplier tube housing angle-adjustment screw; E, optical nose of spectrodensitometer; F, slide adjust screws for varying distance from exciting radiation to photomultiplier tube housing; G. rotating disk. Reprinted with permission from C. D. Ford and R. J. Hurtubise, "Design of a phosphoroscope and the examination of room temperature phosphorescence of nitrogen heterocycles," *Analytical Chemistry* **51**, 660 (1979). Copyright 1979 American Chemical Society.

Ford and Hurtubise (36) designed a phosphoroscope and reflection mode assembly for use with a Schoeffel SD 3000 spectrodensitometer to obtain RTP signals from nitrogen heterocycles adsorbed on silica gel chromatoplates and filter paper. The same instrument can be used to measure RTP from powdered samples. The reflection mode assembly is shown in Fig. 1.3. The assembly permitted the distance from the source exit to the photomultiplier tube to be varied by means of an adjustable slide. In addition, the angle of the photomultiplier tube housing could be adjusted to maximize the reflected RTP striking the lens of the photodetector system. The phosphoroscope assembly was constructed with a variable-speed dc motor (0–12 V) and a rotating disk phosphoroscope. The assembly was designed for use with the modified reflection mode unit of the spectrodensitometer. The phosphoroscope was constructed of thin sheet aluminum which was painted flat black to minimize scattered radiation (Fig. 1.4). In region x (Fig. 1.4), the compound on the solid surface was excited by the source radiation; however, the lens that collected the reflected luminescence and focused it onto the photomultiplier tube was positioned so that no luminescence was detected. As the phosphoroscope rotated region y over the adsorbed compound, the

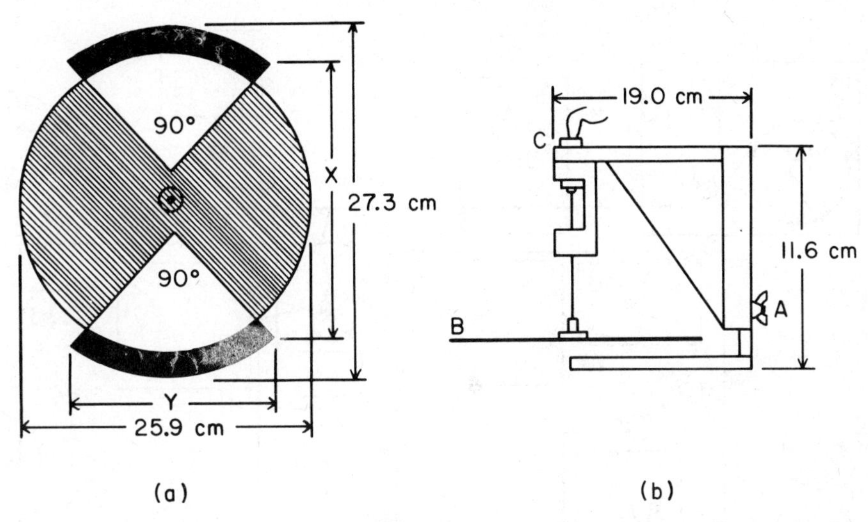

Fig. 1.4. The disk (*a*) and mount (*b*) of the phosphoroscope assembly. A, wing nut for disk height adjustment; B, rotating disk; C, motor. Reprinted with permission from C. D. Ford and R. J. Hurtubise, "Design of a phosphoroscope and the examination of room temperature phosphorescence of nitrogen heterocycles," *Analytical Chemistry* **51**, 661 (1979). Copyright 1979 American Chemical Society.

compound was no longer excited and the detector system detected any delayed luminescence (Fig. 1.4).

Vo-Dinh et al. (83) described an automatic phosphorimetric instrument for RTP measurements with a continuous filter paper device. Figure 1.5 shows a block diagram of the instrumental system. The detection unit of the system was an Aminco-Bowman spectrophotofluorometer. The sample compartment was modified and equipped with a laboratory-constructed rotating mirror assembly for detection of phosphorescence. The rotating mirror and general operating aspects are shown in Fig. 1.6. A diagonally cut section of an aluminum cylindrical rod was used for the mirror and reflecting surface. The well-polished surface permitted good reflection of ultraviolet and visible radiation. Excitation radiation from the excitation monochromator was reflected onto the surface of the filter paper by the reflecting surface (Fig. 1.6). The filter paper moved horizontally across a slit located at the top of the sample compartment. The reflection plane moved into the emission path as the cylindrical reflecting surface rotated, and the phosphorescence emitted by the sample was reflected back into the detection system. The excitation radiation was not observed by the detector during the excitation period, and scattered radiation was decreased substantially by inserting the reflecting cylindrical surface into an Aminco–Keirs phos-

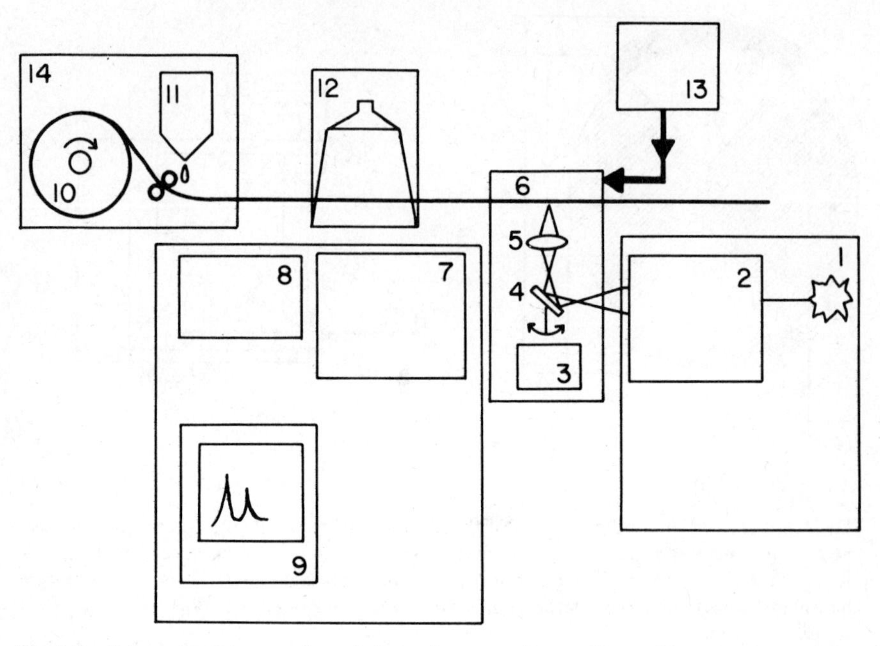

Fig. 1.5. Schematic diagram of an AutoAnalyzer continuous filter with room-temperature phosphorescence detection system. 1, light source; 2, excitation monochromator; 3, rotation motor-phosphoroscope; 4, reflecting surface; 5, optics; 6, filter paper; 7, emission monochromator; 8, detection unit; 9, recorder; 10, filter paper roll; 11, spotting syringe; 12, drying IR lamp; 13, dry-air supply; 14, AutoAnalyzer continuous filter. Reprinted with permission from T. Vo-Dinh, G. L. Walden, and J. D. Winefordner, "Instrument for the facilitation of room temperature phosphorometry with a continuous filter paper device," *Analytical Chemistry* **49**, 1127 (1977). Copyright 1977 American Chemical Society.

phoroscope attachment (Fig. 1.6). At the upper part of the sample compartment an aperature allowed a continuous warm-air flow over the filter paper. A Technicon continuous filter paper roll was used as a solid surface and was drawn into the drying chamber and the cell compartment (Fig. 1.5). The samples were spotted drop by drop with a hypodermic syringe onto the moving filter paper by manually spotting 2 μL of sample solution. After spotting, the filter paper was fed into the drying chamber where the sample remained for about 2 min. After the drying step, the paper was passed over the sample compartment of the spectrophotofluorometer (Fig. 1.5). Two important experimental variables had an influence on the RTP measured with this system. Predrying the sample before measurement and the continuous flushing of dry gas through the sample compartment during measurements were the important variables. Optimal predrying time was between 5 and 10 min for the compounds they investigated. In other work, several

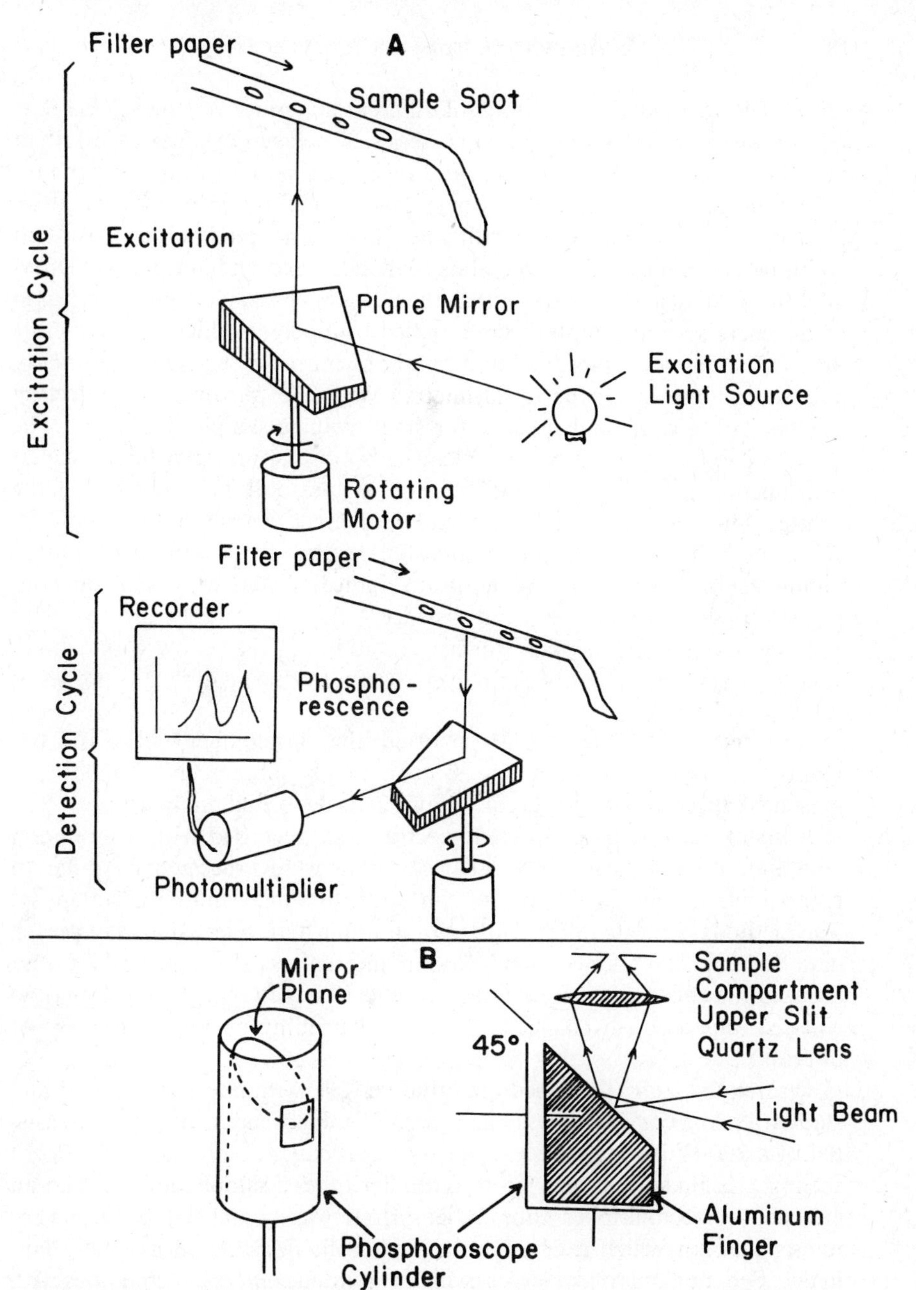

Fig. 1.6. (*a*) Principle of phosphorimetric excitation and detection with the rotating mirror phosphorimeter; (*b*) design of the rotating mirror assembly. Reprinted with permission from T. Vo-Dinh, G. L. Walden, and J. D. Winefordner, "Instrument for the facilitation of room temperature phosphorimetry with a continuous filter paper device," *Analytical Chemistry* **49**, 1127 (1977). Copyright 1977 American Chemical Society.

series of 10 to 15 identical samples of different materials were measured. The relative standard deviation for most series of measurements was less than 5%. The limit of detection was found to be in the nanogram and subnanogram ranges. Later, Lue-Yen Bower and Winefordner (84) designed an improved experimental system with a new filter paper guide which permitted continuous sampling of organic phosphors adsorbed on filter paper. Generally, the analytical data showed that the new system would be very useful in areas where several samples were handled routinely. Walden and Winefordner (85) compared ellipsoidal and parabolic mirror systems in fluorometry and room-temperature phosphorimetry. With the mirrors, the collection efficiency of sample luminescence for small-volume samples increased. The authors emphasized that one problem with most commercial luminescence instruments is that only a small fraction of the total 4π steradians of the emitted luminescence is collected and measured. An instrument with f/4 collection optics collects approximately 0.015π steradians of the total emitted luminescence. Walden and Winefordner indicated that with RTP on filter paper, emission from the front surface was considerably greater than emission levels from the back surface. In addition, they showed that high-collection-efficiency mirrors can be very useful for the analysis of samples of small area.

Goeringer and Pardue (86) reported the development of a silicon-intensified target vidicon camera system for phosphorescence studies and presented applications to RTP of organic salts deposited on filter paper. The instrument allowed time-resolved spectra to be recorded with a minimum scan time of 8 ms/scan. Spectral decay data were processed by different regression methods to obtain rate constants, lifetimes, and initial intensity. The methods permitted the handling of multiple data rates, multiwavelength data for each component, and for an internal standard procedure that reduced the effects of experimental variables. The internal standard method reduced imprecison by factors of 2 to 5, depending on the component of interest.

Warren et al. (78) discussed quantitative RTP with internal standard and standard addition techniques. The experimental measurement system consisted of a 200-W Xe-Hg source, excitation and emission monochromators, a reference beam splitter and photomultiplier tube, a sample module, and an emission photomultiplier tube. The system was controlled by a master microprocessor which ran programs written in BASIC and assembly language. Separate microprocessors, which communicated over serial lines with the master microprocessor, controlled the excitation and emission monochromators. Input and output devices included a teletypewriter for communication between the operator and the master microprocessor, a thermal plotter/printer, and a video monitor for display of experimental data and calculated results.

Su et al. (34) developed a computer-assisted background correction method for RTP. The RTP spectra were obtained by a computer-interfaced spectrometer and then retrieved and corrected against the phosphorescence background emission. The computer system was interfaced with the spectrometer by a laboratory-constructed analog-to-digital converter, a scaler, and a low-pass active filter. The spectra were printed using a digital plotter, and the programs and outputs were printed by a paper printer. The use of background correction by computer data treatment did not yield the desired improvement of the sensitivity; however, spectral quality and the precision of quantitative analysis were improved (8,34).

Su et al. (87) constructed a continuous sampling system which was suitable for RTP, low-temperature fluorescence, low-temperature phosphorescence, and room-temperature fluorescence. Samples of powders, paper disks, and liquids could be handled with the system. The sampling system essentially consisted of a circular chopper disk with 20 equally spaced concentric circular depressions. Twenty copper rings were inserted into the depressions to hold paper disks firmly in the correct position for RTP observation. To the bottom of the copper disk, a smaller-diameter concentric hollow copper cylinder was welded to facilitate the conduction of a cold stream from liquid nitrogen to the copper disk and to the sample for low-temperature studies. A laboratory-built chopper was used to block fluorescence signals from phosphorescence signals.

McAleese and Dunlap (88) investigated the problem of sample sizes larger than the excitation beam in conventional instruments. A loss in phosphorescence sensitivity can occur for samples outside the illumination area. In their work, they modified a spectrophotofluorometer and reduced the size of the filter paper so that complete front-surface illunination was accomplished. Generally improved sensitivity and pecision were observed for the RTP technique.

In addition to the instrument described by Su et al. (87) for low-temperature solid-surface phosphorescence, Gifford et al. (89) constructed and evaluated a thin-layer, single-disk, multislot phosphorimeter for the direct measurement of phosphorescence from separated components on thin-layer chromatograms near liquid-nitrogen temperature. The phosphorimeter can also be used to measure RTP. Later, the authors described the construction and evaluation of an improved thin-layer phosphorimeter (90). In a typical experiment, an aluminum-backed thin-layer chromatoplate or filter paper was affixed with elastic bands to the outside a hollow copper sample drum that could be filled with liquid nitrogen. The rate of rotation of the turntable was controlled by a variable-output transformer and a scanning rate of 3–40 cm min^{-1} was obtainable. Gifford et al. (89) discussed the ability of their phosophoroscope to resolve short- and long-lived phosphorescence. Also, they compared disks of different slot dimensions and showed under what design conditions the phosphorescence intensity became constant.

In this section a general review was given of instruments and accessories for solid-surface lumninescence analysis. Even though significant advances have been made in this area, several improvements can still be made in instrumentation. For example, location of components on a solid surface in an automatic manner, scanning in two dimensions, and digital information such as the area under curves and excitation and emission wavelengths would be very useful features (91). In addition, laser excitation has been used to a limited extent in solid-surface luminescence analysis and evaluations of laser technology is needed in this field. In solid-surface luminescence analysis and other closely related areas of phosphorimetry, Hurtubise (92) has considered new instrumental developments in solid-surface, micelle-stabilized, and solution-sensitized room-temperature phosphorescence.

1.4. THEORETICAL INTENSITY EXPRESSIONS FOR QUANTITATIVE SOLID-SURFACE LUMINESCENCE ANALYSIS

Theoretical equations have been derived for fluorescence and phosphorescence of compounds in homogeneous media which can be used in the quantitative aspects of solution luminescence analysis. These equations are readily derived from the Beer–Lambert law. When dealing with luminescence from solid surfaces, one has to consider absorption, reflection, and transmission from media that diffusely scatter radiation. Several approaches have been developed for describing the optical behavior of diffusely scattering media, and some of those approaches have been applied successfully to solid-surface luminescence analysis. However, much more research remains to be done in experimentally verifying the theoretical expressions.

1.4.1. Theoretical Models for Scattering Media

A general review of theories and equations that have been developed for diffusely scattering media as related to solid-surface luminescence analysis has been reported (8). Some of the equations have their origins from developments in the foggy atmosphere of radiation from stars. Kubelka (93) has published a system of differential equations, frequently described as the Kubelka-Munk equations, that have been used in reflectance spectroscopy and luminescence spectroscopy. Körtum (6) and Wendlandt and Hecht (5) have discussed several aspects of the Kubelka–Munk equations in reflectance spectroscopy.

At least two different models have been reported using statistical approaches in diffuse reflectance spectroscopy (8,94). One model assumes a powdered sample as a collection of plane-parallel layers whose thickness is equal to the average diameter of the sample particles. The other model assumes a collection of randomly shaped and rough-surfaced particles having random orientations compared to the medium surface (8).

Work remains to be done on the theoretical aspects of solid-surface luminescence phenomena as related to quantitative chemical analysis. Very little experimental data have been published to substantiate recent theoretical equations which describe luminescence from compounds adsorbed on solid surfaces.

1.4.2. Theoretical Models for Solid-Surface Luminescence Analysis

Zweidinger and Winefordner (95) derived theoretical intensity expressions for low-temperature phosphorescence for clear rigid solvents, cracked glasses, and snowed matrices. Experimental verification was given for the equations. The intensity expressions have not been applied directly to solid-surface lumninescence analysis; however, with some modifications they should prove useful. The phosphorescence intensity expressions were derived from the theoretical Kubelka–Munk equations and include both absorption and scattering coefficients which are important parameters in describing variations in diffusely scattering media.

Because almost all surfaces used in solid-surface luminescence analysis are highly scattering media, both exciting radiation and luminescence radiation are scattered extensively. In addition, the compound adsorbed on the surface absorbs a fraction of the exciting radiation, and it is possible for the solid surface to absorb exciting radiation. Many reports have appeared on the relationship of the Kubelka–Munk theory to thin-layer densitometry as applied in the transmission-absorption and reflection-absorption modes (46,58,96–99). The concepts in these reports are important because they emphasize various aspects of the Kubelka–Munk theory in densitometry and several of the concepts are applicable to solid-surface luminescence analysis.

Goldman and Goodall (96), Goldman (100), and Pollak and Boulton (50) considered the assumptions of the Kubelka–Munk theory related to solid-surface absorption and fluorescence analysis. An extensive discussion of these assumptions applied to reflected radiation can be found in textbooks on reflectance spectroscopy (5,6). Several of the assumptions of the Kubelka–Munk theory which apply to solid-surface luminescence are listed below.

1. Source or luminescence radiation within the medium propagates only in the forward and background directions perpendicular to the plane-parallel boundary surfaces of the medium (50).

2. Once the radiation reaches the surface, the radiation is scattered in all directions (50).
3. The factors that define the optical response of the medium are k, the absorption coefficient of exciting radiation; s, the scattering coefficient of fluorescent radiation; and x, the thickness of the scattering medium (50,93,100).
4. The direction of the exciting radiation is perpendicular to the surface (50).
5. The medium is assumed to be homogeneous between the boundary surfaces. Theoretically, this means that the absorption coefficient and scattering coefficients are independent of the thickness of the medium (50).

1.4.3. Goldman's Equations

Goldman and Goodall (41,42,96) discussed the densitometric application of the absorptiometry theory of Kubelka and Munk. In fluorescence densitometry, Goldman (100) considered both excitation radiation and fluorescent radiation in scattering media and obtained two pairs of differential equations. One pair corresponded to absorption of exciting radiation in transmitted and reflected directions and the other pair corresponded to fluorescent radiation emitted in the transmitted and reflected directions. With additional mathematical manipulation of the two pairs of differential equations, Goldman obtained two general equations that were very complex. One equation defined fluorescence transmitted through a surface such as a thin-layer chromatoplate, and the other equation represented fluorescence reflected from the surface. Goldman's complex equations can be simplified considerably. For very low-level fluorescence and relatively large values for scattering coefficients for both exciting radiation and fluorescent radiation, Goldman reported the following equations:

$$\frac{I^+}{i_0\alpha} = \frac{1}{3}\,kx\left(1 - \frac{7}{30}\,sx\cdot kx\right) \tag{1.1}$$

$$\frac{J^+}{i_0\alpha} = \frac{2}{3}\,kx\left(1 - \frac{4}{30}\,sx\cdot kx\right) \tag{1.2}$$

where I^+ is the intensity of transmitted fluorescence, i_0 the intensity of initial exciting radiation, α the portion of absorbed radiation converted into fluorescence, k the absorption coefficient of exciting radiation, s the scattering coefficient of exciting radiation, x the thickness of the scattering medium, and J^+ the intensity of reflected fluorescence. The scattering coefficient of fluorescence radiation does not appear in Eqs. (1.1) and (1.2) because of mathematical approximations. The term kx is proportional to the amount of absorbing compound in the exciting beam.

When Goldman derived his fundamental equations, he assumed that the solid surface did not absorb exciting radiation. It was shown by Goodall (101) that common thin-layer chromatographic adsorbents such as silica gel and aluminum oxide absorb radiation in the approximate range 225–340 nm. Goldman's equations have to be modified for solid surfaces that absorb exciting radiation. This has been discussed by Goldman (100), and he showed theoretically that both transmitted fluorescence and reflected fuorescence can yield linear relationships with the amount of fluorescent material adsorbed on the solid surface.

Goldman (100) has considered other theoretical conclusions. For example, at small amounts of fluorescent components, the best analytical mode is the direct measurement of reflected fluorescence if the solid surface is nonabsorbing. If the solid surface absorbs and the fluorescence is at a low level, direct measurement can be performed; however, layer thickness must be considered because the reflected fluorescence is inversely proportional to layer thickness. The transmission mode can be used effectively in situations where the medium absorbs exciting radiation by adding a fixed amount of some inert fluorescent substance to the solid medium. Goldman showed that the amount of fluorescent substance is proportional to $(I^* - I_0^*)/(I_0^*)^2$, which is the linear in I^*. The term I^* is the fluorescence intensity of the fluorescent component of interest, and I_0^* is the background fluorescence of the inert fluorescent material. Measurement of transmitted fluorescence is possible, and the background fluorescence level can be employed to correct the sample readings for layer thickness variation with the expression $(I^* - I_0^*)/(I_0^*)^2$.

1.4.4. Experimental Data Related to Goldman's Equations

Experimental data were obtained by Hurtubise (102) from thin-layer chromatoplates with fluoranthene as a model compound to give experimental support for the simplified Goldman equations (1.1) and (1.2). A spectrodensitometer and aluminum oxide (Al_2O_3) and silica gel (SiO_2) glass-backed chromatoplates with an *n*-hexane mobile phase were used in the experimental work. Experimental values of sx were obtained with the following equation, $T_0 = 1/(sx + 1)$, where T_0 is the transmission of the chromatoplate when kx is zero (6). T_0 was calculated from the equation $A_0 = \log 1/T_0$, and A_0 was determined experimentally. The kx values for fluoranthene were calculated by successive approximations using the general Kubelka–Munk equation for transmission in scattering media (6,96). The transmission equation is

$$T = \frac{b}{a \sinh bsx + b \cosh bsx} \tag{1.3}$$

where T is transmittance of a chromatoplate with an adsorbing material in the scattering medium,

$$a = \frac{s + k}{s} = \frac{sx + kx}{sx} \qquad b = (a^2 - 1)^{1/2}$$

Ideally, it would be advantageous to express kx simply in terms of T and sx; however, this is very complicated theoretically. In the work by Hurtubise (102), T was calculated from the equation $A = \log 1/T$, where A is equal to A_0 plus the experimental absorbance for fluoranthene. In the final calculation step, kx was calculated by successive approximations. It was assumed that the chromatographic material did not adsorb exciting radiation because Goodall (101) showed that silica gel and aluminum oxide absorb very weakly around 340 nm. An excitation wavelength of 370 nm was used by Hurtubise (102). The $I^+/i_0\alpha$ values were calculated from Eqs. (1.1) and (1.2) once sx and kx were obtained. With $I^+/i_0\alpha$ and the corresponding kx values, theoretical calibration curves were plotted for reflected and transmitted fluorescence. In addition, experimental fluorescence calibration curves were obtained for fluoranthene in the reflection and the transmission modes and several comparisons were made with the theoretical curves (102).

The microgram values at which the theoretical curves and experimental curves approximately changed slopes were compared. Data were obtained for fluoranthene on developed and undeveloped aluminum oxide and silica gel chromatoplates to determine if chromatographic conditions had any influence on the reflected and transmitted fluorescence. The data in Table 1.2 show reasonably good agreement between theoretical and experimental values for both developed and undeveloped chromatoplates, whether the fluorescence was measured in the transmission or reflection modes. These results indicate that Goldman's simplified equations can be used to predict approximately when a calibration curve first changes slope. Goldman (100) reported that the range of linearity of reflected fluorescence should be about twice that obtained by transmission in terms of kx. However, the data in Table 1.2 do not support this conclusion. The data indicate that the ranges of linearity of reflected and transmitted fluorescence are relatively close.

Based on Goldman's simplified equations, the theoretical ratio of the

Table 1.2. Comparison of Theoretical and Experimental Approximate Points of First Slope Change in Terms of Microgram of Fluoranthene

Developed Al_2O_3			Undeveloped Al_2O_3			Developed SiO_2			Undeveloped SiO_2		
Plate	Theor.	Exptl.	Plate	Theor.	Exptl.	Plate	Theor.	Exptl.	Plate	Theor.	Exptl.
Transmitted Fluorescence											
1	0.12	0.17	3	0.15	0.17	5	0.16	0.18	7	0.15	0.18
2	0.12	0.18	4	0.11	0.15	6	0.16	0.20	8	0.18	0.18
Reflected Fluorescence											
1	0.12	0.14	3	0.21	0.15	5	0.16	0.22	7	0.19	0.13
2	0.13	0.15	4	0.16	0.14	6	0.14	0.15	8	0.18	0.16

Reprinted with permission from R. J. Hurtubise, "Comparison of experimental and theoretical calibration curves in solid-surface fluorescence analysis," *Analytical Chemistry*, **49**, 2162 (1977). Copyright 1977 American Chemical Society.

Table 1.3. Experimental Ratios of Slopes of Calibration Curves of Transmitted Fluorescence to Reflected Fluorescence

Plate	Developed Chromatoplate	Plate	Undeveloped Chromatoplate
	Al_2O_3		
1	0.39	3	0.26
2	0.36	4	0.51
	SiO_2		
5	0.62	7	0.91
6	0.58	8	0.89

Reprinted with permission from R. J. Hurtubise, "Comparison of experimental and theoretical calibration curves in solid-surface fluorescence analysis," *Analytical Chemistry*, **49**, 2162 (1977).

slopes of a linear fluorescence calibration curve measured by transmission compared to a linear fluorescence calibration curve measured by reflection is 0.50, namely, $\frac{1}{3}\,kx/\frac{2}{3}\,kx$. Table 1.3 compares the slopes of experimental calibration curves. The ratio of the slopes for calibration curves obtained for the developed aluminum oxide and silica gel chromatoplates compares reasonably well with the theoretical ratio. As indicated in Table 1.3, the experimental ratios for the developed aluminum oxide chromatoplates are lower than the theoretical ratio of 0.50. The experimental ratios for the developed silica gel chromatoplates are higher than the theoretical ratio. This suggests that the reflected fluorescence from the aluminum oxide chromatoplates is relatively greater than the reflected fluorescence from the silica gel chromatoplate. The experimental ratio for the undeveloped aluminum oxide plate 3 was 0.26. The small value suggested that fluoranthene did not penetrate the aluminum oxide chromatoplate greatly because with a ratio of 0.26 the reflected fluorescence was substantially greater than the transmitted fluorescence. The ratios for the undeveloped silica gel plates in Table 1.3 are close to 1. This implies that the fluorescence signals measured in the reflected and transmitted modes were approximately the same. It was discovered that with the undeveloped silica gel chromatoplates, the fluoranthene fluorescence at a given amount was substantially greater than its fluorescence on the developed silica gel chromatoplates. This indicated that the fluorescence of fluoranthrene was enhanced on undeveloped silica gel chromatoplates by an adsorption mechanism, or by mechanisms different from that affecting fluoranthene on developed chromatoplates. Apparently, the fluorescence intensity of fluoranthene was great enough that the layer thickness of the

Table 1.4. kx^a and sx Values

Plate	Transmission, kx	Reflection, kx	sx	Transmission, 7/30 $sx \cdot kx$	Reflection, 4/30 $sx \cdot kx$
1	0.080	0.066	8.1	0.15	0.071
2	0.086	0.01	8.6	0.17	0.081
3	0.058	0.047	16	0.22	0.10
4	0.0014	0.0013	43	0.14	0.073
5	0.047	0.058	2.9	0.032	0.022
6	0.041	0.031	6.6	0.063	0.027
7	0.048	0.034	4.9	0.055	0.022
8	0.056	0.050	4.4	0.058	0.029

Reprinted with permission from R. J. Hurtubise, "Comparison of experimental and theoretical calibration curves in solid-surface fluorescence analysis," *Analytical Chemistry*, **49**, 2163 (1977). Copyright 1977 American Chemical Society.

[a] The kx values correspond to the microgram values in Table 1.2.

chromatoplates did not substantially affect the fluorescence intensity. The data indicated that the fluorescence intensity in the reflection and transmission modes is independent of thickness, over a range of layer thicknesses, for strongly fluorescent components. Goldman's simplified equations apparently cannot be used to predict that ratio of relative fluorescence intensity in the transmitted and reflected modes for highly fluorescent components over various ranges of layer thicknesses.

Table 1.4 shows data for kx and sx values for fluoranthene for the conditions described in Table 1.2. The last two columns of Table 1.4 show that the $\frac{7}{30}$ $sx \cdot kx$ and $\frac{4}{30}$ $sx \cdot kx$ products for SiO_2 chromatoplates are smaller than the corresponding products for the Al_2O_3 chromatoplates. These products appear in Eqs. (1.1) and (1.2) and are significant theoretically in determining the points at which a fluorescence transmission calibration curve and a fluorescence reflection calibration curve become nonlinear. The experimental results showed that generally, the linear range for silica gel chromatoplates, in terms of micrograms (or kx), was greater than the linear range for aluminum oxide chromatoplates (102). However, the greater linear range for SiO_2 was not substantial. With the Kubelka–Munk theory, uniform particle size of the scattering medium is assumed. The commercial chromatoplates used in Hurtubise's work probably had a range of particle sizes. Körtum (6) considered the dependence of scattering coefficient on particle size and states it is nearly inversely proportional to the average particle size. The aluminum oxide chromatoplates in Table 1.4 had sx values greater than the silica gel chromatoplates. For both the silica gel and aluminum oxide chromatoplates, the sx values were substantially greater than the kx values. The data indicated that the average particle size for the silica gel chromatoplates was larger than the average particle size for the aluminum oxide chromatoplates. Particle size is a major factor contributing to the smaller $\frac{7}{30}$ $sx \cdot kx$ and $\frac{4}{30}$ $sx \cdot kx$ products for silica gel chromatoplates.

Prosek et al. (103) discussed the advantages of the combination of fluorescence scanning from the far side and near side of chromatoplates for simultaneous measurement of fluorescence in quantitative analysis. In their work, they used the unmodified Goldman equations (100) and showed that experimental calibration curves gave very good fits to the corresponding theoretical calibration curves for transmitted fluorescence, reflected fluorescence, and simultaneous measurement of transmitted and reflected fluorescence. Generally, the simultaneous fluorescence measurements were the most sensitive, and measurement of the reflected fluorescence was more sensitive than transmitted fluorescence. No detailed comparison of experimental and theoretical results was given. However, the results of Prosek et al. (103) are in general agreement with Goldman's theory.

1.4.5. Pollak and Boulton's Equations

A theory for fluorescence from thin-layer chromatoplates based on the Kubelka–Munk theory and an electrical transmission-line model was developed by Pollak and Boulton (50) and Pollak (57,60). Pollak and Boulton (50) used the assumptions of the Kubelka–Munk theory and considered both excitation and fluorescence radiation in scattering media. Pollak and Boulton considered the effect of fluorescence on the propagation of exciting radiation as equivalent to an increase in absorbance, namely, an additional energy loss of the exciting radiation (50). In addition, they reported a differential equation that was essentially identical to an equation describing an electrical transmission line with purely resistive parameters. The electrical transmission line was used as a model to simulate the optical behavior of turbid media in both the absorption and fluorescence modes (46,50). Pollak (56) has discussed the transmission-line model in detail.

Pollak (57,60) reported two complex equations based on the Kubelka–Munk theory and the electrical transmission-line model. The equations described the fluorescence reaching the surface on the same side as the exciting radiation (refection mode), and the fluorescence reaching the side opposite the exciting radiation (transmission mode). He discussed simplified equations which are useful in practical laboratory situations. Media with relatively large absorbance and strong scattering are frequently encountered in solid-surface fluorescence analysis. Pollak (57,60) presented two theoretical equations which can be useful under the conditions of reflected and transmitted fluoescence. It was assumed that the coefficient of fluorescence was relatively small in both cases. The coefficient of fluorescence was defined as the ratio of the intensity of fluorescence excited at any particular point in the interior of the medium divided by the total energy density of the exciting radiation at that point. Eqs. (1.4) and (1.5) define transmitted fluorescence and reflected fluorescence, respectively (57,60). The layer thickness x does not appear in Eqs. (1.4) and (1.5) because it was normalized to 1.

$$I_{FT} = \frac{Fc}{2} \exp(-\gamma_F)(1 - Q_E^2)(1 + Q_F)E_0 \quad (1.4)$$

$$I_{FR} = \frac{Fc}{2} \frac{(1 - Q_E^2)(1 + Q_F)}{\gamma_E + \gamma_F} E_0 \quad (1.5)$$

where the indices E and F refer to excitation radiation and fluorescent radiation, respectively, I_{FT} and I_{FR} are the intensities of the fluorescent radiation emitted in the transmission mode and reflection mode, respectively, c is the amount of fluorescent substance in the illuminated area, E_0 is the intensity of the exciting beam, F is the coefficient of fluorescence of the component of interest, Q is the coefficient of reflectance of a sheet of medium thick enough so that its transmission can be disregarded, and γ is the natural logarithm of the transmittance of a very thin sheet of medium, which has negligible reflectance. Pollak (60) stated that Q^2 is very small, so that chance variation of the term can be neglected. The influence of the terms containing Q was incorporated into a constant correction factor b. Thus Eqs. (1.4) and (1.5) reduced to Eqs. (1.6) and (1.7) (60).

$$\frac{I_{FT}}{E_0} = \frac{Fcb}{2} \exp(-\gamma_F) \quad (1.6)$$

$$\frac{I_{FR}}{E_0} = \frac{Fcb}{2} \frac{1}{\gamma_E + \gamma_F} \quad (1.7)$$

It can be seen in Eq. (1.6) that I_{FT}/E_0 is a function of γ_F, but in Eq. (1.7), I_{FR}/E_0 is a function of γ_E and γ_F. It was concluded by Pollak (60) that the linear range for reflected fluorescence was more limited than the linear range for transmitted fluorescence because of the influence of γ_E. Pollak (57) also concluded that the reflected fluorescence was expected to be greater than the transmitted fluorescence.

1.4.6. General Comparison of Goldman's Equations and Pollak and Boulton's Equations

It would be of considerable interest at this point to make a detailed comparison of the theoretical equations developed by Goldman, and Pollak and Boulton. However, these researchers gave no experimental data to support their theoretical conclusions, and only a small amount of theoretical data was presented by the authors. It appears that the only experimental data obtained to support a theory of solid-surface luminescence analysis was obtained by Hurtubise (102) and by Prosek et al. (103). They both considered Goldman's equations. As considered earlier, Zweidinger and Winefordner (95) derived equations for phosphorescence using the Kubelka–Munk theory and gave supporting theoretical and experimental data which related phosphorescence intensity of species in optically inhomogeneous matrices to various parameters. They discussed mainly snow matrices or densely cracked glasses that were formed at liquid-nitrogen temperature. With some modifications these equations should prove to be applicable to solid-surface luminescence analysis.

Goldman's equations are more readily related to optical theories because they were derived directly from the Kubelka–Munk theory. Pollak and Boulton used the Kubelka–Munk theory, but in addition, they used the properties of an electrical transmission line as a model to simulate the behavior of radiation in scattering media. More work is needed both theoretically and experimentally before a firm conclusion can be made about which approach is more useful in solid-surface luminescence analysis. For example, Goldman (100) predicted that at small amounts of fluorescer the linear range for reflected fluorescence would be approximately twice that of transmission in terms of kx. However, Pollak (60) predicted that the linear range for reflected fluorescence would be more limited than that for transmitted fluorescence because of the influence of γ_E. It seems obvious that there is a need to obtain fluorescence data for several chromatographic systems so adequate comparisons can be made between Goldman's equations and Pollak and Boulton's equations. In addition, little work has appeared that describes theoretical calibration curves for RTP from solid surfaces.

1.5. THEORETICAL ASPECTS IN SOLID-SURFACE LUMINESCENCE INSTRUMENTATION

Many of the instrumental concepts developed for solution luminescence analysis are applicable to solid-surface luminescence analysis (1). Because highly scattering media are used in luminescence from solid surfaces, there are additional instrumental aspects to consider, and instruments designed for solution luminescence are not necessarily the best for solid-surface luminescence analysis.

Several papers have been published which consider theories for the absorption of radiation by components adsorbed on scattering media

(41,42,96, 104). Many of the concepts related to the theories can be applied to instrument design in solid-surface luminescence analysis. Pollak (104) has stated, however, that arguments in favor of flying-spot scanning do not always apply to fluorescence measurements. This results because the fluorescence radiation reaching the detector is a linear function of the amount of substance for analysis at small amounts of the material (50,57). Goldman (100) reached similar theoretical conclusions for small amounts of fluorescent material. This is in contrast to absorption methods in which the photodetector output signal is, in general, a nonlinear function of the spatial concentration of the substance.

For direct fluorescence measurements from a scattering medium, the fluorescing substance stands out as a bright zone on a dark background. Theoretically, the dark background yields almost an ideally flat baseline. However, baseline problems can be obtained from electrical noise and residual fluorescence from impurities in the scattering medium. By integrating the output signal over a long period of time, electrical noise can be minimized (60,104). Because the residual fluorescence from impurities is practically invariant, it would not be diminished by integration. Pollak (60) discussed two integration approaches for fluorescent signals of very low intensities. These were integration on photographic film and photon counting with subsequent integration. Neither approach has found extensive application in luminescence from solid surfaces. Of the two approaches, photon counting with subsequent integration is more appealing because of the difficulty in processing photographic film. With photon counting techniques, it should be possible to obtain detection limits of luminescent components to much lower ranges than can be achieved at present (60).

Direct fluorescence measurements can give nearly an ideally flat baseline; however, optical noise can be a problem. Optical noise is caused by random fluctuations of the optical transfer in the scattering medium. In general, optical noise is the main factor which limits performance in solid-surface luminescence analysis. Electrical noise is important at very low light levels and is generated in the photodetector and preamplifier stages. Pollak (60,104) has discussed several sources of optical noise and these are outlined briefly below.

1. Any specular component of the scattered radiation at the surface of medium varies randomly, and as a result, the intensity of the light entering the medium exhibits random fluctuations from point to point. These fluctuations appear as optical noise in both the transmission and reflection modes.
2. In the reflection mode, it is possible for part of the specularly reflected luminescence to reach the photodetector and cause optical noise. This can be reduced by careful optical design.

3. One serious source of optical noise is caused by local variations in the thickness of the medium.
4. Density fluctuations of the medium and nonuniform particle size can cause optical noise.
5. Special treatment of the medium, such as with chemical reagents, can cause nonrandom changes in optical parameters.
6. Optical noise can also originate in the source radiation through lamp instability.

Theoretically, the signal-to-noise ratio has a definite value which does not depend on the coefficient of fluorescence (F) or the amount of fluorescent component (60) [see Eqs. (1.4 and 1.5)]. Pollak (60) estimated the signal-to-noise ratio for transmitted fluorescence and reflected fluorescence. The relationships are given in Eqs. (1.8) and (1.9).

$$\left(\frac{S}{N}\right)_{FT} \simeq \frac{1}{\delta_\gamma} \tag{1.8}$$

$$\left(\frac{S}{N}\right)_{FR} \simeq \frac{\gamma}{\delta_\gamma} \tag{1.9}$$

The optical noise fluctuations are incorporated in the coefficients γ which were defined with Eqs. (1.4) and (1.5). The term δ_γ represents the rms value of the fluctuations of the γ terms. The optical noise fluctuations are contained in the coefficients γ which were defined with Eqs. (1.4) and (1.5). Pollak assumed that γ_E and γ_F were about equal [see Eqs. (1.7)]. Eqs. (1.8) and (1.9) suggest that the signal-to-noise ratios are both constant and independent of the amount of fluorescer. However, the output signal of the photodetector is still dependent on the stability of the excitation source.

Eqs. (1.4) and (1.5) indicate that increasing the intensity of the excitation radiation will increase the fluorescence intensity of the adsorbed fluorescent component. Nevertheless, this does not mean that the sensitivity will increase, because there may be a corresponding increase in impurity fluorescence from the scattering media. However, the increased output of fluorescent radiation can mask electrical noise and improve sensitivity and accuracy (60).

In addition, Pollak has discussed the effect of nonuniform concentration distribution with depth in a solid surface (61) and the relationship between the dimensions of the scattering medium and sensitivity (62). In the papers published by Pollak, some useful and interesting theoretical ideas for several instrumental aspects of solid-surface luminescence analysis have been considered. However, Pollak presented no experimental data in support of the conclusions. Thus there is a need for experimental proof to substantiate the theoretical, instrumental concepts. Also, because RTP has been analytically useful for several years, more attention to the theoretical and practical aspects of instrumentation is needed. Two areas that should be investigated for both fluorescence and phosphorescence are excitation of the sample and detection of the emitted radiation. Tunable lasers should be beneficial for excitation of the sample, and detector systems with high collection efficiency are important for highly scattering media. Goeringer and Pardue (86) have used a silicon-intensified target (SIT) vidicon system for optical detection in RTP work. Vidicon detector systems are capable of integrating radiation intensity over a period of time because of their charge storage capability, and they are also multichannel detectors.

1.6. PHYSICOCHEMICAL INTERACTIONS IN SOLID-SURFACE LUMINESCENCE

Specific interactions between compounds and solid surfaces that cause fluorescence and phosphorescence remain relatively unexplored. However, several of the conditions needed for enhanced fluorescence and phosphorescence from organic compounds adsorbed on a variety of surfaces have been published. In this section, a general survey of the physicochemical interactions in solid-surface fluorescence and solid-surface phosphorescence will be given with emphasis on solid-surface phosphorescence.

1.6.1. Solid-Surface Fluorescence

Nicholls and Leermakers (105) considered the photochemical and spectroscopic properties of organic compounds adsorbed on surfaces such as silica gel. They discussed various aspects of the spectra of molecules adsorbed on silica gel. In addition, Weis et al. (106) investigated the excimer fluorescence of pyrene and other compounds adsorbed on silica gel. They were interested in using excimer formation as a method for confirming the behavior and orientation of molecules interacting with silica gel surfaces.

Sawicki (14) considered several aspects of the fluorescence of organic compounds adsorbed on filter paper and thin-layer chromatoplates. He discussed the fluorescence of compounds in the wet and dry state on surfaces, excimers, charge-transfer fluorescence, sensitized fluorescence, and photodecomposition. Sawicki used this information for fluorescence analysis in air pollution research. Recently, Hurtubise (7) has summarized Sawicki's work.

Lloyd (107) showed that a neutral dichloromethane solution of acridine injected into a flow-through cell packed with silica gel gave intensified fluorescence. However, the excitation spectrum was essentially the same as that of the neutral molecule, but the emission spectrum was of the protonated form. He concluded that the intensification of fluorescence was due to the protonation of the excited acridine molecules by the acidic silica gel surface. Benzo[*f*]quinoline behaved in the same way as acridine. The excited states of benzo[*h*]quinoline and benz[*c*]acridine were protonated to a reduced extent by the silica gel; thus the fluorescence emission of both protonated and unprotonated forms appeared in the spectra. Lloyd used the experimental information in designing a detector for high-performance liquid chromatography.

Hurtubise (108) obtained excitation and emission spectra from acridine adsorbed on two different brands of silica gel and 30% acetylated cellulose chromatoplates spotted from ethanol and ethanolic 0.1 *M* hydrochloric acid. For both brands of silica gel and the 30% acetylated cellulose chromatoplates, the excitation and emission wavelengths were essentially the same. These data suggested that the same form of acridine was adsorbed on those chromatoplates from either neutral or acidic solution. Excitation and emission spectra were obtained from acridine in neutral and acidic ethanol solutions and the fluorescence wavelength maxima were compared to the fluorescence wavelength maxima obtained from the silica gel and 30% acetylated cellulose chromatoplates. Comparison of wavelengths obtained from solution with the wavelengths obtained from the chromatoplates indicated that protonated acridine was responsible for the fluorescence from the chromatoplates. As discussed above, Lloyd (107) showed that acridine dissolved in 1,2-dichloromethane and passed over silica gel was protonated in the excited state by silica gel, and the protonated acridine emitted fluorescence. The data obtained by Hurtubise also indicated that the protonated form of acridine was emitting fluorescence from both silica gel and acetylated cellulose. However, the chromatoplates used were dried and then the emission data were obtained. Thus it appears that the silica gel and 30% acetylated cellulose can protonate acridine in the "dry" state. With Lloyd's system, the silica gel was "wet" with 1,2-dichloromethane.

Bauer et al. (109) studied the effects of coadsorbed molecules on the fluorescence of pyrene and the dimerization of acenaphthylene adsorbed on silica gel. Alkanols, acenaphthene, glycerol, malonic acid, and 1-adamantanol were used as coadsorbates. The stronger adsorbates apparently displaced pyrene from preferred adsorption sites, reduced the concentration of ground-state associated pairs, and allowed the observation of the "growing in" of pyrene excimer emission.

Lochmuller et al. (110–113) investigated chemically modified silica surfaces with fluorescence spectroscopy. Aminated microparticulate silica gel was derivatized with the fluorescent tag, dansyl chloride, to give a chemically modified silica surface that was investigated by fluorescence spectroscopy (110). In other work it was shown that the fluorescence maxima at room temperature of dansylamide groups covalently attached to the surface of silica particles in dry form and slurried in acetonitrile showed varying degrees of dependence upon excitation wavelength (111). Lochmuller et al. (112) studied the fluorescence of pyrene silane molecules chemically bonded to microparticulate gel at several surface concentrations to determine the proximity and distribution of chemically bound molecules on the native silica gel. Also, a study of the time-dependent luminescence of pyrene silane molecules chemically bound to microparticulate silica was undertaken to assess the distribution of molecules chemically bound to silica and their organization in contact with different solvents (113). The results showed that the organization and proximity of such molecules were controlled mainly by an inhomogeneous distribution of chemically reactive silanol groups on the surface.

In general, the photophysics of molecules adsorbed on surfaces has received relatively little attention (109). In the area of solid-surface fluorescence analysis, there is a need for rather detailed investigations of surface-fluorophor interactions to improve the sensitivity, selectivity, and reproducibility for this area of trace organic analysis. Also, it is important to develop new surfaces that induce strong fluorescence and have minimum fluorescence background.

1.6.2. Solid-Surface Phosphorescence

Several materials have been discovered that induce RTP from adsorbed organic compounds. Little is understood about the specific chemical and physical interactions that cause RTP from organic compounds. In this section, a summary will be given of the more important interactions that have been proposed that yield RTP.

1.6.2.1. Sodium Acetate

Von Wandruszka and Hurtubise (13,23) first introduced the use of sodium acetate as a material that would stimulate RTP from certain organic compounds. In the investigation of the interactions of compounds adsorbed on sodium acetate, comparison of molecular structures and consideration of reflectance, fluorescence, and infrared spectra were used in addition to surface-area data, solvent considerations, and various molecular criteria (23).

Table 1.5 compares the phosphorescence intensities of some of the compounds investigated (23). Ethanol solutions of the compounds were used

Table 1.5. Phosphorescence Intensities of Compounds Adsorbed on Sodium Acetate[a]

Compound	77 K	Room Temp.	Room Temp./77 K
p-Aminobenzoic acid	1000	333	0.333
p-Hydroxybenzoic acid	1040	48	0.046
3-Methyl-4-aminobenzoic acid	232	29	0.125
N,N-Dimethyl-4-aminobenzoic acid	260	42	0.16
Benzocaine	146	—	—
Terephthalic acid	238	0.9	0.0038
Hydroquinone	38	6.8	0.18
Folic acid	132	43	0.32
p-Hydroxymandelic acid	56	6.5	0.116
p-Aminohippuric acid	1200	262	0.218
5-Hydroxyindoleacetic acid	200	32	0.160
5-Hydroxytryptophan	210	39	0.186

Reprinted from R. M. A. Von Wandruska and R. J. Hurtubise, "Room-temperature phosphorescence of compounds adsorbed on sodium acetate," *Analytical Chemistry*, **49**, 2165 (1977). Copyright 1977 American Chemical Society.

[a] 77-K phosphorescence of *p*-aminobenzoic acid arbitrarily set at 1000 units. All compounds 500 ng per 10 mg of NaOAc.

to deposit the compounds on the sodium acetate. The data in Table 1.5 imply that the differences in RTP intensities for a given compound are not due mainly to inherent molecular effects, but rather to differences in rigidity of the absorbed compound. *p*-Aminobenzoic acid appeared to be adsorbed most strongly because it gave the largest room-temperature ratio (Table 1.5). It was found that certain molecular requirements were needed to obtain strong RTP signals. For example, 3-methyl-4-aminobenzoic acid gave a 11.5-fold reduction in RTP compared to *p*-aminobenzoic acid. From the compounds they investigated, it was concluded that the presence of a carboxyl group bonded to the 1 position was one requirement for compounds with the benzene nucleus. Also, attached to the 4-position on the ring, an electron-donating, hydrogen-bonding substituent appeared to be necessary. Two compounds with the indole nucleus showed RTP on sodium acetate, 5-hydroxyindoleacetic acid, and 5-hydroxytryptophan. Analytically useful RTP signals were obtained only when alkaline ethanolic solutions of the compounds were evaporated onto sodium acetate. They also found that the solvent used in adsorbing the compounds on sodium acetate was important in obtaining RTP, but water gave a threefold reduction in RTP for *p*-aminobenzoic acid. Aprotic solvents such as ether, acetone, dimethylformamide, and cyclohexane gave no RTP from *p*-aminobenzoic acid on sodium acetate.

Von Wandruszka and Hurtubise (23) indicated that the adsorption of *p*-aminobenzoic acid on sodium acetate was preceded by partial neutralization with dissolved sodium acetate in alcoholic solutions. The *p*-aminobenzoic acid anion formed had a strong tendency to adsorb on the surface of sodium acetate, forming the sodium salt. This conclusion was supported by the strong RTP of the sodium salt of *p*-aminobenzoic acid when it adsorbed on suspended sodium acetate in an ethanolic solution of the sodium salt of *p*-aminobenzoic acid. Under similar experimental conditions, no RTP signals were observed for *p*-aminobenzoic acid and the sodium salt of *p*-aminobenzoic acid dissolved in acetone or dimethylformamide. Recent work in this laboratory showed that sodium acetate is insoluble in acetone and ether. As discussed, no RTP was obtained from *p*-aminobenzoic acid when these solvents were used to adsorb *p*-aminobenzoic acid onto sodium acetate. Most likely, no RTP was observed with these solvents because partial neutralization of *p*-aminobenzoic acid by reaction with sodium acetate could not occur in solution. For the ethanol solutions of 5-hydroxyindoleacetic acid and 5-hydroxytryptophan, the dissolved sodium acetate did not give the required neutralization of the solutes because the indole compounds are very weak acids. Thus it was necessary to add sodium hydroxide to ethanol solutions of these compounds.

Von Wandruszka and Hurtubise (23) investigated the mode of adsorption of *p*-aminobenzoic acid and other compounds on sodium acetate, talc, and starch using reflectance spectroscopy. The compounds studied did not give RTP on talc and starch. Figure 1.7 shows the reflectance spectra of *p*-aminobenzoic acid adsorbed on sodium acetate, starch, and talc. As shown, there was a blue shift of about 35 nm in the reflectance maximum of *p*-aminobenzoic acid on sodium acetate compared to *p*-aminobenzoic acid on talc or starch. The reflectance spectral results indicated strong interactions between sodium acetate and adsorbed *p*-aminobenzoic acid, suggesting the formation of the sodium salt of *p*-aminobenzoic acid upon adsorption. Other supporting reflectance and fluorescence data were also published (23).

Infrared spectroscopy was also used to study the adsorption of *p*-aminobenzoic acid on sodium acetate (23). Infrared spectra of *p*-aminobenzoic acid, of sodium acetate, and of *p*-aminobenzoic acid adsorbed on sodium acetate were obtained. Strong sodium acetate infrared bands obscured much of the adsorbate spectra, but several observations and conclusions were made. The N–H stretching vibrations of *p*-aminobenzoic acid at 3350–3450 cm^{-1} disappeared for *p*-aminobenzoic acid adsorbed on sodium acetate. This result showed that the bands were shifted to longer wavelengths and broadened due to hydrogen bonding between the amino group of *p*-aminobenzoic acid and the carboxyl group of the sodium acetate surface. *o*-Aminobenzoic acid did not give RTP when adsorbed on sodium

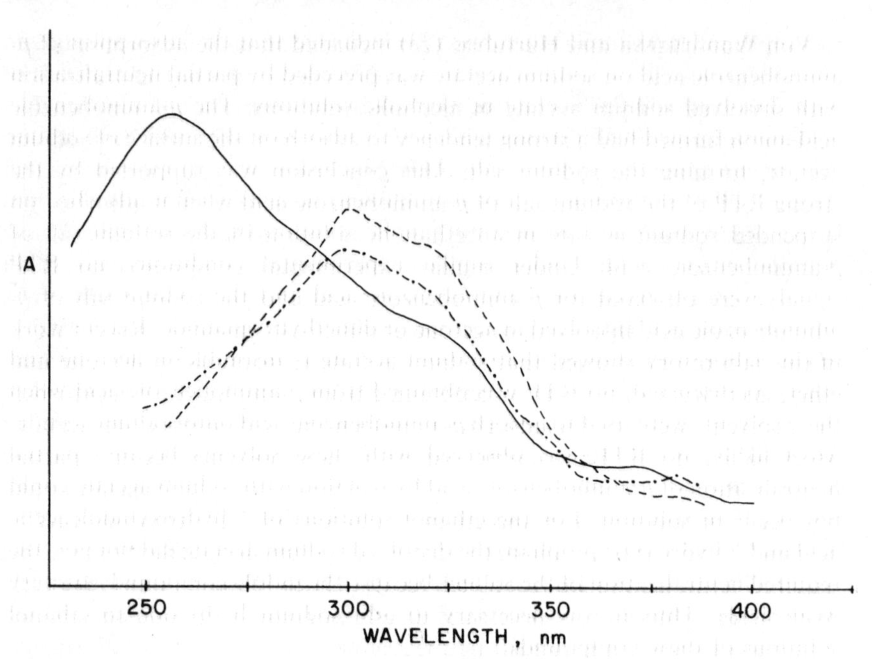

Fig. 1.7. Reflectance spectra of *p*-aminobenzoic acid on sodium acetate (———), starch (– – – –), and talc (–·–·–). Reprinted with permission from R. M. A. von Wandruszka and R. J. Hurtubise, "Room-temperature phosphorescence of compounds adsorbed on sodium acetate," *Analytical Chemistry* **49**, 2167 (1977), Copyright 1977 American Chemical Society.

acetate. Similar infrared experiments were carried out with *o*-aminobenzoic acid adsorbed on sodium acetate, and the *o*-aminobenzoic acid retained its N–H infrared bands. Because of the intramolecular hydrogen bond between the amino group and the carboxyl group in *o*-aminobenzoic acid, it appeared that this compound was not held to the surface by strong intermolecular hydrogen bonding.

Von Wandruszka and Hurtubise (23) assumed that chemisorbed *p*-aminobenzoic acid molecules on the sodium acetate were distinguished from physically adsorbed molecules by their RTP behavior. Only those molecules that strongly and directly interacted with the sodium acetate were held rigidly enough to yield RTP. They postulated that *p*-aminobenzoic acid molecules in the second and subsequent adsorbed layers would not phosphoresce, but would decrease the signal by absorbing exciting radiation and possibly absorbing emitted phosphorescence from the chemisorbed layer. It was determined that the maximum RTP signal was obtained at 6100 ng of *p*-aminobenzoic acid on 10 mg of sodium acetate. They assumed the maximum RTP signal to correspond to complete monolayer coverage of the sodium acetate. Von Wandruszka and Hurtubise (23) used a method developed by Snyder (114) to calculate the surface area occupied by a flatly adsorbed *p*-aminobenzoic acid molecule. They postulated *p*-aminobenzoic acid was adsorbed flatly, based on spectral and other data that they obtained. Then they calculated the surface area of sodium acetate to be 1.8 m^2/g. The previous surface area value was identical to the surface area of sodium acetate obtained from a commercial source. The results they obtained indicated that *p*-aminobenzoic acid was adsorbed flatly on sodium acetate. Additional calculations showed that two sodium acetate molecules were needed to hold one *p*-aminobenzoic acid molecule. A similar investigation was carried out with 5-hydroxyindoleacetic acid, and it was found that three sodium acetate molecules were needed to hold one 5-hydroxyindoleacetic acid molecule.

1.6.2.2. Silica Gel

Silica gel has been shown to be useful for inducing RTP from organic compounds under certain conditions (16,26,36,108,115–120). Ford and Hurtubise (26) tested several brands of silica gel for their ability to yield RTP from benzo[*f*]quinoline. Table 1.6 lists the brands of silica gel investigated. As indicated in Table 1.6, acidic ethanol solutions of benzo[*f*]quinoline adsorbed on some brands of silica gel gave strong RTP signals from benzo[*f*]quinoline. However, benzo[*f*]quinoline exhibited moderate RTP to no RTP on other brands and types of silica gel when ethanol or acidic ethanol solutions were used. Ford and Hurtubise (26) carried out experiments to determine to what extent the RTP enhancement could be attributed to adsorbing benzo[*f*]quinoline on EM silica gel chromatoplates in the protonated form. The hydrochloride of the compound was prepared by passing HCl gas through an ether solution of benzo[*f*]quinoline. The product of the reaction was shown to be the hydrochloride by solution fluorescence spectroscopy. A comparison was made of the RTP relative intensity values for equimolar amounts of benzo[*f*]quinoline and the hydorchloride of benzo[*f*]quinoline spotted from ethanol, and benzo[*f*]quinoline spotted from 0.1 *M* HCl in ethanol onto an EM chromatoplate. The relative RTP signals were in the order 1.0, 1.03, and 11.6, respectively. These data indicated that the RTP enhancement was more than the result of adsorbing the protonated species on the chromatoplate. The HCl probably interacted with the chromatoplate in some manner which allowed strong adsorbate–solid surface interactions. Diffuse reflectance spectra of benzo[*f*]quinoline spotted from ethanol, the hydrochloride of benzo[*f*]quinoline spotted from ethanol, and benzo[*f*]quinoline spotted from 0.1 *M* HCl ethanol solution were obtained from these samples on MN

Table 1.6. Silica Gel Brands Tested as RTP Supports for B[f]Q

Brand	Description	RTP (Neutral)[a]	RTP (Acid)[b]
EM[c]	Aluminum backed TLC chromatoplate	Moderate	Strong
EM	Glass-backed	Moderate	Strong
EM	Plastic-backed TLC chromatoplate	Moderate	Strong
EM	Glass-backed (HPTLC) chromatoplate	Moderate	Strong
EM	Plastic-backed N-HR (TLC) chromatoplate	Moderate	Strong
Brinkmann	Plastic-backed Sil-G (TLC) chromatoplate	None	None
Brinkmann	Glass-backed (TLC) chromatoplate	None	Moderate
S & S[d] Applied Science Labs	Glass-backed Permakotes I (TLC) chromatoplate	None	Weak
EM	Silica Gel 40, column chromatography	None	None
EM	Silica Gel 60, column chromatography	None	None
EM	Silica Gel 100, Column chromatography	None	None
MN[e]	Silica Gel 60, column chromatography	None	Weak
MN	Kieselgel 60 for TLC chromatography	None	Weak
Mallinckrodt	SilicAR TLC-7G chromatography	None	None

Reprinted with permission from Charles D. Ford and Robert J. Hurtubise, "Room temperature phosphorescence of nitrogen heterocycles adsorbed on silica gel," *Analytical Chemistry*, **52**, 657 (1980). Copyright 1980 American Chemical Society.

[a] EM Laboratories.

[b] B[f]Q spotted from 0.1 *M* HCl ethanol.

[c] EM Laboratories.

[d] Schleicher & Schuell.

[e] Macherey, Nagel & Co.

silica gel for column chromatography, an EM silica gel chromatoplate, and a Brinkmann N-HR silica gel chromatoplate. The reflectance spectra indicated that with the neutral benzo[*f*]quinoline sample the neutral form was on the surface of the silica gel samples. For the hydrochloride sample, the spectra also indicated that the neutral form was present. Examination of the ultraveiolet absorption spectra of the neutral compound and hydrochloride in ethanol showed the hydrochloride spectrum to be very similar to the spectrum for the neutral compound. This indicated that the hydrochloride was in equilibrium with a relatively large fraction of the neutral counterpart in ethanol solution. This is reasonable because benzo[*f*]quinoline has a pK_a of 4.75 in a water–ethanol solution (121). All the reflectance spectra for the benzo[*f*]quinoline samples spotted from 0.1 *M* HCl solution showed that the protonated form of the compound was adsorbed on silica gel.

In other experiments, EM silica gel chromatoplates were pretreated with hydrochloride acid by soaking the chromatoplates for 10 s in an acidic water solution. The acid-treated chromatoplates were dried 0.5 h at 110°C prior to use as RTP supports. With the acid-pretreated chromatoplates, enhanced RTP was observed from neutral benzo[*f*]quinoline spotted from ethanol solution. Because the acid studies indicated a change in the silica gel chromatoplates with acid treatment, infrared spectroscopy was used to examine silica gel samples after ethanol or acid treatment. Figure 1.8*a* shows the infrared spectra of column chromatography MN silica gel. The band at about 1870 cm^{-1} is an overtone band of silica gel and the band at 1630 cm^{-1} is attributed to a water deformation band (122). As indicated in Fig. 1.8*a*, there is essentially no difference in the infrared spectra of the acid-treated silica sample compared to the ethanol-treated silica sample. Figure 1.8*b* gives the infrared spectra of EM silica gel samples scraped from an aluminum-backed chromatoplate. The ethanol-treated silica gel sample exhibited board bands at 1560 cm^{-1} and 1720 cm^{-1} in addition to the 1870 cm^{-1} band and an ill-defined band at 1630 cm^{-1}. Upon acid treatment, the 1560 cm^{-1} band disappeared and the 1720 cm^{-1} band became more prominent. The changes in the infrared spectrum with acid treatment were consistent with carboxylate anions being converted to carboxylic acid groups (123). After consulting an EM patent, the patent indicated that the sodium salt of polyacrylic acid is used as a binder in the manufacture of silica gel chromatoplates (124).

The infrared spectra discussed above showed that the form of the binder EM used changed upon acid treatment (Fig. 1.8*b*). In addition, strong RTP was observed only when the chromatoplate was treated with acid. For these reasons polyacrylic acid was investigated as a possible material for inducing RTP. Polyacrylic acid was mixed with column chromatography MN silica gel 60 to give mixtures containing varying percentages of polyacrylic acid. Benzo[*f*]quinoline showed a weak RTP when adsorbed on only MN silica

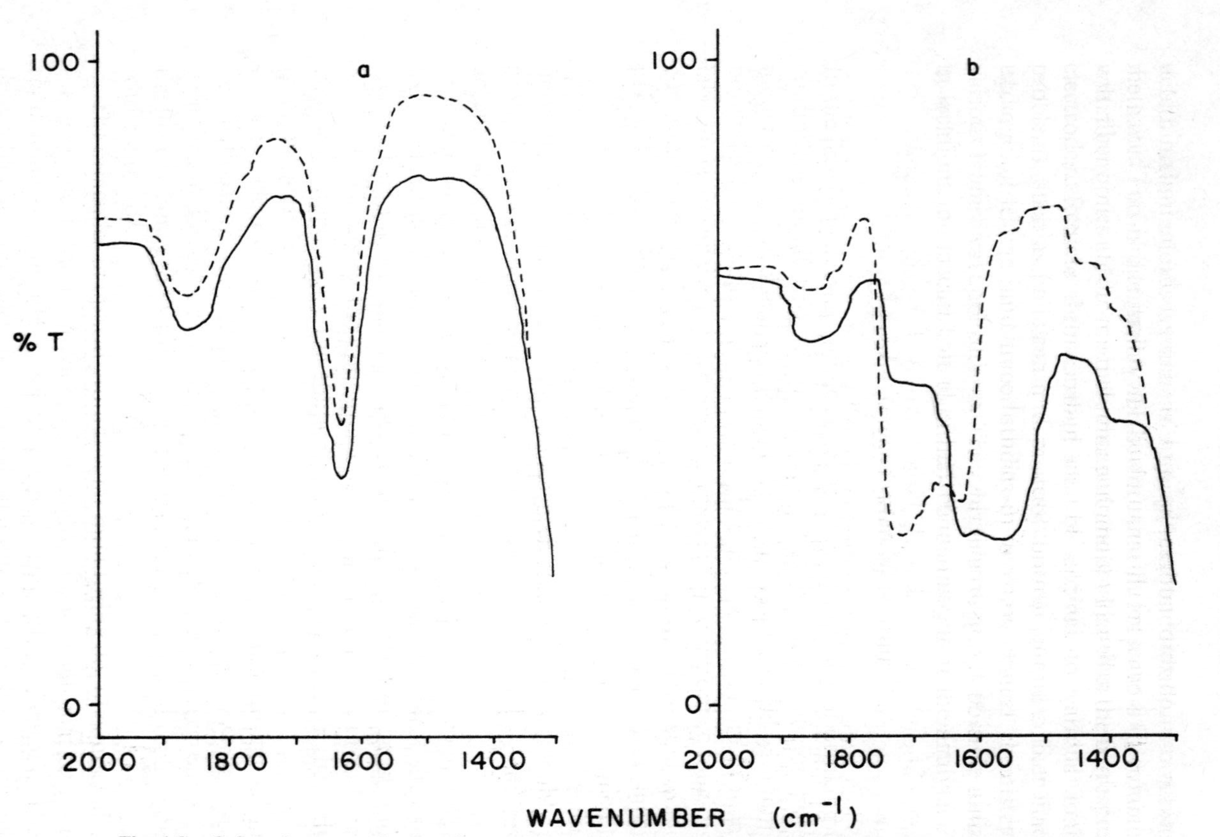

Fig. 1.8. Infrared spectra of ethanol-treated (———), and 0.1 *M* HCl ethanol-treated (----) silica gel sample: (*a*) MN silica gel for column chromatography; (*b*) EM silica gel from a chromatoplate. Reprinted with permission from C. D. Ford and R. J. Hurtubise, "Room temperature phosphorescence of nitrogen heterocycles adsorbed on silica gel," *Analytical Chemistry* **52**, 660 (1980). Copyright 1980 American Chemical Society.

gel (Table 1.6). A fixed amount of benzo[*f*]quinoline from a 0.1 *M* HCl ethanol solution was adsorbed onto several mixtures of MN silica gel and polyacrylic acid with increasing amounts of polyacrylic acid. The RTP signals increased almost linearly to 10% polyacrylic acid and decreased above 20% polyacrylic acid. Also, 100% polyacrylic acid yielded only weak RTP from benzo[*f*]quinoline. Polyacrylic acid–sodium chloride mixtures were also investigated with benzo[*f*]quinoline and similar RTP results were obtained, except that RTP signals were somewhat greater. Deanin (125) reported that modulus in polymers increases when the polymer is mixed with an inorganic powder of higher modulus. Generally modulus measures the resistance to deformation of materials to external forces (126). The polymer is restricted in its ability to rotate and migrate by the powder in the mixture. With polyacrylic acid and silica gel or NaCl mixtures, it appears the silica gel and NaCl serve to restrict the general movement of the polymer matrix, thereby giving the matrix greater rigidity and permitting greater RTP signal intensities from adsorbed compounds.

Ford and Hurtubise (26) proposed that the protonated form of benzo[*f*]quinoline was adsorbed flatly on pure silica gel without binder and that hydrogen bonding between the protonated form and silanol groups was sufficient to hold the molecule rigidly enough so weak RTP could be observed. When acidic polymers or their salts, such as polyacrylic acid, are used as binders for commercial silica gel chromatoplates, other adsorbate–solid surface interactions are implicated. The results from acid studies, luminescence, reflectance, infrared, and binder studies showed that enhanced RTP signals were obtained only when the binder was in the acidic form (26). The presence of carboxyl groups dispersed throughout the silica gel provided sites for strong hydrogen-bonding interaction. Infrared results for the EM chromatoplates showed that the amount of binder was about 5% by weight (26).

Any mechanism of interaction between protonated benzo[*f*]quinoline and EM silica chromatoplates, which contain an acidic polymer as a binder, must account for possible interactions between the adsorbed compound and the binder. The adsorbate solid-surface interaction would involve hydrogen bonding between the π-electron system of the adsorbate and the carboxyl groups of the binder. Because the carboxyl group in the polymer is more strongly acidic than the silanol group, the carboxyl group would be expected to form a stronger hydrogen bond with the π-ring system of protonated benzo[*f*]quinoline and thus hold the molecule more rigidly to the surface. In addition, the NH^+ moiety could form a hydrogen bond with the carbonyl oxygen of the carboxyl group. In other experiments, phenanthrene showed a moderate RTP on EM silica gel chromatoplates when spotted from 0.1 *M* HCl ethanol solution. The results indicated that the excess acid converted carboxylate groups to carboxyl groups, which then interacted with the π-electrons of phenanthrene to hold the compound rigidly enough to obtain RTP (26). Other interactions or conditions are probably responsible for RTP from benzo[*f*]quinoline, and additional research is needed to clarify these aspects.

Ford and Hurtubise (16) earlier proposed that the main interaction causing RTP from terephthalic acid adsorbed on EM silica gel chromatoplates was hydrogen bonding between the surface silanol hydroxyl groups and the hydroxy and carbonyl groups of terephthalic acid. However, because it was found that a salt of polyacrylic acid was present in EM chromatoplates and polyacrylic acid was largely responsible for inducing RTP from benzo[*f*]quinoline, it was important to reexamine the previously reported mechanism for terephthalic acid. Hurtubise and Smith (119) adsorbed neutral, basic (0.1 *M*), and acidic (0.1 *M*) terephthalic acid and coumarin-3-carboxylic acid solutions on Brinkmann Sil-G chromatoplates and Grace column-chromatography silica gel. The Brinkmann Sil-G chromatoplates were known to contain a binder other than polyacrylate (26), and the Grace silica gel contained no binder. The Grace silica gel would induce an RTP signal if the silanol groups were responsible for inducing RTP. Experiments showed that little or no RTP was obtained from terephthalic acid or coumarin-3-carboxylic acid, adsorbed on either support under all conditions tested. The lack of RTP from the two carboxylic acids on the Grace silica gel showed that hydrogen bonding of the carboxyl groups with silanol hydroxyl groups is not the mode of interaction that induces RTP. The results with the Brinkmann Sil-G chromatoplates showed that the combination of binder in the chromatoplates and the silica gel yielded little RTP. Additional experiments with polyacrylic acid–sodium chloride mixtures yielded relatively strong RTP from terephthalic acid. These experiments strongly indicated that the salt of polyacrylic acid in the EM chromatoplates was responsible for inducing RTP from adsorbed terephthalic acid (119). In general, it has been found that silica gel does not give strong RTP from the adsorbed organic compounds which have been investigated. Additional work is needed to determine if other compounds will yield strong RTP from silica gel.

1.6.2.3. Polyacrylic Acid

As discussed in the preceding section, it was determined that polyacrylic acid mixed with silica gel or sodium chloride induced RTP from various organic compounds. Several aspects of the conditions and the interactions needed for RTP from compounds adsorbed on polyacrylic acid salt mixtures have been reported (27,29,30,119,120). Also, Dalterio and Hurtubise (27) investigated several polymers containing polar functional groups as surfaces for inducing

Table 1.7. Relative Intensities (RI) for Terephthalic Acid Anion Absorbed on Polyacrylic Acid–NaCl Mixtures[a]

Polyacrylic Acid–NaCl	RI	Sodium Salt of Polyacrylic Acid–NaCl	RI
Acid	1.3	Acid	1.0
Dianion	2.8	Dianion	2.4

Source: From Hurtubise and Smith (119).

[a] 2 μg of terephthalic acid adsorbed from ethanol–water (1 + 1) solution.

RTP from hydroxyl aromatics. In no case did a polymer alone induce RTP from the adsorbed compound. The polymers examined had to be mixed with an inorganic salt for them to be useful materials for inducing RTP. It was found that polyacrylic acid (secondary standard, mol wt 2,000,000)–salt mixture induced the strongest RTP from the model compounds investigated. RTP analytical data were reported for nine compounds using 1% polyacrylic acid–sodium bromide mixtures (27). The main repeating group in polyacrylic acid is

$$-CH_2-\underset{|}{\overset{\overset{\displaystyle H}{|}}{C}}-COOH$$

Hurtubise and Smith (119) investigated two polyacrylic acid–NaCl mixtures, one containing unneutralized and one containing neutralized polyacrylic acid for their potential to induce RTP from terephthalic acid and the dianion of terephthalic acid. The data in Table 1.7 indicate several modes of interaction of terephthalic acid and its dianion with polyacrylic acid, and with the sodium salt of polyacrylic acid. Terephthalic acid probably forms hydrogen bonds with polyacrylic acid both in the initial "wet" state and final dry state. The dianion of terephthalic acid should react to some extent with polyacrylic acid in the wet state to form terephthalic acid. Polyacrylic acid behaves like an aliphatic carboxylic acid and can be titrated with a solution of sodium hydroxide. It is possible, depending on the extent of the reaction of the dianion with polyacrylic acid, that a mixture of the monoanion, dianion, and the terephthalic acid would remain on the dried surface with various combinations of hydrogen bonds being formed. Terephthalic acid added to the sodium salt of polyacrylic acid could react with this sodium salt to form the monoanion and/or dianion in the wet state. In the dry state, there could be a mixture of the monoanion, dianion, and terephthalic acid with various combinations of hydrogen bonds. For the final situation (Table 1.7) in which the dianion was added to the sodium salt of polyacrylic acid, a relatively high RTP signal was obtained. This is somewhat surprising because the dianion would not give an acid–base reaction with the sodium salt of polyacrylic acid in the wet state and no hydrogen bonds could be formed. The results in Table 1.7 do not allow a simple interaction mechanism to be presented for all the conditions studied. While hydrogen bonding appears to be the important interaction holding the compound rigid for neutral terephthalic acid adsorbed on polyacrylic acid–NaCl, it cannot explain the dianion adsorbed on the sodium salt of polyacrylic acid–NaCl mixture. Niday and Seybold (127) postulated that various salts or sugars packed into filter paper could inhibit internal molecular motions of the phosphorescent compound and thus enhance RTP. This may be one consideration with the terephthalate dianion adsorbed on the sodium salt of the polyacrylic acid–NaCl mixture. However, in other experiments terephthalic acid gave essentially no RTP when adsorbed on sodium acetate. Sodium acetate has been used to induce RTP from certain compounds as discussed in Section 1.6.2.1. If a simple "matrix packing" mechanism were occurring with sodium acetate by protecting terephthalic acid from collision with oxygen, then RTP should have been observed. More work is needed to elucidate the interactions and conditions for RTP from terephthalic acid.

Ramasamy and Hurtubise (28) studied the RTP of benzo[*f*]quinoline and nitrogen heterocycles with polyacrylic acid–salt mixtures under a variety of conditions. Because polyacrylic acid–sodium chloride mixtures induced RTP from several classes of compounds, various aliphatic carboxylic acid–sodium chloride mixtures were investigated for their potential to induce RTP from benzo[*f*]quinoline. Nine 0.5% aliphatic carboxylic acid–sodium chloride mixtures were investigated, but none yielded as strong an RTP signal as benzo[*f*]quinoline adsorbed on 0.5% polyacrylic acid–sodium chloride mixture from an 0.1 *M* HBr–ethanol solution. The previous mixture gave an RTP signal about four times the signals obtained for the aliphatic carboxylic acid–sodium chloride mixtures. It appeared that benzo[*f*]quinoline was entangled in the polymer–salt matrix and thus the molecules were held very rigidly, which allowed strong RTP to be observed. It was also possible that benzo[*f*]quinoline was buried in the matrix and collisions with oxygen were minimized. Sodium chloride or some other salt has to be mixed with polyacrylic acid for strong RTP to be observed from benzo[*f*]quinoline. In addition, an acidic solution of benzo[*f*]quinoline yields stronger RTP signals than a neutral solution of the compound because the protonated form of the molecule can interact more strongly with the surface. Very weak or no RTP was observed for phosphors adsorbed individually on either polyacrylic acid or sodium chloride. Several other inorganic salts (LiCl, KCl, K_2SO_4, and

NaBr) were mixed separately with polyacrylic acid to investigate the effects of the salts on the RTP of the polyacrylic acid–salt mixtures. However, the polyacrylic acid–sodium chloride mixture induced the strongest RTP signal. Deanin (125) discussed the effects of inorganic fillers on the thermal and mechanical properties of polymers. In polymers without a filler, the polymer molecule has a certain freedom to migrate and rotate. In a polymer mixed with an inorganic salt, some polymer molecules are directly adjacent to inorganic particles which have practically no mobility. A polymer molecule lying near such a rigid species is restricted in its ability to rotate and migrate. Deanin (125) stated that the most important effect of fillers on thermal properties of polymers was to reduce the coefficient of thermal expansion of the polymer. This indicates that the mobility and motion of the polymer will be less. With the added rigidity and lower coefficient of thermal expansion for the polymer–salt matrix, the nitrogen heterocycle is an environment with less relative motion and these conditions favor enhanced RTP.

Samples of polyacrylic acid were reacted with different amounts of NaOH to give 25, 50, 75, and 100% neutralized samples of polyacrylic acid. The RTP signal of adsorbed benzo[*f*]quinoline dropped by a factor greater than 10 at 100% neutralization for a 0.5% polyacrylic acid–sodium chloride mixture, but the RTP signal was maximum for an unneutralized polyacrylic acid–sodium chloride mixture. Because the RTP intensity decreased with the percent neutralization, this indicated that some carboxyl groups participated in hydrogen bonding with benzo[*f*]quinoline to anchor the molecules so that RTP could be observed. The results also showed that specific geometric requirements in the matrix were needed to achieve the optimal environment for strong RTP. In other experiments, Ramasamy and Hurtubise (28) showed that the solvent used to adsorb the phosphor onto 0.5% polyacrylic acid–sodium chloride surface was important. For example, 11 solvents were investigated and of the 11 solvents, methanol was the best. This solvent yielded a benzo[*f*]quinoline RTP signal 19.6 times greater than the poorest solvent, chloroform. In addition, for methanol–water solvents, a 70% methanol water (v/v) solvent resulted in the largest RTP signal with benzo[*f*]quinoline. The RTP signal was 3.1 times greater than a sample adsorbed from methanol.

It was found that the amount of polyacrylic acid was important for inducing RTP from nitrogen heterocycles (28). The polyacrylic acid content in polyacrylic acid–NaCl mixtures was varied over a wide range. Relatively large signals were obtained for the nitrogen heterocycles between 0.5 and 1.0% polyacrylic acid. Ethanol was used as a solvent. Beyond 1% polyacrylic acid, the RTP of benzo[*f*]quinoline decreased nonlinearly so that at 90% polyacrylic acid–NaCl, practically no RTP was observed. Figure 1.9 gives the RTP intensity as a function of percent polyacrylic acid from 0 to 1%

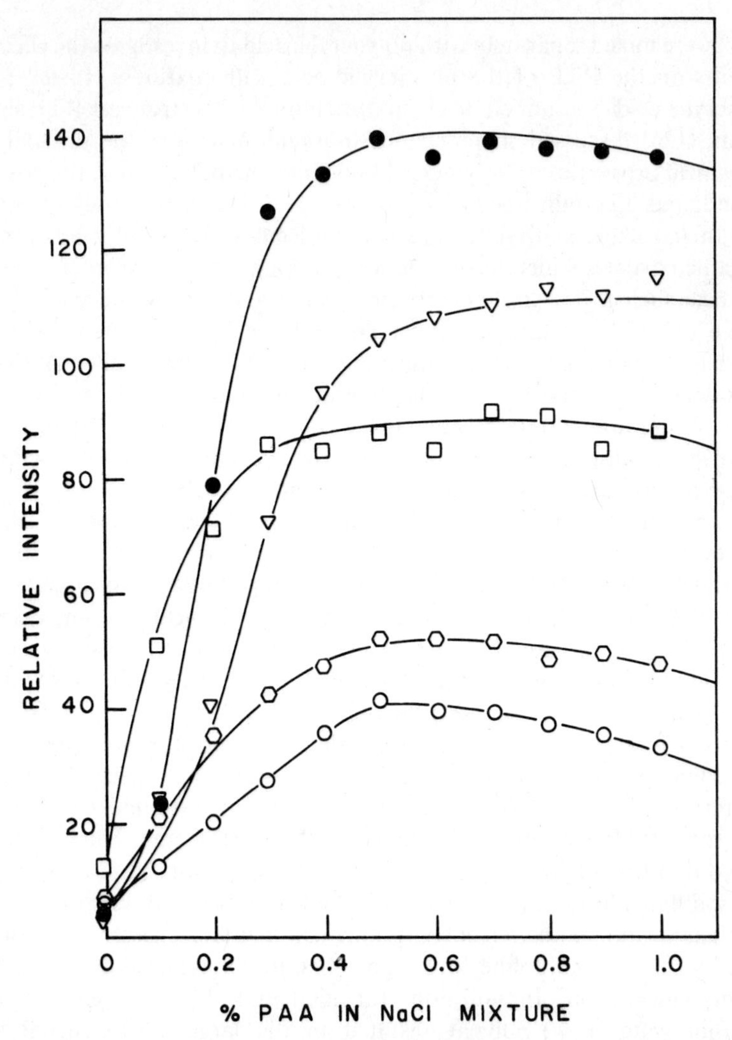

Fig. 1.9. Graphs of room-temperature phosphorescence for nitrogen heterocycles versus percent polyacrylic acid in NaCl mixtures. One hundred nanograms of each phosphor was adsorbed from 0.1 *M* HBr–ethanol solutions: (●) benzo[*f*]quinoline; (▽) 4-azafluorene; (□) phenanthridine; (⬡) 13*H*-dibenzo[*a,i*]carbazole; (O) isoquinoline. Reprinted with permission from S. M. Ramasamy and R. J. Hurtubise, "Matrix and solvent effects on the room-temperature phosphorescence of nitrogen heterocycles," *Analytical Chemistry* **54**, 2480 (1982). Copyright 1982 American Chemical Society.

polyacrylic acid in polyacrylic acid–NaCl mixtures for five nitrogen heterocycles. For all the compounds, the RTP signals increased with polyacrylic acid content and then the RTP intensity reached a maximum signal and stayed approximately constant over a range of percent polyacrylic acid values. As shown in Fig. 1.9, a certain optimal percent polyacrylic acid was needed to obtain maximum RTP signals. The ratio of the number of repeating groups in polyacrylic acid to one phosphor molecule for all the compounds was calculated. For benzo[*f*]quinoline, a value of 1.54×10^3 was obtained. From this calculation, it was clear that many of the carboxyl groups did not interact with the phosphor molecules. The large ratio obtained for benzo[*f*]quinoline implies that a given molecule would be far from its nearest neighbor. This condition would minimize the interaction of nitrogen heterocycle molecules with each other and allow the polyacrylic acid–NaCl matrix to interact effectively with nitrogen heterocycle molecules. In addition to the calculation of the ratio of repeating groups in polyacrylic acid to one phosphor molecule, the ratio was calculated of the number of repeating groups in polyacrylic acid, at the optimal polyacrylic acid concentration, to one molecule of dissolved NaCl. For benzo[*f*]quinoline, a ratio of 3.8 was obtained from this information, it was concluded that there are more repeating groups than NaCl molecules at the optimal concentration of polyacrylic acid. The role of NaCl in the RTP of the nitrogen heterocycles is important because strong RTP is not induced without NaCl. It was postulated that dissolved NaCl initially on the wet surface breaks some of the intra- and/or intermolecular hydrogen bonds of the polyacrylic acid dissolved in ethanol and this allows nitrogen heterocycles to compete for hydrogen bonding with the carboxyl groups.

In related experiments, the RTP of benzo[*f*]quinoline was obtained as a function of percent polyacrylic acid; however, methanol was used as a solvent (28). The solubility of NaCl in methanol is substantially greater than in ethanol. The ratio of repeating groups in polyacrylic acid to one NaCl molecule at the optimal polyacrylic acid concentration was calculated as 0.28, which was smaller than the ratio obtained with ethanol (3.8). The ratio of 0.28 indicated that there was an excess of sodium and chloride ions in solution at the optimal polyacrylic acid content. This condition should favor breaking of intermolecular hydrogen bonds of the carboxyl groups in the initial wet state of the solid surface. As previously discussed, benzo[*f*]quinoline gave a greater RTP signal when adsorbed onto 0.5% polyacrylic acid–NaCl with methanol compared to ethanol. Most likely, with methanol as a solvent and the relatively large amount of NaCl dissolved in methanol, more intra- and/or intermolecular carboxyl hydrogen bonds are broken in polyacrylic acid, and benzo[*f*]quinoline can interact with a larger number of carboxyl groups, which results in greater RTP with methanol.

Allerhand and Schleyer (128) reported a very large spectral shift to lower wavenumbers for the OH stretching frequency of methanol with halide ions in solution. This was attributed to the anion hydrogen bonding with the OH of the methanol. Their results lend support to the concept that Cl^- can interact in solution with the carboxyl groups of polyacrylic acid in addition to OH groups of methanol.

Ramasamy and Hurtubise (120) employed reflectance and infrared spectroscopy to study the interactions of benzo[*f*]quinoline and phenanthrene on polyacrylic acid–salt mixtures. Based on the spectral results, benzo[*f*]-quinoline adsorbed on polyacrylic acid–salt mixtures (0.5% polyacrylic acid–NaCl, 1% polyacrylic acid–NaBr, and 1% polyacrylic acid–KBr) gave neutral benzo[*f*]quinoline and protonated benzo[*f*]quinoline on the surface. Because the protonated form of benzo[*f*]quinoline is analytically more important, the protonated species was considered in more detail. It was found that the hydroxyl groups of polyacrylic acid interacted with the π-electrons of protonated benzo[*f*]quinoline. The NH^+ group could form bonds with the oxygen of either the carbonyl group or the hydroxyl group of polyacrylic acid. Because protonated benzo[*f*]quinoline forms more bonds and presumably stronger bonds than does neutral benzo[*f*]quinoline, it should be held more rigidly by polyacrylic acid than neutral benzo[*f*]quinoline. Hydrobromic acid solutions of benzo[*f*]quinoline were adsorbed onto the polyacrylic acid–salt mixtures; thus the hydrobromide of benzo[*f*]quinoline was adsorbed on the surface. An important consideration is the interactions of the hydrobromide with the surface, particularly bromide ion. The infrared data did not provide direct evidence on bromide, but it appeared that the positive charge of the NH^+ group could be shared by bromide, carbonyl, and hydroxyl groups.

The hydrocarbon analog of benzo[*f*]quinoline, phenanthrene, which also yields RTP from polyacrylic acid–salt mixtures, was studied by infrared spectroscopy (120). With phenanthrene, the most important interactions should be between the π electrons of phenanthrene and the carboxyl groups of polyacrylic acid. The infrared data supported the model that carboxyl groups interacted with π electrons of phenanthrene which would allow phenanthrene to be anchored to the surface.

The external conditions and interactions in the RTP of hydroxyl aromatics adsorbed on polyacrylic acid–salt mixtures were reported by Dalterio and Hurtubise (29). It was discussed previously that sodium chloride is important for inducing RTP from compounds adsorbed on polyacrylic acid. For 4-phenylphenol, a 0.5% polyacrylic acid–sodium chloride mixture induced an RTP signal 2.4 times greater than a similar sample on 0.5% polyacrylic acid–sodium bromide. In fact, NaBr mixtures yielded the lowest RTP of the salts examined, indicating that the heavy-atom effect was unimportant in this particular study of polyacrylic acid–salt mixtures. The polyacrylic acid in various polyacrylic acid–salt mixtures was converted partially or completely to sodium polyacrylate by reaction with NaOH solutions to study the change in RTP of three model compounds as a function of percent neutralization of polyacrylic acid. All three compounds yielded the largest RTP signals on unneutralized 0.5% polyacrylic acid–sodium chloride. Surprisingly, relatively large RTP signals were obtained with 4-phenylphenol and 4,4′-biphenol adsorbed on 75% neutralized polyacrylic acid. In addition, 4,4′-biphenol gave a relatively strong RTP signal adsorbed on a mixture of 100% neutralized polyacrylic acid. However, no RTP was observed from 2-naphthol on any of the neutralized polyacrylic acid samples. From these results it was concluded that the polyacrylic acid coils achieve a conformation in the presence of NaCl which is favorable for inducing RTP from 4-phenylphenol and 4,4′-biphenol when the $COO^-/COOH$ ratio is 3:1. The results from the neutralization studies indicated that for the model compounds examined, the polymer chain conformation is more important for inducing RTP than the number of carboxyl groups present in the polymer chain.

In other experiments, it was shown that enhanced RTP could be obtained from 4-phenylphenol adsorbed on 0.8% polyacrylic acid–NaBr mixture which contained a considerable amount of moisture in the solid matrix (29). The enhanced RTP with moisture present is an unusual result since all previously investigated adsorbents showed optimal RTP with presumably little or no moisture present (7,129,130). Schulman and Parker (19) showed with filter paper that the presence of moisture favors increased quenching of RTP by aiding the transport of O_2 into the sample matrix. For enhanced RTP to be observed with water present, the H_2O molecules incorporated in the polyacrylic acid–salt matrix must change the matrix structure in a fashion that diminishes oxygen quenching of RTP. Additional studies are needed to understand the effect of water on the RTP from polyacrylic acid–salt mixtures. Dalterio and Hurtubise (29) also reported data on the time dependence of RTP intensity for α-naphthoflavone adsorbed 1% polyacrylic acid–NaBr mixture, and compared relative RTP and low-temperature phosphorescence intensities of several model compounds adsorbed on the same mixture.

Solution fluorescence polarization data showed that model hydroxyl aromatics associate in ethanol solutions with polyacrylic acid (29). With the addition of either NaCl or NaBr to ethanol polyacrylic acid solution, the fluorescence polarization increased, with NaBr solution giving the larger fluorescence polarization. The polarization data showed that halide ions in solution interacted with polyacrylic acid molecules by breaking intra- and intermolecular hydrogen bonds between carboxyl groups. This permitted a larger number of carboxyl groups to interact with the phosphor molecules and produced larger fluorescence polarization values. However, there was an inverse correlation between the magnitude of fluorescence polarization with the salt solutions and the magnitude of RTP. In related experiments, the measurement of phosphorescence polarization of 4-phenylphenol on polyacrylic acid–salt mixtures was attempted. However, because extensive depolarization of RTP occurred, no useful data were obtained (29). In general, the area of luminescence polarization as applied to solid-surface luminescence is in need of further investigation.

Diffuse reflectance, fluorescence, phosphorescence, and infrared spectrometry were used by Dalterio and Hurtubise (30) to study the interactions of hydroxy aromatics and aromatic hydrocarbons on polyacrylic acid–salt mixtures. With ultraviolet diffuse reflectance spectrometry the ground states of the adsorbed phosphors were investigated. The lowest excited singlet states of the adsorbed molecules were studied by fluorescence spectrometry, and the lowest excited triplet states were studied by phosphorescence spectrometry. These three spectral techniques were employed primarily to detect interactions of the phosphors with solid supports. Infrared spectrometry was used to investigate interactions of the solid supports with phosphors. In the ultraviolet diffuse reflectance spectral work, the model compounds were adsorbed onto pure NaBr, which served as a reference surface. No RTP was observed from the compounds adsorbed on NaBr. The longest-wavelength diffuse reflectance bands and wavelength shifts relative to the compounds adsorbed on NaBr were obtained for several hydroxyl aromatics and aromatic hydrocarbons adsorbed on 1% polyacrylic acid–NaBr. It has been found for phenol and some substituted phenols that the longest-wavelength absorption band, due to a π–π^* transition, will shift to the red when the phenolic hydroxyl groups act as proton donors in hydrogen-bond formation and to the blue when the phenolic hydroxyl groups act as proton acceptor (131–133). Red shifts in the diffuse reflectance spectra relative to the pure NaBr surface were obtained for 4-phenylphenol and 2-phenylphenol adsorbed on 1% polyacrylic acid–NaBr. The red shift indicated that the hydroxyl groups of the phenols were hydrogen bonded to the matrix by a predominately proton-donating mechanism. Red shifts were also reported for biphenyl and naphthalene adsorbed on polyacrylic acid–NaBr. Because no hydroxyl groups were present in these compounds, the red shifts were most likely caused by intermolecular π-electron hydrogen bonds (OH–π) between the aromatic hydrocarbons and the carboxyl and hydroxyl containing surfaces. Only a small blue shift was obtained for 2-naphthol adsorbed on 1% polyacrylic acid–NaBr. The important feature of this spectral data was the lack of a relatively large spectral shift. From the blue spectral shift, it was concluded that 2-naphthol was behaving either as a proton acceptor or as both a proton donor and proton acceptor (131,132). The different hydrogen-bonding interactions occurring with the phenylphenols and 2-naphthol were partly attributed to different steric and geometric factors of the phosphor fitting into the solid matrix.

Low-temperature and room-temperature fluorescence and phosphorescence spectra for several polycyclic aromatic hydrocarbons and hydroxyl aromatics adsorbed on 1% polyacrylic acid–NaBr were obtained to study the interactions of model compounds in excited singlet and triplet states. The fluorescence λ_{max} values of the model compounds were obtained at low temperature in ethanol glass and at room temperature in ethanol. In general, the room-temperature fluorescence λ_{max} for each model compound in ethanol was red shifted with respect to the low-temperature λ_{max}. In most cases, the room-temperature fluorescence for the model compounds adsorbed on 1% polyacrylic acid–NaBr showed no spectral shift or small shifts compared to the low-temperature fluorescence. The small spectral shifts indicated that the excited molecules were strongly adsorbed on the solid supports and did not reorient themselves extensively from the Franck–Condon excited singlet state. Table 1.8 gives the phosphorescence λ_{max} for several model compounds at low temperature in ethanol glass and at room temperature adsorbed on 1% polyacrylic acid–NaBr. Table 1.8 also lists the LTP λ_{max} of the anions of the hydroxyl aromatics. The LTP λ_{max} of the anions were red shifted by 11–16 nm compared to the LTP λ_{max} of the neutral hydroxyl aromatics. The RTP λ_{max} of 4-phenylphenol and 2-naphthol adsorbed on 1% polyacrylic acid–NaBr from neutral ethanol solutions were not red shifted as much as the LTP λ_{max} of the respective anion (Table 1.8). This was evidence that the triplet emitting species for these compounds on the solid surfaces were the neutral molecules, which would be expected. The RTP λ_{max} of 2-phenylphenol on 1% polyacrylic acid–NaBr was red shifted a greater amount than the LTP λ_{max} of the 2-phenylphenol anion. It was not likely that 2-phenylphenol ionized in the triplet state on the solid surfaces because of the experimental conditions. The large shift was related to the reorientation of 2-phenylphenol on the surface and the Franck–Condon states. Little solvent or phosphor reorientation would occur at low temperature. However, at room temperature, there would be reorientation to an equilibrium triplet state and then phosphorescence would occur from the equilibrium triplet state to the Franck–Condon singlet ground state. With 2-phenylphenol, steric crowding of the phenyl ring probably favors reorientation of the phenyl ring. Generally, the RTP λ_{max} shifts in Table 1.8 showed similar trends with respect to the RTF and diffuse reflectance λ_{max} for the model compounds. The RTP_{max} for the phenylphenols, biphenyl, and naphthalene were red shifted with respect to the LTP_{max}. The red shifts for the phenylphenols suggested increased hydrogen bonding as proton donors in the triplet state.

Table 1.8. Phosphorescence λ_{max} for Model Compounds at Low and Room Temperature[a]

	λ_{max} (nm)						
Compound	LTP[b]	LTP[b] (0.1 *M* OH^-)	$\Delta\lambda$[c]	RTP[d] 1% PAA–NaBr	$\Delta\lambda$[e]	RTP[d] Filter Paper	$\Delta\lambda$[e]
Biphenyl	464			472	+8	475	+11
4-Phenylphenol	477	488	+11	482	+5	483	+6
2-Phenylphenol	463	479	+16	495	+32	508	+45
Naphthalene	506			511	+5	514	+8
2-Naphthol	526	541	+15	520	−6	493	−4
9-Anthracenemethanol	494			497	+3	493	−1

Reprinted with permission from R. A. Dalterio and R. J. Hurtubise, "Interactions of phosphors and solid supports in room-temperature phosphorescence of aromatic compounds," *Analytical Chemistry*, **56**, 339 (1984). Copyright 1984 American Chemical Society.

[a] Phosphorescence λ_{max} values were taken as the most intense band of the corrected phosphorescence spectra. Average of duplicate runs. Overall reproducibility ± 1 nm.

[b] Sample concentrations were between 10 and 50 μg/mL in 100% ethanol. Solutions were frozen at ~77 K by liquid N_2.

[c] $\Delta\lambda = \lambda_{anion,LTP} - \lambda_{neutral,LTP}$.

[d] 200 ng of phosphor was adsorbed from ethanol for all RTP spectra.

[e] $\Delta\lambda = \lambda_{RTP} - \lambda_{neutral,LTP}$.

With 2-phenylphenol, however, apparently substantial reorientation occurs. The RTP_{max} of 2-naphthol was blue shifted compared to its LTP_{max}, indicating that 2-naphthol acted as a proton donor and proton acceptor.

Infrared spectra were obtained for polyacrylic acid alone in KBr, model compounds alone in KBr, and for polyacrylic acid with model hydroxyl aromatics or polycyclic aromatic hydrocarbons in KBr (30). Both the carboxyl O–H stretching frequencies and the carbonyl stretching frequencies of polyacrylic acid were investigated. When hydroxyl aromatics were present with polyacrylic acid, the carboxyl O–H stretching band could still be examined because it was broader and more intense than the phenolic O–H band. The positions of the broad bands were determined from derivative spectra to increase the accuracy of determining band maxima. Shifts in the infrared band corresponding to the polyacrylic acid O–H stretching vibration indicated an increase or decrease in hydroxyl association. The O–H stretching band of polyacrylic acid with 4-phenylphenol present was shifted to a lower wavenumber by 24 cm^{-1} compared to polyacrylic acid alone. The shift indicated a net increase in the hydrogen-bonding association of the polyacrylic acid hydroxyl groups in the presence of 4-phenylphenol. With 2-naphthol–polyacrylic acid mixtures, the polyacrylic acid hydroxyl stretching band shifted to a higher wavenumber by 59 cm^{-1}. This suggested a net decrease in the hydrogen-bonding association of the polyacrylic acid hydroxyl groups. It appeared that the interaction of 2-naphthol with the polymer caused more polyacrylic acid inter- and intramolecular hydrogen bonds to be disrupted than were formed. For the aromatic hydrocarbons biphenyl and naphthalene, mixed with polyacrylic acid, the polyacrylic acid O–H stretching band was shifted to lower wavenumbers by 51 and 78 cm^{-1}, respectively, with respect to the hydroxyl stretching frequency of polyacrylic acid alone. A net increase in polyacrylic acid hydroxyl association was implied with biphenyl or naphthalene present with the polymer. Shifts to smaller wavenumbers were observed for the polyacrylic acid carbonyl stretching frequency when any of the four model compounds were present with polyacrylic acid. The shifts ranged from 3 to 6 cm^{-1} and suggested a net increase in the hydrogen-bonding association of the polyacrylic acid carbonyl groups.

In other experiments, infrared spectra were obtained with a polyacrylic acid–KBr disk placed in the reference beam of the infrared spectrophotometer and polyacrylic acid with 4-phenylphenol- or 2-naphthol–KBr disks in the sample beam (30). The polyacrylic acid hydroxyl stretching band was practically canceled and the phenolic O–H stretching band could be observed for the model compounds. The hydroxyl stretching bands of 4-phenylphenol and 2-naphthol were shifted to smaller wavenumbers by 12 and 17 cm^{-1}, respectively, when mixed with polyacrylic acid, compared to the hydroxyl stretching bands of these compounds alone in the KBr

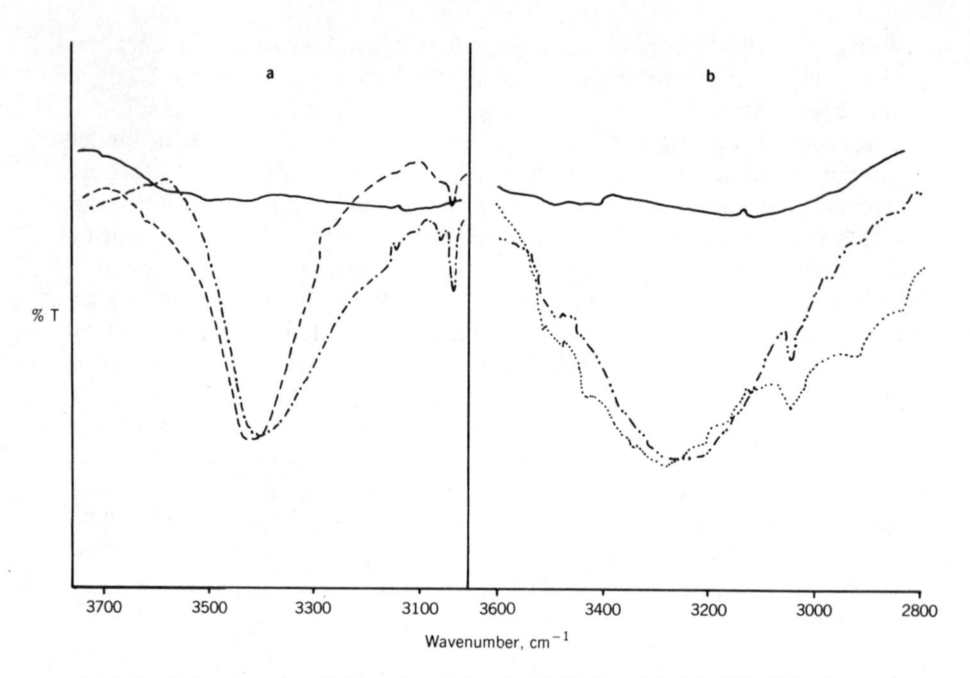

Fig. 1.10. Infrared spectra of (*a*) 4-phenylphenol and (*b*) 2-naphthol in KBr disks: (———) blank, polyacrylic acid in both beams; (– – – –) 4-phenylphenol alone; (—·—) 4-phenylphenol and polyacrylic acid in the sample beam, polyacrylic acid in reference beam; (····) 2-naphthol alone; (··—··) 2-naphthol and polyacrylic acid in sample beam, polyacrylic acid in reference beam. Reprinted with permission from R. A. Dalterio and R. J. Hurtubise, "Interactions of phosphors and solid supports in room-temperature phosphorescence of aromatic compounds," *Analytic Chemistry* **56**, 340 (1984). Copyright 1984 American Chemical Society.

(Fig. 1.10). For 4-phenylphenol and 2-naphthol, the shifts of the phenolic OH stretching bands to lower wavenumbers indicated increased hydrogen bonding for these compounds with polyacrylic acid in KBr disks. The infrared shifts for 4-phenylphenol an 2-naphthol were generally consistent with the reflectance and luminescence data discussed earlier, namely, increased hydrogen bonding for these compounds. For 4-phenylphenol the 12-cm^{-1} red shift suggested increased hydrogen donation, although the results did not preclude the possibility that 4-phenylphenol behaved as both a proton donor and acceptor. The larger red shift for 2-naphthol indicated that the compound could be acting as both a proton donor and a proton acceptor. For example, the O–H frequency for phenol is shifted to a lower frequency when the O–H group acts as both a donor and an acceptor compared to the

O–H groups acting individually as either a proton donor or a proton acceptor (134).

1.6.2.4. Filter Paper

Filter paper is the most widely used surface for inducing RTP from organic compounds. However, relatively little is known about the interactions that are responsible for RTP with filter paper. Some studies have indicated several conditions and interactions that are needed for RTP from adsorbed compounds. The first workers to investigate phosphor interactions with filter paper were Schulman and Parker (19). They considered the effects of moisture, oxygen, and the nature of the support-phosphor interaction using two model compounds. Earlier, Schulman and Walling (116,117) suggested that surface adsorption of phosphorescent compounds to the support inhibited collisional deactivation of the triplet state and restricted oxygen quenching when the sample was dried. Schulman and Parker (19) expanded the foregoing hypothesis by proposing that hydrogen bonding of ionic organic molecules to hydroxyl groups on the solid surface was the primary mechanism of providing the rigid sample matrix for RTP. They showed that effective removal of surface hydroxyl groups from filter paper by silanization reduced the RTP of sodium 1-naphthoate by 90%. In addition, they proposed that moisture acts to disrupt hydrogen bonding and aids in the transport of O_2 into the adsorbent. Their work showed that both moisture and oxygen can independently quench RTP. This was concluded by relative intensity data for humidified argon and oxygen. Some of their data are presented in Table 1.9 and Fig. 1.11 for samples adsorbed on Whatman No. 1 filter paper. In Fig. 1.11, NaBPCA refers to sodium 4-biphenylcarboxylate. In the absence of oxygen, moisture acts by itself as a powerful quencher (Fig. 1.11). At low humidity a moderate quenching effect was noted; however, at high humidity quenching was quite dramatic (Fig. 1.11). Schulman and co-workers (19,113,116,117) concluded that moisture competes with surface hydroxyl groups for hydrogen bonding to the phosphor molecules and ties up hydroxyl groups so that the phosphor is not held rigidly. Quenching by triplet ground-state oxygen occurred in the absence of moisture, but the magnitude of oxygen quenching was facilitated notably by the presence of moisture. This is indicated in Fig. 1.12. The term Q_{O_2} represents the amount of quenching due to just oxygen at a given humidity, whereas the term I_{Ar} represents the phosphorescence intensity in argon relative to 0% humidity in argon. Schulman and Parker (19) stated that moisture must be regarded as the most important contributor to quenching RTP because it can transport oxygen into the sample matrix and allow

Table 1.9. Relative Intensities of Sodium 4-Biphenylcarboxylate Samples in Ar and O_2 as a Function of Relative Humidity

%H[a]	I_{Ar}[b]	Rsd in Ar[c]	I_O[d]	Rsd in O_2[c]	Q_{H_2O}[e]	Q_O[f]
0	100	1.2	70.9	1.5	0	29.1
3.2	98.1	1.1	70.3	1.3	1.9	27.8
8.5	91.6	2.4	61.3	1.9	8.4	30.3
18.8	57.6	1.2	25.5	3.8	42.4	32.1
37.1	12.8	2.6	2.5	14	87.2	10.3
58.3	2.6	7.1	0.4	2.9	97.4	2.2
80.5	0.7	4.3	0.1	17	99.3	0.6
100	0.3	11	0.0		99.7	0.3

[a] Percent relative humidity of gas at 298 K.

[b] Intensity in argon relative to 0% humidity in argon ($I^O{}_r$).

[c] Calculated from triplicate runs and given in percent.

[d] Intensity in oxygen relative to I^O_{Ar}.

[e] $I^O_{Ar} - I_{Ar}$.

[f] $I_{Ar} - I_O$.

collisional deactivation to occur. Alkaline solutions of only two model compounds, sodium 4-biphenylcarboxylate and sodium 1-naphthoate, were used by Schulman and Parker (19). There is a need to study the effects of moisture and of oxygen on the RTP of other compounds adsorbed on filter paper under neutral, acidic, and basic conditions.

Wellons et al. (135) compared the phosphorescence signals of several compounds adsorbed from alkaline solutions onto filter paper at room temperature and at 77 K in a solid matrix. The data obtained gave a measure of the degree of rigidity with which the molecules were held on filter paper compared to a solid solution at 77 K. The authors indicated that the molecules that had the most ionic sites show the greatest rigidity on the surface. Vanillin and 2,4-dithiopyridimine are double-charged species in strongly alkaline solution, and these two compounds gave very strong RTP signals. Most likely ionic interactions were occurring on the filter paper. However, two uncharged compounds, sulfaquanidine and 5-acetyluracil, gave relatively strong RTP and probably hydrogen bonding was occurring with the surface for these two compounds. Lue-Yen Bower and Winefordner (24) investigated the effect of sample environment on the RTP of several polyacrylic aromatic hydrocarbons. They found that the heavy atom effect gave significant enhancement of RTP for the polycyclic aromatic hydrocar-

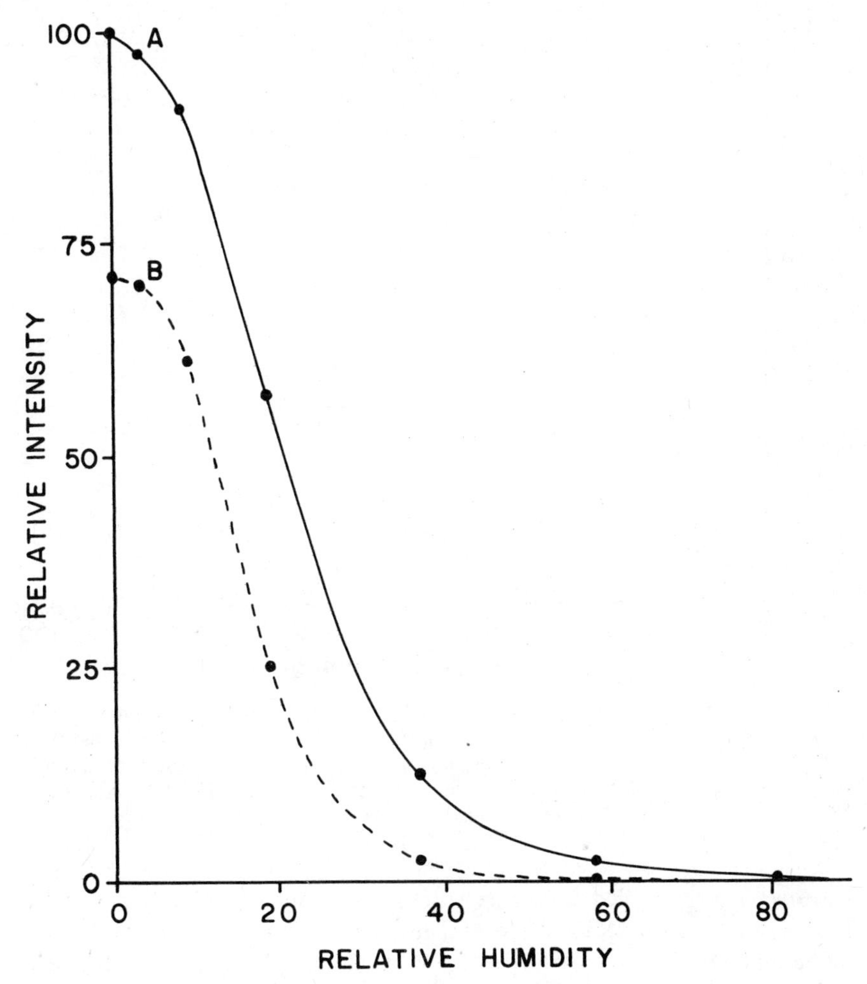

Fig. 1.11. Relative intensities of NaBPCA samples in Ar (curve A, I_{Ar}) and O_2 (curve B, I_{O_2}) as a function of humidity. Reprinted with permission from E. M. Schulman and R. T. Parker, "Room temperature phosphorescence of organic compounds. The effects of moisture, oxygen, and the nature of the support-phosphor interaction," *Journal of Physical Chemistry* **81**, 1935 (1977). Copyright 1977 American Chemical Society.

bons. It was postulated that silver ion formed π complexes with the π electron cloud of the aromatic hydrocarbons. They thus considered the silver ions to be bonded to the molecule and to the functional groups on filter paper. This would then create links between the phosphor molecules and the support and would give the necessary rigidity for observation of RTP. In other work,

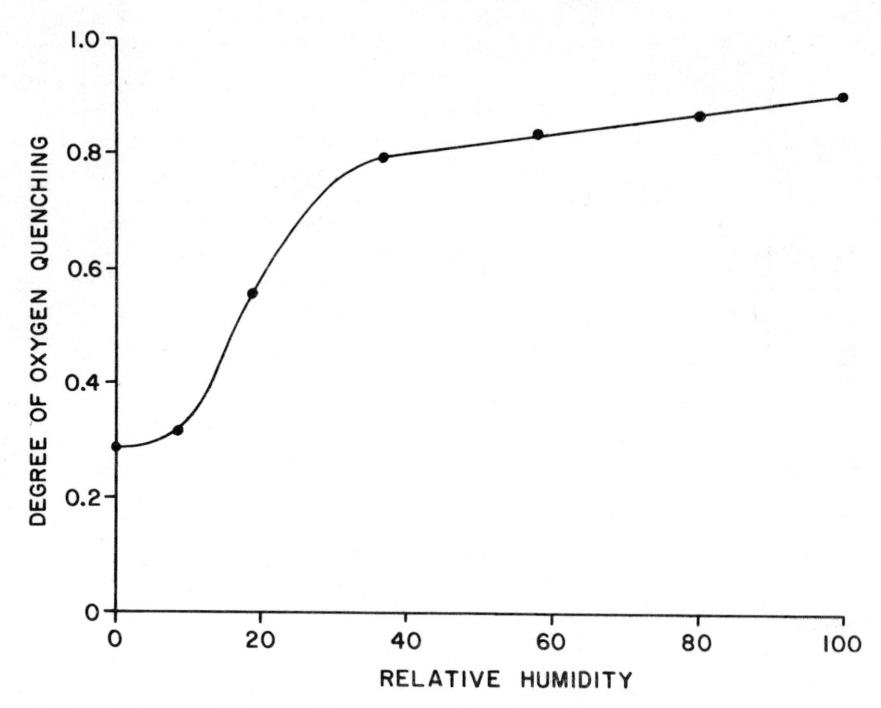

Fig. 1.12. Degree of O_2 quenching represented by Q_{O_2}/I_{Ar} plotted as a function of relative humidity for NaBPCA samples. Reprinted with permission from E. M. Schulman and R. T. Parker, "Room temperature phosphorescence of organic compounds. The effects of moisture, oxygen, and the nature of the support-phosphor interaction," *Journal of Physical Chemistry* **81**, 1936 (1977). Copyright 1977 American Chemical Society.

Aaron et al. (22) studied the consequences of ion-exchange filter paper and of heavy atoms on the RTP of several indoles. Anion-exchange filter paper (Whatman DE-81) gave the largest RTP signals with iodide present. S&S 903 paper treated with diethylenetriaminepentaacetic acid or supporting carboxymethylcellulose resin were not as useful. The presence of a heavy atom was necessary to observe analytically useful RTP signal of indoles, with iodide showing a greater effect than thallium(I). For the particular case of 5-fluorindole with thallium(I) ions present, the use of cation-exchange CM23 carboxymethylcellulose resin on S&S 903 filter paper enhanced the RTP signal about six times compared to Whatman DE-81 filter paper. The authors postulated that the smaller RTP signal on the DE-81 filter paper might be due to the perturbation of hydrogen bonding in DE-81 filter paper by thallium(I) ions. This would result in a decrease in the RTP signal. In addition, they postulated that thallium(I) ions may favor the adsorption of planar 5-fluoroindole molecules at the surface of the CM resin on S&S 903 filter paper by a charge-transfer complex stabilized by hydrogen bonding with the carboxymethylcellulose groups of the paper.

White and Seybold (136) investigated the effect of added alkali halide salts on the room-temperature fluorescence and phosphorescence of 2-naphthalenesulfonate adsorbed on filter paper. They observed a normal external heavy atom effect on the luminescence. The dependence of fluorescence quenching on perturber concentration was described by a modified version of the Perrin equation. The modified Perrin equation gave an excellent fit for all the halogen quenching data. The Stern–Volmer and normal Perrin models adequately represented the quenching data for the lighter halogens but not iodide results. For the phosphorescence results, it was shown that external heavy atoms increased the radiative triplet decay constant more than the competing triplet-state nonradiative constant. In other investigations, Meyers and Seybold (137) reported the effects of external heavy atoms and other factors on the room-temperature fluorescence and phosphorescence of tryptophan, tyrosine, and various derivatives of the previous two compounds. The RTP of tryptophan on filter paper was increased 455-fold, its methyl ester 340-fold, and that of indole 370-fold by the addition of sodium iodide to the surface. Niday and Seybold (127) investigated matrix effects on the lifetime of RTP. They obtained the phosphorescence half-life of 2-naphthalenesulfonate on filter paper with various compounds added to the filter paper. Some of the compounds added to filter paper were NaF, NH_4Cl, H_3BO_3, glycine, glucose, and sucrose. In all cases, with added compound an increase in the phosphorescence lifetime was observed. However, in no situation was the phosphorescence lifetime as long as that observed at 77 K in a rigid mixture of ethyl ether, isopentane, and ethyl alcohol (EPA) in a ratio of 5 : 5 : 2. The authors emphasized that one explanation for RTP is that the matrix holds the adsorbed compound rigid and thereby restricts vibrational motions necessary for nonradiative decay from the triplet state. They concluded that their results were consistent with the above model by assuming that packing the matrix with salts and sugars further inhibited internal motion of the phosphorescent compound. It was considered that the added compounds "plug up" the channels and interstices of the matrix, decreasing oxygen permeability and protecting the phosphorescent molecules from quenching by oxygen.

McAleese and Dunlap (138) proposed a matrix isolation mechanism for RTP from cellulose paper samples based on the swelling property of cellulose. Cellulose undergoes substantial swelling in the presence of strongly polar solvents. The solvents benzene, acetone, propanol, ethanol, methanol, and water were used in their work, and the magnitudes of RTP intensities from various phosphors were compared in reference to the solvent used to adsorb the compound on the filter paper. Swelling of filter paper would favor entry of phosphor molecules into the submicroscopic pores in the paper. After drying the matrix, the molecules could become trapped between cellulose chains and this would provide the necessary rigidity for the phosphor. The authors contended that phosphor immobilization with filter paper could not be explained by a simple hydrogen-bonding mechanism.

Dalterio and Hurtubise (27) studied the effect of sodium halide salts (NaCl, NaBr, NaI) on the RTP intensity of 4-phenylphenol adsorbed on filter paper. NaI on filter paper induced the largest signal. The enhanced RTP with NaI was not likely the result of only heavy-atom enhancement, because NaCl caused about a 10-fold RTP increase over untreated paper. The NaI caused a 17-fold increase in RTP compared to the untreated paper. An individual salt could contribute to matrix packing by which the phosphor molecules would be more rigidly held within the matrix and consequently less prone to collisional deactivation of the triplet state. For the case of NaBr, it was also shown that the maximum RTP of 4-phenylphenol depended on the amount of NaBr present on the filter paper. In contrast to the previous results, it was found that benzo[*f*]quinoline samples spotted on filter paper from 0.1 *M* HBr–ethanol solution containing either NaBr or NaCl did not intensify the RTP compared to 0.1 *M* HBr ethanol solution, but the RTP signals remained about the same for all the samples (28). Ramasamy and Hurtubise (120) utilized reflectance spectroscopy to determine the forms of benzo[*f*]quinoline and quinoline absorbed on filter paper. Comparison of the absorption wavelengths for benzo[*f*]quinoline adsorbed on filter paper from ethanol with the absorption wavelengths in ethanol indicated that the neutral form of benzo[*f*]quinoline was the predominant form adsorbed on filter paper. For a 0.1 *M* acid solution of benzo[*f*]quinoline adsorbed on filter paper, the data showed that the cation of benzo[*f*]quinoline was adsorbed on the surface. Similar results were obtained for quinoline.

Dalterio and Hurtubise (30) used several spectral techniques to explore the interactions of hydroxyl aromatics and aromatic hydrocarbons on filter paper. Using ultraviolet diffuse reflectance spectroscopy, the spectral shifts of the compounds on filter paper were obtained and compared to those of the compounds on NaBr. Red shifts were observed for all the compounds adsorbed on filter paper except for 2-naphthol and 1,2-dihydroxynaphthalene which showed blue shifts. The red shifts suggested that the hydroxyl group of a hydroxyl aromatic was hydrogen bonded to the filter paper by a predominately proton-donating interaction. The red shifts for the aromatic hydrocarbons were attributed to intermolecular π-electron hydrogen bonds (OH–π) between the aromatic hydrocarbons and the hydroxyl group of filter paper. A small blue shift was observed for 2-naphthol on filter paper. With this result, it was speculated that 2-naphthol was acting as both a proton donor and a proton acceptor. Because 1,2-dihydroxynaphthalene contains two adjacent hydroxyl groups that form an intramolecular hydrogen bond, the interpretation of its spectral shift on filter paper is not simple. However, the rather large blue shift observed on filter paper indicated that one or both of the oxygens of the hydroxyl groups of the compound accepted protons from the the hydroxyl groups of filter paper.

With phosphorescence spectroscopy, the phosphorescence λ_{max} values were obtained for several model compounds at low temperature (LTP) in ethanol glass and at room temperature adsorbed on filter paper (30). See Table 1.8. The LTP λ_{max} values for the anions of the hydroxyl aromatics were also recorded. The anions showed red shifts between 11 and 16 nm with respect to the LTP λ_{max} values of the neutral hydroxyl aromatics. The magnitudes of phosphorescence red shifts for 4-phenylphenol and 2-naphthol indicated the triplet emitting species on filter paper were the neutral molecules, which would be expected. However, the RTP λ_{max} value of 2-phenylphenol on filter paper was red shifted a greater amount than the LTP λ_{max} of the 2-phenylphenol anion. It was very unlikely that 2-phenylphenol ionized in the triplet state on filter paper because of the conditions of the experiment. The large red shift was probably due to the reorientation of 2-phenylphenol on the surface and the Franck–Condon states. At low temperature, very little solvent or phosphor reorientation would occur. However, at room temperature, there would be reorientation to an equilibrium triplet state, and then phosphorescence would take place from the equilibrium triplet state to the Franck–Condon singlet ground state (30). Most likely, the steric crowding of the phenyl ring in 2-phenylphenol favors reorientation of the phenyl ring. The RTP λ_{max} for the phenylphenols, biphenyl, and naphthalene were red shifted compared to the respective LTP λ_{max} values. For the phenylphenols, the red shifts suggested increased hydrogen bonding as proton donors in the triplet state. As discussed above, 2-phenylphenol apparently undergoes significant reorientation on the surface, which causes a large red shift. The RTP λ_{max} of 2-naphthol was blue shifted compared to its LTP λ_{max} value, indicating that 2-naphthol acted as a proton acceptor in the triplet state or behaved as both a proton donor and an acceptor (Table 1.8).

Multiple internal reflectance infrared spectroscopy was employed mainly to study the O–H stretching vibration of filter paper with adsorbed model hydroxyl aromatics and salts (30). As discussed above, by adsorbing inorganic salts onto filter paper with certain phosphors, the RTP of the phosphor can be increased substantially. This effect appears to be different from the heavy-atom effect. Table 1.10 gives the hydroxyl stretching frequencies of filter paper alone, with adsorbed NaCl and NaBr, and adsorbed model compounds with salts. In Table 1.10 it can be seen that with NaCl and NaBr adsorbed onto filter paper, the cellulose hydroxyl stretching band shifts

Table 1.10. Multiple Internal Reflectance Infrared Hydroxyl Stretching Band of Filter Paper with Adsorbed Salts and Model Compounds[a]

	Wavenumber (cm^{-1})								
	No Phosphor	4-Phenylphenol	Δcm^{-1}[b]	Biphenyl	Δcm^{-1}	2-Naphthol	Δcm^{-1}	Naphthalene	Δcm^{-1}
Filter paper alone	3282	3304	+22	3251	−31	3258	−24	3266	−16
Filter paper with NaCl[c]	3273	3329	+56	3226	−47	3257	−16	3228	−45
Filter paper with NaBr[d]	3270	3328	+58	3263	−7	3259	−11	3262	−8

Reprinted with permission from R. A. Dalterio and R. J. Hurtubise, "Interactions of phosphors and solid supports in room-temperature phosphorescence of aromatic compounds," *Analytical Chemistry*. **56**, 340 (1984). Copyright 1984 American Chemical Society.

[a] A sample of 5 mg of model compound adsorbed on each of two 14 × 49 mm filter paper sheets. Average of duplicate runs. Overall reproducibility of band maxima $\pm$ 3 cm^{-1}.

[b] $\Delta cm^{-1} = cm^{-1}$ (phosphor on filter paper) − cm^{-1}(no phosphor).

[c] A sample of 2.3 mg of NaCl adsorbed on each filter paper sheet.

[d] A sample of 9.7 mg of NaBr adsorbed on each filter paper sheet.

to lower wavenumbers by 9 and 12 cm^{-1}, respectively, with respect to filter paper alone. This suggested increased hydrogen-bonding association of the filter paper with the salts present. Allerhand and Schleyer (128) showed by infrared spectroscopy that chloride and bromide ions can form hydrogen bonds with the hydroxyl groups of methanol. With filter paper, the ions from the salts may fill in spaces in the matrix and form hydrogen bonds to cellulose O–H groups. The overall result would be a more tightly packed matrix, which would restrict moisture and oxygen penetration and favor RTP. Packing the filter paper matrix could also have the effect of holding the phosphor more rigidly within the matrix and this would favor RTP. With model compounds added to filter paper, additional shifts in the O–H stretching bands were observed; however, the direction of the shifts was different (Table 1.10). For adsorbed 4-phenylphenol, the O–H band of filter paper shifted to larger wavenumbers by 22, 56, and 58 cm^{-1} on filter paper alone, with NaCl and with NaBr, respectively (Table 1.10). The adsorption of 4-phenylphenol onto filter paper probably causes the disruption of some of the hydrogen bonds in the filter paper matrix. This type of shift could occur by replacing stronger O–H hydrogen bonds in the filter paper with weaker hydrogen bonds in the filter paper by formation of weaker bonds to the phosphor. However, reflectance and luminescence data showed that 4-phenylphenol acted as a proton donor. Thus 4-phenylphenol also participated in hydrogen donation to the hydroxyl groups of filter paper. For the adsorption of 2-naphthol, naphthalene, or biphenyl onto filter paper, the filter paper O–H band shifted between 7 and 47 cm^{-1} to smaller wavenumbers, in all cases. This indicated a net increase in hydrogen-bonding association of filter paper O–H groups, presumably by proton donation by the O–H groups of the filter paper to the phosphors. The varying effects on the O–H stretching band of the filter paper were probably influenced by steric factors related to the phosphors fitting into and interacting with the filter paper matrix in different ways. The phenolic O–H stretching bands of 4-phenylphenol and 2-naphthol did not interfere with the observation of the filter paper O–H band, because there was not enough compound on the surface to give an observable band. The infrared results showed that various types of hydrogen-bonding interactions take place between the filter paper and phosphors. One type of hydrogen-bond interaction does not predominate in all cases. The surfaces that induce RTP and the phosphors behave as proton donors, proton acceptors, or simultaneously as proton donors and acceptors.

1.6.2.5. Summary of Conditions and Interactions for RTP

There are four general aspects that should be considered in experiments for RTP. First, the initial solution chemistry can be a factor in yielding strong

RTP. For example, a phosphor that is a neutral molecule, a cation, or an anion in solution prior to depositing the sample on the surface can determine, in some cases, if strong RTP will be obtained. Second, the wet-surface chemistry can yield important conditions that give strong RTP. As an example, the solid surface should be somewhat soluble in the adsorbing solvents, so appropriate wet chemistry can occur. Third, the conditions used to dry the surface, such as the length and temperature of the drying period, are important. Fourth, the properties of the final dried matrix are very important in providing the proper conditions for enhanced RTP. All four aspects could be important in giving strong RTP from an adsorbed phosphor; however, the properties of the final dried matrix are generally the most important.

From the reported results on solid-surface RTP, no one mechanism has emerged that adequately describes the phenomenon of RTP from adsorbed compounds. Indeed, there is probably more than one mechanism with RTP because of the different solids that induce RTP and the variety of experimental conditions that have been used for RTP. It is generally agreed that the phosphor should be held rigidly to prevent vibrational motions which favor nonradiative decay from the triplet state. In one mechanism, the molecules can be held rigidly by hydrogen-bonding interactions between the phosphor and the solid matrix. In addition, the matrix can be packed with salts or other compounds that can prevent internal molecular motions of the phosphorescent molecules. Also, the molecules can become entrapped in the matrix by treating, for example, filter paper with a solvent that swells the paper, and then by drying, the cellulose fibers collapse and entrap the phosphor. Oxygen and moisture are important quenching agents in RTP work, but several conditions have been developed to minimize problems associated with these quenching agents. Heavy-atom materials have been used in many cases to enhance RTP, and it appears that the external heavy-atom effect is operative in these situations.

Considerably more work remains to obtain an adequate model for solid-surface RTP. With an acceptable model for RTP, new surfaces should become available, and the sensitivity and selectivity of the approach should improve even more.

1.7. APPLICATIONS

Numerous applications have been published for solid-surface fluorescence analysis over the years. Over the past 10 years several applications have appeared in solid-surface phosphorescence analysis. It is beyond the confines of this chapter to discuss all of the solid-surface luminescence analysis applications. Other references can be consulted for detailed discussions in solid-surface luminescence analysis. Hurtubise (7) has considered applications in both solid-surface fluorescence and phosphorescence analysis. Vo-Dinh (8) has given a very comprehensive treatment of room-temperature phosphorimetry in chemical analysis in which several RTP applications are treated. Parker et al. (129,130) have given a survey of several analytical aspects and applications in RTP. Because of the availability of the foregoing comprehensive treatments in solid-surface luminescence analysis, only a general survey of some recent applications will be given in this section. However, prior to the discussion of applications, it is important to mention other areas in luminescence that have not been considered in this chapter. Micelle-stabilized room-temperature phosphorescence is an important expanding area of luminescence analysis and has been reviewed by Cline Love and Weinberger (139). Donkerbroek et al. (140) have reported on sensitized RTP in liquid solutions, and Miller (141) has surveyed recent developments in fluorescence and chemiluminescence analysis.

1.7.1. Fluorescence Analysis

Ho et al. (142) investigated the fluorescence enhancement of polycyclic aromatic hydrocarbons on silica gel high-performance thin-layer chromatography plates. Dodecane, Triton X-100, and Fomblin Y-Vac were evaluated for their ability to enhance the fluorescence of polycyclic aromatic hydrocarbons. The procedure for fluorescence enhancement involved dipping the dry developed chromatoplate into a solution of dodecane or Triton X-100 in hexane, or Fomblin Y-Vac in 1,1,2-trichlorotrifluoroethane. The chromatoplate was removed immediately and the silica gel layer became transparent. Of the three materials investigated, Fomblin Y-Vac gave the greatest enhancement, the least spot broadening due to dipping, and permitted measurements throughout the full fluorescence emission region of the samples.

The effects of different thin-layer chromatography stationary phases and surfactant or cyclodextrin spray reagents on the fluorescence of polycyclic aromatic hydrocarbons and dansylated amino acids were reported by Alak et al. (143). Five common thin-layer chromatography stationary phases were investigated and the fluorescence intensity varied appreciably for the compounds on the different stationary phases. The reagents did not affect all the compounds to the same extent, suggesting that qualitative information could be obtained in some situations. The greatest fluorescence increase for a compound spotted on silica gel was for pyrene (47-fold increase) sprayed with sodium cholate.

Zennie (144) obtained greater than 100% recovery using high-performance thin-layer chromatography and fluorescence densitometry for the determination of aflatoxins in spiked corn samples. It was found that

C_{16}–H_{18} free fatty acids on the chromatoplates enhanced the fluorescence of aflatoxin B_1 and aflatoxin B_2. By including glacial acetic acid in the thin-layer chromatography mobile phase, the mobility of the free fatty acids was increased and the positive interference was thus eliminated.

Wollback et al. (145) described a method for the determination of phospholipids of mitochondria following one-dimensional high-performance thin-layer chromatography. Fluorescence from the chromatoplates was used in the quantitative step. Because the method was very sensitive, small amounts of the extracts could be applied to the chomatoplates. The reproducibility for the phospholipid fractions was in the range 8–17%. The method allowed for the determination of cardiolipin, phosphatidyl ethanolamine, phosphatidyl inositol, phosphatidyl choline, and sphingomyelin.

Sackett (146) separated gibberellins in fermentation broths by high-performance thin-layer chromatography and then fluorometrically determined gibberellin A_3 and gibberellin $A_4 + A_7$ using a commercial thin-layer chromatography scanner. The linear range for measurements was from 0 to about 200 μg/mL, and the precision was 3–4% for replicates of a single sample. Fourteen fermentation broth samples were assayed and resolution from potential interferences was discussed.

Surface-sensitized fluorescence was investigated by Seybold et al. (147). With this method a second, sensitizing material is used in the solid matrix to absorb incoming source radiation and transfer the excitation energy to a fluorescent analyte. The practical aspects of the method were illustrated by the example of naphthalene as sensitizer and anthracene as analyte, both adsorbed on filter paper. Under the conditions of the experiment the anthracene fluorescence signal was increased approximately 40-fold. The sensitization approach extended the limit of detection of anthracene about three orders of magnitude.

Hofstraat et al. (148) showed that highly resolved fluorescence spectra of pyrene adsorbed on silica gel chromatoplates could be obtained by using suitable laser line excitation. The spectra were comparable in quality to Shpol'skii spectra. The results obtained were essentially the same as those in the site-selection or the fluorescence line-narrowing technique. The spectra for pyrene were obtained on a dry chromatoplate or on a chromatoplate sprayed with *n*-heptane. Spraying the chromatoplate with *n*-heptane prior to the spectrometric step improved the signal-to-noise ratio by at least a factor of 10.

1.7.2. Phosphorescence Analysis

Onoue et al. (149) used RTP and delayed fluorescence for the analysis of porphyrins on filter paper. Various porphyrin compounds adsorbed on filter paper showed intense E-type delayed fluorescence. The delayed fluorescence was used to determine porphyrin derivatives at the nanogram level. As an application, E-type delayed fluorescence was employed for the determination of chlorophyll *c* from algae.

Low-temperature filter paper phosphorescence was investigated by McCall and Winefordner (82). A conduction cooling bar was designed for the low-temperature measurements and luminescence signals were detected from samples adsorbed on filter paper with a commercial spectrofluorometer. Several analytical figures of merit were compared for various compounds, and generally the low-temperature phosphorescence signals were enhanced by a factor of approximately 10.

Recently, Vannelli and Schulman (150) reported the RTP response, limit of quantification, and linear dynamic range for several pesticides that were adsorbed on filter paper under a variety of conditions. The pesticides that were investigated in some detail were carbaryl, coumaphos, warfarin, morestan, dexon, and benomyl. The analysis of benomyl in fruits by RTP was discussed. The RTP approach was found to be very useful for confirmation of benomyl residues at levels of 7 ppm for apples and 10 ppm for grapes.

As discussed in Section 1.3.2, Su et al. (87) constructed a continuous sampling system which was useful for RTP, low-temperature fluorescence, low-temperature phosphorescence, and room-temperature fluorescence work. The sampling system was used to study the RTP of certain pesticides. Limits of detection, linear dynamic range, and heavy-atom enhancement factors were reported for six phosphorescent pesticides. Analyses of several synthetic mixtures of phosphorescent pesticides were performed without separation by using various degrees of RTP enhancement of the pesticides with substrates and heavy atoms.

Vo-Dinh (151) has discussed the principles and applications of solution synchronous luminescence spectrometry and RTP spectrometry from solid surfaces as applied to air pollution analysis. Both synchronous luminescence and RTP spectrometry can be employed for screening large numbers of unfractionated air samples for an initial ranking for polynuclear aromatic content. Also, both approaches can be used for detailed identification and quantification of certain polynuclear aromatic compounds in samples that have been fractionated prior to analysis. In addition, the RTP approach has been used as a sensitive detection method in a new passive dosimeter that serves as a personnel or area monitor for polynuclear aromatic vapors in the workplace. Vo-Dinh et al. (152) described the field evaluation of simple and cost-effective luminescence techniques for screening ambient air particulate samples. Both synchronous luminescence and RTP were used to estimate the content of polynuclear aromatic species in air particulate extracts obtained at two wood-burning communities. Good agreement was obtained between the screening data and the results acquired by detailed gas chromatography/mass spectrometry and high-performance liquid chromatography.

Senthilnathan and Hurtubise (20) studied the effects of polyacrylic acid on the RTP of 4-phenylphenol, *p*-aminobenzoic acid, 1,2-benzocarbazole, and benzo[*f*]quinoline adsorbed on filter paper. With polyacrylic acid adsorbed on filter paper along with the phosphors, improvements in sensitivity ranged from 26 times to 1.1 times and limits of detection from 100 times to 1.1 times for the samples on filter paper. The relative standard deviations for the samples with polyacrylic acid added were also improved.

Total solid-surface room-temperature luminescence for analysis of mixtures was investigated by Senthilnathan and Hurtubise (153). One polycyclic aromatic hydrocarbon and nitrogen heterocycles were combined to form various binary and ternary mixtures of the compounds. By combining solid-surface room-temperature fluorescence and RTP with selective excitation and emission, the components were determined at the nanogram level. With fluorescence and phosphorescence calibration curves, it was possible to determine all the compounds in a given mixture without isolation of the components. In the mixtures, the smallest amount of material that could be determined was about 2.5 ng.

Dalterio and Hurtubise (154,155) used zeroth and second derivative solid-surface fluorescence and phosphorescence for the identification of mixtures of hydroxyl aromatics adsorbed on filter paper. Correlations were made between the second derivative luminescence wavelengths of standard compounds and the second derivative luminescence wavelengths from mixtures of the hydroxyl aromatics. Several two- and three-component mixtures were investigated by this approach and in all cases the individual components were identified. With a four-component hydroxyl aromatic mixture, three out of the four components were identified (154). In related work the zeroth and second derivative solid-surface fluorescence and phosphorescence spectra were employed for the identification of two hydroxyl aromatic compounds in high-performance liquid chromatography fractions. The method was applicable to solvents from both normal and reversed-phase chromatography systems. This solid-surface luminescence analysis approach demonstrated how fluorescence and phosphorescence could be combined at room temperature for the identification of components in liquid chromatography (155).

Su and Winefordner (21) evaluated several ion-exchange filter papers as solid surfaces and studied the heavy-atom affect for room-temperature phosphorimetry. Drugs, polycyclic aromatic hydrocarbons, and pesticides were used as model compounds. The authors concluded that anion-exchange filter paper was a promising substrate for RTP work. Whatman DE-81 anion-exchange filter paper was an excellent surface for the measurement of drugs and pesticides with iodide ion as an external heavy atom. The authors emphasized that it was important to choose the right sampling procedure of either drying with an infrared heating lamp or with a N_2 stream. They concluded that for background luminescence, limit of detection, and linear dynamic range, ion-exchange filter papers were the best RTP substrates for the compounds they investigated.

Aaron et al. (22) reported on the effects of ion-exchange filter papers and of heavy atoms on the RTP of several indoles. Whatman anion-exchange filter paper yielded the largest RTP signals with the presence of iodide. S & S 903 paper treated with diethylenetriamine–pentaacetic acid or supporting carboxymethylcellulose resin was not as useful. The heavy-atom effect was necessary for analytically useful RTP signals from indoles. Absolute limits of detection between 0.2 and 14 ng showed RTP to be a sensitive analytical technique for the compounds.

McAleese and Dunlap (35) developed a technique for reducing the background emission from filter paper. They found that by exposing filter paper to 285-nm radiation for about 3 h the RTP background emission of filter paper was significantly reduced. In batch experiments, several supports were exposed to illumination by white light from a xenon lamp. After 24 h of exposure, the paper circles were dried in a glovebag for 2 h and then analyzed. Compared to the controls, the background intensities were reduced by an average of 10.3-fold. McAleese and Dunlap (88) discussed the problem of sample sizes larger than the excitation beam in conventional phosphorimeters. They modified the excitation monochromator of a commercial instrument and reduced the size of the filter paper so that complete front-surface illumination could be accomplished. With proper alignment both the RTP of model compounds increased and the relative standard deviation of the RTP measurements improved significantly.

1.8. FUTURE TRENDS

Solid-surface luminescence analysis has shown itself to be very sensitive and selective for organic trace analysis (7,8,129,130). Important advances have been made in solid-surface room-temperature phosphorescence analysis recently. In the future, activity in solid-surface luminescence analysis will probably center around theory, instrumentation, and applications.

The state of analytical theory in solid-surface luminescence is somewhat undeveloped. There is a need for new theoretical equations that adequately describe luminescence reflected or transmitted from solid surfaces as a function of adsorbed luminescent compounds. More research under a variety of experimental and instrumental conditions is needed to establish the validity of recent theoretical equations. Also, it is important to continue to develop a theoretical basis for the phenomenon of solid-surface RTP to improve its sensitivity and selectivity even more. The theoretical basis for solid-surface fluorescence has been somewhat neglected and conditions for optimum quantum yields and selectivity should be pursued. With a firm theoretical basis for both solid-surface fluorescence and phosphorescence, the combined use of both phenomena can be exploited in trace analysis.

In general, commercial and research instruments are acceptable for obtaining good luminescence data. However, several improvements could be made. Positioning of the adsorbed component in the path of the exciting radiation does cause difficulty and can result in errors that exceed 2% (156). More extensive use of computers and digital electronics would minimize this source of error. There has been very little use of laser sources in solid-surface work. The laser work by Hofstraat et al. (148) for site-selection fluorescence from pyrene adsorbed on a thin-layer chromatoplate is one example of the potential use of lasers in solid-surface luminescence analysis. Because highly scattering media are usually employed as solid surfaces, new detection systems should be developed to improve signal-to-noise ratio. Television-type multichannel detectors (image devices) should be useful in this area.

Because of the speed, simplicity, sensitivity, selectivity, and moderate cost of solid-surface luminescence analysis, many applications will continue to appear in the future. The use of solid surfaces should be thought of in general terms and considered as another dimension in chemical analysis. Solid surfaces should be used as solutions are used in chemical analysis. Small samples are easily handled by solid-surface luminescence techniques, and chemical and physical changes can be carried out readily on solid surfaces. These features can be particularly important when dealing with toxic materials or biological samples.

REFERENCES

1. J. D. Winefordner, S. G. Schulman, and T. C. O'Haver, *Luminescence Spectroscopy in Analytical Chemistry*, Wiley, New York, 1972.
2. G. G. Guilbault, *Practical Fluorescence: Theory, Methods and Techniques*, Marcel Dekker, New York, 1973.
3. S. G. Schulman, *Fluorescence and Phosphorescence Spectroscopy: Physicochemical Principles and Practice*, Pergamon Press, Elmsford, N.Y., 1977.
4. J. R. Lakowicz, *Principles of Fluorescence Spectroscopy*, Plenum Press, New York, 1983.
5. W. W. Wendlandt and H. G. Hecht, *Reflectance Spectroscopy*, Wiley, New York, 1966.

6. G. Körtum, *Reflectance Spectroscopy*, Springer-Verlag, New York, 1969.
7. R. J. Hurtubise, *Solid Surface Luminescence Analysis*, Marcel Dekker, New York, 1981.
8. T. Vo-Dinh, *Room Temperature Phosphorimetry for Chemical Analysis*, Wiley, New York, 1984.
9. J. C. Touchstone and M. F. Dobbins, *Practice of Thin Layer Chromatography*, Wiley, New York, 1978.
10. J. C. Touchstone and J. Sherma, Eds., *Densitometry in Thin Layer Chromatography*, Wiley, New York, 1979.
11. J. G. Kirchner, *J. Chromatogr.*, **82**, 101 (1973).
12. S. Samuels and C. Fisher, *J. Chromatogr.*, **71**, 297 (1972).
13. R. M. A. von Wandruszka and R. J. Hurtubise, *Anal. Chem.*, **48**, 1784 (1976).
14. E. Sawicki, *Talanta*, **16**, 1231 (1969).
15. C. R. Sawicki and E. Sawicki, in A. Niederwieser and G. Pataki, Eds., *Progress in Thin-Layer Chromatography*, Vol. III, Ann Arbor Science Publishers, Ann Arbor, Mich. 1972, Chap. 6.
16. C. D. Ford and R. J. Hurtubise, *Anal. Chem.*, **50**, 610 (1978).
17. R. A. Paynter, S. L. Wellons, and J. D. Winefordner, *Anal. Chem.*, **46**, 736 (1974).
18. E. Lue-Yen Bower, J. L. Ward, G. Walden, and J. D. Winefordner, *Talanta*, **27**, 380 (1980).
19. E. M. Schulman and R. T. Parker, *J. Phys. Chem.*, **81**, 1932 (1977).
20. V. P. Senthilnathan and R. J. Hurtubise, *Anal. Chim. Acta*, **157**, 203 (1984).
21. S. Y. Su and J. D. Winefordner, *Can. J. Spectrosc.*, **28**, 21 (1983).
22. J. J. Aaron, M. Andino, and J. D. Winefordner, *Anal. Chim. Acta*, **160**, 171 (1984).
23. R. M. A. von Wandruszka and R. J. Hurtubise, *Anal. Chem.*, **49**, 2164 (1977).
24. E. Lue-Yen Bower and J. D. Winefordner, *Anal. Chim. Acta*, **102**, 1 (1978).
25. R. T. Parker, R. S. Freedlander, E. M. Schulman, and R. B. Dunlap, *Anal. Chem.*, **51**, 1921 (1979).
26. C. D. Ford and R. J. Hurtubise, *Anal. Chem.*, **52**, 656 (1980).
27. R. A. Dalterio and R. J. Hurtubise, *Anal. Chem.*, **54**, 224 (1982).
28. S. M. Ramasamy and R. J. Hurtubise, *Anal. Chem.*, **54**, 2477 (1982).
29. R. A. Dalterio and R. J. Hurtubise, *Anal. Chem.*, **55**, 1084 (1983).
30. R. A. Dalterio and R. J. Hurtubise, *Anal. Chem.*, **56**, 336 (1984).
31. S. Y. Su and J. D. Winefordner, *Microchem. J.*, **27**, 151 (1982).
32. D. L. McAleese, R. S. Freedlander, and R. B. Dunlap, *Anal. Chem.*, **52**, 2443 (1980).
33. R. P. Bateh and J. D. Winefordner, *Talanta*, **29**, 713 (1982).
34. Y. S. Su, D. L. Bolton, and J. D. Winefordner, *Chem. Biomed. Environ. Instrum.*, **12**, 55 (1982).
35. D. L. McAleese and R. B. Dunlap, *Anal. Chem.*, **56**, 600 (1984).
36. C. D. Ford and R. J. Hurtubise, *Anal. Chem.*, **51**, 659 (1979).
37. P. F. Lott, J. R. Dias, and R. J. Hurtubise, *J. Chromatogr. Sci.*, **14**, 488 (1976).
38. P. F. Lott, J. R. Dias, and S. C. Slahck, *J. Chromatogr. Sci.*, **16**, 571 (1978).
39. G. G. Guilbault, *Photochem. Photobiol.*, **25**, 403 (1977).

40. J. F. Lawrence and R. W. Frei, *Chemical Derivatization in Liquid Chromatography*, Vol. 7, Elsevier, New York, 1976, pp. 48–60.
41. J. Goldman and R. R. Goodall, *J. Chromatogr.*, **40**, 345 (1969).
42. J. Goldman and R. R. Goodall, *J. Chromatogr.*, 47, 386 (1970).
43. R. R. Goodall, *J. Chromatogr.* **103**, 265 (1975).
44. V. Pollak and A. A. Boulton, *J. Chromatogr.*, **45**, 200 (1969).
45. V. Pollak and A. A. Boulton, *J. Chromatogr.*, **46**, 247 (1970).
46. V. Pollak and A. A. Boulton, *J. Chromatogr.*, **50**, 19 (1970).
47. V. Pollak and A. A. Boulton, *J. Chromatogr.*, **50**, 30 (1970).
48. V. Pollak and A. A. Boulton, *J. Chromatogr.*, **50**, 39 (1970).
49. V. Pollak and A. A. Boulton, *J. Chromatogr.*, **63**, 87 (1971).
50. V. Pollak and A. A. Boulton, *J. Chromatogr.*, **72**, 231 (1972).
51. V. Pollak and A. A. Boulton, *J. Chromatogr.*, **76**, 393 (1973).
52. A. A. Boulton and V. Pollak, *J. Chromatogr.*, **45**, 189 (1969).
53. A. A. Boulton and V. Pollak, *J. Chromatogr.*, **63**, 75 (1971).
54. V. Pollak, *J. Chromatogr.*, **63**, 145 (1971).
55. V. Pollak, *J. Chromatogr.*, **77**, 245 (1973).
56. V. Pollak, *IEEE Trans. Biomed. Eng.*, **17**, 287 (1970).
57. V. Pollak, *Opt. Acta*, **21**, 51 (1974).
58. V. Pollak, *J. Chromatogr.*, **105**, 279 (1975).
59. V. Pollak, *Opt. Acta*, **23**, 25 (1976).
60. V. Pollak, *J. Chromatogr.*, **133**, 49 (1977).
61. V. Pollak, *J. Chromatogr.*, **133**, 195 (1977).
62. V. Pollak, *J. Chromatogr.*, **133**, 199 (1977).
63. V. Pollak and A. A. Boulton, *J. Chromatogr.*, **155**, 335 (1975).
64. A. A. Boulton, W. Gietz, and V. Pollak *J. Chromatogr.*, **115**, 349 (1975).
65. G. G. Guilbault and A. Vaughan, *Anal. Lett.*, **3**, 1 (1970).
66. G. G. Guilbault and A. Vaughan, *Anal. Chim. Acta*, **55**, 107 (1971).
67. H. K. Y. Lau and G. G. Guilbault, *Enzyme Technol. Dig.*, **3**, 164 (1974).
68. R. L. Zimmerman and G. G. Guilbault, *Anal. Chim. Acta*, **58**, 75 (1972).
69. G. G. Guilbault, "Fluorescencc Analysis on Solid Surfaces," in E. Wanninen, Ed., *Analytical Chemistry: Essay in Memory of Anders Ringbom*, Pergamon Press, Elmsford, N.Y., 1977, pp. 435–452.
70. M. R. Berman and R. N. Zare, *Anal. Chem.*, **47**, 1200 (1975).
71. T. G. Curtis and W. R. Seitz, *J. Chromatogr.*, **134**, 513 (1977).
72. H. Shanfield, F. Hsu, and A. J. P. Martin, *J. Chromatogr.*, **126**, 457 (1976).
73. R. Segura and A. M. Gotto, *J. Chromatogr.*, **99**, 643 (1974).
74. E. Lue-Yen Bower and J. D. Winefordner, *Anal. Chim. Acta*, **101**, 319 (1978).
75. T. Vo-Dinh and P. R. Martinez, *Anal. Chim. Acta*, **125**, 13 (1981).
76. J. L. Ward, R. P. Ward, and J. D. Winefordner, *Analyst*, **107**, 335 (1982).
77. C. G. de Lima and E. de M. Nicola, *Anal. Chem.*, **50**, 1658 (1978).
78. M. W. Warren, J. P. Avery, and H. V. Malmstadt, *Anal. Chem.*, **54**, 1853 (1982).
79. I. M. Jakovljevic, *Anal. Chem.*, **49**, 2048 (1977).
80. F. Abdel Fattah, W. Baeyens, and P. De Moerloose, "Room Temperature Phosphorescence of Some Pharmaceutical Important Imidazoles," in M. A. De

Luca and W. D. McElroy, Eds., *Bioluminescence and Chemiluminescence: Basic Chemistry and Analytical Applications*, Academic Press, New York, 1981, pp. 335–346.
81. J. B. F. Lloyd, *Analyst*, **103**, 775 (1978).
82. S. L. McCall and J. D. Winefordner, *Anal. Chem.*, **55**, 391 (1983).
83. T. Vo-Dinh. G. L. Walden, and J. D. Winefordner, *Anal. Chem.*, **49**, 1126 (1977).
84. E. Lue-Yen Bower and J. D. Winefordner, *Appl. Spectrosc.*, **33**, 9 (1979).
85. G. L. Walden and J. D. Winefordner, *Appl. Spectrosc.*, **33**, 166 (1979).
86. D. E. Goeringer and H. L. Pardue, *Anal. Chem.*, **51**, 1054 (1979).
87. S. Y. Su, E. Asafu-Adjaye, and S. Ocak, *Analyst*, **109**, 1019 (1984).
88. D. L. McAleese and R. B. Dunlap, *Anal. Chem.*, **56**, 836 (1984).
89. L. A. Gifford, J. N. Miller, D. T. Burns, and J. W. Bridges, *J. Chromatogr.*, **103**, 15 (1975).
90. J. N. Miller, D. L. Phillips, D. T. Burns, and J. W. Bridges, *Anal. Chem.*, **50**, 613 (1978).
91. L. R. Treiber, *J. Chromatogr.*, **123**, 23 (1976).
92. R. J. Hurtubise, *Anal. Chem.*, **55**, 669A (1983).
93. P. Kubelka, *J. Opt. Soc. Am.*, **38**, 448 (1948).
94. E. L. Simmons, *Appl. Opt.*, **14**, 1380 (1975).
95. R. Zweidinger and J. D. Winefordner, *Anal. Chem.*, **42**, 639 (1970).
96. J. Goldman and R. R. Goodall, *J. Chromatogr.*, **32**, 24 (1968).
97. H. G. Hecht, *Anal. Chem.*, **48**, 1775 (1976).
98. F. A. Huf, H. J. DeJong, and J. B. Schute, *Anal. Chim. Acta*, **85**, 341 (1976).
99. F. A. Huf, *Anal. Chim. Acta*, **90**, 143 (1977).
100. J. Goldman, *J. Chromatogr.*, **78**, 7 (1973).
101. R. R. Goodall, *J. Chromatogr.*, **78**, 153 (1973).
102. R. J. Hurtubise, *Anal. Chem.*, **49**, 2160 (1977).
103. M. Prosek, E. Kucan, M. Katic, M. Bano, and A. Medja, *Chromatographia*, **11**, 578 (1978).
104. V. Pollak, *J. Chromatogr.*, **123**, 11 (1976).
105. C. H. Nicholls and P. A. Leermakers, in J. N. Pitts, G. S. Hammond, W. A. Noyes, Eds., *Advances in Photochemistry*, Vol. 8, Interscience, New York, 1971, pp. 315–336.
106. L. D. Weis, T. R. Evans, and P. A. Leermakers, *J. Am. Chem. Soc.*, **90**, 6109 (1968).
107. J. B. F. Lloyd, *Analyst*, **100**, 529 (1975).
108. R. J. Hurtubise, *Talanta*, **28**, 145 (1981).
109. R. K. Bauer, P. de Mayo, K. Okada, W. R. Ware, and K. C. Wu, *J. Phys. Chem.*, **87**, 460 (1983).
110. C. H. Lochmuller, D. B. Marshall, and D. R. Wilder, *Anal. Chim. Acta*, **130**, 31 (1981).
111. C. H. Lochmuller, D. B. Marshall, and J. M. Harris, *Anal. Chim. Acta*, **131**, 263 (1981).
112. C. H. Lochmuller, A. S. Colborn, M. L. Hunnicutt, and J. M. Harris, *Anal. Chem.*, **55**, 1344 (1983).

113. C. H. Lochmuller, A. S. Colborn, M. L. Hunnicutt, and J. M. Harris, *J. Am. Chem. Soc.*, **106**, 4077 (1984).
114. L. R. Snyder, *Principles of Adsorption Chromatography*, Marcel Dekker, New York, 1968, p. 199.
115. M. Roth, *J. Chromatogr.*, **30**, 276 (1967).
116. E. M. Schulman and C. Walling, *Science*, **178**, 53 (1972).
117. E. M. Schulman and C. Walling, *J. Phys. Chem.*, **77**, 902 (1973).
118. C. D. Ford and R. J. Hurtubise, *Anal. Lett.*, **13**(A6), 485 (1980).
119. R. J. Hurtubise and G. A. Smith, *Anal. Chim. Acta*, **139**, 315 (1982).
120. S. M. Ramasamy and R. J. Hurtubise, *Anal. Chim. Acta*, **152**, 83 (1983).
121. G. Favaro, F. Masetti, U. Mazzucato, *Spectrochim. Acta Part A*, **27**, 915 (1971).
122. J. P. Peri, *J. Phys. Chem.*, **70**, 2937 (1966).
123. R. M. Silverstein, G. C. Bassler, and T. C. Morrill, *Spectrometric Identification of Organic Compounds*, 3rd ed., Wiley, New York, 1974, pp. 99–102.
124. K. Bruckner, H. Halpaap, and H. Rossler, U.S. Patent 3,502,217.
125. R. D. Deanin, *Polymer Structure, Properties and Applications*, Cahners Books, Boston, 1974, pp. 384–392.
126. L. E. Nielsen, *Mechanical Properties of Polymers and Composites*, Marcel Dekker, New York, 1974, p. 39.
127. G. T. Niday and P. G. Seybold, *Anal. Chem.*, **50**, 1577 (1978).
128. A. Allerhand and P. von R. Schleyer, *J. Am. Chem. Soc.*, **85**, 1233 (1963).
129. R. T. Parker, R. S. Freedlander, and R. B. Dunlap, *Anal. Chim. Acta*, **119**, 189 (1980).
130. R. T. Parker, R. S. Freedlander, and R. B. Dunlap, *Anal. Chim. Acta*, **120**, 1 (1980).
131. M. Ito, *J. Mol. Spectrosc.*, **4**, 125 (1960).
132. G. Nemethy and A. Ray, *J. Phys. Chem.*, **77**, 64 (1973).
133. S. G. Schulman, in E. L. Wehry, Ed., *Modern Fluorescence Spectroscopy*, Vol. 2, Plenum Press, New York, 1976, pp. 245–246.
134. A. Hall and J. L. Wood, *Spectrochim. Acta Part A*, **23A**, 2657 (1967).
135. S. L. Wellons, R. A. Paynter, and J. D. Winefordner, *Spectrochim. Acta Part A*, **30**, 2133 (1974).
136. W. White and P. G. Seybold, *J. Phys. Chem.*, **81**, 2035 (1977).
137. M. L. Meyers and P. G. Seybold, *Anal. Chem.*, **51**, 1069 (1979).
138. D. L. McAleese and R. B. Dunlap, *Anal. Chem.*, **56**, 2244 (1984).
139. L. J. Cline Love and R. Weinberger, *Spectrochim. Acta*, **38B**, 1421 (1983).
140. J. J. Donkerbroek, N. J. R. Van Eikema Hommes, C. Gooijer, N. H. Velthorst, and R. W. Frei, *J. Chromatogr.*, **255**, 581 (1983).
141. J. N. Miller, *Analyst*, **109**, 191 (1984).
142. S. S. J. Ho, H. T. Butler, and C. F. Poole, *J. Chromatogr.*, **281**, 330 (1983).
143. A. Alak, E. Heilweil, W. L. Hinze, H. Oh, and D. W. Armstrong, *J. Liq. Chromatogr.*, **7**, 1273 (1984).
144. T. M. Zennie, *J. Liq. Chromatogr.*, **7**, 1383 (1984).
145. D. Wollbeck, E. V. Kleist, I. Elmadfa, and W. Funk, *J. High Resolut. Chromatogr. Chromatogr. Commun.*, **7**, 473 (1984).

146. P. H. Sackett, *Anal. Chem.*, **56**, 1600 (1984).
147. P. G. Seybold, D. A. Hinckley, and T. A. Heinrichs, *Anal. Chem.*, **55**, 1996 (1983).
148. J. W. Hofstraat, M. Engelsma, W. P. Cofino, G. Ph. Hoornweg, C. Gooijer, and N. H. Velthorst, *Anal. Chim. Acta*, **159**, 359 (1984).
149. Y. Onoue, K. Hiraki, and Y. Nishikawa, *Bull. Chem. Soc. Jpn.*, **54**, 2633 (1981).
150. J. J. Vannelli and E. M. Schulman, *Anal. Chem.*, **56**, 1030 (1984).
151. T. Vo-Dinh, "Air Pollution: Applications of Simple Luminescence Techniques," in L. H. Keith, Ed., *Identification and Analysis of Organic Pollutants in Air*, Butterworth, Woburn, Mass., 1984, Chap. 16.
152. T. Vo-Dinh, T. J. Bruewer, G. C. Colovos, T. J. Wagner, and R. H. Jungers, *Environ. Sci. Technol.*, **18**, 477 (1984).
153. V. P. Senthilnathan and R. J. Hurtubise, *Anal. Chem.*, **56**, 913 (1984).
154. R. A. Dalterio and R. J. Hurtubise, *Anal. Chem.*, **56**, 819 (1984).
155. R. A. Dalterio and R. J. Hurtubise, *Anal. Chem.*, **56**, 1183 (1984).
156. S. Ebel and J. Hocke, *J. Chromatogr.*, **126**, 449 (1976).

CHAPTER

2

TIME-RESOLVED AND PHASE-RESOLVED EMISSION SPECTROSCOPY

J. N. DEMAS

Chemistry Department
University of Virginia
Charlottesville, VA 22901

2.1. INTRODUCTION

Luminescence spectroscopy is an exceptionally powerful and useful tool in numerous scientific areas, including solar energy conversion, physical and analytical chemistry, biology, physics, medicine, and engineering. Luminescence spectroscopy is useful because exceptionally high sensitivity and selectivity can be achieved (1–8).

Traditional luminescence spectroscopy examines both emission and excitation spectra (1). Emission spectra are obtained by measuring the spectral distribution of the emission versus wavelength while holding the excitation wavelength fixed. Excitation spectra are obtained by measuring the emission intensity at a fixed wavelength as the excitation wavelength is varied. Generally, excitation and emission spectra are distorted by instrumental factors such as the variations in light source intensity, detector response, and monochromator and optical efficiencies with wavelength. Correction for these factors produces the fundamentally more useful corrected excitation and emission spectra (1).

In multicomponent systems or systems with complex properties, it has proved particularly valuable to combine both excitation and emission spectra. Even in many complex mixtures, for example, a unique set of excitation and emission spectra can be used to "fingerprint" the mixture and resolve it into its components.

While either excitation or emission spectra alone can provide considerable qualitative, quantitative or fundamental information about a system, the combination of the two provides far more information. If additional unique properties of components can also be simultaneously measured, even more complex systems can be unraveled or greater accuracy and precision can be realized. The use of multiple parameters to characterize a system is called multiparameter analysis. If the multiple parameters can be used to display the data in multidimensional format, the term "multidimensional luminescence measurements" (MLM) has been suggested. Parameters include time and phase resolution, quenching, sensitization, depolarization, circular polarized luminescence, and chromatography (7).

We will deal here with the use of temporal information to enhance luminescence spectroscopy. We will cover two types of time resolution. Generally, one method utilizes a pulsed source and the other a modulated source. For the pulsed source we will be concerned with time evolution of luminescence spectra following excitation with a short-duration excitation pulse. For mixtures of components with different lifetimes or systems in which the nature or environment of luminescent species changes following excitation, the emission properties will vary with delay after excitation. Resolution of emission spectra based on time delay after excitation is called time-resolved emission spectroscopy (TRES) (2–5).

For the modulated source, variations in luminescence properties are detected by variations in phase differences between the excitation and the emission. Longer-lived components exhibit greater phase shifts than do short-lived components; short lifetimes permit a component to track the excitation more faithfully. Resolution of time information on the basis of phase shifts is called phase-resolved spectroscopy (PRS) (2,3,5–8).

Researchers in TRES and PRS tend to be strongly polarized as to which method is best and there are many acrimonious debates in the literature and at meetings between the two schools of thought. In an attempt to justify their position, many researchers tend to oversell the strengths of their method and undersell the weaknesses, which frequently leaves the uninitiated confused. In our opinion the truth lies somewhere in between. Although TRES and PRS are quite different in experimental implementation and data workup, both fundamentally exploit differences in excited-state lifetime, and the information that can be extracted by the two approaches is quite similar. The optimum choice of method will depend on factors such as the nature and depth of detail required, data acquisition speed, simplicity of measurement, and cost. Frequently, in fact, the two methods tend to complement each other. Regardless of whether TRES or PRS is used, the additional degree of freedom adds enormously to the information that can be extracted from complex biological and chemical systems.

In this chapter we discuss both TRES and PRS. We first describe the experimental and theoretical underpinnings of each method and end with a brief discussion of applications. Although we limit our discussion to emission spectroscopy, there are numerous other types of time- and phase-resolved spectroscopy. These include flash photolysis (where one usually monitors excited-state concentration by absorption spectroscopy) (1) and transient grating methods (where excited-state or chemical concentrations or spatial distribution are monitored by the washing out of a transient grating set up in a film sample) (9,10). We also omit thermal spectroscopic methods such as thermal blooming, which monitors sample heating produced by relaxation of excited states or chemical reactions (11,12). Finally, although an extremely powerful tool for studying excited states, we omit time-resolved emission anisotropy (2–4).

The purpose of TRES and PRS is the resolution of emission spectra based on differences in sample decay time. As a general statement TRES uses pulsed light sources, whereas PRS uses periodically modulated sources. The differences between these two approaches can be a thin one, however, when the source is a narrow pulse occurring at a high repetition rate.

Generally speaking, we would like to know the emission spectrum $I(\lambda, t)$ of a sample as a function of wavelength and time following a delta function or zero duration flash. Thus a complete time-resolved (TR) emission spectrum would consist of a three-dimensional plot of emission intensity

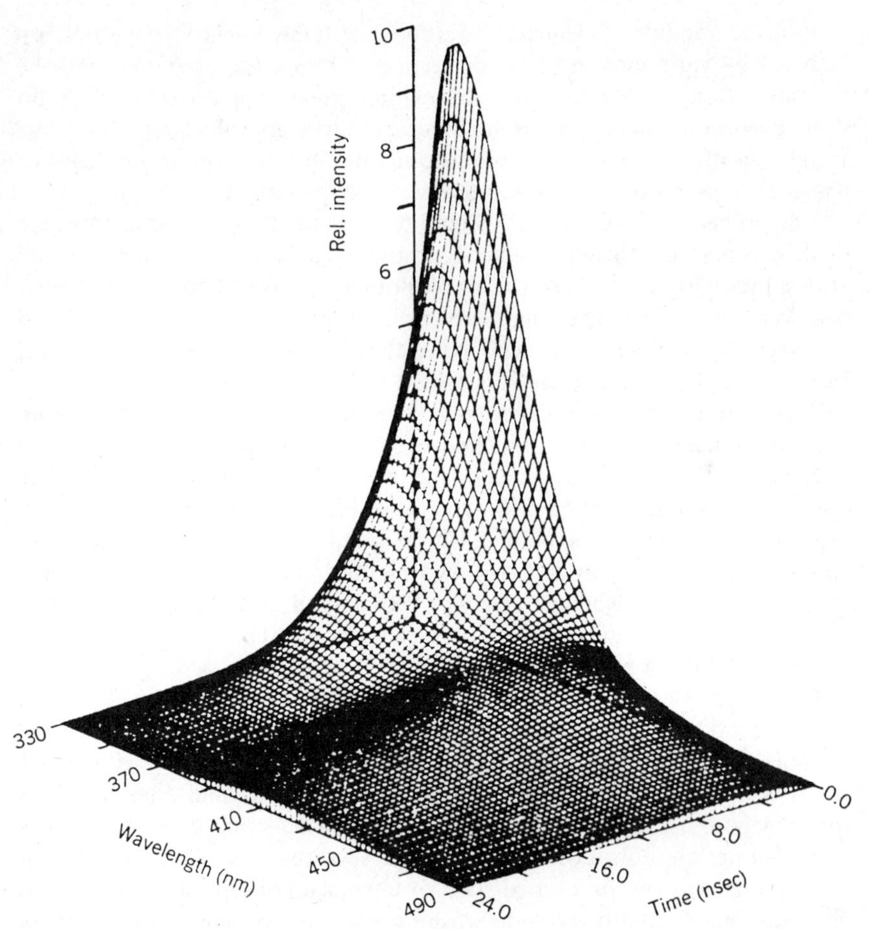

Fig. 2.1 A three-dimensional view of the fluorescence emission of 2-naphthol as a function of time in 1 m*M* phosphate buffer, pH 7.15. Decay curves were collected at 40 wavelengths and deconvoluted. (Reprinted with permission from *J. Phys. Chem.*, **83**, 799. Copyright 1979 American Chemical Society.)

versus wavelength and time. Figure 2.1 shows a complete TR emission spectrum for 2-naphthol at pH 7.15. The relative emission intensity is plotted as a function of emission wavelength and time after excitation. There is a pronounced initial rise in the emission beyond about 420 nm before the inevitable decay. Because of the enormous amount of data required to generate such plots, however, we frequently see more restricted plots. A common alternative format is a series of emission spectra taken for different fixed delays after excitation (see below).

In this chapter we first discuss the theory and experimental implementation of both TRES and PRS. Where appropriate we will compare and contrast the two methods and the information to be gleaned from the two approaches. Finally, we will give a number of examples of the use of TRES and PRS.

Pragmatically, another natural division in experimental methodologies occurs with the length of the sample lifetime. These two ranges are short-lived species such as occur in fluorescences (< 5 μs) and long-lived ones such as arise from phosphorescences (> 10 μs). As we will see, the instrumentation can vary quite dramatically between pulsed and modulation methods and for short- versus long-lived species.

We begin with a description of simple mechanical methods, then cover pulsed methods, and conclude with modulation methods. We will also discuss powerful multidimensional analyses, which are applicable to both pulsed and modulated methods. We will give warnings where the uninitiated may go astray. Where appropriate we include theory.

2.2. MECHANICAL PHOSPHOROSCOPES

The simplest form of TRES involves mechanical chopping to resolve a long-lived phosphorescence from a short-lived fluorescence and scatter. For this application the simple Becquerel phosphorescope or one of its variants serves admirably (1,14,15). Figure 2.2 shows two designs of widely used mechanical phosphorescopes—a straight-through and a right-angle rotating can. In both cases a mechanically rotated shutter alternately illuminates the sample and then, after shutting off the excitation source, opens to permit the detection system to view the sample. If the sample has an afterglow that persists until the detector is illuminated, the detector can only see this long-lived emission. The rotating can phosphorescope is extremely popular on commercial spectrofluorimeters because of its simple, compact construction and because it does not disturb the standard right-angle viewing geometry.

Mechanical phosphorescopes can be made extremely efficient in eliminating sample scatter and fluorescence. Extremely weak phosphorescences can be viewed in the presence of even the most intense fluorescences. For example, the design of Peled et al. (15) is so efficient that the sample could be irradiated by a 450-W Hg lamp with no photographically detectable excitation. Figure 2.3 shows the total luminescence (fluorescence and phosphorescence) of phenanthrene as well as the phosphorescence, which has been completely resolved by means of a mechanical phosphorescope (16).

Mechanical phosphorescopes have several disadvantages. They generally fail if the desired emission is the short one rather than the long one, although

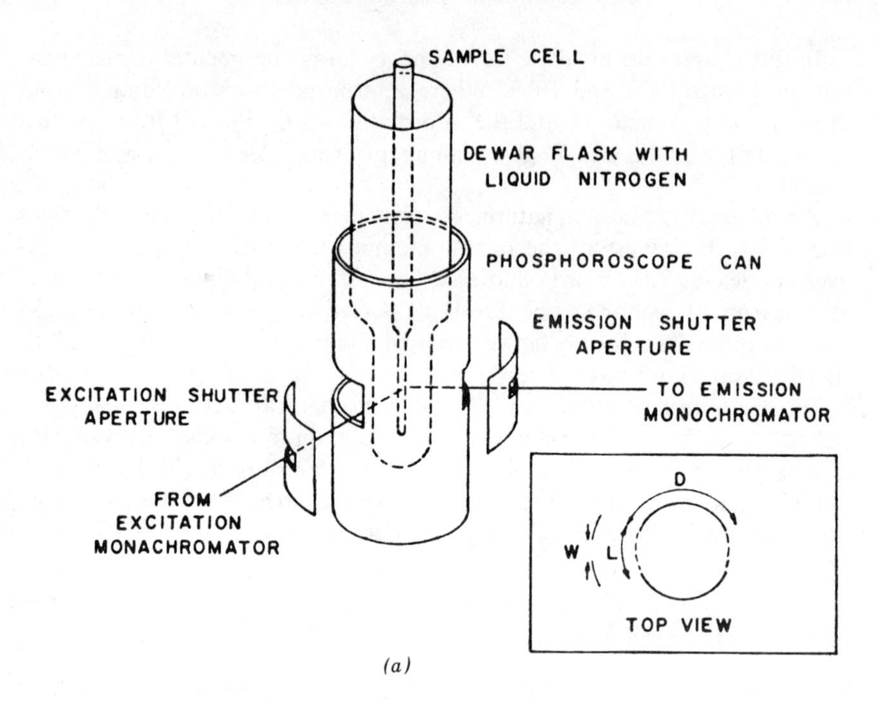

(*a*)

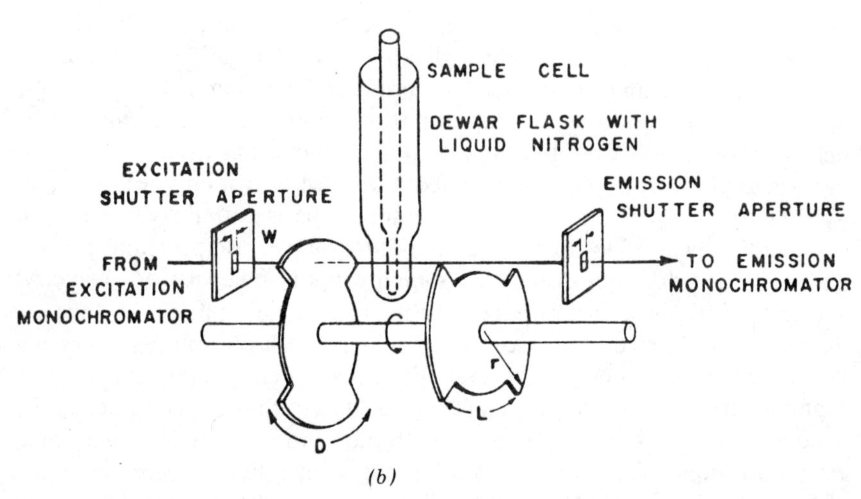

(*b*)

Fig. 2.2 Schematic diagram of (*a*) a rotating can phosphorescope and (*b*) a rotating disk phosphorescope. (Reprinted with permission from *Acct. Chem. Res.*, December, 361. Copyright 1969 American Chemical Society.)

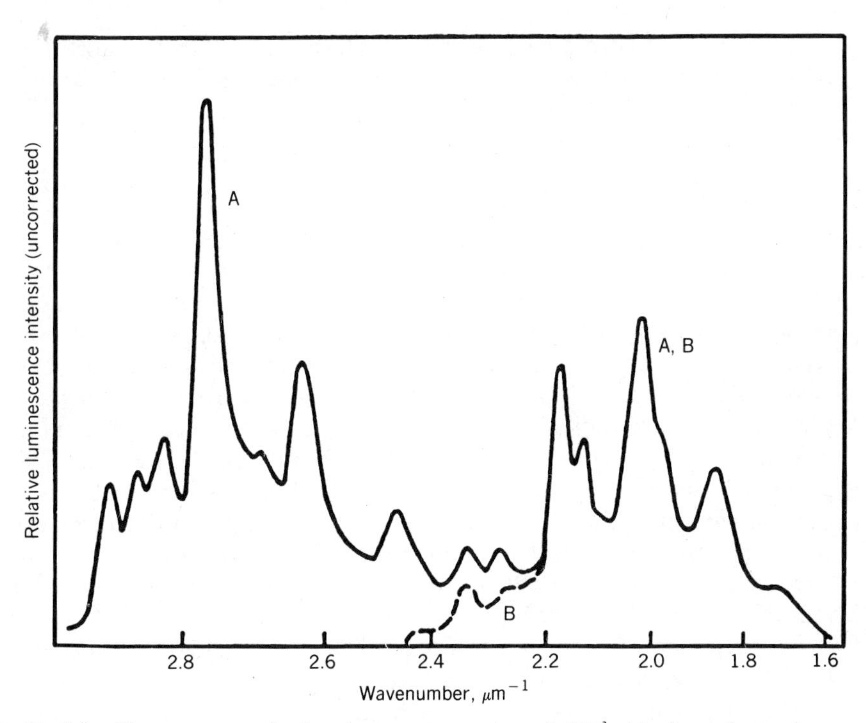

Fig. 2.3. Fluorescence and phosphorescence spectra of 10^{-3} *M* phenanthrene in an ether–isopentane–ethanol glass at 77 K. A, Fluorescence plus phosphorescence; B, phosphorescence resolved with mechanical phosphorescope modulated at 800 Hz. Phosphorescence lifetime is 4.3 s. [Reprinted with permission from *Analyst*, **87**, 664 (1962).]

designs based on unusual blade geometries and phasing can be used to enhance fluorescences relative to the phosphorescences (1,16). Also, because of the mechanical chopping time, only relatively long emissions (>100 μs) can pass through the phosphorescope. All shorter emissions appear to be fluorescences. Pulsed shutters with opening times of less than 5 μs have, however, been used with pulsed lasers to eliminate scatter and fluorescence peaks that would otherwise overload the detection system (17).

Finally, mechanical chopping is largely incapable of resolving more than one long-lived component. Any emission that persists long enough for the shutter to open is detected. If a variable-speed chopper is used and the component lifetimes differ appreciably, a rudimentary resolution can be achieved by varying the chopper speed. At lower chopper speeds, long-lived species will decay less by the time the shutter opens than will short-lived species; this enhances the long-lived component relative to the short lived one. However, it is rarely possible to achieve anything like a quantitative analysis of more than one component in this fashion. Further, the short-lived component is never free of the long-lived one because the long-lived component has its most intense emission at early times.

2.3. PULSED METHODS

By replacing the mechanical excitation system with a flash lamp or pulsed laser, a much faster time resolution system can be achieved. Indeed, with fast laser sources and detectors, time resolution of fluorescences or even separation of scatter and fast fluorescences becomes feasible. Intense inexpensive flash lamps with pulse widths in the region 1–10 μs are available. Laser sources can readily give excitation widths in the low-nanosecond or picosecond range. Thus the luminescence of all species with lifetimes appreciably shorter than the flash width will follow the flash and die out with it. By monitoring the emission only after the flash, the phosphorescence can be detected with very little contribution from the fluorescence. At least one commercially available phosphorescope attachment for a spectrofluorimeter uses this principle.

The detection of the luminescence is frequently achieved by gating the detector or the electronics after the excitation has extinguished. Alternatively, a transient recorder or signal averager can be used to record the entire decay. Single-photon counting methods can also be used for data recording. We briefly discuss these methods.

2.3.1. Gated Methods

Detection gating is much more versatile than mechanical phosphorescopes (18–26). Various regions of the luminescence decay can be selectively gated. This feature is shown in Fig. 2.4 for systems exhibiting both fast fluorescence and slow phosphorescence when excited by a short laser excitation pulse. The indicated decay has a fast fluorescence spike at early times that follows closely the excitation flash and a long phosphorescence tail. By selecting different gate delays and gate widths, the electronics can be used to sample different components of the emission. The fast component during or immediately after the flash can be enhanced by using no delay and matching the gate width to the fluorescence pulse width. The long-lived emission can be enhanced by delaying until the fluorescence has died out. A similar strategy can be used to enhance components of a mixture of phosphorescent species by adjusting the delay and width (14). If one phosphorescence is much longer-lived than any others, complete elimination of the shorter compo-

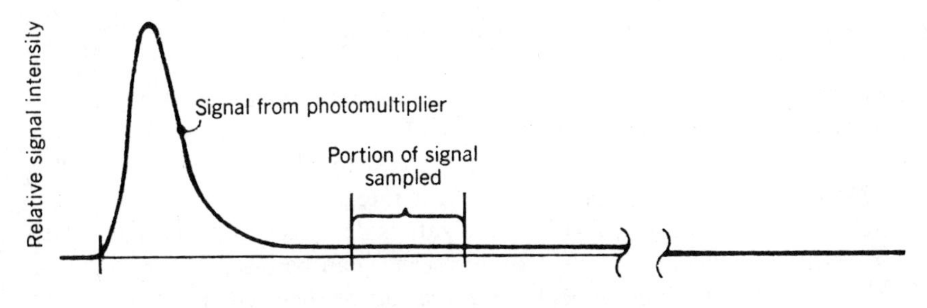

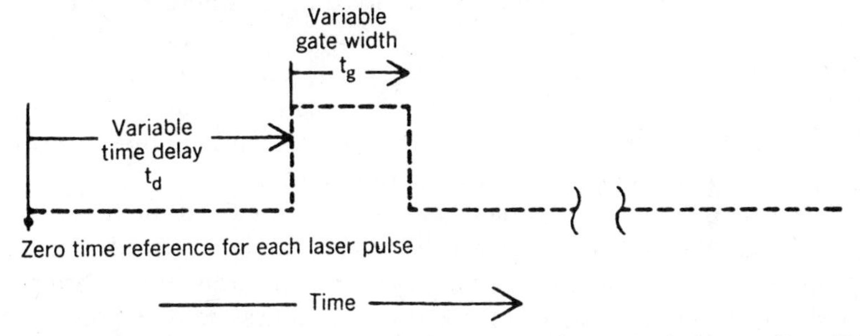

Fig. 2.4. Schematic representation of a variable boxcar sampling gate. t_g is the sampling width and t_d is the delay after excitation before opening the gate. (Reprinted with permission from *Anal. Chem.*, **46**, 1691. Copyright 1974 American Chemical Society.)

nents can be achieved. As with the mechanical phosphorescope, it is not possible to achieve this level of discrimination for the short-lived component.

2.3.1.1. Photomultiplier Gating

Gating of the photomultiplier (PMT) is most commonly done by altering the voltage distribution across the dynode string so as to greatly reduce the PMT gain (18–20). Microsecond gating off/on times, as opposed to subnanosecond gating (18), are particularly easily and inexpensively achieved. Photomultiplier gating is frequently preferred over gating an external amplifier. The reason for this is that high scatter and fluorescence levels can overload the PMT and its dynode string even if the external amplifier is gated; the PMT output may not recover during the desired viewing period. Especially if the PMT is gated at an early dynode, excessive dynode and PMT currents will not flow and recovery will be very fast, which will eliminate any overload problems, even in the presence of intense backgrounds.

2.3.1.2. *Boxcar Integrators*

Boxcar integrators are a common and powerful method of signal enhancement. Basically, a boxcar is a gated integrator. The delays following the start of the signal and the gate width are controllable. The gated signal is smoothed by a filter. For TRES a boxcar integrator can be thought of as an electron gate after the PMT. If the gate is delayed to just after the flash and made very wide, virtually all fluorescence can be discriminated against while virtually all of the phosphorescence can be detected. By using a very short gate that overlaps the excitation, all of the fluorescence can be detected while most of the much-longer-lived phosphorescence is eliminated (14,21). Figure 2.5 shows an elegant use of a boxcar to resolve the prompt fluorescence and the long-lived phosphorescence of 4,4′-dihydroxybenzophenone (21). The $t_d = 0$ spectrum is largely fluorescence. With increasing delay the fluorescence decreases, and for the longest delay the emission is essentially pure phosphorescence.

If a relatively narrow sampling gate width is used and the delay is varied slowly, a complete decay curve for the sample can be obtained. Used in this mode the boxcar functions as a signal-averaging transient recorder. Gate width times on most boxcars are 0.1–1 μs unless sampling oscilloscope methods are used. The use of these recorded transients is described in the next section.

2.3.1.3. *Transient Recording and Multichannel Analyzers*

A more elegant and quantitatively useful procedure is to record the sample decay. This can be done using a digital transient recorder, a boxcar integrator, a signal averager, or a multichannel analyzer (MCA).

Transient recorders are fast analog-to-digital converters coupled to a digital memory. The combination is driven by a logic controller and a clock, which determines the rate of data acquisition. A digital signal averager is a transient recorder in which each succeeding transient is added to the accumulated transient stored in a memory. The signal-to-noise ratio in such averagers increases with the square root of the number of transients added. Transient recorders now have acquisition speeds of >200 million points per second.

The MCA is used in the multichannel scaling (MCS) mode. In the MCS mode the MCA is stepped sequentially through its channels. The MCA dwells on each channel for a fixed period of time set by the MCA time base and accumulates and adds the number of signal counts to the existing count during the dwell period. These counts can be derived from optical signals either by directly counting the single-photon electron pulses from the PMT

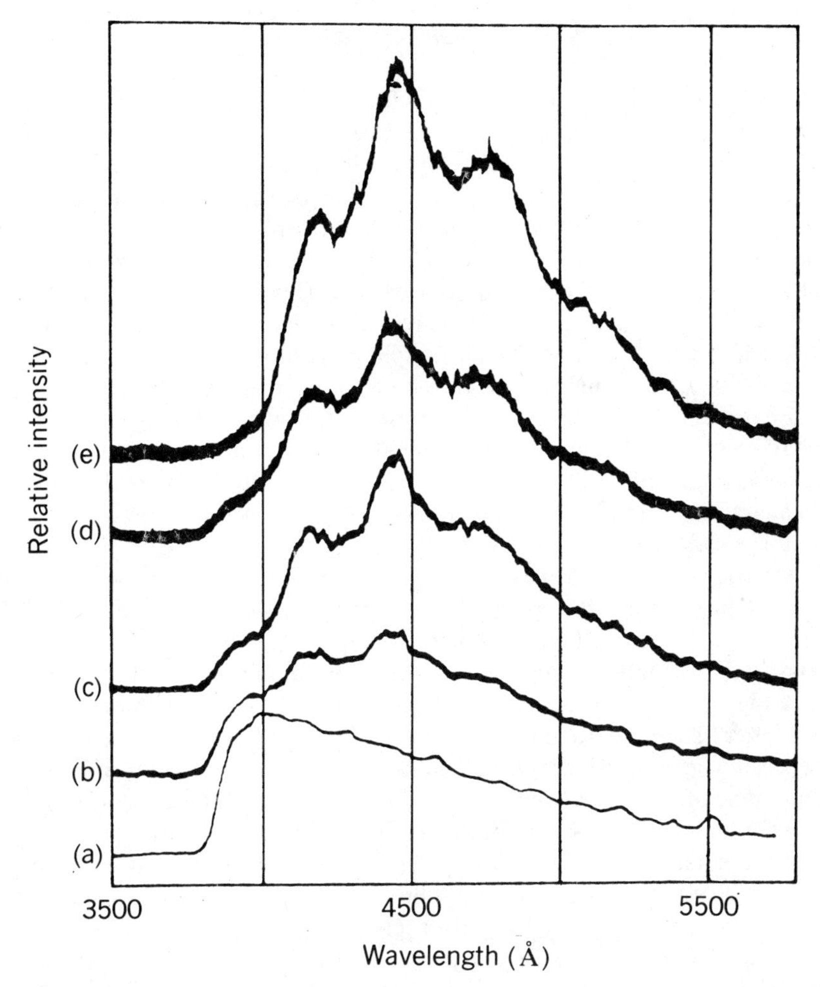

Fig. 2.5. Time-resolved emission spectra from 0.1 *M* 4,4′-dimethoxybenzophenone in benzene, 25°C. t_g = 100 ns. Curve a, t_d = 0 ns; curve b, t_d = 200 ns, *rs* = 5; curve c, t_d = 500 ns, *rs* = 10; curve d, t_d = 1 μs, *rs* = 10; curve e, t_d = 5 μs, *rs* = 50. *rs* is the relative sensitivity versus (a). (Reprinted with permission from *Anal. Chem. 46*, 1693. Copyright 1974 American Chemical Society.)

or by converting the analog signal from the PMT to a proportional count rate by means of a voltage-to-frequency converter. After the decay is recorded, the channel of the recorder or MCA contains a histogram representation of the decay. If the signal-to-noise ratio is inadequate, multiple flashes can be used to accumulate further counts in each channel and perform signal averaging.

2.3.1.4. Sampling Oscilloscopes

Boxcar integrators are very popular for relatively slow phenomena. For faster phenomena transient recorders are still quite expensive. Sampling methods provide a feasible and relatively inexpensive alternative.

In the viewing of very high speed phenomena, sampling techniques have been widely used for viewing repetitive waveforms. Rather than trying to view the entire transient on each occurrence, one acquires and views only a single point from the transient on each occurrence. By selecting different points, one can build up a complete representation of the transient. We will describe briefly a conventional sampling oscilloscope system and then describe how sampling methods are used to record TRES.

Figure 2.6 shows a schematic representation of a sampling oscilloscope (SO). The critical elements are a triggered fast linear time base similar to that of a conventional oscilloscope, a slow-sweep generator, an ultrafast analog comparator, and a sample and hold (S/H). Initially, the slow-sweep generator is set at its lowest value.

On triggering, the fast-sweep generator starts a linear rise. As this ramp crosses the slow one, the comparator senses this crossing and strobes the S/H to latch onto an analog data point. The S/H holds the signal until a new triggering; its output equals the input signal amplitude at the instant of crossing. The sampled point is displayed on a CRT, which functions as an X-Y plotter. The amplitude of the slow sweep is the X input. Since the delay between triggering the scope and the crossing is directly proportional to the amplitude of the slow sweep, the X coordinate of the display is proportional to the time of the sampling on the waveform. After the acquisition of each point, the amplitude of the slow sweep is incremented; this causes a longer delay between triggering and sampling. Thus the transient is sequentially built up pointwise by sampling at progressively longer times. In an SO, the interval between samples is adjustable to permit faithful reconstruction of a waveform. After completion of one sweep, the entire process is repeated.

The rise times for SOs range from 10 to 1000 ps, which makes them ideal for viewing repetitive picosecond phenomena. Since each point is held on the screen until acquisition of the next point, SOs are free of the low-viewing-intensity problem of low-repetition-rate signals of conventional oscilloscopes.

Sampling oscilloscopes lend themselves to efficient boxcar averaging of signals. Instead of rapidly sweeping across the entire waveform, the slow sweep is raised slowly. The result is that the S/H outputs many points for each time delay. Then, by merely smoothing the S/H output, a signal-averaged waveform is obtained. Generally, the sweep is done slowly enough so that the output can be displayed on a mechanical recorder or recorded on

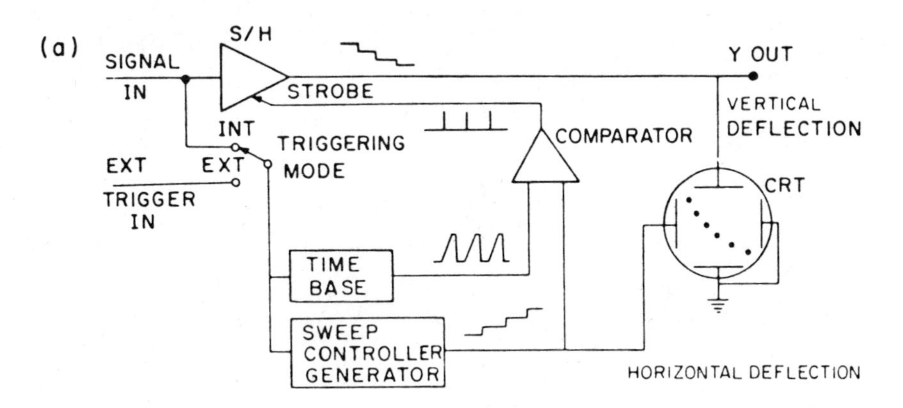

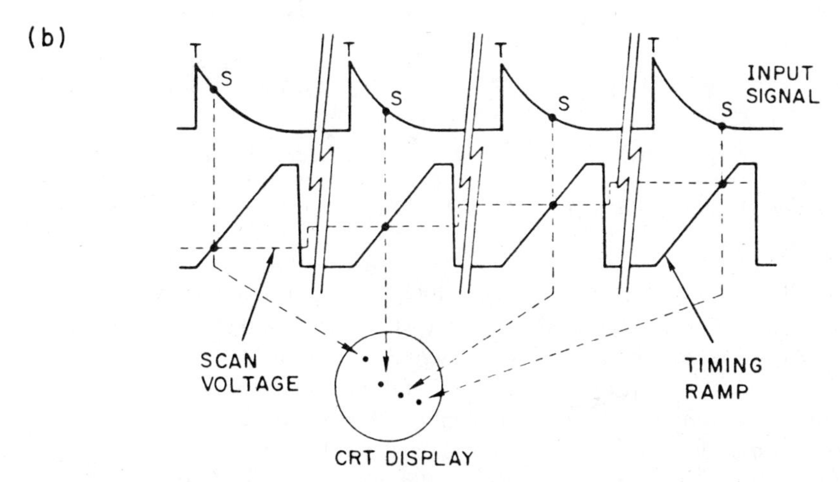

Fig. 2.6. Schematic representation of a sampling oscilloscope (*a*) and relevant waveforms (*b*). (*a*) The time base is triggered on each occurrence of a waveform. The sweep controller output is increased by a fixed amount after each sweep. A CRT display is used as an X-Y point plotter. S/H is triggered once per waveform. (*b*) T indicates the trigger point at which the linear time base sweep is actuated. The S/H is strobed at the sampling points that occur when the scan voltage (---) and linear timing ramp (———) are equal. [Reprinted with permission from J. N. Demas, *Excited State Lifetime Measurements*, Academic Press, (New York; 1983).]

a computer. Computer control is especially useful since the sweep rate and the degree of averaging can be controlled by the computer.

To measure time-resolved emission spectra, the experiment is changed only slightly. The slow sweep is not scanned but rather, is set to a voltage that corresponds to the desired delay after triggering. The emission monochromator is then scanned while the delay remains fixed. The output of the

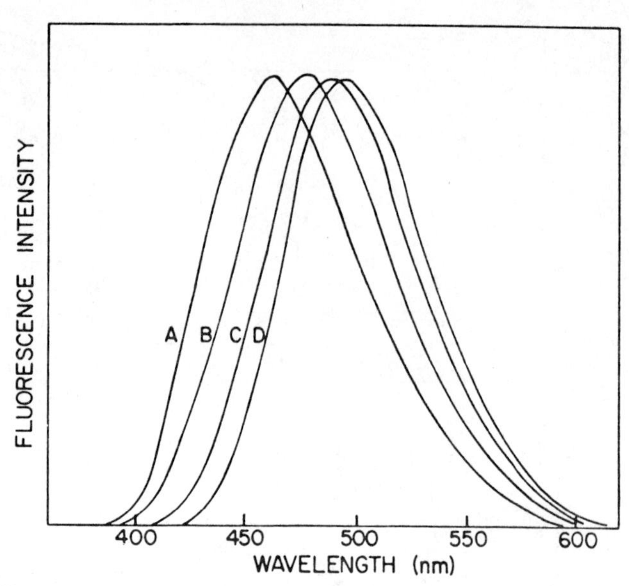

Fig. 2.7. Time-resolved fluorescence spectrum versus delay for 4-aminophthalimide at −70°C in 1-propanol. Curve *A*, 4 ns; curve *B*, 8 ns; curve *C*, 15 ns; curve *D*, 23 ns. [Reprinted with permission from *J. Chem. Phys.*, **54**, 4729 (1971).]

S/H then represents the emission intensity at a fixed delay following excitation. Slow sweeping of the monochromator coupled with filtering of the SO output can be used to affect signal averaging.

Figure 2.7 is a beautiful set of TRES obtained for the solvent relaxation of 4-aminophthalimde in *n*-propyl alcohol as a function of time following excitation. These data were not obtained with a SO, but with a pulsed photomultiplier (PMT) functioning as the sampling gate. The basic sampling principle and the results are, however, similar to those of a SO.

There are also commercial boxcar integrators that do not use SO circuitry but are optimized for slower phenomena. They can be gated at a fixed time delay to measure TRES. Sampling plug-ins are available on some of these boxcars to permit fast acquisition. A number of very fast sampling gates, some that use optical synchronization, have been described, which circumvent the need for a sampling oscilloscope (23–26).

Sampling has several disadvantages. The linearity and dynamic range are limited. One to two decades of signal linearity are the best that can routinely be achieved. For many applications this is adequate, but where a weak long-lived component follows an intense fast one, it is difficult to obtain faithful TRES. Also, the signal must be stable and repetitive, which rules out photochemically unstable systems.

While lacking in the elegance and speed of data acquisition of vidicon- or laser-based TRES, sampling methods are simple and relatively low in cost. Sampling will be more rapid for many spectra than with a flash lamp–based single-photon counting system, and we do not expect sampling methods to become extinct within the near future.

2.3.2. Video Fluorimeters

An extremely elegant, rapid, and powerful method of recording spectra and decays is the video fluorimeter (7,28–32). Instead of recording an emission spectrum one point at a time, a vidicon or silicon-intensified target (SIT) is placed at the focal plane of the emission spectrograph. By scanning the tube to read out the intensities at different points on the surface, the complete emission spectrum can be recorded at once.

Further, a simple refinement permits simultaneous determination of excitation spectra. The sample is excited, not with monochromatic light, but with dispersed light. The dispersion is vertical so that the top of the sample is excited with one wavelength and the bottom with another. The image in the focal plane of the emission monochromator is the dispersed sample emission for different sections of the sample cell. For example, the top of the image may correspond to the image of the top of the sample cell, which, in turn, corresponds to excitation at one wavelength. The bottom of the image would then correspond to the bottom of the cell, which corresponds to excitation at another wavelength. Thus the projection across the detector is the two-dimensional image of the relative intensity as a function of emission wavelength (horizontal) and excitation wavelength (vertical). By digitizing this image one has a complete excitation-emission matrix for the sample. Figure 2.8 shows a typical excitation emission matrix for several aromatic hydrocarbons in low-temperature glasses.

For weak emitters a SIT vidicon provides the ability to integrate the signal over an extended period of time before readout. In this way considerable signal-to-noise enhancement can be achieved. Indeed, a good video fluorimeter seems to have the sensitivity of a good conventional spectrofluorimeter even though it is recording a complete excitation emission matrix compared to the single point on the conventional instrument. Thus a video fluorimeter can record a complete excitation-emission array thousands of times faster than a conventional instrument.

A final advantage of video fluorimeters is that the detector can be gated (7,32). In this way it is possible to obtain a complete emission-excitation matrix in given time delays following an excitation pulse. Figure 2.9 shows a time-resolved emission-excitation matrix for a mixture of the components of Fig. 2.8. The enormous amount of analytically useful information that can

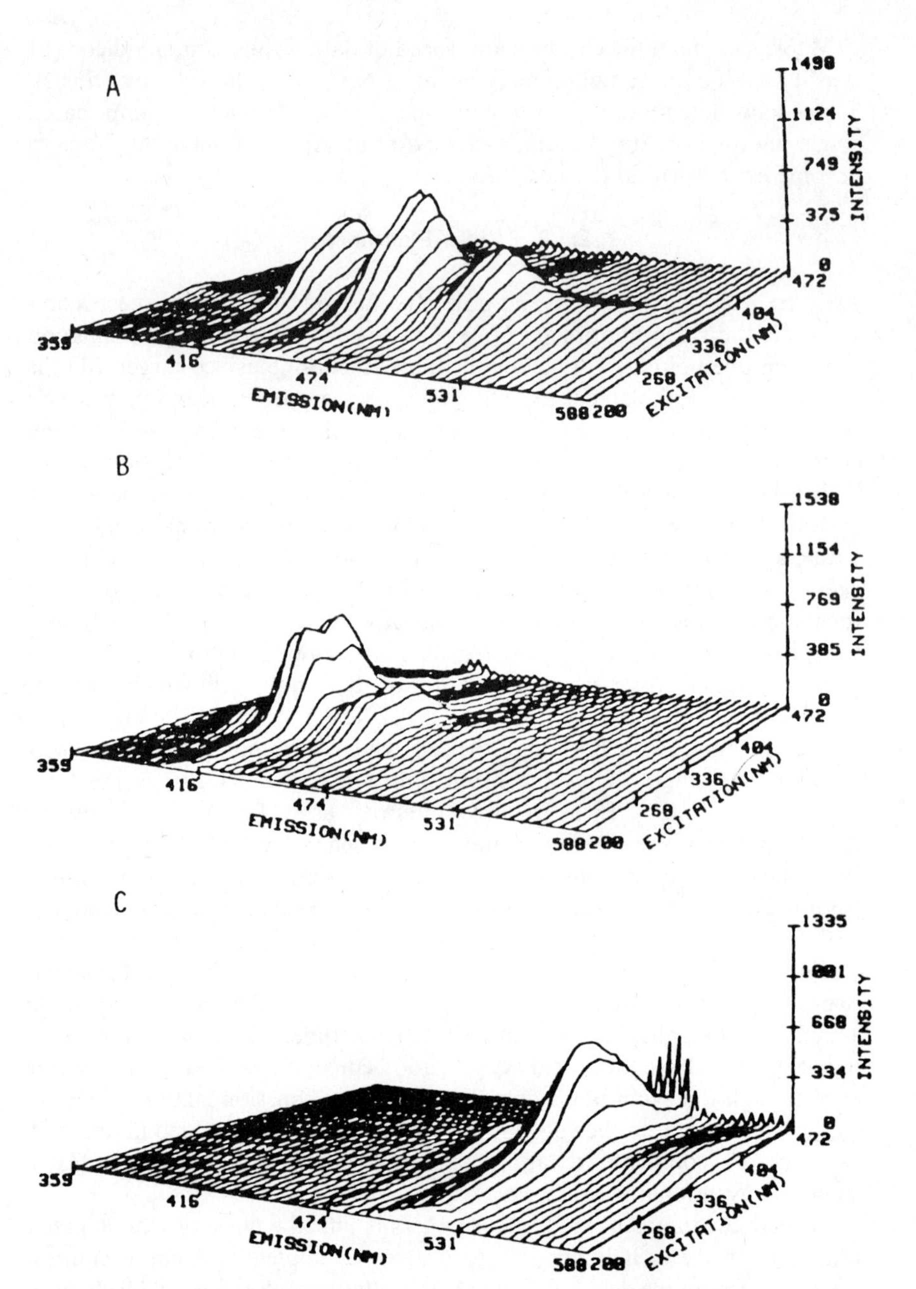

Fig. 2.8. Phosphorescence excitation-emission matrices: (*a*) phenanthrene; (*b*) triphenylene; (*c*) coronene. (Reprinted with permission from *Anal. Chem.*, **54**, 2488. Copyright 1982 American Chemical Society.)

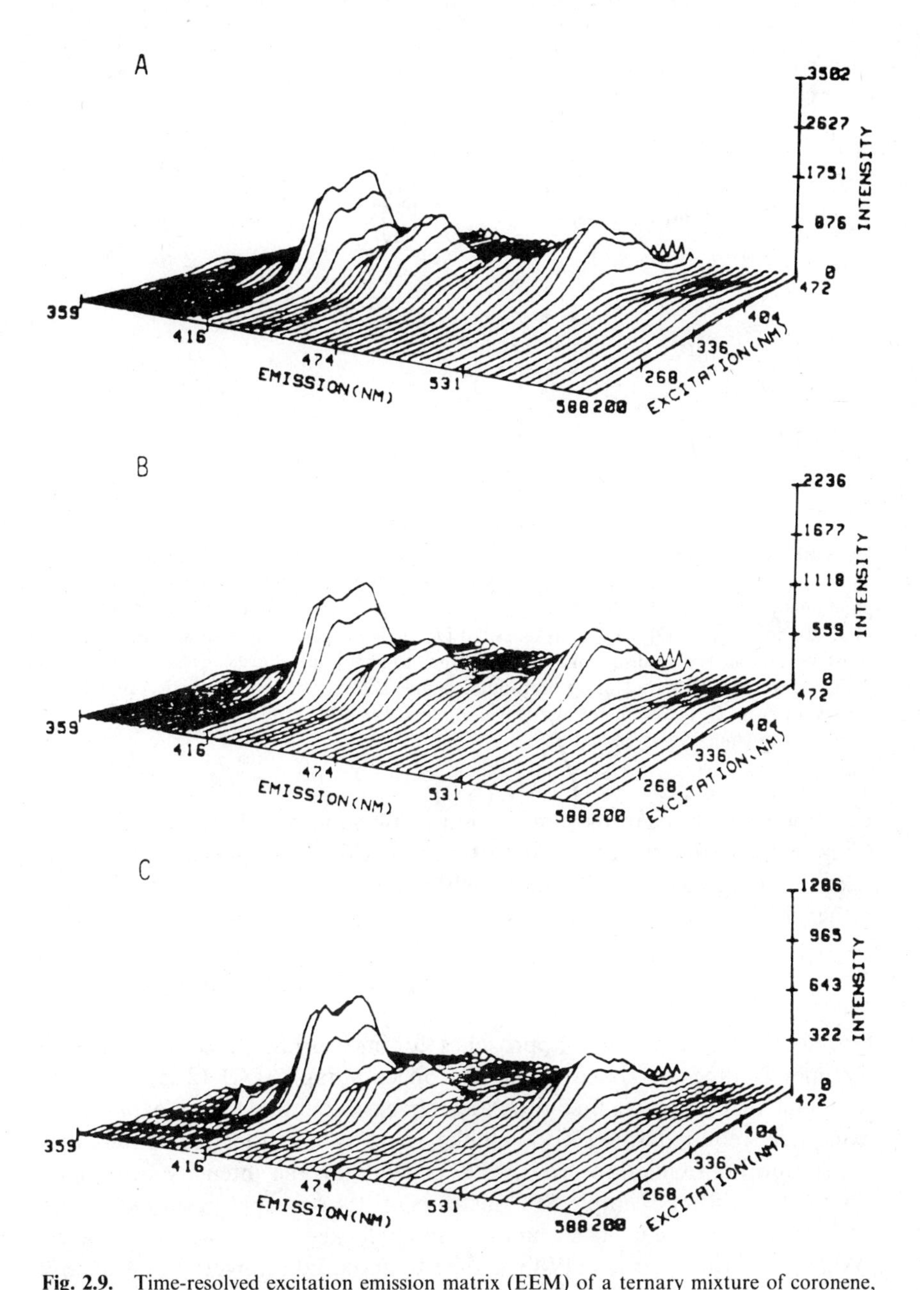

Fig. 2.9. Time-resolved excitation emission matrix (EEM) of a ternary mixture of coronene, phenanthrene, and triphenylene versus delay after excitation: (*a*) $t = 0$ s; (*b*) $t = 5.5$ s; (*c*) $t = 12.5$ s. (Reprinted with permission from *Anal. Chem.*, **54**, 2489. Copyright 1982 American Chemical Society.)

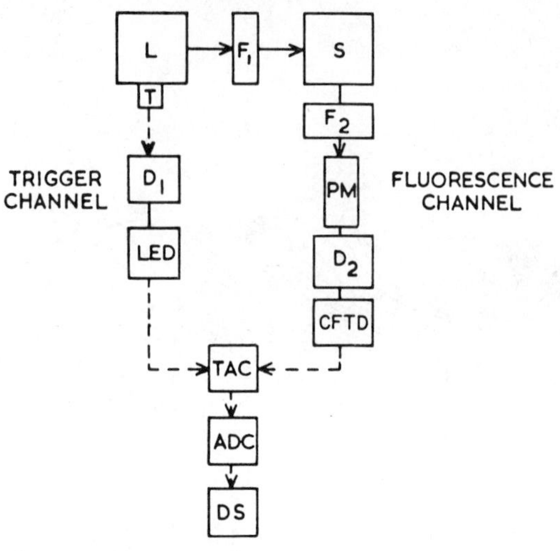

Fig. 2.10. Block diagram of a conventional single-photon counting apparatus. (———) Optical signal; (----) electronic signal; L. excitation source; T, trigger [antenna (fiber optic) and photomultiplier tube, etc.]; S. sample holder; F_1, F_2, filter or monochromator; PM, fast photomultiplier tube; D_1, D_2, delay lines; LED. leading-edge timing discriminator; CFTD, constant-fraction timing discriminator; TAC, time-to-amplitude converter; ADC, analog-to-digital converter; DS, data storage unit (multichannel analyzer or computer). [Reprinted with permission from D. V. O'Connor and David Phillips, *Time-Correlated Single Photon Counting* London; Academic Press, 1984).]

be gathered with such instrumentation is awesome and has yet to be fully utilized. Unfortunately, the gating times on video fluorimeters is typically relatively long (>40 ns), which largely limits their use to phosphorescence rather than fluorescence analyses.

2.3.3. Time-Correlated Single-Photon Counting

One of the most popular approaches for measuring short luminescence lifetimes is time-correlated single-photon counting (SPC) (2–5). As with sampling methods, we first describe the basic method, then explain the extensions required for TRES. Unlike virtually all other techniques that depend on high emission levels, SPC depends on the intensity being low. Figure 2.10 shows a simple SPC instrument. Its principal parts include an excitation source (e.g., flash lamp or mode-locked laser), a start (trigger) PMT, a stop (fluorescence) PMT, a time-to-amplitude converter (TAC), and a multichannel analyzer (MCA) used in the pulse height analysis (PHA) mode. Discriminators are used to shape the pulses to make them compatible with the TAC.

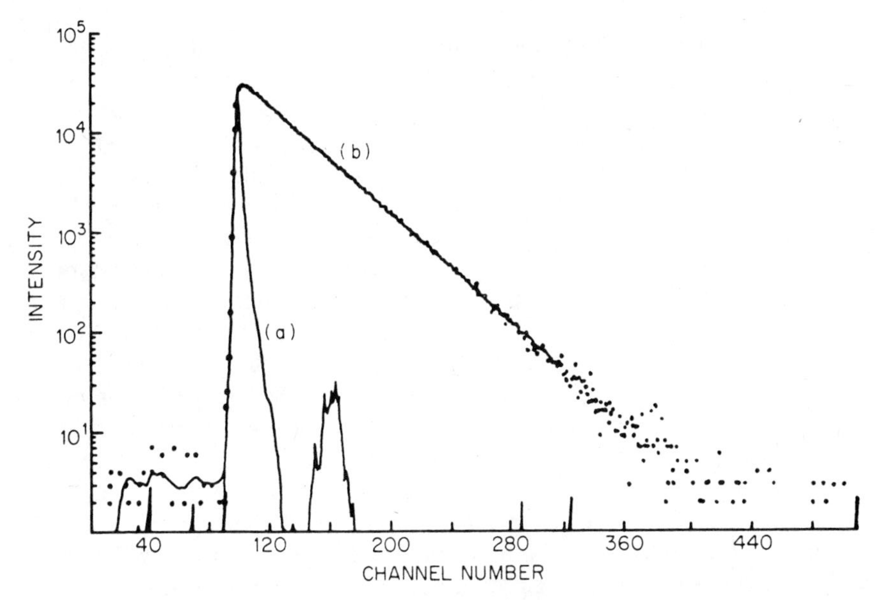

Fig. 2.11. Decay of anthracene (2×10^{-5} *M*) in cyclohexane following laser excitation at 300 nm: (*a*) instrument response function; (*b*) points, observed decay; line, fitted curve. 0.164 ns per channel, τ = 5.23 ns. [Reprinted with permission from D. V. O'Connor and David Phillips, *Time-Correlated Single Photon Counting* (London: Academic Press, 1984).]

The amplitude of the TAC output pulse is directly proportional to the time between the start and stop pulses. In the PHA mode the MCA digitizes the height of the input pulse and increments by one the contents of the memory channel corresponding to this digital value. In operation, the firing of the flash lamp generates a start pulse to the TAC. The sample PMT generates a stop pulse for the TAC when it detects a photon following the flash. The TAC output pulse is presented to the MCA, which determines the memory channel corresponding to the delay time between the start and stop pulses.

For an SPC system to work properly, the probability of a stop pulse occurring during the TAC range must be small (<0.01–0.1 per pulse). Under these conditions the probability of a stop pulse occurring at a specific delay after the flash is directly proportional to the probability of a photon being emitted by the sample at that time. Thus, if many flashes are analyzed, the contents of MCA channels will develop into a histogram having the shape of the sample decay curve. Figure 2.11 shows a set of SPC transients for anthracene. In this case a very short duration laser was the excitation source.

The SPC method has numerous advantages. Since the probability of detecting a photon must be low, excitation intensity and optical efficiency are not as important as with other methods. SPC is, thus, one of the most sensitive decay time methods. Further, system response is not limited by the minimum output pulse width of the PMT; the leading edge of the PMT pulse does the triggering, whereas the width and any tail are discarded. The minimum pulse response is then controlled by jitter in the leading edge of the PMT pulse, the TAC, and the source's width. Therefore, a nanosecond-wide pulse width is easily observed with a PMT having an impulse response several nanoseconds wide. SPC dynamic range and linearity are enormous. With patience, three or four orders of magnitude are routine and five orders are possible (see Fig. 2.11). The data acquisition time is a serious problem with flash lamp systems, but if cost is no concern, the high repetition rate of mode-locked lasers has largely overcome this problem.

Another major advantage of SPC is that the statistics are uniquely defined. SPC data exhibit nuclear counting Poisson statistics (33). The standard deviation of a Poisson distribution equals the square root of the number of counts. These well-behaved statistics permit proper weighting of each point in any data-fitting procedure. This attribute seems to be unique to SPC.

The limitations of SPC include the following: With the available N_2- and D_2-filled flash lamps, excitation is most readily done below 400 nm. These lamps achieve short pulse widths at the expense of intensity (10^8–10^9/flash), so that even with the enormous sensitivity of SPC the intensity can be marginal. Further, flash lamps run at 10^4–10^5 Hz. If only 1% of the pulses are counted, the rate of collection is only 100–1000 counts/sec. To obtain tens of thousands of counts in 512 or 1024 channels may require hours; however, with fewer channels and brighter samples, acquisition is typically done in 5–15 min. Finally, it is more expensive than some other instruments.

Subnanosecond pulse width high-repetition-rate lasers and fast PMTs do permit direct studies of subnanosecond phenomena. But the cost and complexities of operating such systems greatly limit their general applicability. The periodic pulses of synchrotron radiation also make a superb SPC excitation source (34), but the difficulties and expense of gaining access to, and using, such sources largely limits their utility to all but specialized problems.

We turn now to the use of SPC for TRES. The most common procedure is to record a series of complete decay curves at a number of different wavelengths (2–4) (see Fig. 2.1). Once having recorded all the decays, one plots a TRES merely by taking a cross section of the different decays at a fixed delay time. Given the relatively slow acquisition rate of SPC, this procedure is frequently exceptionally time consuming, although the final amount of information about the system can be enormous. For example, if 512 point decays are recorded, one has 512 complete TRES—one for each channel.

A more direct method is to place a single-channel analyzer after the TAC (4,27). The upper and lower thresholds on the analyzer are set to accept only pulses that correspond to a given time window. By counting the number of pulses in this window, one obtains the emission intensity in the window. Then, by scanning the emission monochromator and gating the counts into different bins on the MCA by use of the MCS mode, one obtains the time-resolved emission spectrum for a given delay and window width. The principal advantage of this approach is that it can avoid the need for the MCA, but the price is quite high. Data acquisition is still slow because of the required low duty cycle, and when finished, there is only one complete spectrum rather than many. We therefore do not recommend this procedure as a general rule.

2.3.4. Data Treatment

The workup of the experimental data in TRES can be every bit as important as the method used to gather the data. Data treatment of excitation-emission data from video fluorimeters presents especially severe problems, and readers are referred to the literature (7,30–32). We will describe several common situations, including a simple noninteracting mixture with, and without, significant source distortion. We will also discuss the problem when there is no known functional form for the decay. Also, we will warn about indiscriminate data fitting.

2.3.4.1. Simple Mixtures with No Source Distortion

Typically, for a mixture of noninteracting luminescent materials excited by a source of duration much shorter than the shortest sample lifetime, the observed decay will be a sum of exponentials given by

$$D(t, \lambda) = \Sigma\, I_i(\lambda) \exp(-t/\tau_i) \qquad (2.1)$$

where $D(t,\lambda)$ is the observed emission intensity at wavelength λ and time t following the excitation, $I_i(\lambda)$ the emission intensity of the ith component at $t = 0$, and τ_i the decay time for the ith component. The summation is over all luminescent materials in the sample. The I_i values are related to the relative efficiency of excitation and emission as well as the concentrations of each species. Using such techniques as nonlinear least squares it is possible to fit such decay curves to extract the I_i and τ values (4,5,33). A BASIC program is available for the fitting (5). Alternatively, if the τ values are known, much more accurate estimates of the I values can be obtained.

Fitting parameters to the decay curves provides quantitative information on the components in a mixture, or if the sample is made up of unknowns, the relative contributions and lifetimes of the components at each emission wavelength. To reconstruct a complete set of emission spectra for a mixture requires measuring and mathematically reducing a decay curve at each emission wavelength. While experiments of this type have been done with conventional instrumentation, the data acquisition step alone is quite time consuming.

2.3.4.2. *The Deconvolution Problem*

In pulsed TRES most researchers are confronted with the problem that their minimum instrument response function is no longer short compared to the decay times of the phenomena of interest (2–5). This instrument response is made up of both the finite flash duration and the response time of the detector and its associated electronics. For example, most SPC lamps have durations of 1–5 ns. Even fast PMTs yield an output to an infinitely short light pulse (the impulse response) 1–5 ns in width. Clearly, direct observation of TRES at times comparable to, or shorter than, the instrument response will be distorted by instrumental artifacts. The process of extracting true temporal information from the observed decays and the instrument response function is called deconvolution (2–5).

In the general case, the observed sample decay $D(t)$ is related to the observed excitation $E(t)$ and the sample impulse response $i(t)$ by

$$D(t) = E * i = i * E \tag{2.2}$$

$$E * i = \int_0^t E(t)i(t - x) \quad dx \tag{2.3}$$

where $*$ denotes convolution. The problem then becomes one of extracting $i(t)$ from $D(t)$ and $E(t)$ in this integral equation (2–5).

In the case where the sample impulse is given by a sum of exponentials, deconvolution is straightforward. There are many quite efficient algorithms for deconvolution if $i(t)$ is only a single exponential. Where $i(t)$ is a sum of two or more exponentials, the deconvolution of the K_i and τ values is much more formidable (4,5). The two most popular and efficient algorithms are nonlinear least squares (also known as iterative reconvolution) and the moments methods (4,5,34–36). A FORTRAN program for nonlinear least squares is available (4). If one or more of the excited-state lifetimes is known, the determination of the remaining components becomes both faster and more accurate.

Both methods have been successfully applied to three and four exponentials, although the significance of the parameters for such large numbers of components is subject to question. On this subject we quote from O'Connor and Phillips (4): "It is our opinion based on analyses both successful and unsuccessful of triple exponential decays, that physically valid parameters are extremely difficult to attain, and that all the results of this type of fitting should be carefully scrutinized and checked against other measurements."

2.3.4.3. Global Fitting

Although not yet widely practiced, there is one approach that seems quite attractive. Knorr and Harris (31) have shown that greatly improved accuracy in the fitting of time-resolved spectra can be obtained by globally fitting the entire set of several decay curves to get the preexponential factors for the components as well as their lifetimes. The unknowns in this procedure are a single set of lifetimes and the preexponential factors at each wavelength for each component. This procedure forces a single set of lifetimes onto the data, reduces the number of variables, and greatly increases the precision of the measurements. This approach was shown to work equally well with data that required deconvolution.

As originally reported, the authors used a combination of linear and nonlinear least-squares fitting. They first assumed a set of lifetimes and fit the data by a linear closed-form solution. Improved lifetimes were then obtained by holding the preexponential factors fixed and varying the lifetimes with a simplex algorithm. This completed one iteration and the process was started over with the determination of new preexponential factors. A more efficient algorithm would be a normal nonlinear least squares that varied all parameters simultaneously, such as the Marquardt method (see also refs. 30 and 32).

2.3.4.4. Unknown Decay Models

A more interesting problem arises in cases where not only are the number and lifetimes of the components unknown, but there is not even a known functional form for the decays. This problem arises in solvent relaxation studies where suitable theories for the decays have not been derived. Fortunately, this problem is not as serious as it might at first seem. In the case of these systems the real question is not what the decay times are, but rather, what the temporal dependence of $i(t)$ versus wavelength is. Put in this way the problem is actually simpler than the deconvolution problem, with a defined function where the fit parameters must have physical significance.

The extraction of the shape of $i(t)$ depends on the following observation. $i(t)$ will be a smooth, well-behaved decay, which can generally be well fitted by a sum of several exponentials either with floating lifetimes or fixed ones. If the lifetimes are fixed, the fitting is equivalent to expanding the function with a fixed basis set (4). The deconvolution is extremely fast if fixed lifetimes are used, although some skill in the selection of the optimum basis set is required. Once $i(t)$ is obtained, time-resolved emission spectra are then readily reconstructed at each wavelength. A FORTRAN program for carrying out these deconvolutions is given in the literature (4). Alternatively, if the lifetimes are parameters in the fit, no physical significance is attached to these values.

2.3.4.5. *A Caveat on Multicomponent Analysis*

The above-described fitting or deconvolution of multiple exponential decays is a potential mine field. Although mathematically straightforward, it is frequently overlooked that the fitting of a sum of two or more exponentials is intrinsically a poorly poised system, especially if two of the lifetimes are close to each other. To demonstrate this point, we show two apparently quite dissimilar functions (5,37):

$$I(t) = 7500 \exp(-t/5.5) + 2500 \exp(-t/8) \qquad (2.4a)$$

$$I(t) = 2500 \exp(-t/4.5) + 7500 \exp(-t/6.7) \qquad (2.4b)$$

The lifetimes for the two decays are quite different, and the shorter-lived component is the major contributor in the first case and a minor component in the second. Yet, on a linear scale these quite dissimilar functions are visually indistinguishable over the range 0–50 in t. Further, both functions are so close to each other that the differences between the two curves are less than the Poisson noise on the signals had they been derived from single-photon counting. This result demonstrates convincingly that some data sets are so poorly poised that meaningful results cannot be reliably obtained by curve fitting. Although if a full statistical treatment is carried out with valid weighting factors, the analysis will give estimates of the parameters and it will be clear that the system cannot be evaluated with good accuracy. As a final point it should be stressed that if one or more of the lifetimes is known accurately, the fitting of the preexponential factors and any remaining lifetimes can be done with far greater precision.

The problem becomes much more serious where deconvolutions are involved. See the quote in Section 2.3.4.2 from O'Connor and Phillips on multiexponential fitting (4).

2.4. PHASE SHIFT METHODS

Another powerful and increasingly important method of time resolution involves modulated sources and phase shift measurement (2–8,38). We first describe the mathematics of the phase shift method for τ measurements and then show the extension required for time resolution by the method of phase resolution. Phase shift methods generally do not use pulsed sources but excite the sample with a sinusoidally varying excitation. The sample's lifetime is deduced either from the time or phase shift between the excitation waveform and the sample emission or by the degree of modulation of the emission.

Phase shift methods evolved earlier than pulsed methods for short-lifetime measurement (3,8,38). The technology for measuring small phase shifts of continuous wave signals even in the presence of high noise levels was well worked out, as was high-frequency optical modulators. Indeed, the standard modulator is the Debye–Sears acoustic modulator, which dates back to 1932. Thus the phase shift method was the workhorse of fluorescence decay-time measurements. With the advent of intense pulsed laser sources and single-photon counting methods, phase shift methods have unjustifiably fallen into some disrepute. In fact, of the traditional lifetime measurement methods, the phase shift method is still capable of measuring lifetimes shorter than those measured by any of the conventional methods, such as single-photon counting. Further, recent advances in instrumentation suggest continued and expanded use of the phase shift approach (3,8). Improvements in technology and the recent introduction of a commercial phase fluorimeter have added greatly to their acceptance.

2.4.1. Theory of Single-Component Systems

If a luminescent sample is sinusoidally excited, the luminescence is sinusoidally modulated. The sample luminescence lags or is phase shifted behind the excitation source. The degree of phase shift is related to τ. This shift arises from the finite sample lifetime, which delays luminescence after excitation. We assume that the sample is excited with a sinusoidal waveform of the form

$$E(t) = B(1 + M \sin \omega t) \tag{2.5a}$$

$$\omega = 2\pi f \tag{2.5b}$$

where $E(t)$ is the time dependence of the excitation source, ω the angular modulation frequency of the excitation, f the modulation frequency (Hz), B is related to the intensity of the exciting light, and M is the degree of modulation of the exciting light and can assume values of 0–1, which correspond to 0–100% modulation, respectively. After a steady state in the emission waveform $D(t)$ has been achieved, one then finds

$$D(t) \propto \sin(\omega t - \phi) \tag{2.6a}$$

$$m = 1/[1 + (\omega\tau)^2]^{1/2} \tag{2.6b}$$

$$m_{obs} = mM \tag{2.6c}$$

$$\phi = \tan^{-1}\omega\tau \tag{2.6d}$$

where ϕ is the phase shift between $E(t)$ and $D(t)$ and m_{obs} is the observed degree of modulation of $D(t)$ and is related to the intrinsic sample modulation m and the measurable degree of source modulation M.

To measure lifetimes with a phase shift instrument one measures either the phase shift ϕ or the degree of modulation m_{obs}. Then Eq. (2.6*c*) or (2.6*d*) is used to calculate τ. Both methods have been used either separately or simultaneously to complement each other.

Figure 2.12 shows phase shifts and degrees of modulation versus τ for $f = 10$ MHz and $M = 1$. It is clear that the optimum frequency for a lifetime measurement using either the ϕ or m method is when τ is on the order of $1/\omega$. It is thus desirable to be able to select several modulation frequencies depending on the range of lifetimes under study.

In the past, Debye–Sears modulators did not lend themselves to simple frequency changes, as they depended on the resonance frequency of a tuned cavity. At least for molecules absorbing in the red region such as chlorophylls, light-emitting diodes have turned out to be excellent sources (39), and electrooptical modulators can modulate lasers at much shorter wavelengths. Operating frequencies of 100–500 MHz can be achieved yielding low-picosecond time resolution (40).

The availability of high-intensity light sources and efficient light modulators currently makes the phase shift method an easily implemented, quick, and reliable method for obtaining τ's, but it suffers from several disadvantages. The indirect nature of the phase measurement can conceal system complexities that may be more apparent using flash measurements. A single phase shift measurement gives no warning of the presence of several components with different lifetimes; any number of components still gives a pure sinusoidal decay, but with a phase and degree of modulation which is a

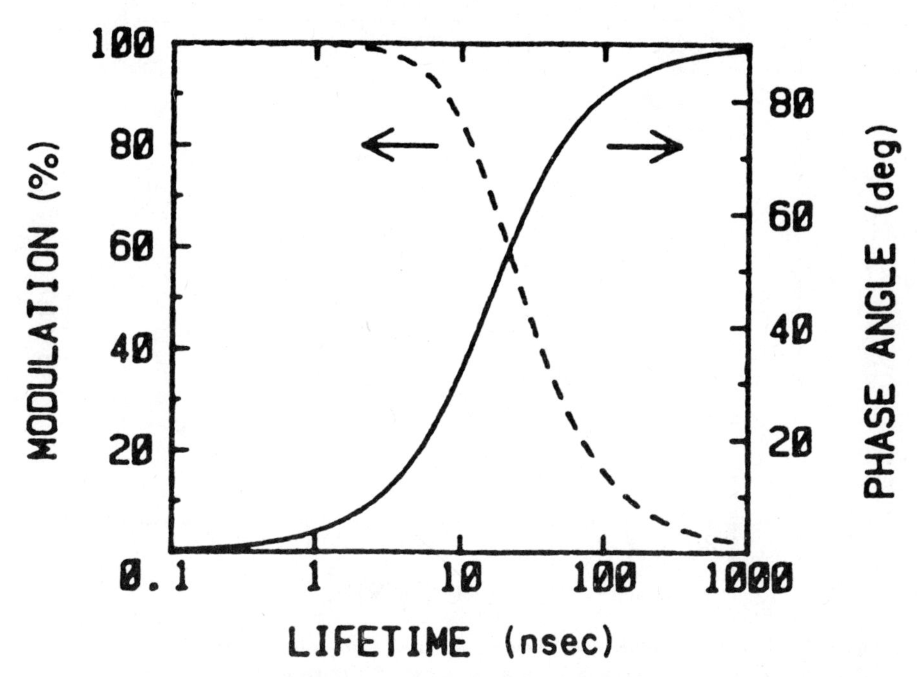

Fig. 2.12. Theoretical phase shift (———) and percent modulation (---) of a luminescent sample versus sample lifetime. Note the nonlinear lifetime scale. The excitation source is 100% modulated at 10 MHz. [Reprinted with permission from J. N. Demas, *Excited State Lifetime Measurements* (New York: Academic Press, 1983).]

vector weighted average of all the components. The flash method, however, yields nonlinear semilogarithmic plots of multiexponential decays. In a suitably modified form, however, multiple exponential resolution is feasible by phase measurements.

2.4.2. Theory of Multicomponent Systems

If the sample emission is made up of two or more overlapping sinusoidal waveforms with different amplitudes and phases, it is possible to extract quantitative spectra. For N components the ac portion of the emission waveform is given by

$$D(\lambda, t) = \Sigma I_i(\lambda) m \sin(\omega t - \phi_i) \tag{2.7}$$

where the subscripts refer to the ith component.

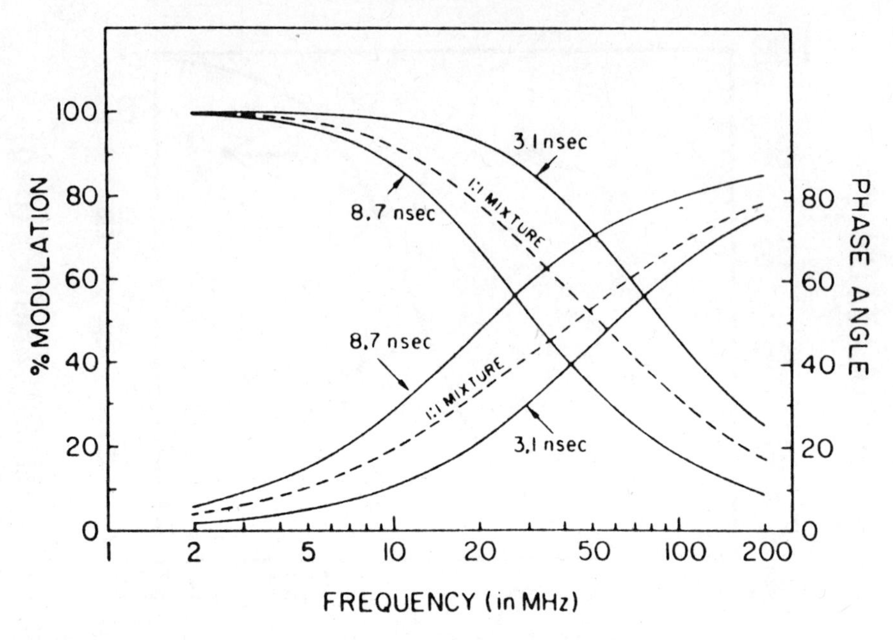

Fig. 2.13. Simulated multifrequency phase and modulation curves for single exponential decays of 3.1 and 8.7 ns and for a 1:1 mixture of the two components. [Reprinted Ref. 8, p. 55, by courtesy of Marcel Dekker, Inc.]

The observed phase shift ϕ and the degree of modulation m_{obs} for the mixture is given by

$$\tan \phi = S/G \tag{2.8a}$$

$$m_{\text{obs}}^2 = S^2 + G^2 \tag{2.8b}$$

$$S = \Sigma f_i m_i \sin \phi_i \tag{2.8c}$$

$$G = \Sigma f_i m_i \cos \phi_i \tag{2.8d}$$

where the f_i, m_i, and ϕ_i are the fractional contributions to the emission, the degree of modulation, and the phase shifts for the ith component, respectively. The summations are over all components.

The far greater complexity of even a binary mixture is revealed in Fig. 2.13, which shows both the degree of modulation and the phase shift for a 1 : 1 mixture. In this case the lifetimes are 3.1 and 8.7 ns. The data for each of the pure components are shown for comparison.

2.4.3. Phase-Resolved Spectroscopy

In PRS of a mixture one exploits the differences in lifetimes and thus phase angles of the components to separate the emission contribution of different components [Eq. 2.6]. This ingenious idea appears to have originated with Veselova and co-workers (41,42) in the early 1970s, but its full applications have only recently been realized (6,42–45). The phase resolution or separation of the luminescence signal is performed with a lock-in amplifier. Synchronous detection in a lock-in amplifier is equivalent to multiplying the waveform by a periodic square wave with the same frequency as the modulation and then filtering the rectified signal to yield its average dc value (6). The amplitude of the square wave is +1 for every half-cycle beginning at a detector phase shift of ϕ_D and -1 for every half-cycle beginning at $\phi_D + 180°$. ϕ_D is the phase shift introduced by the lock-in amplifier and is generally variable over at least 360°. The resultant dc output of the lock-in is given by

$$I(\lambda) = \Sigma A_i(\lambda)\cos(\phi_D - \phi_i) \tag{2.9}$$

where the A_i are intensities weighted by the intensity of each component and the loss of ac emission intensity given by Eq. (2.6). Again the summation is over all the components. The principle of PRS is that there is an instrumental phase angle setting that will quadrature or null out one of the components. For the ith component this condition is satisfied when $\phi_D = \phi_i \pm 90°$. This phase nulling will completely eliminate the dc contribution of the ith component. However, if the lifetimes of the other components are different from that for the ith component, they will be attenuated but not completely suppressed.

If a system has only two components of different lifetimes, proper adjustment of ϕ_D permits complete elimination of one of the two components. Similarly, by readjusting the ϕ_D, the other component can be suppressed while leaving the first. In this way the emission spectra of both components can be extracted.

Figure 2.14 shows a particularly impressive example of phase resolution. The two components POPOP and dimethyl-POPOP differ in excited-state lifetime by only 0.2 ns (1.0 ns versus 1.2 ns). The extraction in this case was accomplished using a modulation frequency of 30 MHz, which is much lower than optimum for such short-lived species.

A two-component spectrum is relatively easy to resolve, especially if the lifetimes of one or both of the species is known so that the proper ϕ_D values can be set. The problem becomes much more severe if there are more than

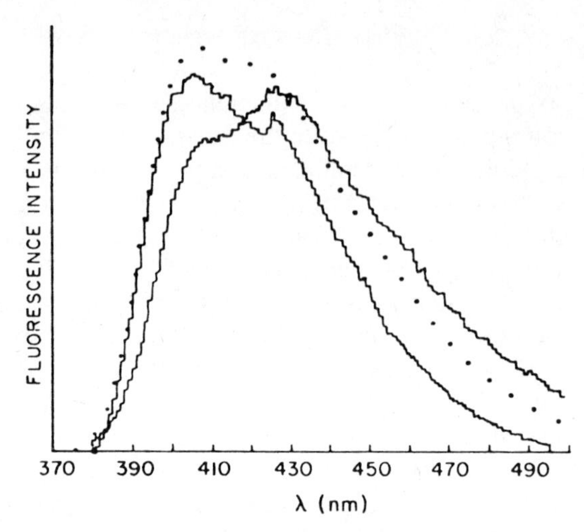

Fig. 2.14. Phase-resolved spectrum of POPOP and dimethyl-POPOP in methanol. The dotted line is the total emission spectrum. The solid lines are the phase-resolved spectra of each component with the high-energy emission from POPOP and the low-energy one from dimethyl-POPOP. λ_{ex} = 355 nm; $\Delta\lambda_{ex}$ = 1 nm; $\Delta\lambda_{em}$ = 8 nm; f = 30 MHz; 10^{-6} M concentration. (Reprinted with permission from ref. 46.)

two components and none of the lifetimes are known. Even if the ϕ_D values for complete suppression of one component is known, suppression of this component simplifies but does not completely resolve a spectrum. The phase-resolved spectrum is made up of a weighted sum of any other components, and direct extraction of these other components is not trivial.

For analytical applications of PRS, where complete spectra are not required, it has been shown that the relative concentrations of two species is best obtained, not by phase nulling, but by determining the emission intensities for several different ϕ_D values (6,43–45). A set of simultaneous equations analogous to Eq. (2.7) was set up. The resultant set of simultaneous equations was solved to yield the concentration of the components. Alternatively, if more measurements are made than were required, the overdetermined set of linear equations can be solved to yield much more accurate K values than would be obtainable by direct phase nulling.

If one were going to extract phase-resolved spectra by the multiple-phase-angle approach, however, the process of carrying out the necessary measurements at the different phase angles for each emission wavelength would be rather tedious with a conventional phase fluorimeter. There are, however, several computer-controlled lock-in amplifiers on the market. A completely computerized spectrofluorimeter coupled to an interfaced lock-in would permit automated data collection at a number of wavelengths and ϕ_D values.

Resolution of the different components could then be achieved relatively easily.

The analysis above assumes that the lifetimes of the components are known. In an unknown mixture, this is rarely true. Under these conditions, it becomes essential to extract an accurate lifetime before attempting spectral fitting. In principle, it is possible to extract these lifetimes from both the degree of modulation and the phase shift. In practice, with a single-frequency measurement, the errors involved are often unacceptably high because the system is too poorly poised. However, if phase shift and degree of modulation measurements are made at several different frequencies, the system again becomes overdetermined, and much more accurate lifetimes can be obtained by such methods as nonlinear least squares (8,40,47,48).

2.4.4. PRS Instrumentation

Instrumentation for carrying out phase shift lifetime determinations has been described extensively and will not be repeated here (3,8,38). A phase-resolved spectrometer can be built from a phase shift lifetime instrument by little more than adding an emission monochromator and possibly a lock-in amplifier.

For low-frequency operation (<200 Hz) such an instrument is quite reasonable. At high frequencies, more extreme measures have been necessary. Traditionally, high-frequency lock-in amplifiers have not been available for the megahertz frequencies required for fluorescence measurements. To circumvent this problem, the high frequency has been down-converted to a lower frequency by beating the high-frequency signal with a slightly different one in a nonlinear mixer. A sum-and-difference frequency are produced. This sum-and-difference frequency contains all the phase and amplitude information present in the original high-frequency signal. If the difference frequency falls in the operating range of a lock-in amplifier, the difference can be processed just as if a high-frequency lock-in had been available.

The nonlinear mixing is usually done in the photomultiplier, an approach pioneered by Spencer and Weber (49). The PMT gain is varied by modulating the dynode voltages by applying an RF signal to the dynode string. The primary and sum frequencies are eliminated with a simple low-pass filter. The difference frequency is usually in the range 20–100 Hz. This cross-correlation method is widely used in homemade instruments and in the commercially available SLM 4800.

There is, however, a commercially available lock-in amplifier that will work to 50 MHz, which is higher than that used on most phase fluorimeters. We have shown that this lock-in performs at near the photon statistical quantum noise limit in a phase fluorimeter.

Typically, the modulators are water/alcohol-filled Debye–Sears modulators driven with quartz crystals (3,8,38). These modulators are nontrivial to tune. Further, the operating frequency cannot easily be changed, and generally the frequency range of operation is rather limited. They do, however, have a relatively large acceptance angle and aperture. They are thus particularly convenient with arc lamp sources and monochromators.

Pockel cells and acoustooptic (AO) modulators, especially when used with laser excitation sources, are very attractive modulators. Inexpensive AO modulators work to 40 MHz with high modulation efficiency.

Figure 2.15 shows a homemade phase-resolved spectrofluorimeter based on an AO modulator and commercially available lock-in amplifiers. Spectra taken with this system are presented later. This system represents only very minor changes to a conventional Raman instrument. The AO modulator worked to 40 MHz. Depending on the lifetimes involved and the frequency used for separation, either a standard low-frequency lock-in or a 50-MHz PAR 5202 lock-in amplifier were used. Even though the instrument was analog, its performance was near the quantum photon statistics theoretical limit.

One of the most intriguing ideas in phase fluorimetry is to replace the sinusoidally modulated source with a high repetition rate pulsed source having a very short duration (8,47,51). Suitable sources have been synchrotron radiation and mode-locked lasers, which can give pulses in the range nanoseconds to 100 ps. PRS does not care what the source looks like as long as it has a frequency component at the frequency to be detected. However, a pulse train of very narrow pulses can be decomposed into its component frequencies. If the pulse train has a frequency f, the fundamental and the higher harmonics will contain appreciable energy. The narrower the pulses, the higher the frequencies available. The fundamental and the harmonics can each be used to excite the sample and extract information.

For example, the ADONE synchrotron source at Frascati has a pulse width of 2 ns and repetition rate of 2.88 MHz. As pointed out by Gratton and Lopez-Delgado (51), such a source has considerable power in fundamentals and harmonics to well over 500 MHz, where the power is only half the value at the fundamental. In reality, it is the response of the PMT that sets the upper-frequency limit. Thus, exciting the sample with a pulsed source of this type is equivalent to exciting the sample with dozens of different sinusoidal sources. Merely by extracting phase and amplitude information at a number of these frequencies, a large set of data is directly available. The beauty of this approach is that a single source behaves as though it were many.

Gratton et al. (47) have successfully built an instrument that capitalizes on this frequency distribution. In this instrument the half-efficiency frequency point was about 100 MHz, limited by the PMT response. Figure 2.16 shows

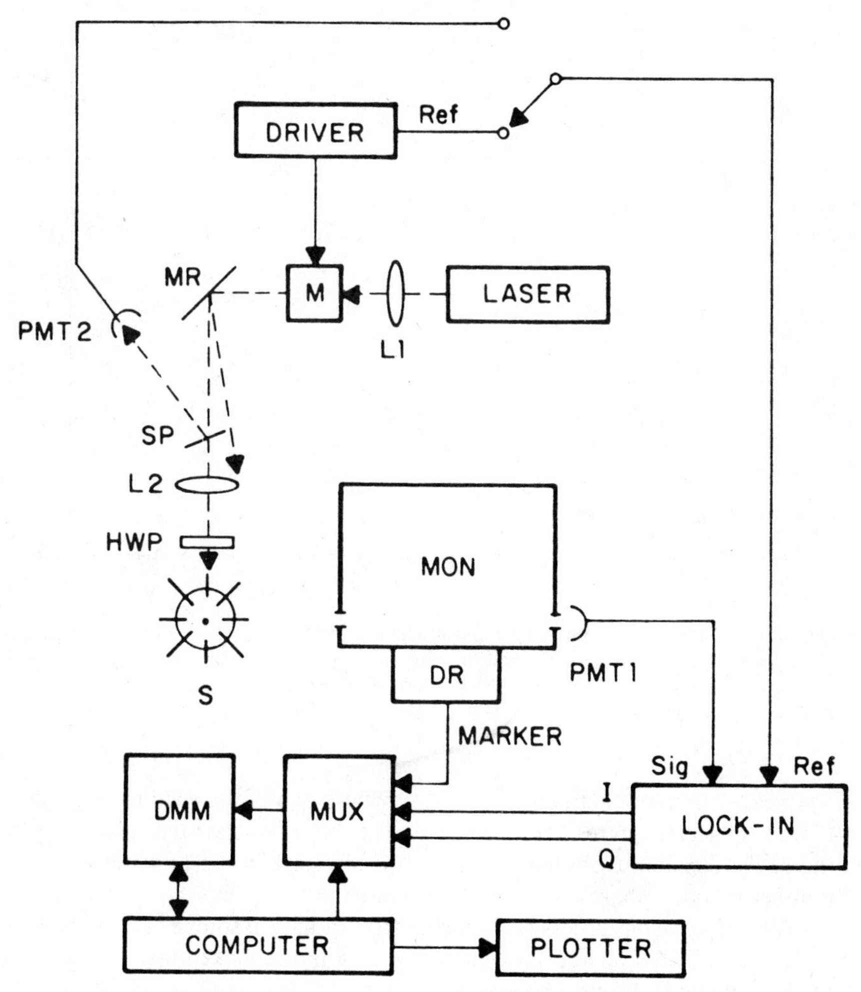

Fig. 2.15. Schematic representation of a computerized phase-resolved spectrometer. LASER, ionized argon laser; PMT1, signal phototube; PMT2, reference phototube; L1, 10-cm focusing lens for AO modulator; L2, defocusing lens or neutral density filters; DRIVER, modulator oscillator/driver; M, AO or electrooptical (EO) modulator; HWP, half-wave plate for EO modulator; MON, monochromator; DR, monochromator wavelength drive; LOCK-IN, lock-in amplifier; MR, mirror; SP, microscope slide beam splitter; MUX, analog multiplexer; S. Sample; DMM, digital multimeter. (Reprinted with permission from *Anal. Chem.*, **57**, 538. Copyright 1985 American Chemical Society.)

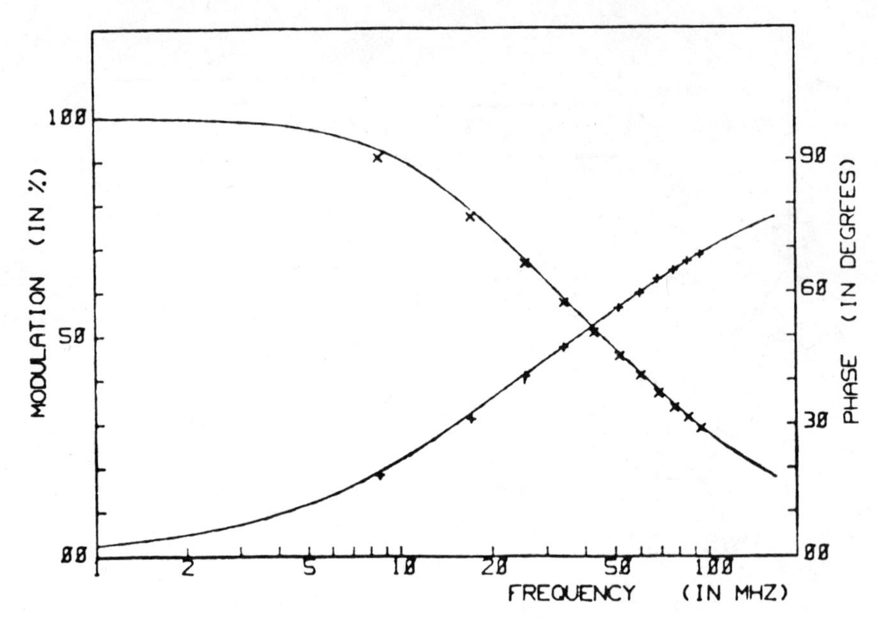

Fig. 2.16. Multifrequency phase (+) and modulation (X) data for tryptophan at 20°C, pH 9.25. Solid lines correspond to the best fit using two exponential components: $\tau_1 = 3.193 \pm 0.026$ ns, $\tau_2 = 9.00$ ns, and $f_1 = 0.396 \pm 0.005$. [Reprinted with permission from *Rev. Sci. Instrum.*, **55**, 492 (1984).]

a typical two-component data set and fit where one of the τ values was fixed. Nonlinear least-squares fits are shown. The excellent statistics, especially in the ratio, can be attributed to the fixing of one lifetime and the use of a number of both phase and amplitude points.

Not everyone has access to a synchrotron radiation source. The same type of data could also have been achieved with a mode-locked laser, which is a readily available, if expensive, item.

The disadvantage of the pulsed light source method is that it does not readily lend itself to spectral recording at more than one frequency. Each point on the curves of Fig. 2.16 appears to represent a relatively tedious experiment. Although by working at only one frequency, changeovers would not be required and routine PR spectra could be obtained. Hopefully, as the technology grows, simpler procedures and data reduction methods will be developed.

McGown and co-workers have also carried out multicomponent analysis using combinations of amplitude versus phase angle and wavelength (6,43–45). These measurements represent a version of PRS. In their current form, however, the computational and demanding measurement requirements limit these approaches to an analysis of well-defined mixtures rather than for exploratory time-resolved spectroscopy.

We suggest that the best approach to phase-resolved TRES would be similar to that proposed for pulsed TRES. One would measure several complete PR spctra at different ϕ_D values and perhaps different modulation frequencies. Then by a global least-squares analysis, simultaneously fit all the spectra using one or more emission spectra with different lifetimes. Although computationally quite complex, modern "microcomputers" can now rival or exceed the performance of older mainframe and minicomputers and the time scales to complete the necessary computations appear much less formidable than they once did.

2.5. APPLICATIONS

Applications of TRES and PRS to fundamental and practical problems are becoming increasingly prevalent. The added degree of experimental freedom gained by time resolution in analytical problems can make a difficult or impossible chemical resolution problem straightforward. In the elucidation of molecular or excited state structure, time resolution provides insights that can frequently be gained by no other method. The following discussion provides a few examples of the utility of TRES and PRS. As the instrumentation grows in sophistication and performance and more machines become commercially available, we expect the field to grow explosively in terms of both routine and imaginative new uses. In particular, we expect dramatic improvements in PRS as new instrumental and mathematical tools are developed.

The area of the applications of TRES and PRS is so broad that it is necessary to limit the discussion. We have attempted to provide important and useful examples from several different areas, but obviously, this selection must be prejudiced to some extent by our own interests and backgrounds. We also present some additional experimental and mathematical information.

2.5.1. Analytical Chemistry

Generally, the low level limits of luminescence analytical measurements are determined by the sample blank. This blank is usually made up of Mie, Rayleigh, or Raman scattered excitation light and solvent or impurity luminescences. A similar problem confronts the Raman spectroscopist who examines the weak spectrally shifted scattered radiation. Because of the intrinsic weakness of the Raman spectrum, impurity or intrinsic sample luminescence can totally obscure it. Different luminescence materials can frequently have different emission lifetimes. Scattered radiation can be treated as a zero-lifetime emitter. These differences in emission lifetimes can then be used to separate the background from the desired sample signal. TRES and PRS have proved exceptionally useful in both areas.

2.5.1.1. *Time-Resolved Emission Spectroscopy*

Time resolution can be extremely efficient at removing an undesired component if the emission lifetimes are sufficiently different. A particularly powerful and widespread example is the virtually complete separation of nanosecond fluorescences, interferents, and scattered excitation light from millisecond to second phosphorescences by means of mechanical phosphorescopes (1,14,16). This approach has been the basis of numerous analytical methods based on low-temperature phosphorescence; sensitivity is greatly enhanced by removing scattering and fluorescence interference. The need for mechanical choppers has limited this approach with phosphorescences to lifetimes longer than about 100 μs.

More recently, pulsed phosphorescence methods have greatly improved time resolution and sensitivity. Time resolution down to about 10 μs can be achieved using flash lamps, while laser excitation sources extend time resolution down to the nanosecond region. This increased time resolution is, however, gained only by the added complexity of gating the detection system.

Figure 2.9 shows a particularly beautiful example of simultaneous multidimensional resolution of a three-component hydrocarbon mixture. Each plot is the relative emission intensity as a function of excitation and emission wavelength for different delay times following a pulsed excitation. The enhancement of the longer-lived triphenylene and corenene with increasing delay times versus the shorter-lived phenanthrene is clearly evident.

Raman spectroscopy can be greatly enhanced by time resolution. Figure 2.17 shows one of the earliest examples of pulsed time-resolution Raman spectra. In this case the Raman band is the 992-cm^{-1} benzene Raman band in solution with 1.3×10^{-9} *M* Acridine Orange, which has a lifetime of about 4 ns. The substantial enhancement is clearly noticeable. Figure 2.18 shows a more recent example using a 16-ns fluorescence of rubrene in benzene. Figure 2.19 shows a biological example for the Raman scattering of adenosinemonophosphoric (AMP) with added tryptophan, which has a 4.5-ns fluorescence. The enhancement in this case is about 3.

It is important to note that in using pulsed methods the separation of scattering from luminescence is more difficult than removal of luminescence from scatter. The reason for this is that if the sample lifetime is longer than

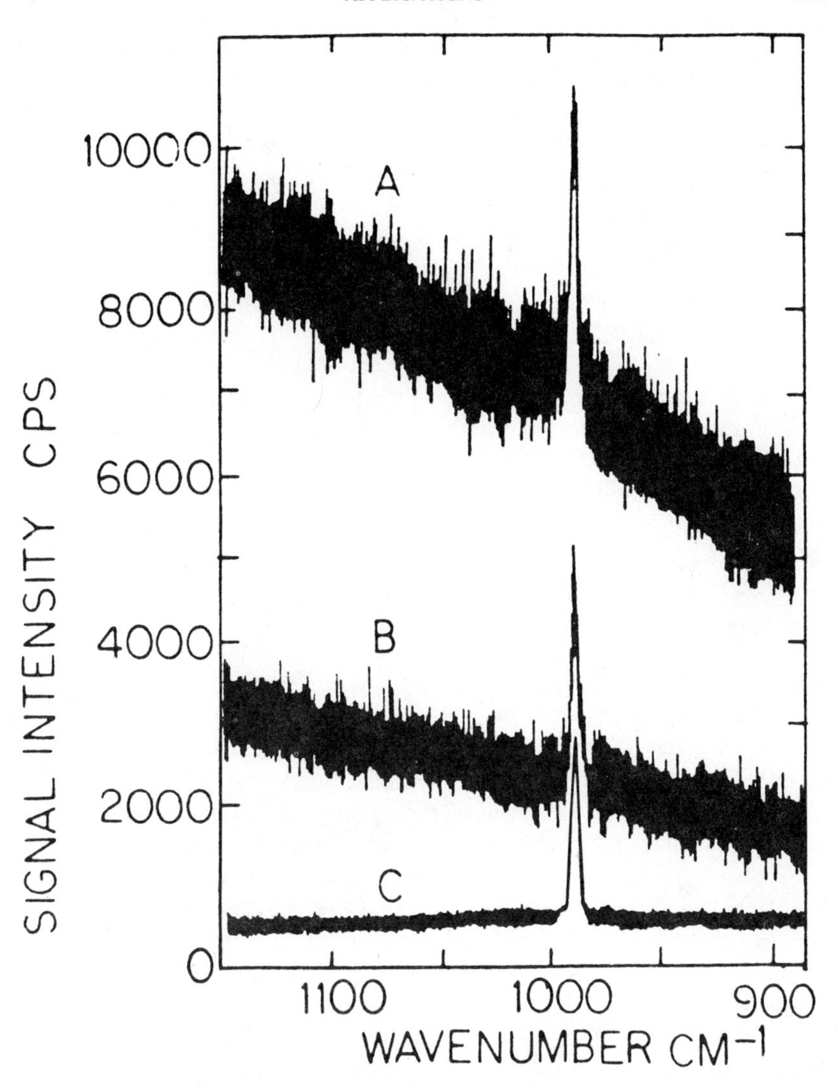

Fig. 2.17. Mode-locked laser Raman spectra of benzene doped with Acridine Orange. *A*, No fluorescence rejection; *B*, with fluorescence rejection; *C*, neat benzene under same conditions as *B*. (Reprinted with permission from *Anal. Chem.*, **46**, 220. Copyright 1974 American Chemical Society.)

the flash, there will always be a time where the sample is still emitting but the scatter is eliminated. However, no matter how long-lived the emission, there will always be a contribution of the emission to the scatter. This accounts for the nonzero backgrounds in all of the Raman spectra.

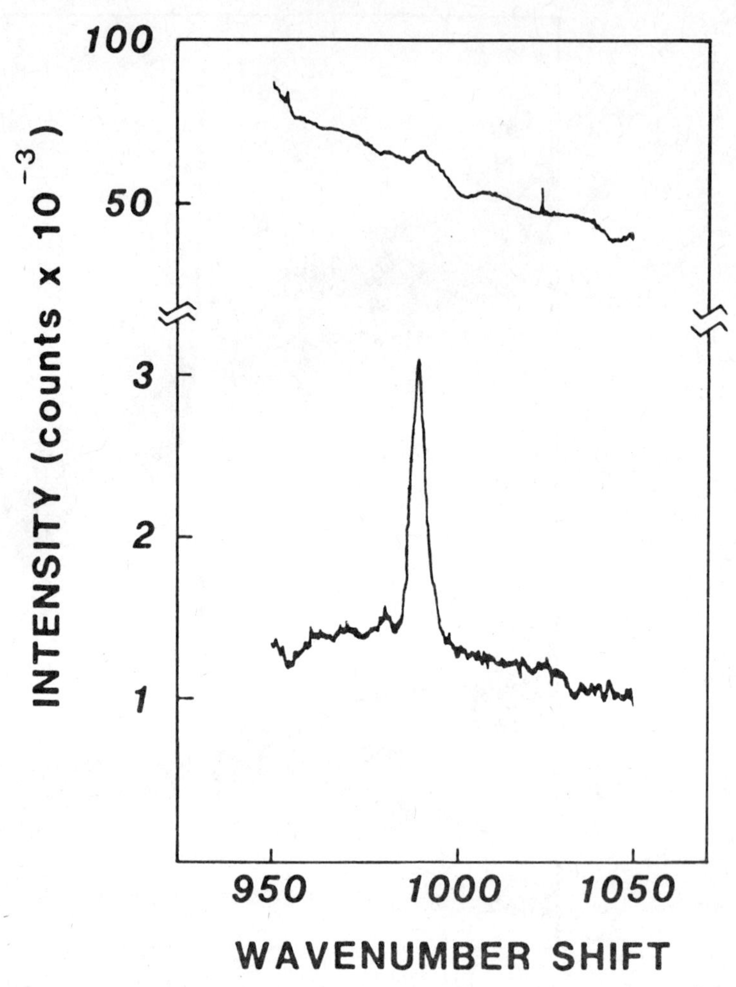

Fig. 2.18. Time-resolved rejection of fluorescence from Raman spectra: 10^{-5} *M* rubrene in benzene. Top curve represents continuous detection; bottom curve taken with high-repetition-rate gated photon counting scheme. Both spectra obtained with 100 mW average power and 1 s integration time (bandpass = 2 cm^{-1}). (Reprinted with permission from *Anal. Chem.*, **54**, 636. Copyright 1982 American Chemical Society.)

2.5.1.2. *Phase-Resolved Spectroscopy*

Phase-resolved spectroscopy (PRS) has also proved successful in examining long-lived phosphorescences. For long-lived organics in particular, low-cost mechanical choppers and low-frequency lock-in amplifiers can be used to

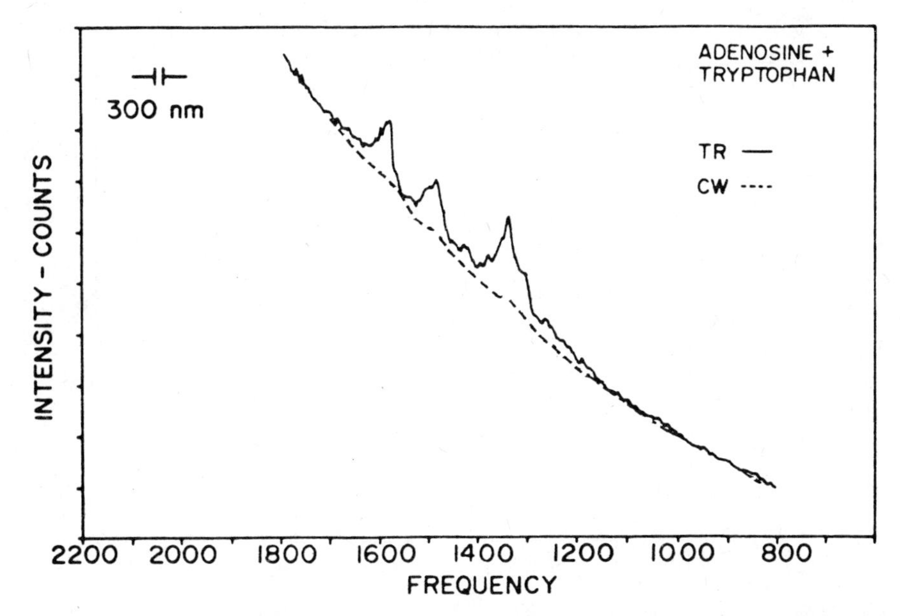

Fig. 2.19. Time-resolved rejection of fluorescence from Raman spectra: 10^{-2} *M* AMP in the presence of 10^{-6} *M* tryptophan. Solid curve represents time-resolved detection; dash curve corresponds to continuous detection. Both spectra were obtained with 2 mW average power and 1 s integration time. Counts have been normalized to emphasize the enhancement in the Raman signal above the background (bandpass = 10 cm^{-1}) (23). [Reprinted with permission from *Anal. Chem.*, **54**, 637. **Copyright 1982 American Chemical Society.]**

increase the performance. Figure 2.20 shows a phase-resolved phosphorescence spectrum (52). The very strong fluorescence and scattered radiation from the sample in Fig. 2.20a is virtually completely suppressed by phase resolution (Fig. 2.20b). It should be pointed out that where mechanical phosphorescopes can be used, traditional PRS will never be as satisfactory at reducing background and noise. Standard PRS only hides the undesired emission, it does not eliminate the fluorescence and scattered photons. Thus, although this suppressed component does not contribute to the dc signal, it does add to the noise. Further, as the magnitude of the scattered component increases, the difficulty of setting and maintaining a stable-phase null increases (50).

PRS also proves quite useful in enhancement of luminescence and Raman in the presence of background. Figure 2.21 shows the spectra of several systems that we selected to evaluate PRS. We selected rhodamine 6G, which has a 3-ns fluorescence, $[Ru(bpy)_3]^{2+}$ (bpy = 2,2′-bipyridine), which has a 375-ns phosphorescence, and the water Raman scattering band. All three

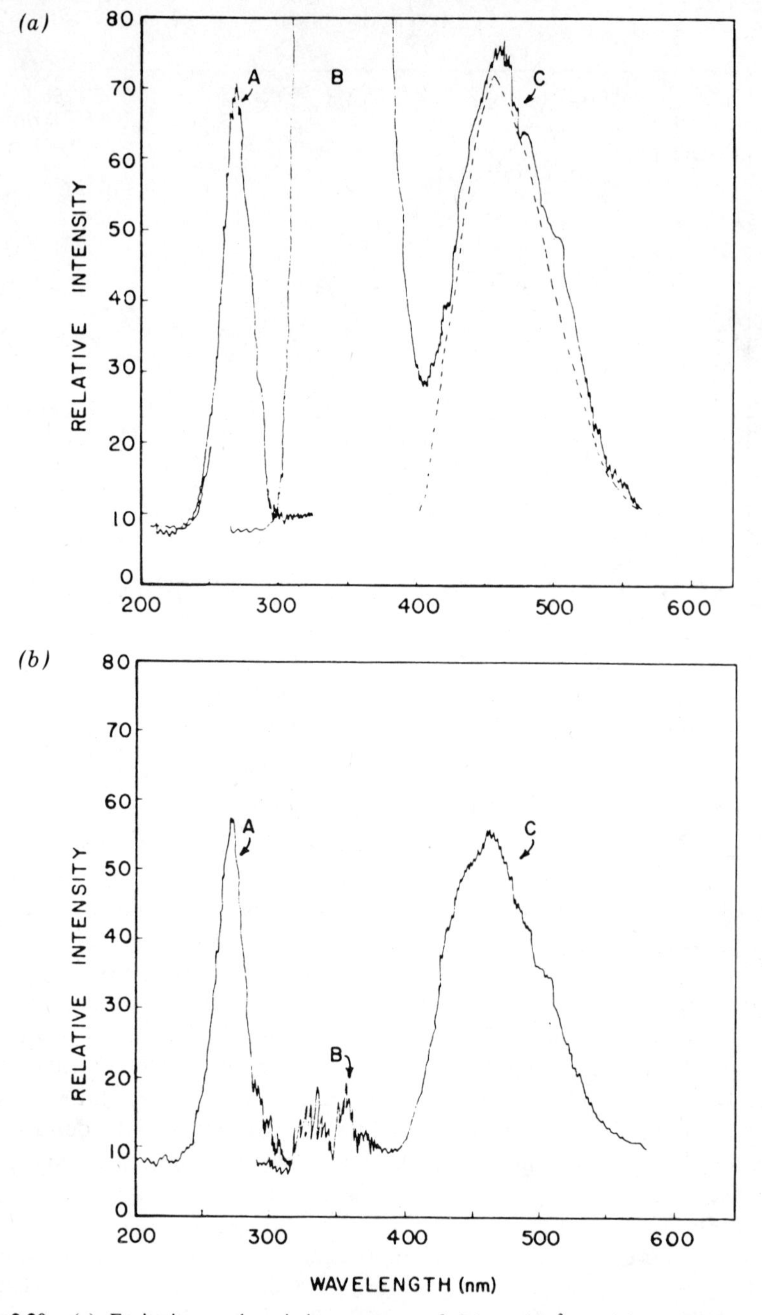

Fig. 2.20. (a) Excitation and emission spectra of 2.1×10^{-3} *M* 2-bromobiphenyl. *A*, Excitation spectrum; *B*, fluorescence emission peak. *C*, phosphorescence emission peak. (---) Phosphorescence emission spectrum of 2-bromobiphenyl standard. (b) Phase-resolved excitation and emission spectra of 2.1×10^{-3} *M* 2-bromobiphenyl at 50 Hz. *A*, Excitation spectrum; *B*, residual fluorescence emission; *C*, phosphorescence emission spectrum. (Reprinted with permission from *Anal. Chem.*, **46**, 1204. Copyright 1974 American Chemical Society.)

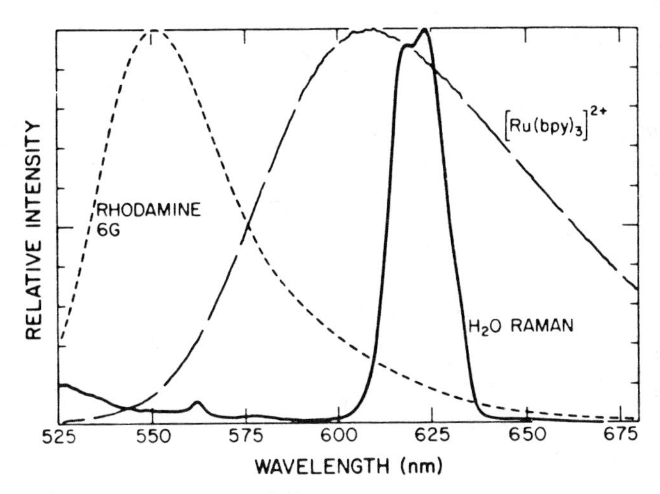

Fig. 2.21. Luminescence spectrum of aqueous 400-nm rhodamine 6G (---) and of 130 μ*M* $[Ru(bpy)_3]^{2+}$ (—— ——). Raman spectrum of pure water (———). Relative intensities cannot be compared. (Reprinted with permission from *Anal. Chem.*, **57**, 540. Copyright 1985 American Chemical Society.)

spectra overlap strongly and the range of lifetimes spanned that expected for analytically useful materials and interferents.

Data were taken with the instrument shown in Fig. 2.15. Figure 2.22 shows extraction of the Raman spectra of water in the presence of an intense phosphorescence, while Fig. 2.23 shows suppression of strong water bands in the luminescence of $[Ru(bpy)_3]^{2+}$. These results were obtained with a low-frequency lock-in amplifier and demonstrate clearly the potential of PRS in such areas as room-temperature phosphorescence. These results show one advantage of PRS over gated TRES. In PRS there is complete symmetry in the equations, and it is just as easy to resolve the short- or the long-lived component. In gated TRES, however, gating can reduce the contribution of the long-lived component, but not completely suppress it.

Examples of resolution of the 3-ns rhodamine fluorescence from the water Raman and water Raman from the rhodamine fluorescence were also presented (50). For resolution of the short-lived species, it was necessary to work at 40 MHz.

Recently, we have demonstrated the utility of phase-resolved spectroscopy for the resolution of a 170-μs phosphorescence of uranyl ion from the Raman and fluorescence (53) of a phosphoric acid analytical media (54). Detection limits of about 2 ppb have been achieved, which is one to two orders of magnitude below that achievable without phase resolution. We add, however, that much more elaborate TRES methods using a N_2 laser pumped dye

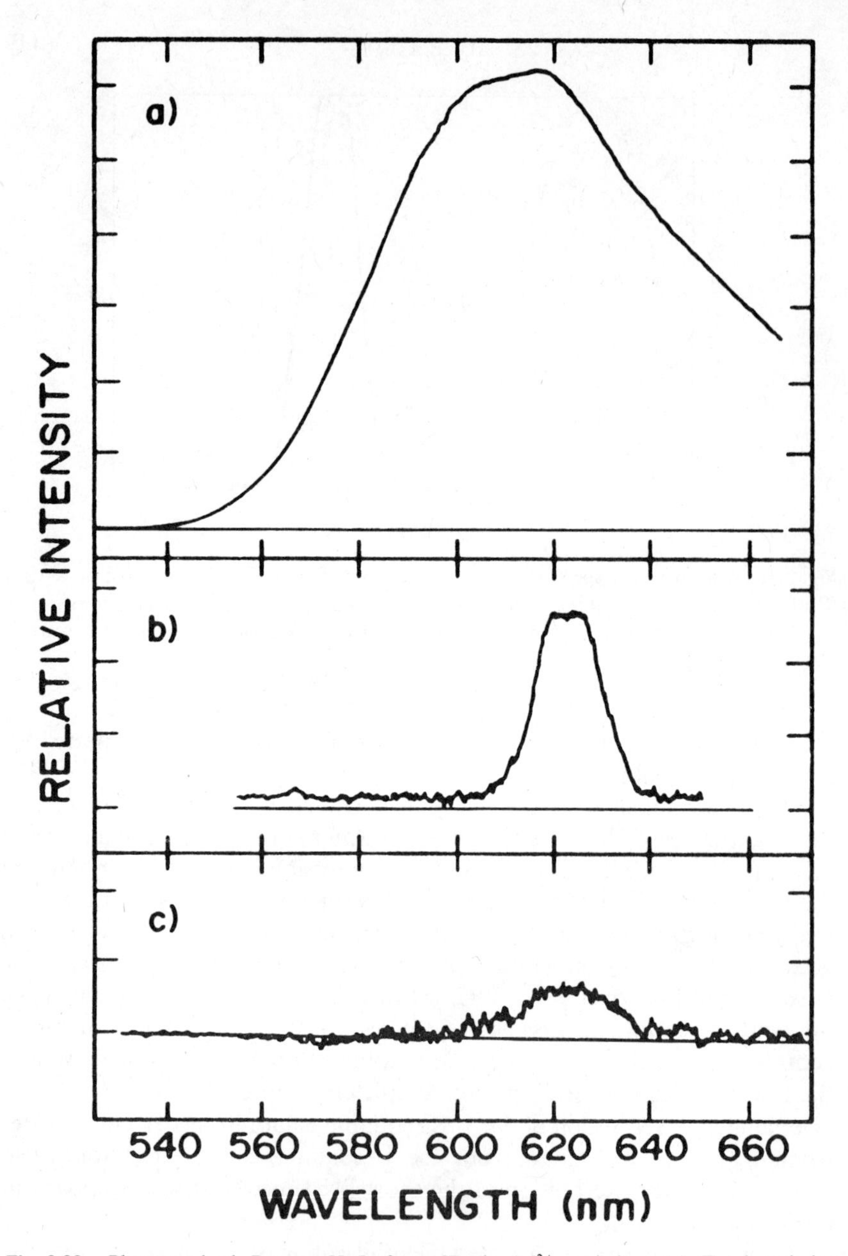

Fig. 2.22. Phase-resolved Raman H_2O from $[Ru(bpy)_3]^{2+}$ emission. (*a*) Total emission spectrum (2 kHz, 0.3-s time constant) of a 5 μM aqueous $[Ru(bpy)_3]^{2+}$ solution (1/25 the concentration of the sample of Figure 2.21). The water Raman band is the bump at 620 nm. (*b*) Phase-resolved (200 kHz, 1-s time constant) water Raman of the sample of part (*a*). (*c*) Phase-resolved (200 kHz, 10-s time constant) water Raman spectrum of a 130 μM $[Ru(bpy)_3]^{2+}$ solution [25 times more concentrated than that in part (*a*)]. (Reprinted with permission from *Anal. Chem.*, **57**, 542. Copyright 1985 American Chemical Society.)

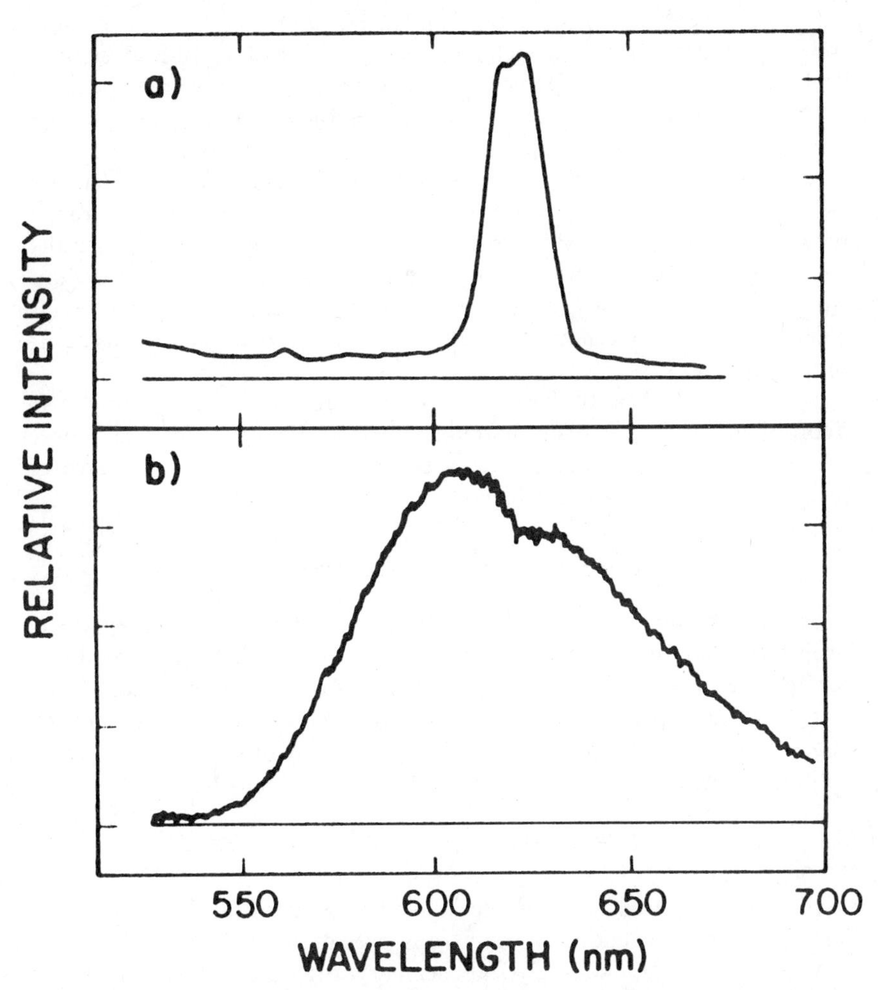

Fig. 2.23. Phase-resolved $[Ru(bpy)_3]^{2+}$ emission from H_2O Raman scattering. (*a*) Total emission spectrum (2 kHz, 0.3-s time constant) of a 0.2 μM $[Ru(bpy)_3]^{2+}$ solution (1/25 the concentration of Figure 2.22*a*). The dominant band at 620 nm is the water Raman band, while the emission is the broad underlying continuum. (*b*) Phase-resolved (200 kHz, 1-s time constant) $[Ru(bpy)_3]^{2+}$ from the sample of (*a*). The small dip at 620 nm is due to a slight overcompensation on the phase adjustment. (Reprinted with permission from *Anal. Chem.*, **57**, 542. Copyright 1985 American Chemical Society.)

laser yield detection limits that are another two to three orders of magnitude lower than that achieved by PRS (55).

We have also recently demonstrated that methods related to PRS can eliminate quenching errors in luminescence spectroscopy (56). We have shown that errors from quenching effects could lead to 10-fold errors in concentration but can be reduced to a few percent by phase methods.

Phase-resolved spectroscopy has successfully been applied to the simultaneous analysis of up to four components by means of a direct determination of the lifetimes of the components (6,43–45). It also has been used for measurements at more than one emission wavelength in order to reduce the number of unknowns. While each measurement is easier than collecting SPC data, the actual collection sequence is rather tedious. Additionally, the measurement does not readily lend itself to automation for complete spectral determinations, although it appears to be quite useful for simple analyses.

A particularly beautiful example of phase-resolved spectroscopy is the fluoroimmunoassay of phenobarbital (45). The lifetime difference between the two fluorescein chromophores was about 100 ps in 4 ns, yet excellent precision in the analysis is achieved.

2.5.2. Excited-State Dynamics

Examples abound of using both TRES and PRS to probe excited-state properties and dynamics. Examples range from excited-state acid–base reactions, energy transfer, state identification, spectral resolution, conformational motion in macromolecules, and solvent relaxation. Interested readers should consult several recent books (2–4) for many more examples than we will discuss here.

2.5.2.1. Excited-State Acid–Base Reactions

TRES has been used to study the dynamics of the intramolecular excited-state acid–base reactions of numerous species in such diverse areas as physical, biological, and analytical chemistry. Figure 2.1 shows an excellent example of the decay kinetics of the excited-state acid–base reaction of 2-naphthol. In this case the flash duration was significant in comparison to the sample decay times; it was therefore necessary to deconvolute the decay curves. The data clearly show the conversion from the higher-energy emitting unprotonated form to the lower-energy protonated form during the decay process.

2.5.2.2. Solvent and Environmental Relaxation

Figure 2.7 shows a very fine example of a solvent relaxation study of 4-aminophthalimide (4-AMP) in 1-propanol. The pronounced lowering of excited-state energy with increasing time following the flash is due to the

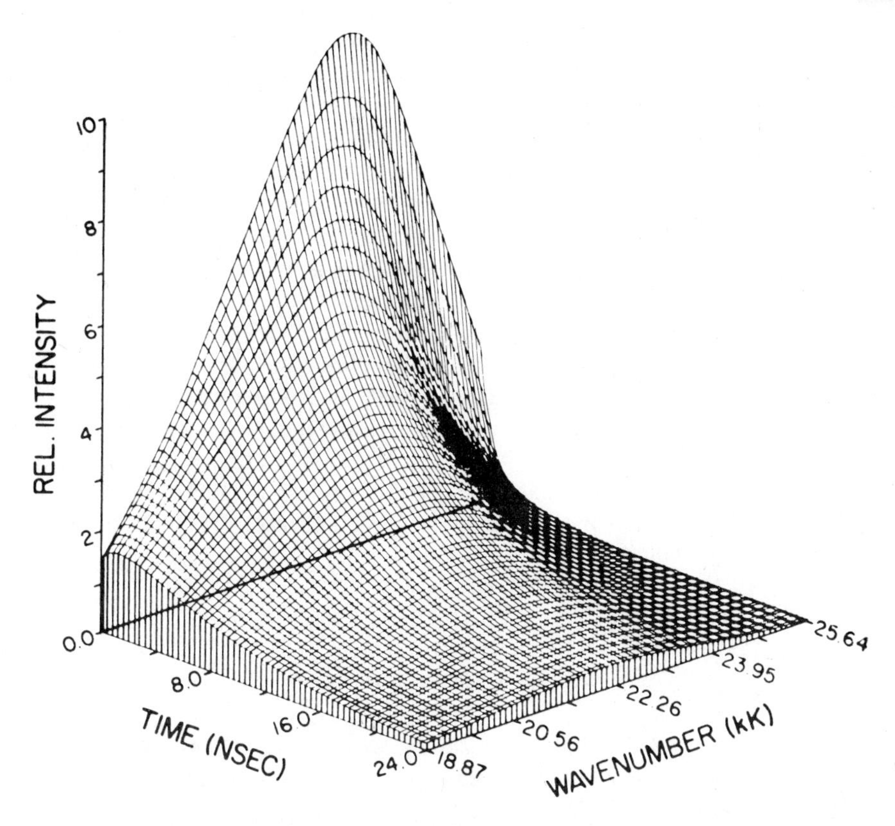

Fig. 2.24. Time- and wavenumber-dependent fluorescent intensities for TNs-labeled egg lecithin vesicles. [Reproduced from *Biophys. J.*, 16, 571 (1976). By permission of the Biophysical Society.]

relaxation of the excited molecule and its solvent environment. The low temperature of the measurement makes the solvent more viscous and thus makes the time scale for the environmental relaxation, coupled with molecular rearrangements, occur on a time scale comparable to the excited-state lifetime. The high-energy emission at early times corresponds to the unrelaxed species. The lower-energy emission at later times corresponds to the molecule after it and its environment have relaxed (3,27).

If the temperature is too high or too low, a time-independent emission is observed. At low temperatures, the media is so viscous that no molecular relaxation occurs during emission and only the high-energy unrelaxed emission is seen. At high enough temperatures the solvent is so fluid that the molecules relax in a time scale that is short compared to instrument resolution; thus all the emissions occur from the low-energy relaxed form.

Figure 2.24 shows a biological example of TRES. The plot is for the

2-*p*-toluidinylnaphthalene-6-sulfonic acid (TNS)-labeled egg lecithin vesicles. As with the 2-naphthol, deconvolution was necessary. The explanation is similar to that for the 4-AMP. The higher-energy initially excited form relaxes to lower-energy forms on a time scale comparable to the excited-state lifetime. The pronounced rise in the emission at lower emission energies is a clear indication of this relaxation (57).

2.5.2.3. Solid-State and Excimer Effects

Figure 2.25 shows excited-state emission spectra as a function of time for pyrene bound to silica gel (58). The kinetics provide insight into the organization of pyrene on the surface as well as the decay of the exciplexes formed. In contrast to solution behavior, the pyrene monomer and excimer decays are independent and there is no buildup of the excimer emission. The result clearly shows that the excimer is formed immediately on excitation and requires no diffusional motion of molecules. It seems quite likely that the packing on the silica surface is such that ground-state complexes actually exist.

2.6. CONCLUSIONS

We have described the theory, instrumentation, and applications of TRES and PRS. We have also described some representative examples of the use of these methods. Which method is best will depend on the particulars of a given researcher's interests, experimental expertise, and financial situation. We have attempted to spell out some of the factors that can aid in selecting one method over another.

We feel that both TRES and PRS are methodologies very much in their infancy. We expect both areas to grow enormously in importance in the next decade. Further, we think that both PRS and TRES will hold stable and valuable niches in the arsenal of tools available to the luminescence spectroscopist, and indeed complementary use of both methods is likely to become more common. As experimental methods mature the distinction between the two methods are, in fact, likely to blur considerably.

We expect significant experimental and mathematical enhancements to both methods to occur and to add to their value. In particular, multichannel TACs will greatly enhance single-photon counting methods by permitting the acquisition of more than one stop pulse per flash. The problem of resolution of spectra in PRS is not dissimilar to the problems encountered in interpreting the output of phased-array radars—an extremely "hot" area at the moment. Thus, in addition to improved experimental methods, we may well see enhancements in radar providing powerful new mathematical tools

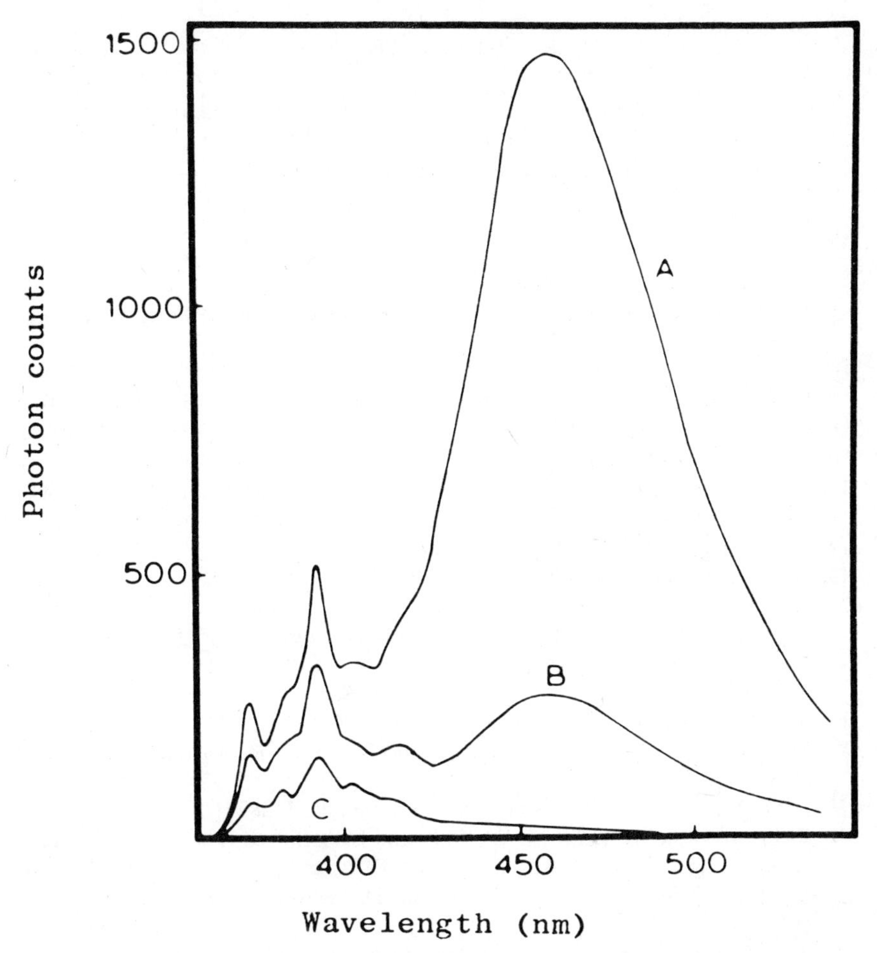

Fig. 2.25. Time-resolved fluorescence spectra of pyrene adsorbed on silica gel (24 mg of pyrene per gram of SiO_2) observed at (*A*) 7–52 ns, (*B*) 108–162 ns and (*C*) 347–404 ns after a flash. [Reprinted with permission from R. B. Cundall and R. E. Dale, Eds., *Time-Resolved Fluorescence Spectroscopy in Biochemistry and Biology* (New York: Plenum Press, 1983).]

for PRS. In summary, we look forward to a very exciting and stimulating decade for both methods.

ACKNOWLEDGMENTS

We gratefully acknowledge the support of the National Science Foundation (Grants 82–06279 and 86–00012) and the donors of the Petroleum Research Fund administered by the American Chemical Society.

REFERENCES

1. C. A. Parker, *Photoluminescence of Solutions*, Elsevier, Amsterdam, 1968.
2. R. B. Cundall and R. E. Dale, Eds., *Time-Resolved Fluorescence Spectroscopy in Biochemistry and Biology*, Plenum Press, New York, 1983.
3. J. R. Lakowicz, *Principles of Fluorescence Spectroscopy*, Plenum Press, New York, 1983.
4. D. V. O'Connor and David Phillips, *Time-Correlated Single Photon Counting*, Academic Press, London, 1984.
5. J. N. Demas, *Excited State Lifetime Measurements*, Academic Press, New York, 1983.
6. L. B. McGown and F. V. Bright, *Anal. Chem.*, **56**, 1400A (1984).
7. I. M. Warner, G. Patonay, and M. P. Thomas, *Anal. Chem.*, **57**, 463A (1985).
8. D. M. Jameson, E. Gratton, and R. D. Hall, *Appl. Spectrosc. Rev.*, **20**, 55, (1984).
9. T. S. Rose, R. Righini, and M. D. Fayer, *Chem. Phys. Lett.*, **106**, 13 (1984).
10. M. D. Ediger, R. P. Domingue, and M. D. Fayer, *J. Chem. Phys.*, **80**, 1246 (1984).
11. N. J. Dovicki and J. M. Harris, *Anal. Chem.*, **53**, 106 (1981).
12. C. A. Carter, J. M. Brady, and J. M. Harris, *Appl. Spectrosc.*, **36**, 309 (1982).
13. W. R. Laws and L. Brand, *J. Phys. Chem.*, **83**, 795 (1979).
14. J. D. Winefordner, *Acc. Chem. Res.*, 361 (December 1969).
15. S. Peled, U. El-Hanany, and S. Yatsiv, *Rev. Sci. Instrum.*, **37**, 1649 (1966).
16. C. A. Parker and C. G. Hatchard, *Analyst*, **87**, 664 (1962).
17. M. M. Broer, C. G. Levey, and W. M. Yen, *Rev. Sci. Instrum.*, **54**, 76 (1983).
18. R. G. Bennett, *Rev. Sci. Instrum.*, **31**, 1275 (1960).
19. M. L. Bhaumik, G. L. Clark, J. Snell, and L. Ferder, *Rev. Sci. Instrum.*, **36**, 37 (1965).
20. F. De Marco and E. Penco, *Rev. Sci. Instrum.*, **40**, 1158 (1969).
21. R. E. Brown, K. D. Legg, M. W. Wolf, and L. A. Singer, *Anal. Chem.*, **46**, 1690 (1974).
22. R. P. Van Duyne, D. L. Jeanmaire, and D. F. Shriver, *Anal. Chem.*, **46**, 213 (1974).
23. T. L. Gustafson and F. E. Lytle, *Anal. Chem.*, **54**, 634 (1982).
24. A. J. Alfano, F. K. Fong, and F. E. Lytle, *Rev. Sci. Instrum.*, **54**, 967 (1983).
25. R. E. Russo and G. M. Hieftje, *Anal. Chim. Acta*, **134**, 13 (1982).
26. R. E. Russo and G. M. Hieftje, *Appl. Spectrosc.*, **36**, 92 (1982).
27. W. R. Ware, S. K. Lee, G. J. Brant, and Peter Chow, *J. Chem. Phys.*, **54**, 4729 (1971).
28. D. W. Johnson, J. A. Gladden, J. B. Callis, and G. D. Christian, *Rev. Sci. Instrum.*, **50**, 118 (1979).
29. I. M. Warner, M. P. Fogarty, and D. C. Shelly, *Anal. Chim. Acta*, **109**, 361 (1979).
30. I. M. Warner, G. D. Christian, E. R. Davidson, and J. B. Callis, *Anal. Chem.*, **49**, 564 (1977).
31. F. J. Knorr and J. M. Harris, *Anal. Chem.*, **53**, 272 (1981).
32. Chu-Ngi Ho and I. M. Warner, *Anal. Chem.*, **54**, 2486 (1982).
33. P. R. Bevington, *Data Reduction and Error Analysis for the Physical Sciences*, McGraw-Hill, New York, 1969.

34. I. Isenberg, *J. Chem. Phys.*, **59**, 5696 (1973).
35. I. Isenberg and E. W. Small, *J. Chem. Phys.*, **77**, 2799 (1982).
36. E. W. Small, L. J. Libertini, and I. Isenberg, *Rev. Sci. Instrum.*, **55**, 879 (1984).
37. A. Grinvald and I. Z. Steinberg, *Anal. Biochem.*, **59**, 583 (1974).
38. F. W. J. Teale, in R. B. Cundall and R. E. Dale, Eds., *Time-Resolved Fluorescence Spectroscopy in Biochemistry and Biology*, Plenum Press, New York, 1983.
39. I. Moya, in R. B. Cundall and R. E. Dale, Eds., *Time-Resolved Fluorescence Spectroscopy in Biochemistry and Biology*, Plenum Press, New York, 1983.
40. H. P. Haar and M. Hauser, *Rev. Sci. Instrum.*, **49**, 632 (1978).
41. T. V. Veselova and V. I. Shirokov, *Bull. Acad. Sci. USSR, Phys. Ser. (Engl. Transl.)*, **36**, 925 (1972).
42. T. V. Veselova, A. S. Cherkasov, and V. I. Shirokov, *Opt. Spectrosc.*, **29**, 617 (1970).
43. F. V. Bright and L. B. McGown, *Anal. Chim. Acta*, **162**, 275 (1984).
44. L. B. McGown and F. V. Bright, *Anal. Chem.*, **56**, 2195 (1984).
45. L. McGown, *Anal. Chim. Acta*, **157**, 327, (1984).
46. G. W. Mitchell and R. D. Spencer, *Application Brief*, Vol. 1, No. 1, SLM Instruments, Urbana, Ill., 19 .
47. E. Gratton, D. M. Jameson, N. Rosato, and G. Weber, *Rev. Sci. Instrum.*, **55**, 486 (1984).
48. G. Ide, Y. Engelborghs, and A. Persoons, *Rev. Sci. Instrum.*, **54**, 841 (1983).
49. R. D. Spencer and G. Weber, *Ann. N.Y. Acad. Sci.*, **158**, 361 (1969).
50. J. N. Demas, and R. A. Keller, *Anal. Chem.*, **57**, 538 (1985).
51. E. Gratton and R. Lopez-Delgado, *Nuovo Dimento B*, **56**, 110 (1980).
52. K. F. Harbaugh, C. M. O'Donnell, and J. D. Winefordner, *Anal. Chem.*, **46**, 1206 (1974).
53. J. N. Demas and W. M. Jones, in preparation.
54. R. Kaminski, F. J. Purcell, and E. Russavage, *Anal. Chem.*, **53**, 1093 (1981).
55. B. A. Bushaw, in W. S. Lyon, Ed., *Analytical Spectroscopy*, Elsevier, Amsterdam, 1984.
56. J. N. Demas and W. M. Jones, and R. A. Keller, *Anal. Chem.*, **58**, 1717 (1986).
57. J. H. Easter, R. P. De Toma, and L. Brand, *Biophys. J.*, **16**, 571 (1976).
58. W. R. Ware, in R. B. Cundall and R. E. Dale, Eds., *Time-Resolved Fluorescence Spectroscopy in Biochemistry and Biology*, Plenum Press, New York, 1983.

CHAPTER

3

FIBER OPTICAL FLUOROSENSORS IN ANALYTICAL AND CLINICAL CHEMISTRY

OTTO S. WOLFBEIS

Institute of Organic Chemistry
Division of Analytical Chemistry
Karl Franzens University
Graz, Austria

3.1. INTRODUCTION

3.1.1. Analytical Aspects of Sensors

Continuous sensing of analytes is a matter of growing interest by virtue of the real-time nature of most sensors. Increasing concern about environmental quality, and—in a more and more cost-conscious world—considerable personel savings in comparison to manual off-line methods contribute to the desirability of sensors. Hence tremendous efforts have been devoted to the development of various sensing devices for use in analytical and clinical chemistry.

By definition, a sensor is a device that is able to indicate continuously and reversibly the concentration of an analyte or a physical parameter. Thus a pH meter as well as a nonbleeding pH paper strip may be called true sensors since they act continuously and fully reversibly. However, certain devices that are able to measure the concentration (or activity) of biomolecules have also been called sensors, although they do not sense continuously but rather allow only a single determination. Some of these "sensors" have been included into this chapter, but they have usually been referred to as probes rather than as sensors.

In view of various requirements imposed on sensors for various purposes, their complexity can vary considerably. An optical sensor for groundwater studies, for example, can be a simple glass fiber that guides a laser beam to the fluorescent analyte and guides its fluorescence back to a detector. However, most molecules of analytical interest require more sophisticated instrumentation, particularly when they are not fluorescent by themselves. In most cases, the sensor consists of an analyte-sensitive coating, a transducer (if necessary), an amplifier, a data acquisition and processing unit, and a data output or display.

A great variety of transducers sensitive to heat, light, anions and cations, gases, electron transfer, mass changes, conductance, absorbance, and light depolarization are known and has already been applied for analytical purposes. These prove to be of particular utility when responding to products of biocatalytic processes and to binding–unbinding phenomena.

Various sensing principles compete successfully with each other (Table 3.1). The most widely used electrochemical sensing method at present is potentiometry, and pH sensing glass electrodes have been used over decades for the determination of pH as such, but also for carbon dioxide, which reversibly changes the pH of a bicarbonate buffer, and ammonia, which affects the pH of an ammonium ion buffer. More recently, pH electrodes have been applied as transducers for bioprocesses in which protons are produced or consumed.

Ion-selective electrodes (ISEs) have tremendously widened the field of application of potentiometry, although some are of limited long-term stability. The ion-sensitive field-effect transistor (ISFET) is often regarded as the second-generation ISE and presents a logical merger between the latter and solid-state integrated circuits. Devices are now in use for the direct in-vitro and in-vivo analysis for blood electrolytes, mainly hydrogen ion, sodium, potassium, calcium, and chloride. ISEs are available for various other analytes and may also act as transducers for enzyme, substrate,

Table 3.1. Sensor and Transducer Types Other Than Optical Ones Exploited in Analytical and Clinical Chemistry

Sensor Type	Principle	Typical Sensing Application	Typical Tranducer Application
Glass electrode	Potentiometry	pH	CO_2, NH_3, enzyme electrodes, titrimetry
Platin electrode	Potentionmetry	Redox status	—
Field-effect transistors	Potentiometry	Ions and gases	Enzymes, substrates, antibiotics, immunosystems
Ion-selective electrodes	Potentiometry	Anions and cations	Titrimetry, enzymes
Polarographic electrodes	Amperometry	Transition and heavy metal cations; as detectors in HPLC	Enzymes, substrates
Clark electrode	Amperometry	Oxygen, halothane	In enzyme electrodes
Piezoelectric crystal	Piezoelectric mass determination	Gases, volatile liquids	—
Conductance sensor	Conductivity	—	Enzyme-catalyzed reactions
Thermistors	Measurement of reaction enthalpy via its highly negative temperature coefficient	—	In chemical and biochemical reactions accompanied by consumption or production of heat

inhibitor, coenzyme, or even bacterial electrodes. The palladium gate metal-oxide semiconductor FET (Pd-MOSFET), on the other hand, is sensitive to hydrogen and other gases, such as ammonia in the ppm range.

Amperometric sensors have been applied mainly for oxygen determination and as transducers for oxygen-producing or oxygen-consuming enzymatic reactions. Polarography using the dropping mercury electrode is a useful method for determination of toxic trace elements in the sub-ppb range, owing to the sensitivity of the method, although its popularity is decreasing because of the toxicity of mercury.

3.1.2. Fiber Optical Sensors

Recent advances in optoelectronics and fiber optical techniques have led to an exciting new technique called optical sensing. Depending on the origin of the optical signal, the devices are classified as absorbance (or reflectance), fluorescence, phosphorescence, Raman, or infrared sensors. This chapter deals with the largest group, fluorescence sensors. Some other optical sensor types have also been included because they freqnectly have very similar working principles or because they represent interesting principles that await their extension to fluorimetry.

The literature surveyed here is thought to be complete to the end of 1986, with a few references from 1987. Particular attention has been paid to the patent literature. Various short reviews (1–12) cover the subject of optical sensing from different points of view. Some are confined to measurements of physical parameters (5), chemical (1,2,6,7,12) or clinical analytes (8, 10, 12), or remote sensing only (5,11), whereas others discuss mainly optical problems (3). This chapter is intended to cover all aspects of chemical fluorosensing except for problems associated with signal amplification, signal processing, and data storage. Hence it comprises sections as different as those on chemistry, mathematics, and optical hardware. In each instance emphasis is given to practical applications rather than to a purely theoretical treatment of the subject.

The dramatic progress that has been achieved in the past 10 years is as result of joint efforts of various specialists in their particular fields: Fluorescence spectroscopy has become an established spectroscopic method because of its sensitivity, selectivity, and versatility. Numerous organic (13), bioorganic (14), and inorganic analytes (15) have been shown to be fluorescent under irradiation or to be determinable via a fluorogenic reaction. The communication industry has provided low-priced fiber optics which allow the transmission of optical signals over large distances, even in the UV and IR. Powerful lasers have become available as ideal light sources in fluorimetry, although their prices are still a limiting factor for their use in commercial instrumentation. Visible-light-emitting diodes (LEDs) and sufficiently sensitive photodetectors have become available at low prices which allow their use in simple and safe instrumentation. Finally, new methods in chemometrics, along with powerful and small-sized calculators and microprocessors which allow data storage and rapid data processing, even in cases of complex signal-to-concentration relations, have significantly contributed to the state of the art.

Doubtless, the strongest impact on fiber optical sensing has come from the optical fiber technology, an outgrow of the communication industry. The first sensors designed to collect information via fiber optics relied on the fact that alterations in a specific physical property of a medium being sensed would cause a predictable change in the light *transmission* characteristics of a fiber. Acoustic waves, acceleration, strain, position, and magnetic field are some of the physical properties measured with these initial sensors.

The field of application of fiber sensors in analytical chemistry greatly increased when fluorescence spectroscopy was coupled with the fiber optic technique (6,7). As a result, sensing no longer became restricted to measuring physical properties that change the transmission of a fiber, but can be extended to numerous organic, inorganic, clinical, and biomedical analytes and parameters.

Various names have been given to fluorescence optical sensors. Lübbers and Opitz introduced both the word "optrode" (16a) (from "*opt*ical elect*rode*") and "optode" (8,16b) (from the Greek: the optical way). Optrode has become popular in the Anglosaxon countries, and optode in continental Europe. Both expressions stress the fact that the signal is optical rather than electrical. Remote sensing with fiber optics was named (6) RFF (remote fiber fluorimetry). RFF emphasizes the fact that sensing can easily be performed over wide distances. Another abbreviation that has been used is FOCS (fiber optical chemical sensor).

3.1.3. Advantages of Fiber Optical Sensing

Depending on the field of application, optical fiber sensing can offer some of the following advantages over other sensor types:

1. Optical sensors do not require a reference signal as is required in all potentiometric methods, where the difference of two absolute potentials is measured. The need for reference electrodes makes potentiometric instrumentation relatively costly. Moreover, the liquid–liquid junction between the two electrolytes is very sensitive to perturbations and can be considered to be the weakest "link" in all potentiometers.

2. The ease of miniaturization allows the development of very small, light, and flexible fiber sensors. This is of great utility in the case of minute sample volumes and in designing small catheters for invasive sensing in clinical chemistry and medicine. By now, fluorescence optical sensor heads much smaller than any electrochemical sensor (including field-effect transistors) can be manufactured.

3. Low-loss optical fibers allow transmittance of optical signals over wide distances, typically 10–1000 m, and even larger distances seem feasible when use is made of amplifiers currently used in optical telecommunication. Remote sensing makes it possible to perform analyses in ultraclean rooms, when samples are hard to reach, dangerous, too hot or too cold, in harsh environments, or radioactive. Fibers appear to be resistant to radiation doses in the order of 10^7 rad or more.

4. Because the primary signal is optical, it is not subject to electrical interferences by, for example, static electricity of the body or surface potentials of the sensor head. Fibers do not present a spark or fire hazard and a risk to patients since there are no electrical connections to the body. Complete electrical isolation is required when sensors are used in strong electric fields.

5. Analyses can be performed in almost real-time since no sampling with its inherent drawbacks is necessary. Response times as short as 1 ms have been reported for an oxygen sensor based on fluorescence quenching.

6. Since several fiber sensors placed in different sites can be coupled to one fluorimeter via a chopper, the method allows multiple analyses with a single central instrument. Placement of the spectrometer at a central location remote from the sensor head (which experiences varying experimental conditions) makes routine maintenance checkout possible, assures that calibration will be preserved, and consequently renders the instrument more reliable.

Coupling of small sensors for different analytes to produce a sensor bundle of small size allows simultaneous monitoring of various analytes by hybrid sensors without crosstalk of the single strands.

7. In many cases the sensor head does not consume the analyte in a measurable rate, as, for instance, in the case of polarographic electrodes. This fact is of particular advantage in case of extremely small sample volumes. Moreover, fiber optical sensing is a nondestructive analytical method.

8. Optical sensors have been developed which respond to chemical analytes or physicochemical parameters for which electrodes are not available.

9. A fiber optic can transmit much more information than can an electrical lead. High information density can be achieved since the optical signals can differ with respect to wavelength, phase, decay profile, polarization, or intensity modulation. Thus one single fiber may guide green and red light in one direction, and blue and yellow light into the other. Therefore, a single fiber can, in principle, guide a huge number of signals simultaneously. In practice, a single fiber may be used to assay several analytes at the same time because different analytes or indicators can respond to different

excitation wavelengths. Moreover, the respective emissions are also spectrally different. Time resolution along with spectral selection offers a particularly fascinating new technique in fiber optical sensing and can make superfluous the need for hybrid sensors.

10. Fiber sensors with indicator or reagent layers are less prone to inner filter effects caused by interfering analytes with similar absorption. This is due to the high optical density of the indicator layer, which may also serve as a filter to the underground emission from the sample.

11. Sensors based on dynamic fluorescence quenching have a useful dynamic range often wider than that of electrochemical sensors. For instance, the fiber optical oxygen sensor (Section 3.6.2) shows much better precision than does the Clark electrode in the oxygen pressure range below 200 torr.

12. Many sensors are simple in design and can easily be replaced by substitute parts, even when manufacturing the sensor head requires relatively complex chemistry. This opens the way for disposable sensors.

13. Most fiber sensors can be employed over a wider temperature range than electrodes and are steam-sterilizable.

14. Optical sensors can offer cost advantages over electrodes, particularly when a single spectrometer is used in combination with several sensors. One of the possible raw materials is sand, an abundant source.

15. The feasibility of fluorescence immunoassay (FIA) using fiber optics has already been demonstrated. This provides a promising principle for nonradioactive determination of antigens and eventually, antibodies. Only very few electrochemical methods are suitable for this purpose.

3.1.4. Disadvantages of Fiber Optical Sensing

Notwithstanding a number of advantages over other sensor types, fiber sensors exhibit the following disadvantages:

1. Ambient light interferes in many cases. Although this plays no role when the sensor is applied in a dark environment, it can result in a major limitation in performing fluorescence titrations with plain fibers (Section 3.6.6). Therefore, they must either be used in dim or dark surroundings, or the optical signal must be encoded so that it can be resolved from ambient background light. It is therefore advisable that the optical system of the sensor be isolated from the sample by a suitable filter.

2. Sensors with indicator phases (Section 3.2.2) are likely to have limited long-term stability because of photobleaching or wash-out. Signal drifts can be compensated for by relating the signals obtained under two different excitation wavelengths or ratioing the signal to the intrinsic Raman scatter.

Photobleaching efficiency increases with increasing irradiation intensity. Consequently, a powerful laser should be used only when necessary, for instance, when long optical cables with their considerable attenuation make lasers indispensible.

3. Since in indicator phase sensors (Section 3.2.2) analyte and indicator are in different phases, there is a mass transfer necessary before a steady-state equilibrium—and consequently a constant response—is established. This, in turn, limits response times for analytes with small diffusion coefficients. This situation makes it desirable to keep the volume of the analyte-accessible indicator phase much smaller than that of the sample in order not to dilute the sample volume.

4. Sensors with immobilized pH indicators as well as chelating reagents have limited dynamic ranges as compared to electrodes since the respective association equilibria obey the mass action law. The corresponding plots of optical signal versus log of analyte concentration are sigmoidal rather than linear as in the case of the Nernst relation.

5. The fiber optics used at present have impurities of a spectral nature that can give a fluorescence and Raman background. Low-priced (plastic) fibers are confined to wavelengths between about 380 and 800 nm, whereas UV light is efficiently transmitted by rather expensive quartz fibers only. The intensity losses in very long fibers are further complicated by spectral attenuation and change in the numerical aperture as a function of fiber length (Section 3.3.3).

6. Commercial accessories of the optical system are not optimal yet. Stable and long-lived light sources, better connectors, terminations, optical fibers, and inexpensive blue LEDs are needed.

7. More selective indicators have to be found for various important analytes and the immobilization chemistry has to be improved so as to achieve both better selectivity and sensitivity.

8. Many indicators suffer a reduction in sensitivity after immobilization or when dissolved in a polymer. In particular, dynamic quenching efficiency is frequently drastically diminished. Consequently, the respective conventional fluorimetric method will be much more sensitive than the fluorosensor method.

3.1.5. Potential Fields of Fiber Sensor Applications

Fiber optical fluorosensors can be expected to be of utility in analytical chemistry whenever the sample cannot be brought into the fluorimeter. They may replace electrodes and even ion-selective electrodes in the determination of pH and related ionic analytes, but seems to be particularly be suited for the determination of gases such as oxygen, carbon dioxide, ammonia, and methane. This is due mainly to the fact that fluorosensors in combination with fiber optics are far tougher, cheaper, and smaller than the corresponding electrodes. Remote fluorosensing can be applied to various analytical problems, such as pollution and process control, jet and rocket machines undergoing testing, and tracer studies in geology, clinical chemistry, and various biomedical applications. Their ruggedness makes possible utilization in locations too inaccessible to either the instrument or the analyte.

3.1.5.1. Groundwater Monitoring

In the face of increasing public concern about the quality of drinking water, continuous monitoring of groundwater has become a major aspect of modern analytical chemistry. Rather than digging a well field with numerous boreholes large enough to admit sample collectors which are brought to a laboratory for analysis, it has been proposed (6) to introduce long-distance communication-grade fibers down to the groundwater level and to monitor pH, chloride, sulfate, uranium, organic pollutants, and tracer substances using the corresponding optrodes. Several fibers can then be coupled to one spectrometer at a central location up to 1 km remote (Fig. 3.1). The ability to make up to 50 unattended in-situ measurements using a reasonably priced centralized fluorimeter system has been discussed (17) and should result in acceptable economy.

The much smaller diameter of the fiber hold (typically 1.2 cm) allows the use of small boreholes. It has been stated (4) that at typical sites such as chemical or nuclear plants, savings in drilling costs can be as much as $500,000. Moreover, the sample can be studied in situ and in real time, with almost no opportunity for contamination of a container or the well itself from outside.

For nuclear waste repositories, environmental monitoring of nuclear installations, or study of underground nuclear tests, the hazards associated with the samples can be avoided by leaving them safely underground. Even if the fiber is damaged by radiation, it is easier and cheaper to replace than the whole monitor.

The principles of optical groundwater monitoring do, of course, also hold for geological tracer studies using highly fluorescent markers such as rhodamine 6G, or pH gradient studies using pH indicators. They are detectable in boreholes in picomol quantities when lasers are used as excitation light sources.

3.1.5.2. Pollution Monitoring

Airborne laser fluorosensors offer a rapid, real-time, and efficient method for the continuous control of seawater and air pollution. The possibility of

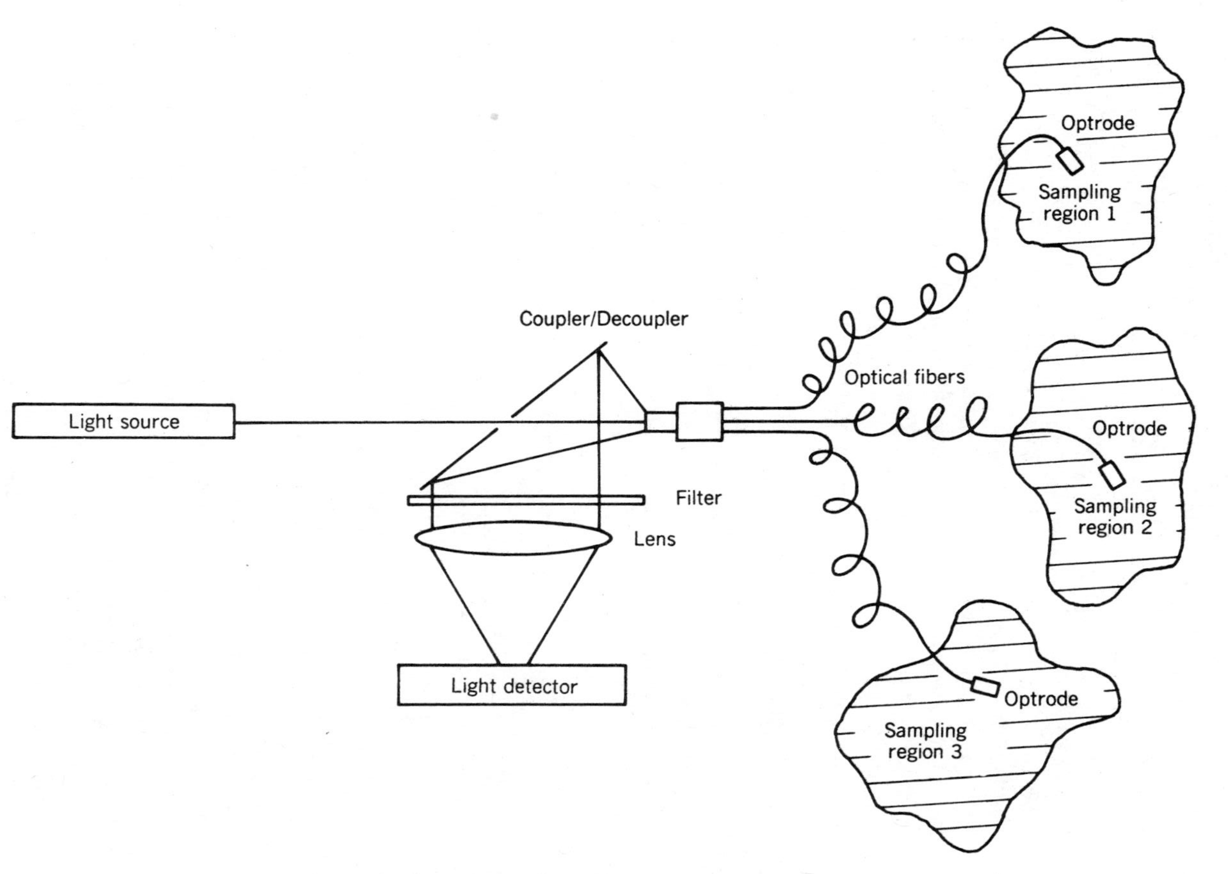

Fig. 3.1. Schematic diagram of a remote fiber sensor.

remote sensing of environmental parameters has been studied by several groups (18,19). Increasing efforts were directed to the detection of polycyclic aromatic hydrocarbons (PAHs) in the environment which are by-products of many new coal-processing factories, including coal-liquidification plants. Similarly, mineral oil spills released by ships can be distinguished from seawater background by this technique and can even be classified into subgroups (see Section 3.6.9).

Airborn laser fluorosensor measurement of chlorophyll concentration (phytoplankton population) have also been applied to water-mapping studies both in inland waters and in the ocean. Abundance, distribution, and biological activity in natural waters can easily be monitored over wide areas and even at various depths.

Air pollution sensing for industrial applications and environmental research can conveniently be performed with a network of specific optrodes hooked up to a central measurement station by optical fibers. The sensors may be instantaneous concentration sensitive or be designed for cumulative measurements of the total integrated exposure. Sensors have been developed (20) for formaldehyde, ammonia, nitrogen oxides, chloroform, hydrogen sulfide, and reactive hydrocarbons. Again, chemical analysis at multiple locations can be made from a central station using fibers that can readily be multiplexed to the station.

3.1.5.3. Process Control

Production efficiency and product quality are critically dependent on process control. More and more on-line sensors are replacing classical sampling techniques, and analytical chemistry is emigrating from the laboratory to the factory. In view of the costs resulting from a failing chemical reaction and its environmental consequences, there are extreme reliability requirements on process control instrumentation.

On the other side, most sensors for continuous process control are faced with a harsh environment. Typically, a reliability of better than 99% is required under extreme and rapidly varying temperatures, high noise and vibration, and substantial chemical exposures, but with a minimum of servicing and maintenance at rather long intervals. Furthermore, the instrument is expected to stay accurate over a prolonged period, despite the absence of recalibration or even checkup (11,21).

Clearly, process control is becoming a major field of application for fluorosensing. The development of long-range, low-cost, and high-performance fibers along with the design of optosensors for the most important chemical and physical parameters has provided a new dimension of continuous monitoring, since the instrument can remain in the benign environment of the laboratory, where service and calibration is no longer a problem.

3.1.5.4. Remote Spectroscopy

Instrumentation for the remote determination of fluorescence spectra is commercially available [e.g., from Oriel (Stamford, CT) or Guided Wave, Inc. (Rancho Cordova, CA)]. It is intended for use in radioactive areas, fermentation systems, high-voltage areas, explosive and dangerous areas, biologial hazard stations, and in marine, river, and reservoir locations. Remote fluorimetry is not as sensitive as conventional fluorimetry because of light attenuation by the fiber and considerably higher background.

Fibers have also been used to measure the intrinsic fluoresence of HPLC effluents (22). Remote fiber fluorimetry has been applied in combustion measurements and will provide a simple means for studying chemical processes in flames and combustion gases (23). Plain fibers have also found application for the measurement of fluorescence of reagent strips in substrate-labeled fluorescence immunoassay (24).

3.1.5.5. Titrimetry

Although not a continuous method, titrimetry is still one of the most widespread tools in analytical chemistry. Recent work demonstrates that fiber fluorosensors can easily replace electrodes in various titrimetric procedures, including acid–base titrations, argentometry, bromometry and iodometry, complexometry, or redox titrations (Section 3.6.7). There are two ways to perform this. In the first one, a fluorescent indicator is added to the solution to be titrated, and the relative change in fluorescence during titration is followed using a bifurcated fiber. This method can offer considerable cost reductions since optical fibers along with indicator solutions are much less expensive than electrodes. The optical and electrochemical detection systems are of comparable price.

In the second method, the analyte-sensitive indicator is immobilized at the end of the fiber to give an analyte-selective optrode, which is dipped into the solution during titration. It may be predicted that the endpoints of the majority of standard titration methods can be determined by fiber optical methods with the same precision as with electrochemical methods, although the full potential of fiber optical titrimetry is not yet fully exploited.

3.1.5.6. Biosensors

Possibly the greatest field of application is sensing clinically and biochemically important analytes such as blood gases, electrolytes, metabolites, enzymes, coenzymes, immunoproteins, and inhibitors. Sensors responding to biomolecular parameters are frequently called biosensors (9,10,25). Fluorosensors for blood analytes are being used and will be used in vivo as

Table 3.2. Potential Biomedical Sensor Applications

Application	Mode	Examples
Critical care	In vivo	O_2, CO_2, temperature, K^+
Chronic maintenance	In vitro	Glucose, therapeutic drugs
Acute diagnosis	In vitro	Blood chemistry, abnormal function, infection, diagnostic metabolites

sensors for the continuous monitoring of the critically ill and as devices for testing blood samples in vitro. Continuous measurements of critical parameters which give warning of life-threatening trends such as pH, oxygen, carbon dioxide, and blood pressure are well established in principle. One may expect from fiber optics to see continued improvements in biocompatibility, signal stability, ease of calibration, and sterilization. A selection of potential biomedical sensor applications is compiled in Table 3.2.

Within the last decade clinical practitioners have gradually moved from the diagnosis of established disease and toward presymptomatic prognosis and preventive measures. Continuous tests for electrolytes, total protein, urea, glucose, creatinine, cholesterol, triglycerides, and others in blood or urine are the subject of intense research and development to produce devices capable of automatically recording results by using computerized data logging and output systems.

Aside from sensors for these analytes which are present in relatively high concentrations, there is now a substantial demand for biosensors for substrates being present in considerably lower concentrations. It is assumed that in these cases fluorosensors will be of particular utility by virtue of the sensitivity of fluorimetry. Most likely, several fluorimetric methods known for low-concentration analytes such as hormones, steroids, thyroid function constituents, and pregnancy markers will be adapted to continuous fluorosensing.

The area of therapeutic drug monitoring is another one of rapidly increasing commercial significance, along with early detection of infection disease and of the various forms of cancer. Table 3.3 demonstrates the enormous increase in the number of clinical tests performed in one year.

A highly significant trend is now visible which will lead to the decentralization of clinical testing away from hospitals, ideally into the patient's home. This will produce a huge market for simple and inexpensive but reliable equipment in the next 10 years. It was predicted (26) that the home/self-care product market in the United States will increase from $2.5 billion in 1980 to $15.0 billion in 1990.

Table 3.3. Expansion of Clinical Testing Markets in the United States[a]

Test Discipline	1983[b]	1984[b]
Infectious diseases	51	112
Therapeutic drug monitoring	27	56
Autoimmune disease	14	30
Plasma proteins	14	28
Cancer tests	2	5
Others	11	27
TOTAL	119[c]	258[c]

Source: BBI/Information Resources Intl., Inc.

[a] Nonisotopic tests only.

[b] Millions of tests.

[c] Minimum values.

Existing fluorosensors for oxygen and pH will be able to fulfill some of these needs. The corresponding sensors may also be utilized to quantify all enzymatic activities which are accompanied by a change in pH or oxygen partial pressure. Typical clinically important analytes, the respective enzyme, and the detected species are given in Table 3.4.

Fluorescence methods will also be applicable to biosensors other than enzyme optodes. Thus the interaction of antibodies with antigens, a process known to be of outstanding selectivity, can be followed by fluorimetry (see Section 3.7.3.). Evanescent wave sensors with immobilized antibodies on the waveguide surface, fluorescence polarization studies of labeled binding partners, fiber optic decay measurements, or combination thereof, offer numerous possibilities. Bilirubin, steroids, albumin, and enzymes have all been assayed by antibody-binding methods, which, however, have not yet been transferred to the fiber optical sensing technology, but readily could be.

In summary, the opportunity exists for fiber fluorosensors to participate in a rapidly growing segment of the market for clinical analyses, although there may be tough competition from other sensor instrumentation, such as electrodes, FETs, piezoelectric devices, Fourier-transform-IR spectroscopy, calorimetry, and dry-reagent chemistry. Probably, an interdisciplinary approach to the solution of existing hurdles will be essential.

3.2. OPTICAL SENSOR TYPES

A variety of devices has been called fluorescence optical sensors, although their sensing ability relies on quite different principles. One common property,

Table 3.4. Clinically Important Organic Analytes, Enzymes for Use in Their Determination, and the Species That Can Be Detected with One of the Existing Fluorosensors

Analyte	Biocatalyst	Enzyme Catalog No.	Detected Species
Acetaldehyde	Xanthine oxidase	1.2.3.2	H^+
Acetylcholine	Acetylcholinesterase	3.1.1.7	H^+
Adenosine monophosphate	Rabbit muscle slice	—	Ammonia
Arginine	Arginase plus urease	3.5.3.1, 3.5.1.5	Ammonia
Cephalosporins	*C. freundii*	—	H^+
Creatinine	Creatinine deaminase		Ammonia
L-Cystein	*Proteus morganii*	—	H_2S
Ethanol	Alcohol dehydrogenase	1.1.1.1	H^+
	Alcohol oxidase	1.1.3.13	Oxygen
Glucose	Glucose oxidase	1.1.3.4	Oxygen
Glutamine	Porcine kidney slice	—	Ammonia
	Glutamate decarboxylase	4.1.1.15	CO_2
	E. coli	—	CO_2
Insulin	Antibody plus catalase	1.11.1.6	Oxygen
Nystatin	Yeast cells	—	Oxygen
Oxalate	Oxalate decarboxylase	4.1.1.2	CO_2
Penicillin	Penicillinase	3.5.2.6	H^+
	Penicillin amidase	3.5.1.11	H^+
Phenylalanine	Phe—ammonialyase	4.3.1.5	NH_3
Tyrosine	Tyr decarboxylase	4.1.1.25	CO_2
	L-Tyrosine decarboxylase	4.1.1.25	CO_2
Urea	Urease	3.5.1.5	NH_3, H^+ or NH_4^+

however, is their full reversibility. Nonreversible devices are included in this chapter but will be called "probes". In surveying the optical sensors known so far, one may differentiate between various sensor types which will be discussed briefly in the following sections.

3.2.1. Plain Fiber Sensors

Plain fiber sensors ("bare-ended" fiber sensors) consist of a fiber (an "optical wire") through which exciting light is guided to the analyte having a native fluorescence. Emitted light is guided back to a detector through the same fiber. Fluorescence may as well be guided back through a second arm of a bifurcated fiber (bundle). The exciting light may be considered as the interrogating pulse, and fluorescence as the encoded answer.

Plain fiber sensors utilize the intrinsic fluorescence of the analyte, or an added indicator, and do not have a reagent phase at the end of the fiber. They are referred to as "first generation sensors", as opposed to those having an analyte-specific chemistry at their ends. A typical schematic of this concept is shown in Fig. 3.1. Light from a light source is focused into an optical fiber and guided to the sampling region. The fluorescence of the analyte is guided back through the same cable. Scattered light having wavelength λ_1 is separated from fluorescence having λ_2 by using a spectral sorter such as a dichroic filter.

An instrument for remote sensing with fibers is available from the Oriel Co. (Stamford, CT 06902). It is operated with a xenon flash lamp of mean dissipation 1 W at 9 Hz and 2 μs pulse duration. The "optode fibers" available can be used for measurements of fluorescence in liquids. Although they may as well be used in clear samples, they are specifically designed for use in strongly absorbing, viscous, or turbid samples, and for measurement of immersed solids (e.g., measurement on underwater plants in vivo).

In the instrument, fluorescence is collected at an angle of 0° and guided back via the output bundle of a bifurcated fiber. The detection limit is 1 ppb rhodamine 6G in water (twice the noise level). In an alternative version shown in Fig. 3.2 the collection of light is at 90° to the incident excitation light. A 45° rear surface mirror backed onto a silica prism allows a convenient probe size to be retained. The significance of this conventional geometry is to keep the background light level low in comparison to the fluorescence to be measured. High sensitivities can be achieved if the ambient light (nonpulsed light) is kept to a minimum.

Typical representatives of plain fiber sensors are those for metal cations with intrinsic fluorescence such as uranyl cation (27) or rare earth cations (6). Fluorimetry with plain fibers is usually not sensitive but diplays extremely long-term stability and simplicity. Remote fiber fluorimetric studies using tracer dyes such as rhodamine B and the application of fibers in place of electrodes to follow the course of titrations (Section 3.6.7) represent other fields of application.

A promising new technique for continuous monitoring of drugs also

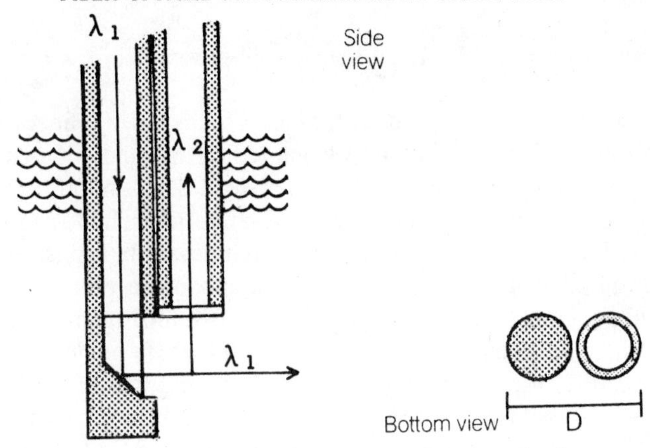

Fig. 3.2. Schematic of the end of a bifurcated fiber optic for measuring fluorescence emission under 90° excitation. The exciting beam (λ_1) enters the sample volume through the first fiber, will be reflected, and produces a fluorescence which leaves the sample through the second fiber (λ_2).

makes use of plain fibers. Here an optical catheter is introduced into a blood vessel and the intrinsic fluorescence of blood components or a drug such as adriamycin or methotrexate is monitored (28). Although these methods usually do not yield absolute analyte concentrations due to fluctuations in the fluorescence background of the serum, they do sufficiently correctly reflect relative concentration changes. The method can, of course, be applied to other bioliquids as well, but drug monitoring in serum seems to be the field with the largest potential. Human blood serum contains a number of fluorescent molecules. Its intrinsic fluorescence at above 400 nm results mainly from contributions from nicotinamide adenine dinucleotide and its phosphate, enzyme-bound pyridoxal phosphate, flavin mononucleotide, and bilirubin (29), Since all these have different decay times (14), time-resolved fluorosensing (30,31) offers a method for their continuous in-vivo monitoring.

Plain fibers have also been applied to monitor the concentrations of algae (by virtue of their intrinsic chlorophyll fluorescence) and their growth in natural waters (32). Although the fluorescence from living algae is rather complex and affected by a variety of factors (see Section 3.6.9), it offers an additional attractive field of application for remote fluorimetry using plain fibers.

Fiber optic sensors are especially useful for systems with poor optical access. Concentrations of Na and OH radicals which serve as analytically important reaction parameters in combustion studies have already been determined (23) and other important radicals, including CH, CN, C_2, CH_2,

N, NH, and NH_2, will certainly be detectable by the same method. A fluorescence spectrometer for multielement analyses by inductively coupled plasma emission has been described in a patent (33). Emission is guided to the detector via a plain fiber optic interposed between the source and the detector.

Plain fibers served to guide excitation light to the detector cells of chromatographic instrumentation. A high-sensitivity laser-induced detector has been designed for use in capillary column liquid chromatography. Its probe volume as viewed by the fiber end was only 3 pL (22).

An electron beam propagation, optical diagnostic system was developed (34) which uses optical fiber links to observe luminescence produced by the electron beam interaction with air or fast scintillators. Fiber optics may also expand the range of direct-reading spark optical emission spectrometers (35a) and have been applied to remote spectrometric analysis of alkali elements by dc arc emission (35b).

A liquid-core optical fiber of 250 μm i.d. has been used as a detection cell for iodine. Carbon disulfide was used as a core material having a high index of refraction and a low surface tension. A funnel-shaped glass was applied for efficiently coupling light emission into the aperture of the hollow fiber. Within a 5-m fiber cell, 0.1 μg/mL iodine (equal to 10 ng of iodine) could be detected (35c).

Plain fiber sensors are the ones that are most easily subject to interferences from ambient light. Various approaches have been made to separate the analytically useful light from background light, so to increase the efficiency of fluorescence excitation. A simple arrangement that is useful in optical sensing of analytes with intrinsic fluorescence is shown in Fig. 3.2.

Plain fiber sensors are simple in design and manufacture. They can be applied, however, only when the analyte is fluorescent or a suitable reagent is found. The sensor type can hardly be applied to mixtures of several fluorophores with similar spectral properties, except when additional discriminations such as time resolution, two-photon excitation, or sequential excitation are employed.

3.2.2. Indicator Phase Sensors

Rather than adding the indicator solution each time to a nonfluorescent analyte, it has become common practice to immobilize indicators on solid supports. In contact with the analyte, the fluorescence of the indicator phase can indicate the concentration. The sensor layer is usually attached to the end of a fiber optic that guides excitation light to the indicator phase, and guides fluorescence to the detector. Alternatively, the indicator phase, when immobilized on the surface of the fiber core or even on a glass slide, may be excited by the field of an evanescent wave (Section 3.2.5). This type of fibers is referred to as "second generation fibers".

Despite several shortcomings, indicator phase sensors are certainly the most elegant ones. Although all fiber devices used for fluorosensing have sometimes been referred to as "optrodes," the reagent-phase sensors really deserve this name since they are the only ones that can compete—with respect to specificity—with electrodes (the source of one half of the word "optrodes").

The first fiber optical sensor with an indicator phase responding to a chemical analyte was described in 1974 (36). An oxygen-sensitive luminophor (silica gel–adsorbed trypaflavin) was attached to the tips of a single or bifurcated fiber and covered with an oxygen-permeable membrane. Luminescence is quenched by oxygen according to the Stern–Volmer equation (see Section 3.5.3) to give an oxygen-dependent signal. In the case of using single-fiber optics, exciting and emitted light were separated by time resolution by applying a phase-sensitive fluorimeter. A reference sensor not in contact with the sample was used to account for photobleaching and temperature effects. This first fiber optical fluorosensor comprises practically all components of a modern instrument.

Typical experimental arrangements are shown in Fig. 3.3. In part (a), a porous glass ball is cemented onto the end of a small single fiber. The indicator is immobilized on the ball, which is exposed to the sample. Exciting and emitted light are guided through the same fiber and are then separated in an optical decoupler. In part (b), the reagent phase is attached to the end of a bifurcated fiber (37,38). Light travels from the source to the common end of a bifurcated *single* fiber. Reflected or fluorescent light is guided to the detector through the second fiber.

Alternatively, input light and output light may also be guided through different strands of a bifurcated fiber *bundle*. Excitation and emission light travels through different fiber strands, which makes the optical separation easier. In contrast to the single-fiber technique, there is no strong background present which may result from impurities in, and Raman scatter from, the excitation light fibers. In Fig. 3.3c, the indicator is immobilized on the end of the fiber core and excited by the evanescent wave. A mirror and a black cover at the tip reflect both excitation and emission beam and prevent ambient light from entering the fiber (17,57).

For certain applications, such as gas and water flow studies, it is not always necessary to apply fibers. A simple arrangement useful for monitoring the oxygen tension of a flowing gas is shown in Fig. 3.4. In this instrument (39) the indicator (fluoranthene) is excited by a UV discharge lamp and fluorescence is detected by a photocell after it has been separated from the

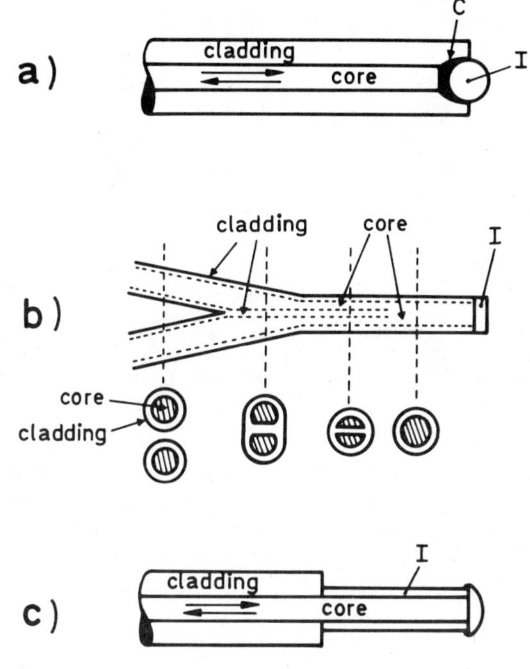

Fig. 3.3. Types of indicator phase sensors. (*a*) A glass ball is cemented onto the end of a single fiber, with an indicator layer immobilized on the ball. Fluorescence is guided back through the same fiber, through which exciting light is guided to the ball. (*b*) An indicator phase is attached to the common end of a bifurcated fiber. (*c*) The indicator is immobilized directly onto the fiber core.

exciting light by a UV filter. This arrangement is not suitable for colored or turbid gases, which, however, can be studied by the same principle, applying the front-face technique as shown in Fig. 3.5.

Three different types of indicator-phase fluorosensors with quite different working principles have been described so far. The first type is based on spectral changes of immobilized indicators undergoing ground-state reactions with analytes. Typical representatives are those based on metallochromic reactions, changes in acid–base equilibria, *static* fluorescence quenching, and competitive binding.

The second type of indicator phase sensors comprises those with immobilized indicators, the fluorescence of which suffers *dynamic* quenching. Since the quenching process occurs during the lifetime of the first excited singlet state, fluorescence is likely to become more efficient the longer the lifetime of the indicator will be. Generally, it is observed that fluorescence quenching of immobilized indicators is much less efficient than quenching of

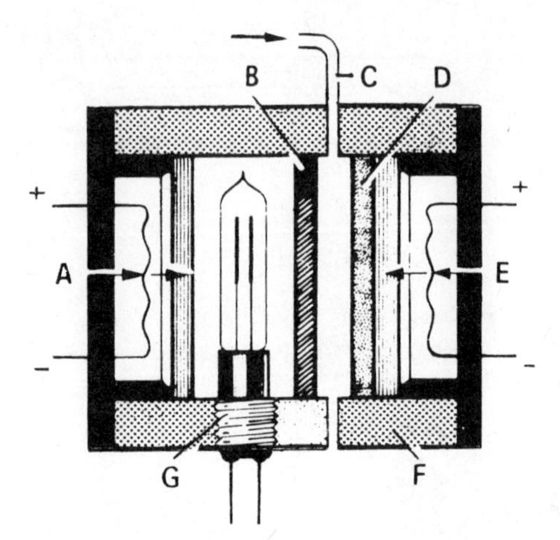

Fig. 3.4. Fluorosensor for continuous measurement of oxygen tension. A, Reference photocell and filter; B, ultraviolet transmitting filter. C, inlet for calibration gases and for gases sampled remotely; D, disk of porous vycor glass treated with fluoranthene; E, photocell with ultraviolet-absorbing filter; F, light-tight sintered metal cylinder for diffusion sampling; G, ultraviolet glow lamp. (From ref. 39 with permission.)

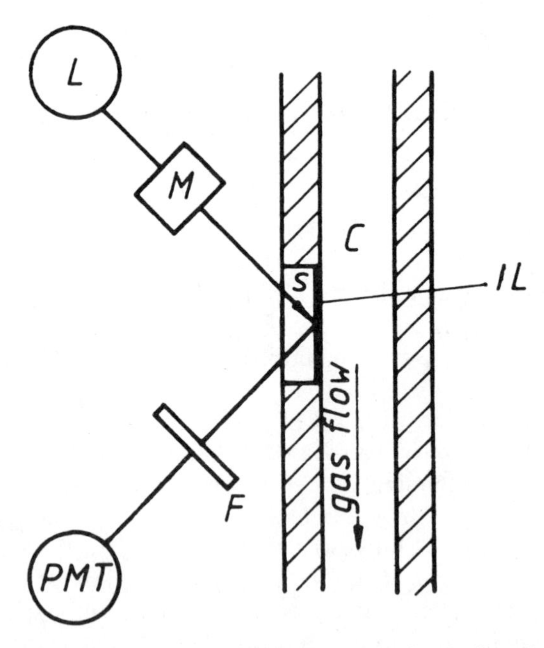

Fig. 3.5. Fluorosensor for continuous monitoring analytes by the front-face technique. The sensor may also be applied to flow-injection analysis.

indicators dissolved in fluid solution, certainly because of the limited mobility and accessibility of an immobilized fluorophore. Fluorosensors based on dynamic quenching by gases have rapid response times. Thus the response of the oxygen sensor shown in Fig. 3.5 for 90% of the total signal change is much less than 20 msec.

Its seems likely that sensors based on dynamic quenching will have a bright future. Dynamic quenching is frequently encountered with analytes for which no other simple optical methods are known. Since dynamic quenching is fully reversible and signal recovery occurs within a time much shorter than that of other sensor types, they seem almost ideally suited for continuous sensing.

The third group of optrode sensors is based on spectral changes of indicators due to physical or chemical perturbations. These include effects of temperature, solvent polarity, pressure, or viscosity. A typical example is provided by the fluorescence optical temperature sensor based on the reversed temperature dependence of the two main groups of the emission lines from europium-doped gadolinium oxysulfide that has been sequestered on the tip of a fiber (41,42).

To allow continuous sensing, the interactions of analyte with indicator have to be fully reversible. This is the case for all known sensors based on dynamic quenching or acid–base equilibration. Unfortunately, some sensors based on chelate formation between metal ions and indicators act irreversibly because the complexation constants are frequently so large that complexation is practically irreversible. This can make the indicator layer insensitive to concentration changes after it has been heavily loaded with the metal ion.

Indicator phase sensors are fairly simple in design and usually do not require excess maintenance. They are expected to have a broad field of application, since numerous fluorescent indicators are known, all of which can—in principle—be used in combination with fiber optics to produce new fluorosensors.

Reagent—phase sensors are almost the only optical ones that can be used for continuous in-vivo sensing of analytes lacking an intrinsic absorption or fluorescence, such as H^+, CO_2, or O_2. They have another advantage—of not being perturbed by inner filter effects. When, in conventional fluorimetry, a second substance is present in solution, having a measurable absorption at the excitation wavelength, the observed emission intensity will be reduced because of partial absorption of either exciting or emitted light by the interfering substance. This would result in too low an intensity. Hence when the intrinsic fluorescence of an analyte or a fluorogenic reaction is followed, the calculated analyte concentration will be too low. In a quenching experiment, on the other hand, the calculated concentration will be found too high.

These interferences are practically not observed with optrode sensors. The sensing layers are very thin (0.01–50 μm) and have high absorbances. Therefore, even when a fraction of an interfering substance penetrates the sensor layer, its fluorescence will remain small. The background fluorescence of the sample that may be stimulated by exciting light passing the sensing layer may be supressed by covering the sensor layer with a colored but nonfluorescent material acting as a filter. Suitable materials are Fe_2O_3, Fe_3O_4,TiO_2, or azo dyes of appropriate absorption maxima. Alternatively, the indicator itself may filter off the sample fluorescence, interfering with the sensor signal.

The convenience of not adding indicator solution to the analyte (as with reservoir sensors) is somewhat compensated for by the photolability of most indicators, in particular under laser excitation. Usually, indicators with fluorescence quantum yields above 0.9 are desirable, not only because of strong emission, but also because they do not tend to undergo efficienty intersystem crossing to the triplet state, which is the photoreactive state.

Fluorosensors often have a limited working range (i.e., the sloped part of the titration curve). When use is made of indicators with overlapping pK_a values, such as fluorescein, the range is considerably enlarged, but then pH/intensity plots no longer obey a simple dissociation curve. Unfortunately, no two-step-chelating indicators are known that would allow an enlargement of the dynamic range of cation sensors (Section 3.6.6).

A serious source of error can result from the fact that in these sensors it is the fraction of an immobilized indicator, or the ratio of acid and conjugate base of a pH indicator, which is measured rather than the analyte itself. This fraction or ratio is then mathematically related to the analyte concentration. Dissociation equilibria as well as complexation equilibria which are known to be governed by thermodynamic constants are subject to interferences by ionic strength, concentration effects, solvent composition, and—in the case of complexation—pH. As a result, a pH indicator will display a different reading for solutions of the same pH but different ionic strength, simply because of the ionic-strength-induced shift in the indicator's pK_a. Electrodes, in contrast, measure true activities via a diffusion potential.

3.2.3. Reservoir Sensors

Only a limited number of inorganic analytes display a native fluorescence that is sufficiently intense to enable sensitive quantitation. Also, all organic substances lacking a π-electron system are not subject to direct fluorimetry. It has therefore become necessary to make analytes fluorescent or to let them undergo a fluorogenic or fluoroquenching reaction. In reservoir sensors this is achieved by bringing the indicator solution into contact with the analyte

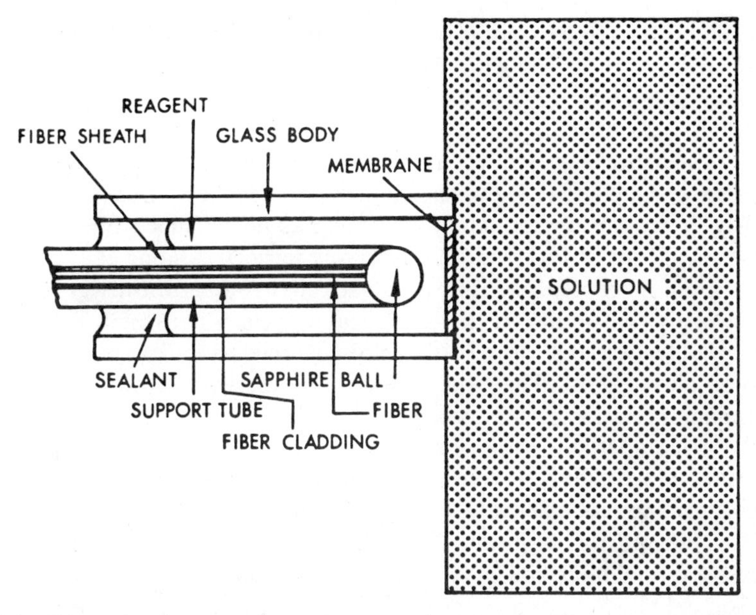

Fig. 3.6. Schematic of a reservoir continuous-sensing optrode. The reagent solution is in contact with the sample solution at the membrane. Light arriving at the terminal of the fiber is focused into the contact area at the membrane. Fluorescence is more efficiently coupled back into the fiber, since the sapphire ball acts as a collimating lens. (From ref. 17 with permission.)

solution in a volume viewed by the fiber. The indicator solution is contained in a reservoir situated close to the end of the fiber.

A typical scheme of a continuously sensing reservoir optrode is shown in Fig. 3.6. The tip of a fiber optical light guide is surrounded by a reagent solution which is in contact with the sample solution via a membrane. A sapphire ball at the end of the fiber acts as a lens, providing an apparent large diameter for observation of the sample through the fiber and effectively refocusing fluorescence into the fiber.

The fluorogenic reaction takes place at the membrane surface, or close to it. The resulting fluorescence is collected by the lens and guided back to the light detector. Again, because of the minute size of the tip, the usage rate of reagent can be as small as approximately 1 mL per month (!), thus making long-lived optrodes practical. However, any processes that may perturb the required mass transfer of analyte and reagent are potential sources of error.

Typical representatives of reservoir optrodes are those for uranyl ion (UO_2^{2+}) and polychlorinated organic compounds. In the former case, the weak intrinsic emission of UO_2^{2+} is greatly enhanced by complexation with phosphate at low pH. This situation is realized by slow addition of dilute phosphoric acid to the sample solution. In the latter case, an alkyl halide such as chloroform is reacted with alkaline pyridine to give a fluorescent dye (see Section 3.6.9).

Reservoir optrodes will always represent an interesting alternative to indicator phase sensors when the analyte does not exhibit native fluorescence. Obviously, however, fluorogenic reactions that proceed very slowly or require drastic conditions, such as high temperature or concentrated acid, are of little value in reservoir sensors. Among the advantages over plain fiber sensors, improved selectivity and sensitivity are most noteworthy. In addition, the analytical wavelength (which is frequently predicted by available laser lines) can be governed to some extent by a proper choice of reagents. The potential of the reservoir sensor principle for chemiluminescence sensors is evident.

When compared to indicator phase sensors based on dynamic quenching, it can be noted that reservoir sensors are more sensitive. This is because all indicators are less efficiently quenched when immobilized in the indicator phase than in fluid solution. For example, the quenching constant of 6-methoxyquinolinium ion by chloride in fluid solution (i.e., in the reservoir sensor) is 133 M^{-1}, while in the immobilized state (i.e., in the indicator phase sensor) the quenching constant is smaller than 1 M^{-1}.

The limited lifetime of reservoir optrodes is certainly disadvantageous, as is the required mass transfer to yield fluorescent products. Manufacturing as well as maintanance of these sensors is more laborious, costly, and delicate than that of simple fiber sensors, thus requiring more skilled personnel.

3.2.4. Phosphorescence Sensors

Apparently, the first luminescence sensor even applied to continuous measurement of an analyte was a room-temperature phosphorescence sensor (43). Use was made of the phosphorescence of dyes such as trypaflavin adsorbed on silica gel, which is most efficiently quenched by molecular oxygen. This is probably due to the extremely long lifetimes of phosphorescent states. The quenching process is practically complete at oxygen tensions above 10^{-3} mm, resulting in an analytical range of 10^{-6}–10^{-3} mm oxygen. This sensitive method eventually led to the discovery of the Kautsky effect in photosynthesis and the first unambiguous detection of singlet oxygen formed by photosensitation.

The method has also been applied by others (44–47). Aside from kieselgel, other supports, such as cellophane, paper, or plexiglass, have been used. Unlike polycyclic aromatics, the dyes used by Kautsky and others have excitation wavelengths beond 480 nm, thereby making them LED-excitable.

In addition, the Stokes shifts observed in phosphorescence are much larger than in fluorescence. This can minimize interferences from scattered and Raman light and makes easier spectral separation of exciting and emitted light.

An even more sensitive oxygen sensor (48) is based on the finding that at temperatures of $-70°C$ or lower, the phosphorescence of adsorbed dyes such as various rhodamines, trypaflavin, uranin, hydroxypyrenetrisulfonate, or umbelliferone is instantaneously transduced into rapidly decaying fluorescence after the addition of the minutest amounts of oxygen, typically 1×10^{-11} mol to a volume of 50 mL. As an example, the slowly decaying orange phosphorescence of silica gel–adsorbed trypaflavine is quenched and an intense flash of green fluorescence is observed when traces of oxygen are admitted. The effect is the same under vacuum, hydrogen, or nitrogen. Working ranges are 10^{-8}–10^{-5} mm of oxygen, which corresponds to roughly 10^{-9}–10^{-12} mol of oxygen in a volume of 50 mL. The visually observed detection limit was reported to be as low as 1.2×10^{-12} mol.

Except for a few cases, it has always been the practice to measure phosphorescence at liquid-nitrogen temperature or lower. The need for working at these temperatures prevents the design of optical sensors for ambient-temperature samples. This disadvantage has been overcome by recent developments (49) in room-temperature phosphorimetry (RTP). Thus a sensor for oxygen may be obtained (50) by incorporating 1-bromonaphthalene into β-cyclodextrin, which, in turn, is embedded into a matrix of polyvinylbromide plus plasticizer. The quenching of RTP and fluorescence by molecular oxygen is similar to the effect observed in solution (51). Unfortunately, the sensitivity of the RTP toward oxygen is much smaller in polymer solution than in liquid solution.

The indicator requires UV excitation at 293 nm, but the observed Stokes shift is so large (232 nm) that spectral separation of emission becomes very easy. Bromophenanthrene has a longer excitation wavelength but is less sensitive toward oxygen.

The RTP of cyclodextrin-embedded bromonapthalene is quenched by oxygen but enhanced by bromoalkanes (51) and halothane (a frequently employed anesthetic). This effect offers a method for continuously sensing halothane (50). The enhancement can be explained by the heavy-atom effect of halothane (1,1,1-trifluoro-2-bromo-2-chloroethane), which favors intersystem crossing to the phosphorescent state. Parallel to the increase in RTP a decrease in fluorescence intensity at 375 nm is found. The ratio of the intensities of the fluorescence and phosphorescence bands is therefore an even more sensitive parameter for the halothane concentration than is measurement of fluorescence or phosphorescence alone.

3.2.5. Evanescent Wave Sensors

The evanescent wave sensor has found application in both absorption and fluorescence sensors. Probably it has its greatest potential for probing immuno reactions since it is capable of detecting fluorescent molecules adsorbed on, or located within, several tens of nanometers of a surface (52,53). The technique utilizes an interfacial evanescent, standing electromagnetic wave to excite fluorescence near an optical interface. The fundamentals of the technique will be outlined briefly here (see also Section 3.3.3).

Light is totally reflected at the interface between an optical dense medium and an optically rare one, when the angle of reflection is larger than a critical angle β. The relation between the two refraction indices and β is given by

$$\sin \beta = n_2/n_1 \tag{3.1}$$

with n_1 and n_2 being the index of refraction of core and cladding, respectively. Interestingly enough, light is not instantaneously reflected when it reaches the interface. Rather, light penetrates to some extent into the optically rare phase. More precisely, the amplitude of the electric field does not drop abruptly to zero at the boundary but has a tail that decreases exponentially in the direction of an outward normal to the boundary. This phenomenon is referned to as an *evanescent wave*.

Figure 3.7 shows how light evanesces from the phase with n_1 into the phase with n_2. The displacement of incoming and reflected ray (D) is called the Goos–Hänchen shift. Provided that the reflection angle is close to the critical angle β, this shift can be calculated as

$$D = d_p \cos \beta \tag{3.2}$$

The depth of penetration (d_p) is defined as the distance within which the electric field of the wave falls to $1/e$:

$$\mathbf{E} = E_0 e^{-z/d_p} \tag{3.3}$$

with **E** being the amplitude of the electric field at depth z. d_p depends on the wavelength of the light and the refraction indices of the two media:

$$\lambda/d_p = 2\pi n_1 [\sin^2\beta - (n_2/n_1)^2]^{1/2} \tag{3.4}$$

Typically, d_p ranges from 50 to 1200nm for visible light, which is more than the thickness of a reagent or protein layer immobilized on a surface, as shown

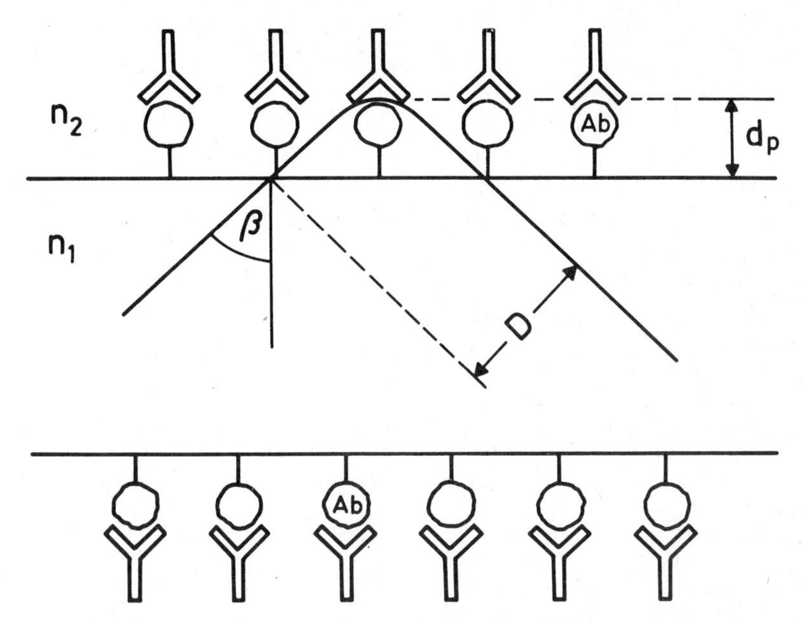

Fig. 3.7. Total internal reflection of light at the interface between optical media having refraction indices of n_1 and n_2, demonstrating that light is evanescing into the phase with n_2. d_p is the penetration depth, D the Goos–Hänchen shift. Because light penetrates into the phase with n_2, it can be used to probe immobilized molecules in this phase, for instance an antibody (Ab).

in Fig. 3.7. Light protruding into this phase is able to induce the fluorescence of the fluorophore, or to be absorbed, or to be scattered.

According to the Maxwell equations, a standing sinusoidal wave perpendicular to the reflecting surface is established in the guiding medium, whereas in the second medium the electric field amplitude shows an exponential decay [Eq. (3.3)]. This is shown schematically in Fig. 3.8. In contrast to the electromagnetic field of normal light, the evanescent field comprises a longitudinal component, too.

From Fig. 3.9 it is evident that the field amplitude depends on the polarization of the incident light. Hence the effective layer thickness d_e is different for an incident beam being polarized parallel from that of an incident beam being polarized perpendicular to the plane of incidence. For the two directions of polarization of the incident wave the effective thickness d_p (parallel) and d_p (perpendicular), respectively, can be calculated from the refraction indices. Obviously, the geometry of the optical arrangement in evanescent wave sensors is rather sensitive to wavelength and polarization effects.

Light penetrating the second medium will be attenuated according to the

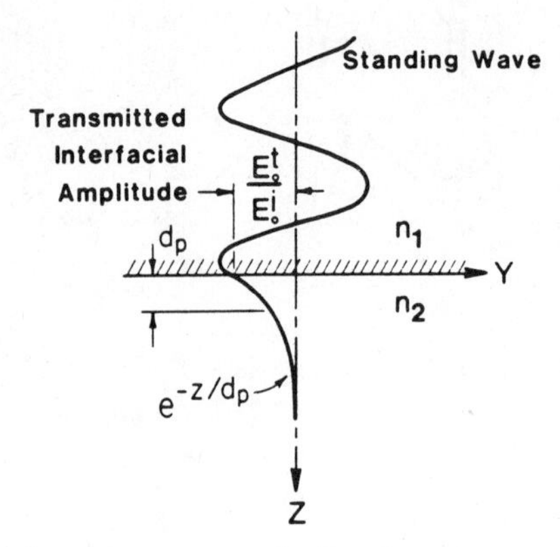

Fig. 3.8. Schematic of the surface wave at the interface of two optical media showing the standing-wave pattern in the medium having n_1, and the exponential decay of the interfacial electric field amplitude ("evanescent wave") into medium having index n_2. [From ref 58 with permission of Elsevier Publ. B.V. (Amsterdam).]

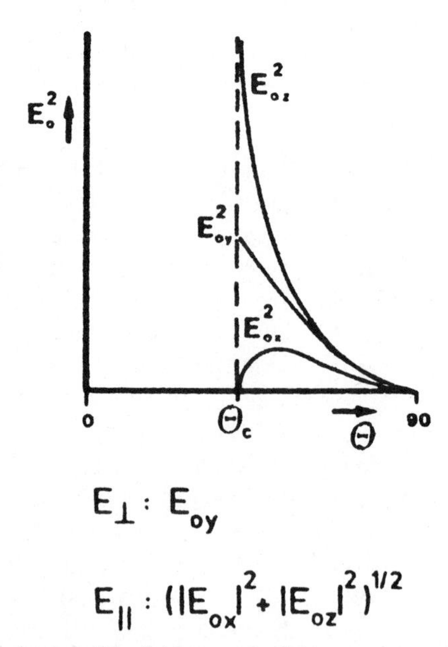

Fig. 3.9. Qualitative behavior of the field strength of the evanescent wave as a function of the angle of incidence ($n_1 = 1.51$, $n_2 = 1.00$). $E_{\parallel}$ and $E_\perp$, respectively, indicate the direction of polarization of the incident light. [From ref. 58 with permission of Elsevier Publ. B.V. (Amsterdam).]

Beer–Lambert law. This is the basis of absorption sensors utilizing evanescent waves (54,55). Sensitivity can be enhanced by combining the evanescent wave principle with multiple internal reflections. The number of reflections (N) is a function of the length (L) and thickness (T) of waveguide and the reflection angle β:

$$N = L \cot \beta / T \tag{3.5}$$

The situation is considerably more complex for evanescent wave–induced fluorescence (55–61), partly because of the different refractive indices of exciting and emitted light, differences in critical reflection angles, and polarization of emitted light. Evanescent wave sensors have been applied successfully to measure the fluorescence of dyes and indicators in solution (56,57,80), proteins (55,58,59), and of labeled antigen/antibody couples (55).

Evanescent wave-type sensors offer the typical advantages of attenuated total reflection spectroscopy. Interface layers that are too thin to support optical propagation as well as monodispersed microspheres can easily be probed. The interface can be between the surface of a fiber (58,61), the face of a prism, a simple glass slide, the surface of a thin-film waveguide, and the material of interest. In these cases a sample film which can be considerably thinner than 1 μm is formed on the surface of a fiber or a slab of glass that acts as waveguide. Both the cylindrical and slab waveguide with evanescent coupling to the sample offer the advantage that the effective pathlength can be made relatively long.

By interposing a 50-nm layer of silver between the glass surface and the sample, the intensity of the evanescent field can be increased by one to two orders of magnitude. The incident field couples to the electrons in the metal layer, an effect that is called surface plasmon resonance. The depth of penetration into the sample is a function of the incidence angle only and is not affected by the presence or absence of a metal layer.

An optical fiber sensor for detecting chemical or physical parameters has been described in a patent (54). It comprises a single fiber having a core section with a cladding whose absorption of the evanescent wave is analyte sensitive. It also has a transmission section surrounded by an analyte-insensitive cladding. The ends of the fiber are connected to the light source and detectors, respectively. Evanescent wave spectrometry along with time-resolution techniques (61a) will considerably improve the selectivity of the method.

3.2.6. Absorption and Reflection ("Photometric") Sensors

These sensors are based on changes in the absorption or reflectance of an analyte or indicator. Although not immediately within the scope of this

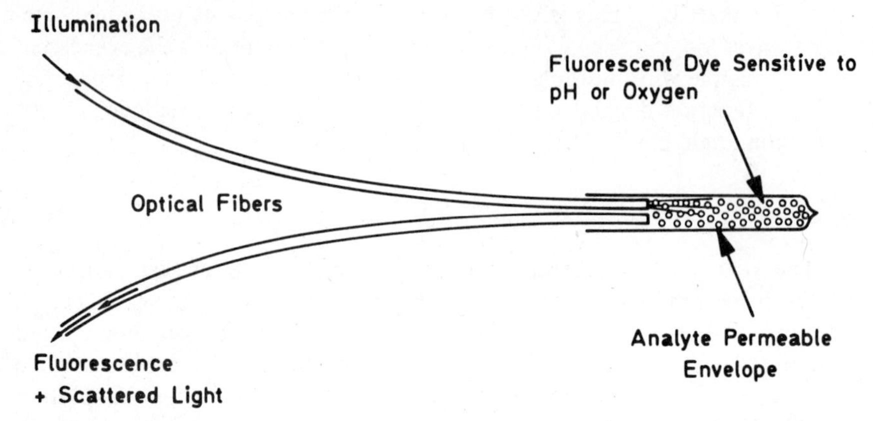

Fig. 3.10. Schematic of a thin fiber optical catheter suitable for invasive sensing. For measurement of pH, the fluorescent dye was replaced by an absorption indicator and the ratio of reflected light at two different wavelengths measured. (Redrawn from ref. 106.)

chapter, photometric sensors will be discussed briefly, since many of the features and problems associated with absorption sensors also hold for fluorosensors.

Several types of photometric pH sensors have been described. Harper appears to have been the first to immobilize pH indicators on porous glass and related materials (61b) to obtain reusable pH optosensors. The immobilization technique involves derivatization of the siliconous surface of glass with a silicone-type reagent such as aminopropyltriethoxy–silane. The aminopropyl–glass was subsequently converted to an arylamine–glass, which is a most useful material for azo coupling reactions. Indicators such as bromocresol purple, crystal violet, methyl red, phenol red, and phenolphthalein were linked to the glass surface in this way. The resulting glass sensors were shown to respond to pH values in the range 1.1–10.2.

A flexible fiber optic sensor 0.4 mm in diameter has been presented by Peterson et al. (62). It is based on spectral changes of the indicator phenol red with pH. Microspheres of polyacrylamide-covered dye and smaller polystyrene microspheres for light scattering are packed in an envelope of cellulosic dialysis tubing at the end of a pair of fibers (Fig. 3.10).

Green light (560 nm) and red light (600 nm) from a tungsten lamp or LED inject light into the illuminating fiber. Light returning from the tip through the other fiber is selected into two wavelengths by a cycling filter wheel. The ratio of the green light intensity to the red light intensity (which is pH independent and can be used as a reference) is then converted to pH. This principle is interesting in that it is based on a ratio measurement that can account for fluctuations in the exciting light intensity, photobleaching, leaking, and temperature effects.

A fiber optic–based sensor for bioanalytical absorbance measurement was used to follow serum absorption changes (63). The end of a 0.6-mm-i.d. fiber was connected to a specially constructed needle that acts as a cannula, and the sample is aspirated with a syringe. Aluminum foil reflects back into the fiber the light not absorbed. A study on the influence of optical path length and stray radiation on the analytical characteristics of the sensor as well as the results of a preliminary experiment involving the measurement of bilirubin in serum demonstrated the advantages of fiber optical sensing of small sample volumes. The need for repetitive sampling makes this particular method somewhat cumbersome and it remains open whether bilirubin can also be assayed in whole blood (with its high absorbance).

It is evident that in addition to pH sensing with indicator dyes, complexation of metal ions with metallochromic dyes may be utilized to devise photometric sensors for metal ions. Also, binding of antigens to antibodies (or vice versa) with one partner being labeled with a suitable dye offers a potential field of application for photometric sensors. While a few absorbance-based sensors have been reported, some of which were coupled to the evanescent wave technique (56,57), they have never become as popular as fluorosensor, probably for the following reasons:

1. Diffuse reflection photometry is much less sensitive than is fluorimetry, resulting in the need for large indicator zones. Only if the analyte is present in high concentration, as, for instance, hemoglobin and oxyhemoglobin, absorbance or diffuse reflectance is the method of choice (see Section 3.7.1).
2. Spectral shifts induced by complexation are usually more expressed in fluorescence than in absorption, a fact that can contribute significantly to better spectral differentiation.
3. When, in a chelate–ligand couple, fluorescence is associated with only one of the two species, the emission intensity is an absolute measure for the analyte concentration. Extensive background correction as in photometry is not required.

Another interesting principle for use in absorbance-based sensors (and probably also for fluorosensors) is represented by an albumin sensor an based on the color changes of immobilized bromcresol green that accompanies its binding to albumin (64).

pH-sensitive dye membranes together with appropriate enzymes that produce protons during their catalytic activity have been applied as transducers in the determination of penicillin G, glucose, and urea (65). In combination with LEDs as small and unexpensive light sources, and photodiodes as light detectors, these sensors compare favorably with electrochemical ones. They are designed for use in optically transparent samples, but are of little value for application in whole blood unless reflection is measured.

Ruzicka and Hansen (66a) have introduced light-reflectance-based optosensors to flow injection analysis. pH values in the range 4–10 were shown to be determinable in a detector cell consisting of a nonbleeding pH indicator strip attached to the end of a bifurcated fiber. The determination of ammonia and urea (via urease) was enabled by a different technique, which involves admixing a pH indicator (bromothymol blue), whose spectral changes with pH are monitored by a fiber optic, to the sample stream.

3.2.7. Other Types of Optical Sensors

Aside from light absorption, fluorescence, and phosphorescence, there are additional effects that can be utilized for optical sensing. These include bioluminescence and chemiluminescence, diffuse reflection, ellipsometry, light scattering, refractometry, and Raman emission.

It can be expected that chemiluminescence sensors will provide a powerful means for continuous monitoring. They offer advantages over fluorosensors in that they do not require an excitation light source and spectral separation of exciting and emitted light. Since all chemiluminescence reactions consume some kind of reagent, it is likely that they are similar in design to the reservoir optrodes.

The first chemiluminescence sensor ever described—which could readily be coupled to a fiber optic—makes use of the chemiluminescence that is produced when tetrakis(dimethylamino)ethylene reacts with molecular oxygen (66b). Reaction takes place in a reagent reservoir into which oxygen is allowed to diffuse through a Teflon membrane. Steady-state luminescence is proportional to the partial pressure of oxygen.

Bifurcated fibers or fiber bundles have been used to measure diffuse light reflection by erythrocytes in the bloodstream. The method allows the continuous determination of oxygen saturation (67) and cardiac output (68). This is described in more detail in Section 3.7.1.

Ellipsometry is an optical technique that exploits the reflection of polarized light from surfaces in order to measure surface monolayers of, for example, biomolecules. It is receiving increasing attention as a tool for investigating biological interactions at liquid–solid interfaces (69). It could thus prove invaluable for the detection of binding processes between biomolecules which otherwise are difficult to follow. As a matter of fact, ellipsometry requires large, complex apparatus and does not lend itself to miniaturization or implantation, unless polarization-maintaining fiber optics will be used. This may limit its capability for in-vivo probing.

Fiber optic sensors have also been used to collect spontaneous Raman emissions (70), and hollow fibers have served for inducing both spontaneous Raman (71) and coherent anti-Stokes Raman emissions (72). Raman emissions generated in a flame have been transmitted away from the hostile environment to a remote spectrometer (73). A fiber optic illumination device was developed for samples that were sensitive to a focused laser beam (74).

A fiber optic Raman probe was described in which both the exciting laser light and the Raman scatter are guided by optical cables (75). The technique requires no alignment of sample with input beam or collection optics. Various technical improvements resulted in a Raman signal up to nine times better than that of a conventional liquid sampling system. Applications to liquids (such as acetonitrile), liquid solutions of transfer ribonucleic acid, low-temperature samples (carbon tetrachloride, acetonitrile), and electrochemically generated radical cations (chlorpromazine) were described.

The most likely application of fiber optic Raman sensors will be for routine or remote sampling of liquids. A valuable aspect is its simplicity. The spectra may be obtained by an unskilled operator with no special cuvettes and no attention paid to sample alignment. The low specifity in complex samples is perhaps disadvantageous.

The refractive index of a liquid or polymer changes with an increasing fraction of other substances being dissolved. Thus refractometry can be used to continuously sense a gas or liquid that has a definite Nernst distribution between polymer and sample. The method is not very selective since all compounds diffusing into the polymer layer will change the index of refraction. It may be useful when encountered with samples having only one polymer-soluble analyte, or in determination of the sum of polymer-soluble compounds. Typical representatives are sensors for (chlorinated) hydrocarbons in air or groundwater. These compounds do readily diffuse from the aqueous phase into a lipophilic polymer whose change in the index of refraction is monitored. Alternatively, a hydrophilic material may be extracted from a lipid phase into a hydrophilic polymer.

Seifert at al. (76a) have published some work on an integrated optical biosensor, in which a differential refractometer of high resolution is employed. It was used to monitor enzyme activities by measuring the refractive index of a substrate converted by an enzyme. In preliminary experiments it was shown that immunoassays can be performed by this technique as well.

Mechanical sensors are based on light-intensity changes following the displacement of a reflecting of fluorescent spot. A typical representative is the pressure sensor, in which a fluorescent surface is held in place at the terminal end of an optical fiber. A flexible bellow or cantilevered mirror varies its position with changing pressure (10,25). A pressure sensor for intracranial use is commercially available from Ladd Research (76b).

3.3. INSTRUMENTATION IN FLUOROSENSING

A fiber optical fluorosensor essentially consists of a light source, optical filters (if required), a fiber light guide including light couplers and decouplers, a sensing zone or layer (if the analyte is nonfluorescent), and a light detector which transforms the optical signal into voltage. This signal is then amplified, processed, and displayed or printed out. The particular instrumental design is, of course, different for the various demands imposed on sensors, but the principle is generally always the same. The components of the optical system needed for fluorosensors as well as the advantages and disadvantages of various hardware components will be discussed in this section.

3.3.1. Light Sources

The following light sources have been used in fluorimetrically sensing chemical analytes: lasers; xenon lamps; hydrogen, deuterium, mercury, and halogen lamps; and light-emitting diodes (LEDs). Lasers have so far been applied exclusively for research purposes, not for commercially available instrumentation. This is because they are relatively costly and do require heavy power supply devices, whereas halogen lamps, gas discharge lamps, and LEDs do not.

Lasers are unique light sources in that they provide high-intensity pulses of very small bandwidth. They are available for a variety of wavelengths, the pattern of which can considerably be enlarged by frequency doubling or tripling (Table 3.5). They are of particular utility when combined with very long fibers, where light losses are large. Since light emission from laser sources is almost unidirectional, the beam is more easily coupled into the fiber than is light emitted by any other source. Moreover, they lend themselves to time-resolved fluorimetry in the nanosecond and picosecond time regime. Time resolution will experience its most useful application in fluorosensing of complex mixtures, wherein one component has a decay distinctly different from that of the other components, and in sensors based on dynamic quenching.

Nitrogen gas and krypton fluoride lasers with their strong pulsed emissions at 337 and 249 nm, respectively, are usually not used in combination with fiber optics because of their strong light attenuation in the UV. They have found application in airborne laser fluorosensorrs for detecting oil pollution (see Section 3.6.9). Such short excitation wavelengths are required

Table 3.5. Laser Types Most Suitable for Fluorescence Optical Sensing

Laser Type	Lasing Wavelength (nm)	Useful Other Wavelengths (nm)[a]	Typical Power[b]	Price
Argon ion	488, 514	—	1 W CW	Fair
Neodym-YAG	1060	530, 353	200 W CW	Medium
Helium-neon	633, 543, 1153	—	5-30 mW CW	Very low
Nitrogen	337	—	1 MW/peak	Fair
Krypton	350.7–356.4,[c] 406.7–415.4, 476.2, 482.5, 521, 531, 568, 647, 676, 752, 799	—	100 mW, 200 mW,[d] 500 mW[e]	Medium
Copper vapor	510, 578	—	10 W quasi-linear, 100 kW/peak	High
Excimer	Near UV	Various	0.1–1.0 MW per peak	Low

[a] After frequency conversion.
[b] CW, continuous wave.
[c] Several lines.
[d] For the 531-nm line.
[e] For the 647-nm line.

since certain significant polycyclic aromatic hydrocarbons display fluorescence emission under far-UV excitation only. Because of the power of the laser and with the help of sophisticated mathematical methods for background discrimination, pollutant-type materials can now be detected both at night and in full daylight (77).

Other useful laser types are listed in Table 3.5. They cover a wide spectral range, including the near UV. Notwithstanding the utility of UV lasers in airborne sensing of pollutants, they are of limited interest for use in fiber sensors. Generally it is observed that UV excitation results in higher background signals from both the fibers and the samples, so that it has become practice to use longwave-excitable indicators only. Moreover, UV-transparent fibers are relatively expensive.

Semiconductor lasers (not listed in Table 3.5) are fairly cheap but exhibit emission lines above 700 nm only. Because of the poor intensity, frequency doubling is very difficult, but recently has been shown to occur.

Dye lasers cover the complete spectral range of interest in fluorosensing but are rather expensive since they are usually pumped by another laser. They are promising light sources for analytes or indicators with spectral maxima that are not efficiently excited by other laser lines, and also because of the rather short pulses they can produce. Thus the spectral range between 370 and 460 nm and the one between 540 and 620 nm are not fully covered by nondye lasers. They have an additional advantage in being tunable over a much greater range than any other lasers operating in the UV or visible.

Aside from the costs of lasers that are prohibitive for their use in commercial instrumentation at present, the only limitation of lasers in combination with fiber optics is the stimulated Raman emission of fibers at high irradiation density. This effect is preferentially encountered when single fibers are used and plays no important role in fluorosensing with bifurcated fiber bundles.

An interesting feature of certain laser types results from the fact that their emission lines closely match the vibrational absorption band of certain analytically interesting compounds. For instance, an erbium laser has been applied as a light source for sensing methane in the atmosphere, owing to the coincidence of the $^2\nu_3$ R(6) line of methane and the erbium:YAG laser emission centered at 1644.9 nm (78,79).

Xenon lamps, either continuous-working or pulsed, offer another useful light source. The spectral range, extending from the UV (where intensity is low) to the IR, is almost ideal for fluorosensing. The emission maximum is broad, ranging from 400 to 500 nm, and is pretty well suited for fluorescence excitation. Krypton gas discharge lamps are also available, but the spectral distribution of the emission is less favorable. There is a relatively constant emission over the entire visible region, but with the strongest bands between 800 and 1000 nm, which is outside the range of interest. As a result, the IR part of the emission of both lamp types has to be shielded from the fiber and sensing layer by a heat protector.

Continuously burning xenon and krypton lamps have several drawbacks in that they are expensive, have limited lifetimes, and like lasers, require more electrical power and supply equipment than do other conventional light sources. Pulsed xenon lamps are preferable because of longer lifetimes, lower price, and less power consumption. On the other hand, the light output of pulsed lamps varies from pulse to pulse because of so-called arc wandering. This means that the arc within the gas bulb does not travel the same way in each flash, which may result in nonideal focusing of light into the end of the fiber. The problem may be overcome by signal averaging.

Despite this, pulsed xenon lamps proved to be very useful light sources. Small-sized lamps are available from, for example, Noblelite Ltd. (Cambridge, UK) or EG&G (Salem, MA) having lifetimes of up to 10^9 flashes (depending on the flash energy). They are run at frequencies between 0.1 and 10 Hz, with flash duration of about 10 ns. Although producing a lot more light energy than halogen lamps, they do not produce much photobleaching because of the short pulses. Small xenon lamps are still relatively expensive, but several fibers can be attached to one small tube without using a rotating wheel so that one lamp is sufficient for several sensors.

Neon lamps are among the cheapest light sources at present. Various minilamps with spectral maxima between 350 and 700 nm are available (from, e.g., Elite Elektronik, Steinhoring, FRG, or ERG, Göttingen, FRG). Although their light intensities are small, the long lifetimes of up to 20,000 h as well as the low power required for their operation makes gas discharge lamps most interesting light sources. They are run at 200–220 V, do not require extremely high starting voltage (and therefore a costly power supply), and do not produce much heat. In combination with fibers and interference filters, neon lamps form another type of useful light sources for technical instrumentation.

Halogen lamps and tungsten filament lamps are cheap but are of limited utility in practice. They produce a lot of heat, have short lifetimes, and low intensity in the spectral range below 450 nm, which drops more than proportionally when the voltage is lowered. This makes the shortwave part of the emission particularly sensitive to voltage fluctuations of the electric power source. On the other side, halogen lamps are easily operated at low dc voltage, the heat produced may be used for thermostatization of an instrument, and the spectral range above 500 nm is of sufficient intensity to excite even weak fluorescers.

Hydrogen lamps, and especially, deuterium lamps are the most useful sources for UV excitation. The range 190–400 nm is covered very well with low power requirements, typically 5–50 W. They have low visible output, which provides little stray light when fluorescence is observed in this spectral range. There is, however, a general tendency to perform fluorosensing outside the UV excitation range because of considerably higher prices of equipment for UV spectroscopy.

Light-emitting diodes (LEDs) are promising new light source types, although available from the IR down to the blue only. They are the light source of choice for application along with long-wave absorbing indicators. Thus LEDs have found application in pH sensors using bromcresyl violet as an indicator (64), in related transducers (62), and in a pressure sensor (80).

The lack of suitable violet and ultraviolet LEDs prevents their use in optrode sensors having indicators with less than 450 nm excitation. LEDs with λ_{max} 480 nm and 100 nm bandwidth at half-height are the most shortwave LEDs at present. They are available from a few companies, such as Sanyo Corp. (Moriguchi, Osaka, Japan), Siemens AG (Munich, FRG), and Telefunken (Heilbronn, FRG).

LEDs are run at low voltage and current, typically 2.5–3.5 V at 20 mA, and are therefore ideal for portable sensor instrumentation. They produce small light intensities (about 2 millicandela), but their extremely long lifetime, spectral constancy, and ruggedness makes them extremely interesting light sources. Their intensities can be modulated in the kHz to MHz frequency domain, thus enabling their use in phase-fluorimetric lifetime-based sensors. The small size of LEDs and lack of heat production offer additional advantages. They may be placed very close to the sensor layer, thereby eliminating the need for light guides as well as heat-protecting devices. It is hoped that the spectral range can soon be extended into the violet and near-UV range, a spectral region where a manifold of interesting indicators is known to absorb light.

3.3.2. Optical Couplers

The separation of fluorescent light from interfering light (mainly Raman emission, Rayleigh scatter, and second-order light resulting from wavelength doubling) is usually achieved with the help of monochromators. Effective separation is of particular importance when single fibers are used to guide both the exciting and emitted light. Single fibers have advantages because of size limitations (e.g., in catheters), lower costs, and slightly better collection of fluorescence. The use of a single fiber for illumination and collection not only eliminates problems caused by fiber alignment variations and modal distortions within the fiber, but maximizes the overlap between the illumination and observation volumes and their subtended solid angle.

To couple light most efficiently into a single fiber and to couple the returning beam into the spectral sorter, various optical couplers have been described. Essentially, light separation is accomplished by four methods. The first makes use of a simple beam splitter which separates the outgoing and returning light by their direction. The method is inefficient.

The second method uses a dichroic mirror: a mirror that reflects light of a certain wavelength only but transmits all other wavelengths. This method is more efficient and relatively simple, but is inflexible in that each mirror is suitable for a definite wavelength only. Hence changes in the spectral properties of the system require an exchange of the mirror.

The third method uses a mirror with a small hole in it to separate the highly collimated excitation beam (usually, a laser) from the divergent returning beam. This is shown in Fig. 3.11. The beam goes through the hole

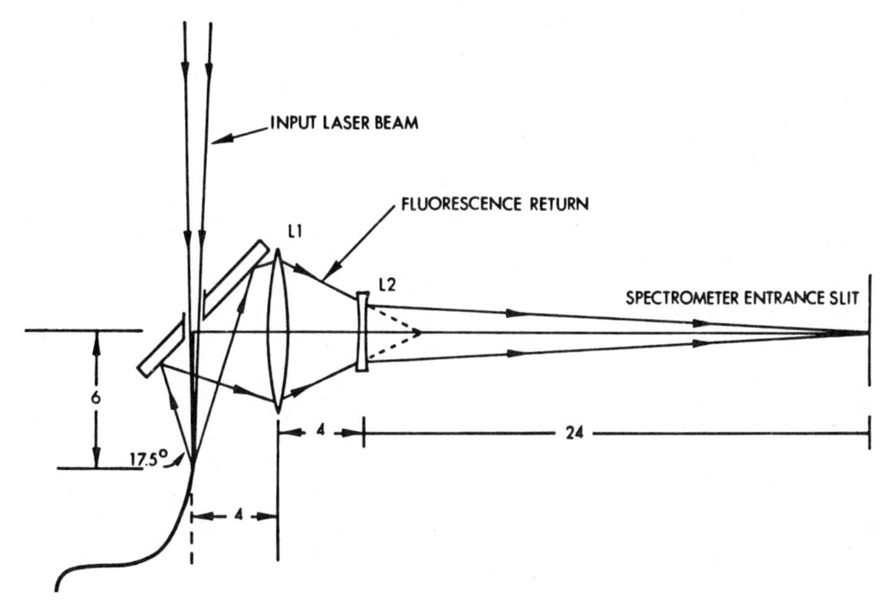

Fig. 3.11. Design of a geometric coupler having a small hole in a mirror, through which a narrow laser beam is focused into the fiber end. Returning light with higher divergence strikes the mirror and is reflected into the spectrometer entrance slit after passing lenses L_1 and L_2. Dimensions are in centimeters. (From ref. 6 with permission.)

in the mirror into the fiber. The returning fluorescence leaves the fiber with a certain numerical aperture, strikes the mirror, and is turned into a collection lens followed by a focusing lens that directs the light in the spectral sorter or a photodetector.

The separation of concentric outgoing and returning beams can also be accomplished by using a small prism and collection lenses, as shown in Fig. 3.12. Here the prism turns the focused excitation beam into the fiber, while the divergent returning light is collected by the lens system. Because of its small size the prism causes essentially no obscuration and thus no degradation of signal intensity.

In the fourth method a single fiber is divided into two rods as shown in Fig. 3.3*b*. The end of one rod is coupled to the light source, the other to the detector. Although only 50–90% of the total fluorescence from the common end is led to the detector by this approach, it allows a simple and vibration-insensitive separation of fluorescence from primary light.

Aside from efficiently coupling the excitation beam into the fiber it is necessary to couple fluorescence most efficiently into the end of the fiber. In conventional arrangements, as shown is Fig. 3.13, only a small fraction of

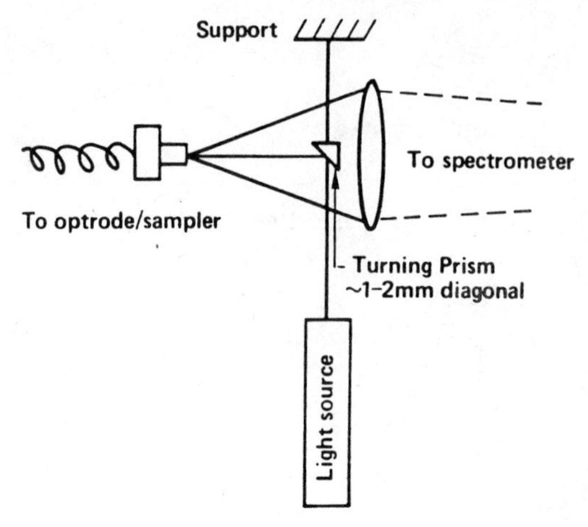

Fig. 3.12. Optical design of geometric coupler, in which a small prism reflects a thin laser beam into the fiber. Again, the returning light leaves the fiber with higher divergence and can easily be separated. (From ref. 25 with permission.)

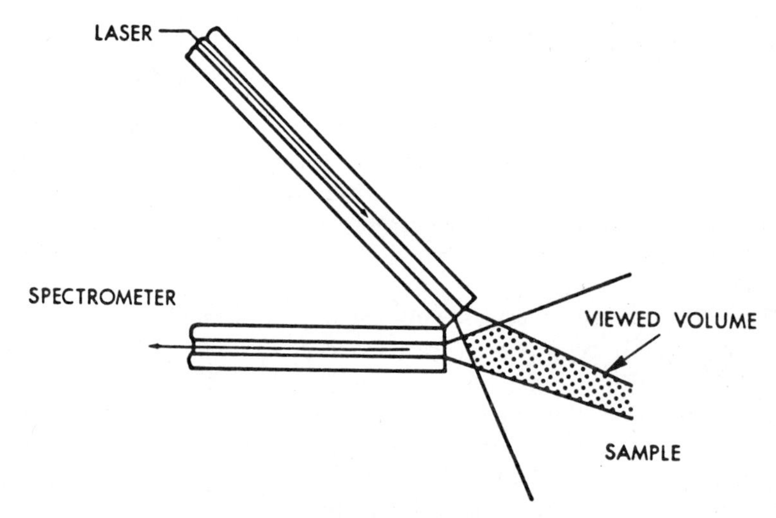

Fig. 3.13. Double-fiber terminal for remote fiber fluorimeter.

the total emission enters the second fiber because of a small sample volume that is viewed by both the "laser fiber" and the "spectrometer fiber." The cone angle overlap is better when single fibers are used, but the effectively viewed fluorescence is still only a minor part of the total emission (Fig. 3.14).

The efficiency of light collection by single fibers can easily be improved by

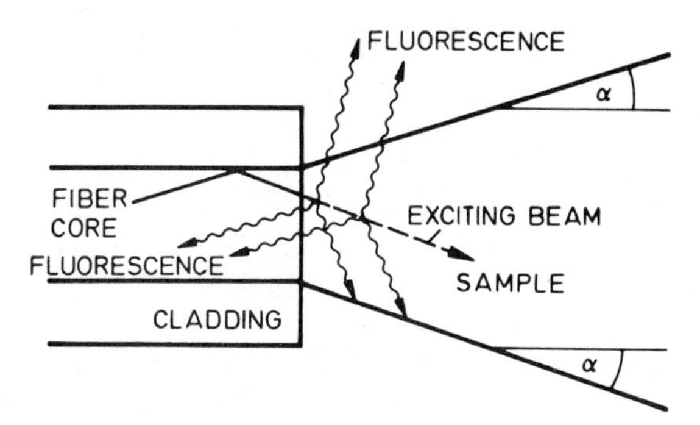

Fig. 3.14. Schematic of the fluorescence produced at a single-fiber termination.

various methods (21). Thus a 3-mm-diameter sapphire ball can be interposed between fiber tip and sample (Fig. 3.15). The sapphire (focal length 0.5 mm) acts as a lens and not only increases the amount of sample that is irradiated, but also increases the amount of fluorescent radiation from the sample that makes its way back into the fiber. Also, the impervious sapphire ball can protect the fiber tip from caustic media. The improvement in sensitivity is about a factor of 2. The sapphire ball may be replaced by a ground lens to better focus the input and output beam. The price for such a lens is estimated to be hundreds of dollars, whereas a sapphire ball costs about $4.

In the capillary optrode shown in Fig. 3.16, a further increase in sensitivity is achieved, owing to reflections of both fluorescence and excitation light in the internally silvered capillary tube. This creates a guiding effect for light that increases the observed sample thickness and simultaneously collects more of the emitted signal.

The above-mentioned fiber terminations are passive in that they invoke no specific response of the media to the incident radiation. They would have application in highly characterized and relatively uncomplicated sample regions such as groundwater or air pollution control, where only one fluorophore is studied.

When fiber bundles are used in place of single fibers, additional losses occur because the detection fibers do not collect the entire fluorescence produced by the illuminating fibers. The losses increase with the distance between the fiber bundles as a result of less cone angle overlap. Most efficient coupling is guaranteed as long as the source of fluorescence (the sample) is large in relation to the field of view of the detection bundle. Imaging optics such as lenses may be used to accomplish more efficient coupling when the sample is too small or too unaccessible.

The size of the fiber optic bundle affects both the amount of light shining on the sample and the amount collected. As a rule of thumb, a 3-mm fiber

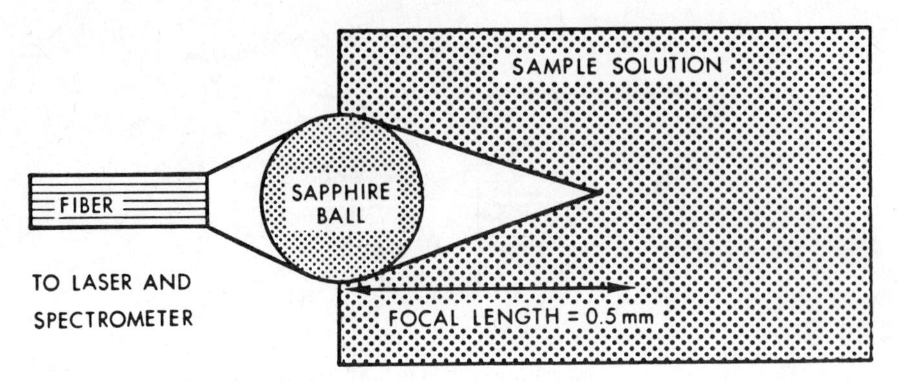

Fig. 3.15. Sapphire ball optrode.

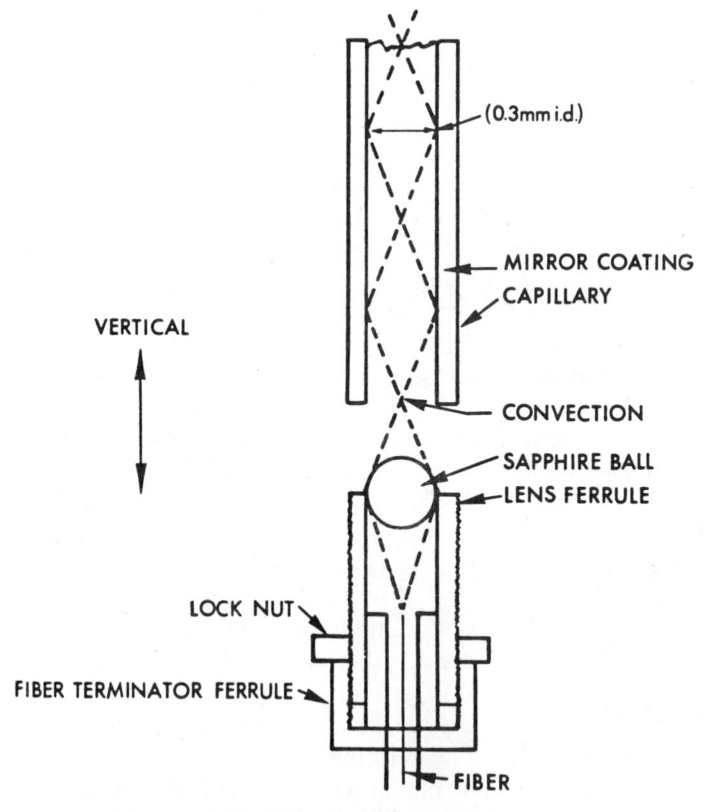

Fig. 3.16. Capillary optrode.

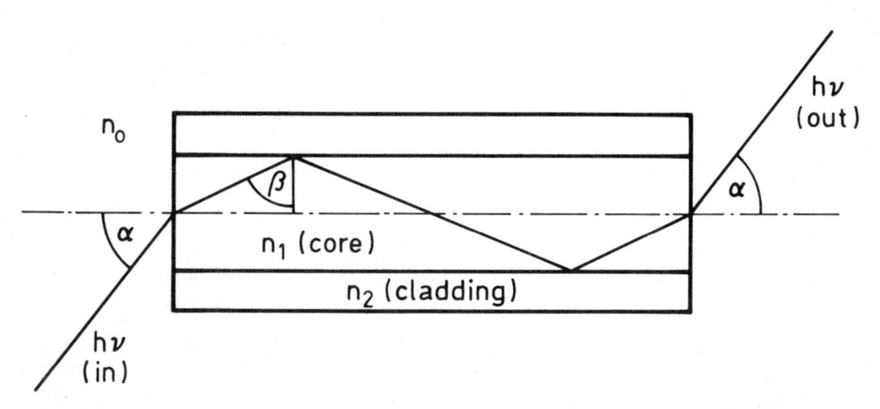

Fig. 3.17. Path of light passing through a section of an optical fiber. The indices of refraction of the surroundings, the core, and the cladding are denoted by n_0, n_1, and n_2, respectively. For air, n_0 is 1.

bundle will produce about 20 to 80 times the fluorescence signal of a 1-mm fiber bundle.

3.3.3. Light Guides

Most fluorosensors have been used in combination with fiber optical light guides. These can also act as heat protectors and enable the focusing of light on very small sensing layers. The number of lenses and mirrors is reduced to a minimum, and lens aberrations, light losses, and time required for setup and alignment are thus reduced.

A light guide is able to transport light over large distances in thin, flexible cables with low losses. The geometrical path of light passing through a section of a fiber is sketched in Fig. 3.17. If the path of the light enters the fiber at an angle smaller than a critical angle α, light will be totally reflected inside the fiber and guided away. Total reflection in the material of higher index of refraction occurs at the interface to the material of lower index of refraction. Instead of α, the fiber may also be characterized by the numerical aperture (N.A.), which is related to α by

$$\text{N.A.} = n_0 \sin \alpha \tag{3.6}$$

The N.A. describes the cone within which light is accepted and guided by the fibers. When n_0 is one (i.e., under air), N.A. is simply the sine function of the critical angle α.

Outside the fiber the angle between the incident source beam and the fiber

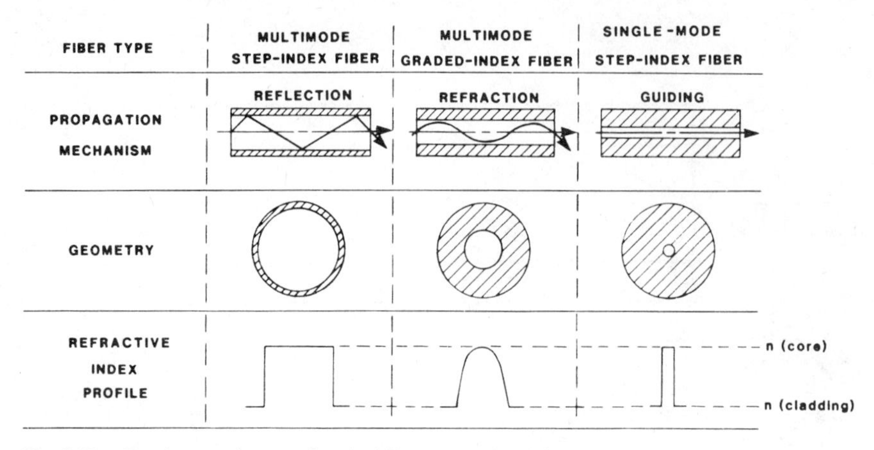

Fig. 3.18. Fundamental types of optical fibers. [Reprinted from *Fiber Optics* (J. C. Daley, Ed.) with permission of CRC Press, Inc., Boca Raton, FL 33431 (1984).]

axis is larger than it is within since the index of refraction of the waveguide material is larger than that of the medium outside the entrance face. Fibers are usually coated with a plastic layer as the fiber is drawn. This plastic coat protects the fiber from abrasion and chemical attack. Within the light guide the profile of the refraction index may be uniform, have a smooth gradient, or a discontinuous "step" profile.

The three fundamental types of optical fibers are shown in Fig. 3.18. The core diameters for multimode graded fibers and step-index fibers are approximately 50–250 μm. The peak difference in refractive indices of the core and the cladding is between 1 and 2%. This difference determines the N.A. of the fiber, which in turn determines the amount of coupled light the fiber can accept from one optical source.

Within the light guide, light must travel parallel to, or at a small angle to, the axis of the guide. This means that the light reaches the walls of the fiber at a large angle of incidence (with respect to the normal to the interface). Transverse gradients or discontinuities in the index of refraction serve to reflect the waves.

To be transmitted by a fiber the light must enter the fiber within a given cone angle. This cone angle depends on the critical angle of total reflection within the fiber. This angle (β in Fig. 3.17) is defined as the smallest angle of incidence, in the medium of greater index, for which light is totally reflected. This means that only when light in an optically dense medium approaches the boundary of a less dense medium at an angle greater than the critical one will it be totally reflected.

This angle β is defined by $\sin \beta = n_2/n_1$ [Eq. (3.1)], and the half-angle α of the acceptance cone is related to the refraction indices by

$$n_0 \sin \alpha = (n_1^2 - n_2^2)^{1/2} \tag{3.7}$$

Thus the N.A. can also be described by

$$\text{N.A.} = (n_1^2 - n_2^2)^{1/2} \tag{3.8}$$

The N.A. of fibers can vary from 0.18 to 0.66. It might appear that the latter would be the better choice because of the greater cone angle. This is true only if one does not intend to go lower than 400 nm. It is also somewhat more difficult to couple in and out of the fiber with the larger cone angle.

Since most multimode fiber optics rotate the light beam in a helical manner, they act as depolarizer. This can help to minimize errors attributed to polarization effects, but limits the capability of fibers for use in fluorescence polarization experiments. There are, however, certain types of fibers which are capable of sustaining a given polarization over lengths of meters, but in most fibers light that is incident in a plane-polarized form is quickly converted into elliptical polarized light. Recently, however, the Hitachi Cable Co. has started offering single polarization monomode fibers which sustain polarization over distances as large as 1000 m. The effective wavelengths are 630, 850, and 1300 nm.

When entrapped in a fiber, light propagates in different modes. The mode of propagation is determined by the characteristics of the incident beam, the waveguide, the angle of light entry, and light wavelength. The fundamental mode (i.e., one with electromagnetic field minima only at the walls of a light guide) as well as two higher-order modes are illustrated in Fig. 3.19. The fields have an exponentially decreasing value when they evanesce the waveguide. This has already been shown in Section 3.2.5. Outside the waveguide the higher-order modes have a greater intensity at a given distance than do the lower-order modes.

The evanescing tail represents an electromagnetic field that oscillates at the frequency of the incident light but does not propagate through the second medium. The field is strong only near the interface between the two media, where it is able to induce the fluorescence of an absorbed indicator or analyte.

It is obvious that mode 1 in the fiber travels through the light guide faster than do higher-order modes since it has a shorter path to traverse in the fiber. This effect is known as mode dispersion. Another dispersive phenomenon is material dispersion, which is due to the wavelength dependence of

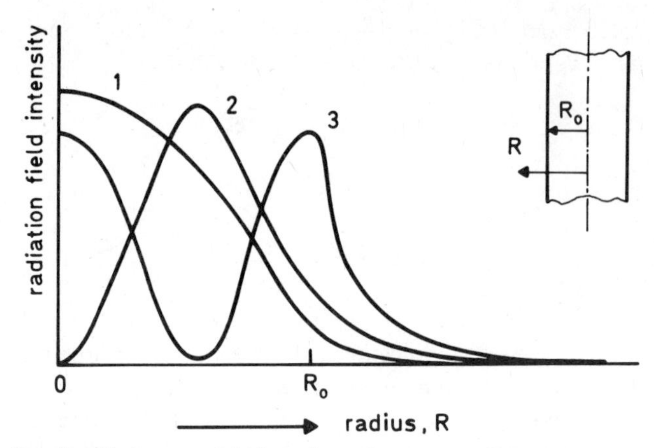

Fig. 3.19. Relationship between field intensity and transverse distance from the sample of a symmetric waveguide illustrated for the first three modes. (Redrawn from ref. 7.)

the refraction index. Both modal dispersion and material dispersion are important effects to consider in using waveguides with pulsed light sources. The effect of dispersion is to alter the pulse shape and width.

Most fiber cores are drawn from either fused silica glass, compound glass, poly(methyl methacrylate), polystyrene, or a suitable liquid. Fused silica fibers transmit in the range 220–1300 nm but cost four times as much as glass fibers, which transmit only in the range 380–1300 nm. However, for most applications the length of the fiber is limited to a few meters, a length where light attenuation is usually not an important factor to consider. The maximum operating temperature of glass fibers is far above 100°C, whereas plastic fibers should not be heated to more than 140°C. This is more than sufficiently high to allow steam sterilization.

Liquid light guides function in a manner similar to fiber optics. The difference is that the liquid light guide consists of tubings filled with a transparent fluid plugged at both ends with a polished silica gel plug. Typically, liquid light guides transmit in the range 270–720 nm. They are therefore a good in-between choice if near-UV guiding is required, since fused silica fibers cost around two to three times as much.

Although total reflection implies no light attenuation, small losses occur due to absorption in the medium, discontinuities and defects in the reflecting surface, and reflections at the surfaces when the light enters and leaves the medium. Also, when light is internally reflected, it penetrates the lower index phase (air or cladding) a short distance. Hence dust particles, fingerprints, and surface scratches cause light scattering and absorption, resulting in light attenuation. The most frequently used parameter to characterize the attenu-

Table 3.6. Angle of Aperture 2α of Fused Silica Fibers

Fiber Length (mm)	2α at λ = 254 nm	2α at λ = 546 nm
150	36 ± 5°	62 ± 5°
250	32 ± 5°	55 ± 5°
500	27 ± 5°	42 ± 5°
1000	22 ± 5°	35 ± 5°
1800	17 ± 5°	30 ± 5°

ation of light when it passes over a certain length of fiber is the decibel (dB). One measures the ratio of light intensities at the beginning (I_0) and the end (I) of the lightguide. The attenuation, expressed in dB, is calculated according to $-10 \log I/I_0$, usually per kilometer. Fibers with losses between 0.01 and 3 dB/km are state of the art, and lower-loss fibers are under development.

The scattered intensity is inversely proportional to the fourth power of the wavelength. As a consequence, not only is the sky blue, but transmission losses are considerably higher for indicators with blue or violet fluorescence than for those with yellow or red fluorescence. The λ^{-4} law is not very important with short fibers but is further support for designing sensors having analyte-sensitive indicators of analytical wavelengths far in the visible or even infrared.

It has been observed (81) that the limiting factor on the sensitivity of a single-optical-fiber fluorimeter system is the intrinsic fluorescence of the fiber, and to a lesser extent, other optical components, giving rise to background fluorescence in the spectral region of interest. This imposes a severe limit on the system performance of a single-fiber fluorosensor. Thus when 1 W of 514-mm light was launched into several fiber types, the sensitivity of remote dye concentration measurement was restricted to 1 part in 10^6, which is four orders of magnitude worse than that of the dual-fiber system. Among the fiber types investigated a "superwet" UV-grade PCS fiber exhibited significantly lower fluorescence than did a similar length of "dry" silica-cored fiber with low OH absorption, or any one of several doped silica fibers that were tried.

Visible fluorescence emission has been observed from heat-treated GeO_2-doped silica fibers in a hydrogen atmosphere (82). Absorption bands with maxima centered at 440 and 630 nm have been found in silica fibers after irradiation with α and γ rays (83) and again are less significant in high-OH fibers. It is particularly noteworthy that γ irradiation of silica can also result in increased fluorescence in the region 640–720 nm with a maximum at around 670 nm.

Aside from the intrinsic fluorescence, Raman bands at 490 and 600 cm^{-1} are present in silica quartz glass, presumably as a result of breaks in Si–O bonds. By doping the glass with GeO_2 or B_2O_3 (84), or by thermal treatment (85), a reduction in Raman intensity is accomplished. Despite this, the stimulated Raman scatter (SRS) becomes very intense when the input power of (pulsed) laser light into fibers exceeds certain limits, typically 100 mW. Since this is more than the energy required in most fluorosensors, the incoming light from laser sources is attenuated by inserting gray filters into the light path. In remote fiber fluorimetry over large distances the SRS effect can seriously hamper sensitivity.

In a Raman spectroscopic study of the absorptions induced during the drawing and irradiation of fibers (86), the fluorescence and Raman emissions of silica fibers were recorded. The maxima of the fluorescence bands in the red region were different in both cases. The presence of emission bands in the same region as that observed in the fibers was demonstrated. The major sources of extrinsic absorption and scattering losses in fluorozirconate glasses and fibers were identified as transition element impurities and hydroxide contamination from humidity (87).

Jones and Spooncer (88) have described a useful method of referencing out adventitious intensity variations in an optical fiber intensity-modulated sensor ("shutter sensor") by measuring the transmitted light intensity at two wavelengths; one carries the signal information and the other is used to normalize the intensity. A similar procedure (i.e., ratioing two light intensities) was applied in the fiber optic pH and oxygen sensors developed by Peterson et al. (37,62).

Both single fibers and fiber bundles have been applied in fluorosensing. Single fibers are cheap and can be made very small, but this necessitates relatively complex instrumentation to separate input and output signals. Single-fiber sensors have lower sensitivity because of considerably higher signal background levels.

Bifurcated fiber are more voluminous but allow a more homogeneous illumination of large sensor layers. Care has to be taken that at the common end of the two arms, the bundles are statistically mixed so that there is most effective overlap in the cones of input and output fibers. Coherent bundles, in which individual fibers are laid up in correct orientation so that both ends are arranged in an identical pattern, are more expensive and less efficient.

3.3.4. Light Detectors

Accurate and sensitive determination of photon flux is a prerequisite in precise and sensitive fluorosensing. In each instance the choice of the proper light detector will be a trade-off between costs and sensitivity. The various light detectors used in practice, including their advantages and disadvantages, are described in this section.

There are three basic types of photosensitive devices in use today: the photoemissive type, the photovoltaic type, and the photoconductive type. The photoemissive one includes photomultiplier tubes (PMTs) and phototubes. A PMT consists of a photoemissive surface called a photocathode, secondary emission electrodes called dynodes, and an anode with a potential difference set up between the various electrodes. When light impinges the photoemissive surface of the photocathode, an electron will be ejected that is attracted by the potential difference between photocathode and dynode. Striking the dynode will cause secondary emission of electrons, an effect that is repeated throughout a series of dynodes until finally a large charge is available at the anode as the output signal. For example, a PMT having 10 stages and a secondary emission factor of 4 per stage has a gain of 4^{10} ($= 1.05 \times 10^6$). Thus only one photoelectron emitted by the photocathode results in 10^6 electrons (or a charge of 1.68×10^{-13} C) at the anode in the form of a pulse having a width of several tens of nanoseconds.

Such PMTs, because of their high intrinsic signal-to-noise ratio, fast response, and high gain are ideal for the measurement of very low light levels. Noise can result from thermal electrons ejected from both the photocathode and the dynodes, and from the glass material hit by cosmic radiation. The thermal noise can be greatly reduced by cooling to $-30°C$, and pulse height discriminators can eliminate noise from glass and dynodes. Obviously, PMTs cannot be operated easily in a radioactive environment.

PMTs and entire photocounting units are available in compact packages also containing preamplifiers and discriminators. The spectral response depends mainly on the nature of the photoemissive layer in the photocathode. For the analytically useful spectral range encountered in fluorimetry (350–800 nm) cesium-doped layers are most useful. A number of them, having wide and flat response, are commercially available. Gallium arsenide cathode PMTs have good near-IR response and a nanosecond time response. Size, price, and possibly also the high voltage required for the operation may be limiting factors for their use in certain instrumentation, where photodiodes are prefered because of considerably smaller size and lower costs. PMTs are, however, very popular in scientific instrumentation such as fluorimeters and single-photon counters, since they are most sensitive and the rise time can be as short as 1.5 ns.

The photovoltaic light detector is a self-generating type and is presented by silicon photodiodes. When a silicon *p–n* junction is operated under zero externally applied voltage and low levels of incident light, the *p–n* junction will generate a current proportional to the light intensity ("photovoltaic effect"). This photocurrent will divide between the diode internal junction resistance and the parallel load resistance. The voltage measured across the load resistor can be related to intensity by various methods.

The major advantages of photodiodes are small size and low price. They exhibit good stability, a wider dynamic range than PMTs, need lower power requirements, but are not as sensitive as PMTs. Limited response linearity seems to be eliminated in newer photodiode types. The spectral response increases steadily upward from 400 nm to reach a maximum at about 900 nm. This makes photodiodes most suitable for fluorimetry in the visible.

The photoconductive light detector types are further classified into photoconductive junction type and photoconductive bulk-effect type. The former utilizes the photoconductive properties of *p–n* junctions and includes phototransistors and PIN ("positive-intrinsic-negative") photodiodes. Cadmium sulfide and cadmium selenide photoconductive cells represent the latter. When light shines on the CdS layer, the resistance decreases as the light level increases.

Photoconductive diodes are available that respond to wavelengths between 400 and 900 nm with good sensitivity, with a normal operating bias of 90 V. Linearity over five to nine decades of light intensity can reasonably be assumed for guard-ring-structured (low dark current) planar diffused photodiodes. Photoconductive cells can be used for prolonged periods unless exposed to high humidity and strong UV light. PIN silicium photodiodes are ideally suited for detecting ultrashort light pulses on the order of 50–100 ps in the range 300–1100 nm.

The advantages and disadvantages of the photovoltaic and photoconductive diodes are important in designing fluorosensors and should be reviewed with respect to the particular application. Thus the photoconductive mode produces a high sensitivity at the longer wavelengths, an extended frequency response due to lower junction capacity, and a photocurrent that is linear with irradiance over an extended range. The photovoltaic operating mode results in a low photodiode-generated noise, lower sensitivity at the red end, and a shorter frequency response.

3.4. SENSING LAYERS

The majority of analytes of interest in fluorosensing are nonfluorescent or have an emission too weak or too far in the UV to be of practical utility. It has therefore become necessary to apply fluorescent indicators, to label molecules with suitable fluorescent tags, or to perform fluorescence quenching analysis. Good indicators and tags are characterized by high fluorescence quantum yields, long-wave excitations and emissions, and selectivity. On the other hand, it should be kept in mind that under these experimental circumstances it is the concentration of an indicator or its complex that is measured, not the analyte itself.

In conventional fluorimetry, a definite amount of indicator is added to the sample of interest. The change in intensity is then measured and, eventually, compared with data from a calibration curve. In reservoir sensors the absolute fluorescence intensity is monitored by viewing the reaction zone with a plain fiber. In an optrode sensor, the indicator is immobilized on a solid support, and its fluorescence as a function of the analyte concentration is followed. While the preparation of suitable sensors is more costly and laborious than working with plain fibers, optrode sensors offer the advantages of high selectivity and good reproducibility. A precise addition of reagent is no longer required, since it is always the same amount of indicator that is immobilized.

The preparation of suitable sensing layers is one of the critical steps in the development of optrode sensors. Indicator, polymeric support, glueing materials, and immobilization chemistry should be chosen carefully in order to produce a sensing layer as sensitive as possible. These aspects will be discussed in this section.

3.4.1. Indicators

Fluorescent indicators have been described for practically all inorganic analytes (15). Organic and bioorganic molecules, unless fluorescent by themselves, can be made fluorescent under mild conditions using a label or a derivatization reagent (89). An indicator suitable for fluorosensing with an indicator phase sensor should possess a chemical function that enables its immobilization on a rigid support. It turned out to be practicable to apply simple immobilization chemistry under mild experimental conditions in order not to change the physicochemical properties of the support. No extensive chemistry should be performed once the indicator is immobilized.

Photochemical stability is another important feature. While indicators used for a single fluorimetric determination of an analyte are under illumination for the short time of a measurement only, an immobilized indicator at the end of a fiber will be under photoexcitation over weeks or even months. Photodecomposition of indicators is a main reason for the limited long-term stability of optical sensors at present. Trypaflavin and the benzoin-boron complex, for instance, have been proposed as most sensitive oxygen indicators, but their photolability makes them useless in practice.

The excitation wavelength of an indicator should favorably be beyond 420 nm to allow the use of inexpensive fibers. Ideally, it should be at 480 nm or higher to make it LED excitable. The fluorogenic reaction of the indicator should occur under conditions prescribed by the nature of the sample. Hence analytically useful reactions which proceed in strong acid or base only are of little significance for continuous fluorosensing with optrodes. They may be applied to resérvoir sensors to a limited extent, but certainly not to in-vivo sensors.

This short discussion demonstrates that despite numerous known fluorogenic reactions, only a limited number are of practical value in fluorosensing. In most cases it will also become necessary to chemically modify an indicator recognized as suitable for a given purpose so to furnish it with a chemical function suitable for immobilization. This modification should not change the physicochemical constants of the indicator, in particular not its pK_a value and its binding constants.

3.4.2. Polymeric Supports

An indicator is usually immobilized on a rigid, optically transparent polymeric support, which then can be attached to the end of an optical fiber. Alternatively, the glass of the fiber may serve as the solid support, although its specific surface is rather small. Immobilization can be achieved by physical, chemical, or electrostatic methods.

The choice of the polymeric support is governed by the requirements predicted by the indicator–analyte interaction. Fluorescent indicators for nonpolar (gaseous) quenchers such as oxygen or sulfur dioxide can be dissolved in a suitable polymer to form a thin sensor layer, through which the quencher can freely diffuse, whereas ionic substances cannot. Thus the polymer also serves as a preselection membrane and thereby improves the specifity of the sensing membrane.

The choice of polymer has a pronounced effect on the sensor performance, in particular on the response time. For gaseous analytes the response will be governed by the diffusion coefficients of gases, their solubility, and their permeability. The respective data for various gases have been compiled by Yasuda and Stannett (90). Among the various polymers studied, silicone is unique because of its excellent solubility for oxygen and also CO_2. It seems noteworthy that water has a high permeability through practically all materials studied, even the most hydrophobic ones, although its solubility in polymers varies over a wide range and can be very small. Although these authors have compiled a considerable amount of material on numerous polymers, it should be noted that copolymers and mixtures of polymers occasionally do not display the properties that may be expected by averaging the data for the pure polymers.

Cox and Dunn (91) have studied oxygen diffusion in polydimethylsiloxane and determined the diffusion coefficient (3.55×10^{-5} cm^2/s) as well as the activation energy. The diffusion coefficient was found to be independent of both the oxygen concentration and fluorophor concentration over wide pressure and temperature ranges. A similar study was performed on silicone rubber filled with small-weight fractions of fumed silica. It was found that the diffusion coefficient is reduced.

Attention should be paid to the intrinsic fluorescence of polymers. Polystyrenes, under UV excitation, exhibit strong blue fluorescence, which can represent a significant blank. Polyethylenes also fluoresce blue or violet, but with much less intensity. Poly(vinyl chloride) and silicones usually do not fluoresce under excitation with light of wavelengths above 300 nm.

Except for silicone rubber, plasticizers are added throughout to organic polymers to make them softer and more gas permeable. Among these, esters of phthalic acid are fluorescent by themselves, whereas esters of phosphoric acid and aliphatic dicarboxylic acids (such as adipic acid) are not.

Polymers are not good solvents for most indicators, so that they tend to crystallize after awhile, or to separate on the polymer surface as a thin film. Silicone rubber is a particularly bad solvent for most large molecules. This problem can be overcome in two ways. One consists of the deposition of a thin layer of indicator molecules on the surface of small polymer beads (37,46). In this form the dye molecules are easily accessible to the analyte, and quenching is much more efficient than in bulk polymer solution. The other approach consists of a chemical derivatization of the indicator by introducing hydrophobic side chains. This results in a several-hundredfold improvement in solubility with practically no loss in indicator sensitivity (92).

Ionic indicators do not readily dissolve in most usual polymers, and charged analytes do not penetrate lipophilic polymers. In such a situation, the indicator has to be made analyte accessible by immobilizing it at the interface between polymeric support and sample solution.

The most frequently employed solid supports for covalent surface immobilization are glass in various forms, cellulose, and polyacrylamides. Glass has the advantage of not swelling in aqueous solution and is easy to handle. Beads of controlled porous glass can provide the desired high specific surface. This material is available in various forms, even with modified surfaces, so that immobilization is particularly simple. On the other hand, it is sensitive to shaking or stirring since it tends to be ground into a fine powder. Plain glass is, of course, more rigid but lacks a high specific surface, so that it cannot be loaded with a sufficient amount of indicator. As a compromise, porous glass beads may be sintered on a glass support to give a rugged material with high specific surface (93).

Cellulose-type membranes have the advantage of being easily penetrated by aqueous samples, a fact that results in distinctly shorter response times than when the sample is in surface contact with a membrane only. Cellulose membranes of thickness as small as 6 μm are commercially available (Cuprophane). These thin membranes require rather careful handling and are easily attacked by bacteria, in particular after they had been brought into contact with serum samples.

Cross-linked polyacrylamide and polymethacrylamide form mechanically fairly stable supports which are easily handled and chemically modified but lack the good permeability of cellulose. Unlike glass and cellulose, they tend to swell in water, which may cause an initial decrease in fluorescence signal when the dry layer is brought into contact with an aqueous sample.

Powdered cellulose, dextrans, and agarose have often been used as supports for enzyme immobilization, and a variety of chemically modified polysaccharides is commercially available. In the first pH sensor described in detail, fluorescein isothiocyanate was immobilized on amino-modified cellulose (94).

Poly(vinyl chloride) (PVC) and polystyrene are supports that can provide a fairly hydrophobic sensor surface but are not easily furnished with functional groups. Poly(chloromethyl styrene) is available from various sources and can be used to immobilize amines and alcohols. Carboxy-modified PVC may be used to immobilize amines or, via spacer groups, other functional groups. A survey on the most frequently used polymeric support materials suitable for covalent immobilization is given in Table 3.7.

Anion- or cation-exchanging membranes offer an attractive alternative for the surface immobilization of charged molecules. The membranes are available in constant quality. A major advantage of this material results from the fact that immobilization occurs only at positions that are accessible to the indicator solution and, consequently, to the analyte as well. Therefore, high indicator blanks resulting from analyte-unaccessible indicator fractions are unlikely to occur. Unfortunately, polystyrene-derived ion-exchange membranes exhibit a strong intrinsic blue-green fluorescence, which can interfere seriously.

Because of the small exclusion size of ion-exchange beads, small indicator molecules as well as analytes can easily diffuse to the binding sites positioned in the innermost regions of ion-exchange beads. Large molecules, such as polysaccharides and proteins, in contrast, do not have access to the binding sites and hence cannot produce adverse interferences. For instance, no "protein error" was observed in the measurement of serum pH with an indicator electrostatically immobilized in the interior of a ion-exchange bead (95). The surface-potential-sensitive indicator rhodamine B has been immobilized on cation-exchanging polystyrene beads simply by dipping them into a methanol solution of the dye (96).

Table 3.7. Frequently Applied Methods to Modify Polymer Surfaces, Reagents Required Therefore, and Types of Molecules That Can Be Attached to the Surface

Polymer	Reactive Group on Surface	Produced by Reaction with:	Suitable for Coupling to:	Coupling Reagent[a]
Agarose: *see* Cellulose				
Cellulose	Aminoethyl	(i) BrCN; (ii) ethylenediamine	Carboxylates, proteins, sulfonic acids	EDC, DCC via acid chlorides
	Epoxy	Epichlorhydrine/NaOH	Amines, including spacers	—
	Carboxymethyl	Chloroacetic acid	Amines	DDC, EDC
	Mercaptomethyl	Homocystein thiolactone	Thiols, acids	—
	Dichlorotriazinyl[b]	Cyanuric chloride	Alcohols, amines, thiols	—
Glass (kieselgel, silica gel, quartz, porous glass	Aminopropyl	Triethoxysilylpropylamine	Acids, acid chlorides	DCC, EDC; acid chlorides
	Chloroethyl	Triethoxysilylpropylchloride	Alcohols, phenols, amines, carboxylates	—
	Mercaptopropyl	Triethoxysilylpropanethiol	Thiols, acids	—
	Long-chain amine	*N*-(2-aminoethyl)-3-aminopropyltrimethoxysilane	Acids, proteins	Via acid chloride or EDC

Table 3.7. (*Continued*)

Polymer	Reactive Group on Surface	Produced by Reaction with:	Suitable for Coupling to:	Coupling Reagent[a]
	Epoxide	Glycidyloxypropyltrimethoxysilane	Carboxylates, amines, thiols, phenols, alcohols	—
	Vinyl	Triacetoxyvinylsilane	Strong nucleophiles	—
	Methacryloyl	3-methacryloxypropyltrimethoxysilane	Strong nucleophiles	—
Poly(acrylic acid)	Carboxyethyl	—	Amines, proteins	EDC, DCC
	CH_2–CH_2–CO–Cl	$SOCl_2$, $POCl_3$	Alcohols, amines, thiols	—
Polyacrylamide	Carboxyethyl	(i) strong alkali; (ii) strong acid	Amines, proteins	EDC, DCC
	Long chain amine	Long-chain diamine	Carboxylic acids, proteins	
Poly(ethyl acrylate)	Carboxyethyl	(i) strong alkali; (ii) strong acid	Amines, proteins	DCC, EDC

Polystyrene[c]	Chloromethyl	Chloromethylation or by copolymerization with vinylbenzyl chloride	Amines, alcohols, carboxylates, thiols	—
	Ammonium ion	—	Anionic species[d]	—
	Sulfonic acid	Sulfonation	Cationic species[e]	—
(Acrylic acid)/(vinyl chloride) copolymer	Carboxyethyl	—	Amines, proteins	EDC, DCC
	CH_2–CH_2–CO–Cl	$POCl_3$	Alcohols, amines, thiols	—

[a] EDC is 1-ethyl-3-(3-dimethylaminopropyl)carbodiimide; DCC is dicyclohexyl carbodiimide.
[b] This material may be further reacted with glycine to give a carboxymethyl derivative good for coupling via the spacer group to amines and proteins.
[c] Amino-modified polystyrene is commercially available.
[d] Anion-exchanging support.
[e] Cation-exchanging support.

3.4.3. Immobilization Techniques

Three methods are important for the preparation of sensing layers: mechanical, electrostatic, and covalent immobilization. Excellent reviews cover all aspects of the chemistry and physics of immobilized reagents and dyes, proteins, and even whole cells (97–104). The methods known from protein chemistry (probably the most thorougly studied field of immobilization techniques) may of course also be applied to problems associated with dye immobilization.

Mechanical (physical) immobilization involves inclusion of molecules in a sphere which they cannot leave. Thus proteins can be included in the interior of polyurethane, and indicators may be entrapped in capsules of only a few manometers in diameter (105). Although not a physical method in a true sense, lipophilic indicators dissolved in a lipophilic polymer are practically not washed out by aqueous samples because of a Nernst distribution that strongly favors the presence of the indicator in the lipophilic phase, and thereby remain "immobilized."

Chemical immobilization is performed by covalently binding the indicator to a polymer surface. Both mechanical and covalent immobilization have been applied to enzymes, antigens, antibodies, and even whole bacteria. Indicators, however, are mostly immobilized by chemical or electrostatic methods. Frequently, the dye is distanced from the matrix by a spacer molecule.

To covalently bind an indicator to a polymer surface, one or two activation steps are usually required to make the reagents undergo a facile room-temperature reaction. The first step involves the modification of the *polymer* to provide it with a sufficiently reactive function. A similar procedure may be required when the *indicator* does not possess chemical functions suitable for immobilization. Polycyclic aromatic hydrocarbons and many metal chelates, for instance, are devoid of functional groups suitable for covalent binding. Therefore, they have to be furnished with sufficiently reactive functional groups. Table 3.7 summarizes the most frequently employed methods for covalent binding of indicators to various polymeric supports.

Surface modification of quartz, kieselgel, silica gel, and conventional glass is achieved almost exclusively with reagents of the type $(RO)_3Si$-R, with R being ethyl or methyl and R being 3-aminopropyl, 3-chloropropyl, 3-glycidyloxy, vinyl, or a long-chain amine. The resulting materials are easily reacted with the indicator or peptide to be immobilized. Because of the widespread use of immobilization techniques, controlled porous glass with various types of organo-functional extension arms is commercially available.

Cellulose (linear chains of 1,4-linked β-glucose organized in fibers of a high degree of crystallinity) can be surface modified by reaction with cyanogen bromide, followed by treatment with a long-chain diamine. The nasty cyanogen bromide may be replaced by cyanuric chloride or epichlorohydrin. The products thus obtained can be reacted with acid chlorides in an aprotic solvent, or with carboxylic acids or even proteins using a carbodiimide as a coupling agent.

Carboxymethylcellulose (CMC) is obtained from cellulose and chloroacetic acid and can serve as a support for amines using similar immobilization methods. CMC and aminoethylcellulose are commercially available in various modifications. After several years of experience with covalent immobilization it is recommended to prefer standardized (i.e., commercially available) support material over self-prepared material whenever possible. The former material is available in continuous quality, whereas the material produced in a laboratory on a small scale varies from preparation to preparation.

Agarose (alternating 1,3-linked β-D-galactose and 1,4-linked 3,6-anhydro-α-L-galactose) is similar in its properties to cellulose. It requires the same surface chemistry and does not offer particular advantages over cellulose.

Polyacrylamide is of low reactivity but can be activated by surface saponification with strong alkali hydroxide. Subsequent acidification yields a material having surface carboxyl groups which are capable of covalent binding to amines. Carboxy-modified polyacrylamide, cross-linked poly(acrylic acid), carboxy-modified PVC, and carboxy-modified polystyrene are commercially available and are capable of coupling to amine-type indicators. Also, long-chain diamines acting as spacer groups may be linked to the surface.

Polymers such as poly(vinyl alcohols). polyglycols, or the strongly basic poly(ethylene imines) are disadvantageous in view of strong swelling with, or even solubility in, water. This limits their utility in case of aqueous sample solutions.

Methods other than covalent immobilization have been discovered in the last years. Thus Peterson (106) has found that nondiffusible forms of pH-indicating dyes are obtained by emulsion copolymerization of the dyes with acrylic monomers. In a typical experiment, radical polymerization of phenol red with aqueous acrylamide in the presence of emulsifier and toluene under nitrogen gave microspheres that are useful in optical pH sensing.

A simple method to immobilize indicators on cellulose and related material is based on the creation of an interpenetrating network within the cellulose backbone (107). The cellulose membrane is simply soaked in an aqueous solution of poly(ethylene imine) or a long-chain diaminoalkane. Within the cellulosic material the amine forms a network, to which an indicator is attached by reaction via its chloride, isothiocyanate or sulfochloride. As a result, the indicator-loaded amine chain can no longer be washed out from the cellulose material, since it is anchored in the cellulose network.

Table 3.7 summarizes the commonly applied methods to provide a polymer surface with a reactive group suitable for immobilization, the reagents required therefore, and the functional groups that can be coupled to such a surface. For a detailed description of the various coupling procedures, reference is made to several useful reviews and monographs (100–103).

Immobilization of indicators and reagents usually requires more drastic conditions than those required for enzyme immobilization. In addition, the chemistry is not confined to reactions in aqueous solutions, so that a broader range of coupling procedures becomes available. However, as a general rule it may be said that once the indicator is immobilized, the chemistry should be kept as limited as possible and confined to mild reactions giving very high yields. Ideally, no further step is necessary after the indicator has been reacted with the modified sensor layer.

Fluorescent indicators should not be covalently bound to polymers by azo coupling because azo dyes do not fluoresce, except for certain metal complexes of *ortho*-hydroxyazo compounds. Attention should also be paid to not using a chelating functional groups for the immobilization chemistry, which would render the reagent useless.

Immobilization of most dyes results in a small change in their spectral characteristics. In addition, there are frequently considerable shifts in pK_a values, binding constants, and in particular, quenching constants. The changes reflect the various interactions between neighboring dye molecules on the polymer surface, interactions between dyes and polymer surface, and electronic effects of the convalent bond. In each instance it is therefore required to determine the respective data of the immobilized dye rather than to use the values determined in solution. The large differences in the quenching constants of dissolved and immobilized indicators also explain why reservoir sensors (with an indicator in fluid solution) will more sensitively respond to a quencher than an indicator phase sensor (with an immobilized indicator), despite the same indicator being used.

3.5. RELATIONS BETWEEN ANALYTE CONCENTRATION AND FLUORESCENCE SIGNAL

While most of the sensors described so far rely on the phenomenon of fluorescence or phosphorescence, the relations between emission intensity and analyte concentration are based on quite different principles, depending on whether the method involves an interaction of an analyte with an indicator in the ground state or in the excited state. Two equations are of special importance for the description of the relations between analyte concentration and emission intensity. One, developed by Parker (108), relates the fluorescence of an indicator dye to its concentration. The other, developed by Stern and Volmer (109), describes the relation between emission intensity and quencher concentration. Both will be discussed here in more detail, along with specific refinements.

3.5.1. Fluorescent Analytes

Plain fiber sensors utilize the intrinsic fluorescence of analytes. In fact, they are a combination of front-face fluorimetry and fiber optical waveguiding. As in conventional fluorimetry, fluorosensing can be performed under either right-angle illumination or front-face illumination. The first situation is found in fiber optical fluorescence detection in HPLC when the two ends of the two fibers guiding the exciting and fluorescent light, respectively, are directed toward the sample at an angle of 90° to each other (22). In fluorosensing, however, front-face fluorimetry is the more common situation since both single fibers and bifurcated fibers are usually immersed in the sample solution and remitted light is collected.

The rate of emission of fluorescence (F) is, by definition, equal to the rate of light absorption, multiplied by the value of quantum yield (ϕ_f). Since the amount of absorbed light (I_a) is the difference between incident light I_0 and transmitted light I_t,

$$F = I_a\phi_f = (I_0 - I_t)\,\phi_f \tag{3.9}$$

With $I_t/I_0 = e^{-\varepsilon cd}$ (Beer–Lambert law), one can see that

$$F = I_0(1 - e^{-\varepsilon cd})\,\phi_f \tag{3.9}$$

with ε being the molar extinction coefficient at the excitation wavelength, c the concentration of the fluorophore, and d the optical depth. A factor k that takes into account the geometrical arrangement of the instrument has to be included in Eq. (3.10).

Expanding the exponential term, we get, with ln ε = 2.3 log ε,

$$F = I_0\phi_f[2.3\varepsilon cd - (2.3\varepsilon cd)^2/2! + (2.3\varepsilon cd)^3/3! \ldots]k \tag{3.11}$$

which rearranges to

$$F = 2.3I_0\phi_f\varepsilon cd[1 - 2.3\varepsilon cd/2! + (2.3\varepsilon cd)^2/3! \ldots]k \tag{3.12}$$

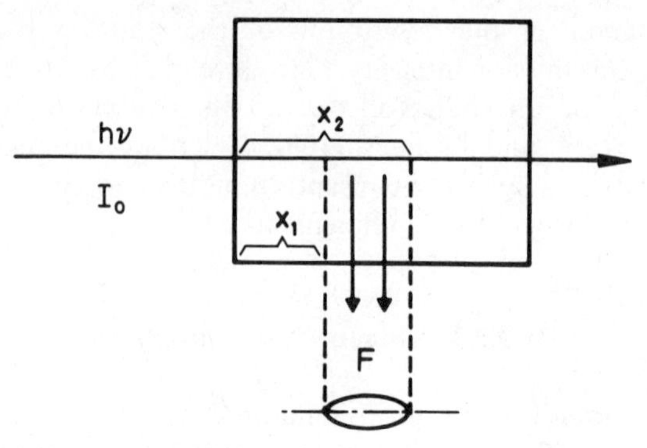

Fig. 3.20. Demonstration of the inner filter effect. Light of intensity I_0 is attenuated by absorption after traversing a distance x_1. The lens will therefore see a fluorescence F that is less than predicted by the relation $F = kI_0\varepsilon cd\phi_f$.

For weakly absorbing solutions, for which εcd is small, Eq. (3.12) simplifies to Parker's law:

$$F = 2.3 I_0 \phi_f \varepsilon cdk \tag{3.13}$$

Again, the constant k accounts for the fact that only a fraction of the total emission is observed.

This equation holds with sufficient precision for most fluorosensors applied in optically thin media such as groundwater and air. Unless d can be made very small, it is not applicable to analysis of strongly absorbing samples such as blood. Therefore, the majority of sensors for clinical analyses are of the indicator phase type (Section 3.2.2.).

Since, in practice, most samples exhibit considerable absorption at the wavelength of excitation, the additional terms in Eq. (3.12) cannot be neglected. In this case the calibration curve will suffer a negative deviation from linearity because of the so-called inner filter effect. It may result from strong absorption by the analyte itself, or by another light-absorbing substance present in the sample. The errors introduced by this effect are in the order of $-100A\%$ up to $A = 0.3$ (with A being the absorbance of the sample). The error will become even larger at higher absorbances.

Errors resulting from the inner filter effect depend on how deep exciting light has to penetrate the sample until reaching the region viewed by the luminescence detector. When, for instance, exciting light of intensity I_0 leaves a fiber to excite the sample contained in a compartment as shown in Fig. 3.20, and a second fiber observes fluorescence through a hole of width $(x_2 - x_1)$ as in Fig. 3.20, the actual excitation light intensity I at the center of the compartment will be smaller than I_0 because of partial absorption over distance x_1. The correction factor to be applied to the observed fluorescence intensity F to give the value F° that would have been observed in the absence of the inner filter effect is now (110)

$$F^\circ/F = \frac{2.3A\,(x_2 - x_1)}{10^{-Ax_1} - 10^{-Ax_2}} \tag{3.14}$$

A is the absorbance per centimeter. A related expression has been presented by others (111).

Lutz and Luisi (112) have proposed another method for correcting inner filter effects resulting from absorption of both exciting light and emitted light. Basically, it consists of a measurement of fluorescence intensity at two different points along the diagonal of the cell. Unlike with former methods, the two points are corrected simultaneously for absorption of exciting and emitted radiation without the necessity of reading the optical density of the solution, and with simple data elaboration.

Consequently, the use of too concentrated a solution will result in a severe distortion of spectra and calibration graphs, since much of the fluorescence is already produced in a region x_1 out of view of the photodetector. Almost the same result will be produced, independent of whether the analyte is present in high concentration, or an absorbing chromophore different from the analyte is consuming light.

A second type of inner filter effect is observed when there is considerable overlap between the longest-wave absorption and the shortwave part of the emission. At high concentration this can lead to a complete loss of the shortwave fraction of the emission. It is therefore desirable to have indicators with large Stokes shifts.

The mathematics for a situation that is complicated because of the presence of other absorbing but not fluorescent species, such as blood, has been developed by Eisinger and Flores (113). An absorbance correction factor f_a was introduced which is used to correct right-angle cell fluorescence intensities for finite sample absorbance. The factor can be calculated for two limiting geometries for the emission: In the focused emission, light is collected from the center of the cell only (Fig. 3.20, with x_2 being very small). In the defocused emission, light is collected from the whole cell face with uniform efficiency (Fig. 3.20), with x_1 small and x_2 large). Actual conditions can be expected to fall somewhere between these limits.

Fluorosensing is usually performed in combination with fiber light guides under frontal illumination. Fortunately, inner filter effects are in some ways less serious with this method. Although the excitation spectra may become distorted at high absorbances, the emission spectra will only be scarcely so. A major advantage of front-face fluorimetry results from the fact that unlike in

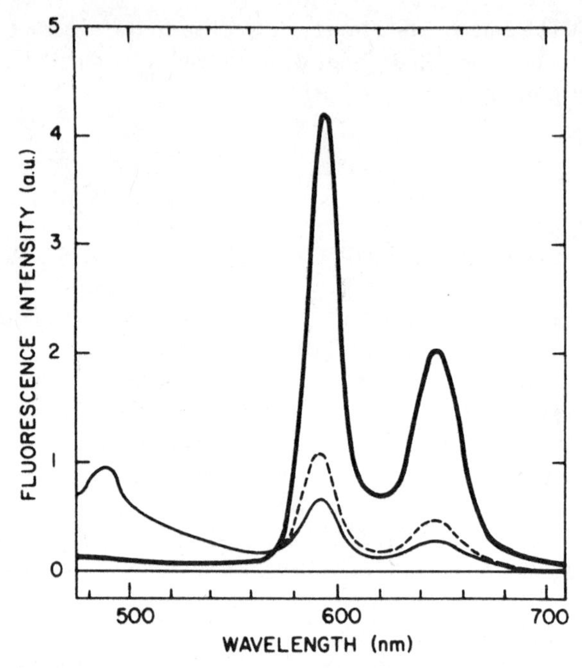

Fig. 3.21. Comparison of the emission spectra of zinc protoporphyrin (ZPP) obtained by the front-face (FF) and right-angle (RA) techniques taken with the same instrumental gain and slit widths. The samples contain hemolyzed blood, ZPP, and 2% of a detergent. (From ref. 113 with permission of Academic Press, Orlando, FL 32887.)

right-angle fluorimetry, the exciting light can be totally absorbed by the sample, which in turn leads to higher emission intensities. There are no losses of excitation light as in right-angle fluorimetry.

Thus, when a catheter is inserted into a blood vessel, the exciting light will be totally absorbed within a few micrometers. Only a fraction of the absorbed light will reach the fluorophore, whereas most of the light will be lost to hemoglobin. The fraction of light that is exciting a fluorophore is given by the ratio between the absorbance of the fluorophore (A_f) and the total absorbance of the sample (A_t). The fraction is, of course, the same in the right-angle and front-face geometries, but with much less light absorption in the former.

Figure 3.21 shows the fluorescence spectra of zinc protoporphyrin (ZPP) in hemolyzed blood, measured by the right-angle (RA) and front-face technique (FF). ZPP was excited at the Soret band (at 420 nm), where the absorbance of the FF sample was 162 cm^{-1}, and the ratio of the absorbances A_{f/A_t} was about 0.04. The sample solutions used in the RA cell were obtained

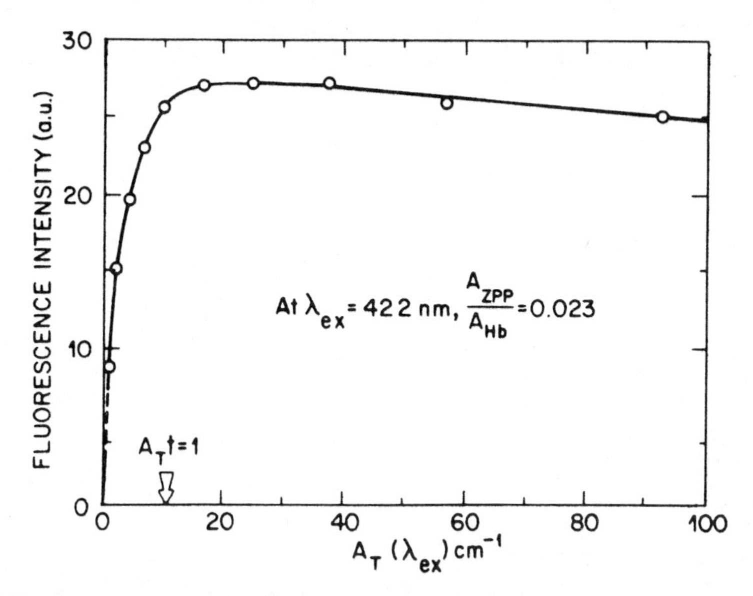

Fig. 3.22. Demonstration of the almost independence of the front-face fluorescence intensity of zinc protoporphyrin (ZPP) in hemolyzed blood upon the sample absorbance at 422 nm at high absorbance. (From ref. 113 with permission of Academic Press.)

by diluting that used in the FF cell to absorbances of 0.2 and 0.4 cm^{-1}, respectively. The corresponding spectra are similar in shape after subtraction of some background. The Raman peak at 495 nm and the background fluorescence caused by the detergent Triton X-100 are seen only in the RA spectrum. The example nicely demonstrates the advantages of FF fluorimetry over RA fluorimetry in strongly absorbing sample solutions.

Unlike fluorimetry under right-angle observation, front-face fluorimetry results in fluorescence intensity almost independent of analyte concentration at high optical densities (113). In this case all incident light will be absorbed shortly after it has entered the sample compartment. A representative example is again provided by the fluorimetric determination of ZPP in whole blood with its strong absorption at the excitation wavelength of ZPP. Figure 3.22 demonstrates that the fluorimetric signal measured by the front-face technique is practically independent of the sample absorbance. Obviously, the technique (with fibers or not) is a most useful analytical technique for samples with high absorbance.

When a bifurcated fiber is immersed in a fluorescent solution the situation is somewhat different. The total signal intensity will depend on the numerical aperture (N.A.) of the fibers, on the overlap of the viewed light cones, and on the penetration depth of the exciting light. Assuming a bundle of fibers in an

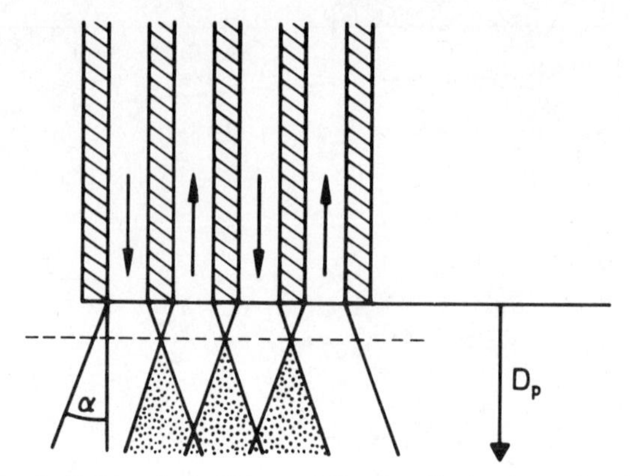

Fig. 3.23. Schematic of the end of a bifurcated fiber bundle demonstrating the lack of overlap in the cone angles below a critical distance from the fiber end.

arrangement as shown in Fig. 3.23 one can see that in the region above the dashed line, fluorescence will be induced by input bundles but will not be seen by the output bundles. It is this limited overlap in the light cones that makes it necessary to place a sensing indicator layer a certain distance away from the end of the fiber bundles. The greater the N.A. of a fiber, the smaller the required distance will be.

The penetration depth of the excitation beam (D_p) is a second important factor, particularly in the case of small sample volumes. When the sample layer is thinner than the distance that the excitation light would be able to penetrate, the signal will be smaller than that of a sample larger than the penetration depth of the light. When, as in Fig. 3.23, the penetration depth is allowed to increase, the observed fluorescence intensity will increase because of more efficient cone overlap. It will reach a limiting value when the sample has a thickness larger than the penetration depth of exciting light. Thus the optimum penetration depth will be mainly determined by the absorbance of the solution.

Figure 3.24 shows how the fluorescence signal observed depends on the penetration depth available and the absorbance of the solution. A bifurcated fiber bundle, consisting of two arms (i.d. 2 mm) with around 75 single strands of quartz fibers of N.A. 0.20 was immersed into solutions of 1-hydroxypyren-3,6,8-trisulfonate in 0.001 *N* NaOH. At low absorbances the signal depends strongly on D_p, over a much longer distance than in highly absorbing solutions. It should be noted that in contrast to the front-face technique, the signal intensity at high penetration depth (indicated by an arrow) is not concentration-independent up to absorbances of 25, although it becomes so at higher optical densities.

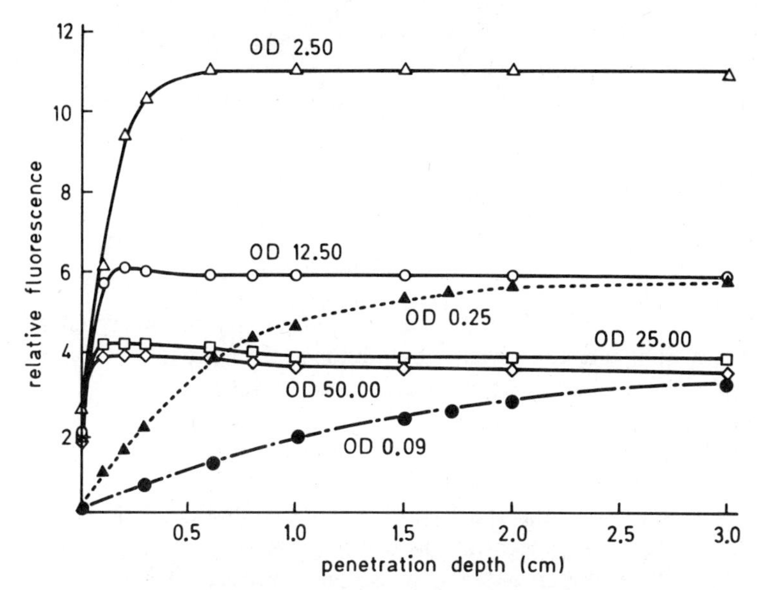

Fig. 3.24. Plot of fluorescence intensity of a dye solution versus penetration depth, which was varied by lifting the end of the fiber bundle stepwise from a black bottom. At low optical density the increase in signal with penetration depth is much more expressed than in strongly absorbing solution. The figures give the absorbances per centimeter of the solutions.

Since sensors are usually immersed directly into the fluid sample, fiber optical sensing does not suffer from interferences from excitation light scattered at the surface of cuvettes. The laws of front-face fluorimetry proved to be of particular advantage in performing fiber optical fluorescence titration with dissolved indicators. Fiber sensors may also be used in performing studies on binding equilibria of biomolecules and of their interactions by titration. Since binding equilibria may differ considerably in dilute and whole blood, front-face fluorimetry should be the method preferred over right-angle fluorimetry of dilute samples.

The mathematics of a situation where bifurcated fiber optics are used for excitation and front-surface fluorescence collection in a 32-μL flow-through cell was described (114). The inner filter effect correction function is formed by deconvoluting the fiber optic/cell light function with mathematical descriptions for primary and secondary absorbance effects. The absorption-corrected fluorescence of quinine sulfate was found to be linear with concentration, despite primary self-absorption values extending to 3. The cell design of the fiber optic fluorimeter eliminates the collimation requirements of other techniques and thereby improves light throughput and signal-to-noise ratio.

3.5.2. Immobilized Indicators

In sensors with immobilized indicator phases as analyte-sensitive zones an equilibrium is established between the concentrations of analyte ([A/]), reagent (indicator) ([R]), and associated complex ([AR]). Assuming a 1 : 1 stoicheiometry, the equilibrium is described by

$$K = [AR]/[A][R] \tag{3.15}$$

with K being the binding constant (equilibrium constant). The analyte concentration can thus easily be obtained by measuring the ratio of associated chelate and reagent concentrations:

$$[A] = [AR]/K[R] \tag{3.16}$$

The limitation of the ratio approach is that both AR and R must be present in sufficient concentrations to be measured with adequate precision. When the ratio between these species becomes larger than 100 or smaller than 0.01, the precision will be poor. As a consequence, the dynamic range of sensors based on ratio measurements will be limited.

In order to utilize the full analytical range of fluorimetry, it is therefore advisable to measure the concentration (fluorescence) of one partner of Eq. 3.15) only. This becomes possible in sensors where the total amount of immobilized indicator remains constant. The total concentration of immobilized indicator or reagent ([C]) is the sum of the concentrations of free and complexed molecules:

$$[C] = [R] + [AR] \tag{3.17}$$

Since

$$[R] + [R][A]K = [C] \tag{3.18}$$

it follows that

$$[C] = [R](1 + K[A]) \tag{3.19}$$

Introducing Eq. (3.19) into Eq. (3.15) and subsequent rearrangement shows that

$$[AR] = K[A][C]/(1 + K[A]) \tag{3.20}$$

Provided that there is linearity between absorbance (or fluorescence intensity) and chelate concentration ([AR]), the resulting signal will depend only on the analyte concentration, since both K and [C] are constants.

Figure 3.25 shows the corresponding signal response versus analyte concentration for the case in which the optical parameter is proportional to [AR]. Such a situation is, for instance, encountered when a nonfluorescent ligand combines with an ion to form a fluorescent complex. As long as [A] is much smaller than $1/K$, the response will be proportional to [A]. As concentration increases, the response becomes curved, reaching a limiting value when [A] is much bigger than $1/K$. This indicates saturation of the reagent phase with the analyte. Typical representatives of this sensor type and performance are those for beryllium(II) (115) and aluminum(III) (116), based on immobilized morin.

If the fluorescence of the free ligand rather than that of the complex is to be measured, Eq. (3.19) has to be rearranged to give

$$[R] = [C]/(1 + K[A]) \tag{3.21}$$

It is obvious that Eqs. (3.19) and (3.21) are equivalent to the Stern–Volmer relation (29) (see Section 3.5.3). The graph shown in Fig. 3.25b describes such a situation with an optical parameter proportional to [R]. In this situation an increase in analyte concentration leads to a decrease in signal, reaching its minimum at reagent saturation.

Typical sensors for which this relation is valid are represented by those in which an immobilized reagent is quenched by the analyte such as a transition metal ion. Unfortunately, most chelating reagents tend to complex more than one species of metal ion, a fact that renders many sensors unspecific. 8-Hydroxyquinoline, for instance, complexes a manifold of bivalent metal ions at near neutral pH, which limits its otherwise excellent properties for use in fluorooptical sensing.

Other examples of sensors for which relation (3.20) is valid are the pH sensors with immobilized indicators where only the dissociated form is fluorescent, whereas protonation ("complexation") leads to "quenching." Thus when the fluorescence of the indicator anion is exclusively measured, Eq. (3.19) is exclusively valid, with $-\log K$ being the pK_a value, [C] the amount of immobilized indicator, and [R] the fraction of the dissociated form at the actual pH.

Specifically for the situation where a pH indicator is immobilized, a similar mathematical approach can be made to relate fluorescence intensity with pH (95). When an immobilized fluorophore is immersed into a solution,

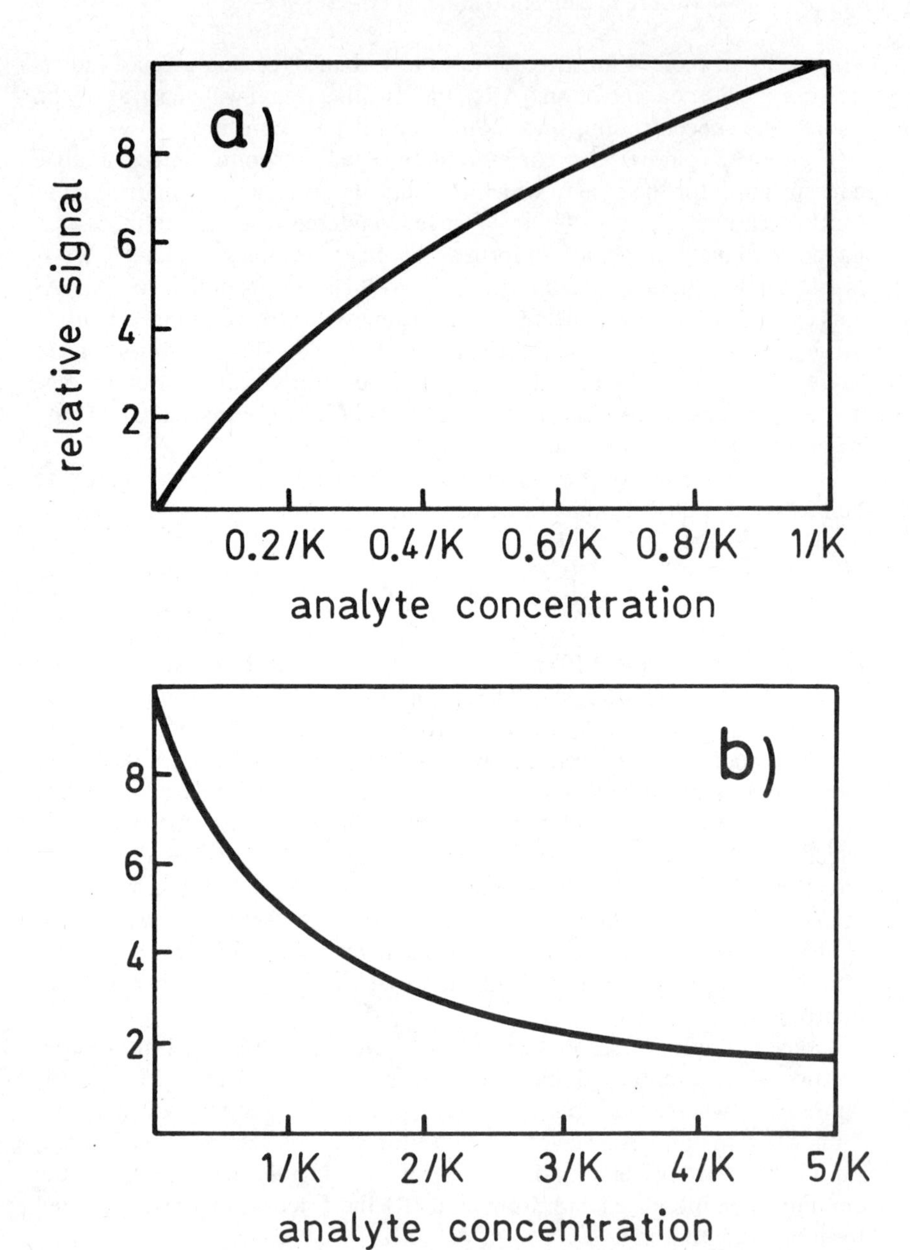

Fig. 3.25. Relation between relative fluorescence intensity, binding constant, and analyte concentration (*a*) when the reagent forms a fluorescent complex with the analyte and (*b*) when the fluorescence of an indicator is statically quenched (e.g., by complexation with a transition metal ion). (Redrawn from ref. 7.)

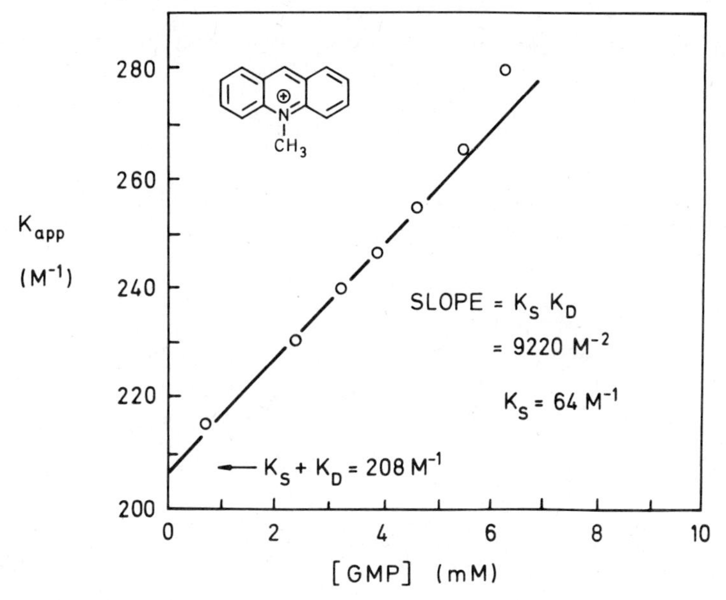

Fig. 3.26. Quenching of 10-methylacridinium chloride by guanosine monophosphate. (Redrawn from ref. 123.)

a prototropic equilibrium will be established in the immobilized phase according to the actual pH value. The equilibrium constant is represented by

$$K = [H^+][F^-]/[HF] \tag{3.22}$$

where $[H^+]$ is the *activity* of hydrogen ions and [HF] and $[F^-]$ are the numbers of moles of the acid and base form of the fluorophore, respectively. Because the total number of immobilized fluorophore in the reagent phase ([C]) is fixed,

$$[C] = [F^-] + [HF] \tag{3.23}$$

Introducing $[HF] = [C] - [F^-]$ into Eq. (3.22) gives $[F^-] = K([C] - [F^-]/[H^+]$, which can be rearranged to $[F] = K[C]/(L + [H^+])$. When $[F^-]$ is the only species that is excited or is fluorescent, intensity I will be proportional to $[F^-]$ by a factor k_1:

$$\mathrm{I} = k_1 \mathrm{K}[C]/(K + [H^+]) \tag{3.24}$$

If, however, both the HF and F^- forms are excited and fluorescent, the concentration of HF can be shown to be, by analogy to the derivation of Eq. (3.24),

$$[\mathrm{HF}] = ([\mathrm{H}^+] + [\mathrm{C}])/(K + [\mathrm{H}^+]) \tag{3.25}$$

which again is related to the measured fluorescence by a factor (k_2). The total intensity observed is therefore

$$F_t = k_1[\mathrm{F}^-] + k_2[\mathrm{HF}] \tag{3.26}$$

where k_1 and k_2 represent constant factors that relate fluorescence intensity to the concentrations of F^- and HF, respectively. Combination of Eqs. (3.24) to (3.26) gives the expression for the pH-dependent fluorescence intensity of an indicator which is excited at a wavelength where both the acid and conjugate base absorb:

$$F_t = [\mathrm{C}](k_1[\mathrm{C}] + k_2[\mathrm{HF}])/(K + [\mathrm{H}^+]) \tag{3.27}$$

Since in all pH fluorosensors the absorbance (A) is relatively high, even in the case of a thin indicator layer, the correction factor given in Eq. (3.14) which accounts for the inner filter effect has to be introduced into Eqs. (3.24) and (3.27). With $x_1 = 0$ and $x_2 = 1$ this factor is given by $(1 - 10^A)/2.3A$.

Assuming excitation at a wavelength where only the conjugate base absorbs (a situation that is very desirable in practice on grounds of simplicity of the optical system), we obtain the final expression for the reation between observed fluorescence intensity and proton concentration:

$$F_t = k \cdot K[\mathrm{C}](1 - 10^A)/A(K + [H^+]) \tag{3.28}$$

Quantitative relationships between the luminescence signal and the analyte concentration are more complex in surface luminescence analysis than in dilute solution work. However, it has been stated (117) that all the theoretical treatments (118–120) make fairly drastic, even unrealistic assumptions about the nature of the solid medium and the optical conditions.

3.5.3. Dynamic Fluorescence Quenching by One Analyte

Quenching of fluorescence obeys the Stern–Volmer law, provided that quenching is exclusively dynamic ("collisional"). The relation between quencher concentration [Q] and measured fluorescence intensity F is given by

$$F^0/F = \tau_0/\tau = 1 + K_d[Q] = 1 + k_q\tau_0[\mathrm{Q}] \tag{3.29}$$

with F^0 being the fluorescence intensity of the indicator in the absence of any quencher and K_d the overall ("Stern–Volmer") quenching constant. τ_0 and τ are, respectively, the lifetimes in the absence and presence of quencher. K_d can be shown to be the product of the lifetime τ_0 of the fluorophore in the absence of quencher and the bimolecular quenching constant k_q.

Fluorescence can consequently be expected to decrease with (1) increasing analyte (= quencher) concentration, (2) increasing indicator decay time, and (3) increasing k_q. Since k_q is viscosity dependent, quenching efficiency can be governed to some extent by proper choice of solvent viscosity. An increase in temperature makes collisional quenching more efficient, whereas ground-state complexation reactions usually have a negative temperature coefficient.

The bimolecular quenching constant k_q is related to the collisional frequency Z by a modified Smoluchowski equation, according to which

$$k_q = 4\pi\gamma NDR/1000 \tag{3.30}$$

where γ is a factor that accounts for the quenching efficiency of a collision. Its maximum value is 1 for the case where each collision results in quenching. N is Avogadro's number, D the sum of the diffusion coefficients of fluorophore and quencher, and R the collision radius. The sum of the radii of fluorophore and quencher is usually taken as R. The term $N/1000$ converts molarity into molecules per cubic centimeter.

Diffusion coefficients may be obtained from the Stokes–Einstein relation

$$D = k_b(T/6)\pi\eta R \tag{3.31}$$

where k_b is the Boltzmann constant and η is the solvent viscosity. This equation shows why quenching is viscosity dependent. Equation (3.31) frequently underestimates the diffusion coefficients of small molecules.

Stern–Volmer constants can be obtained from plots of F_0/F or τ_0/τ versus [Q], which should result in a straight line of slope K_d and an intercept of 1. It is useful to note that $1/K_d$ is the quencher concentration at which half the maximal fluorescence can be observed (i.e., when $F_0/F = 2$). Linearity between [Q] and F_0/F is occasionally lost when Q is present in high concentrations because of both static and dynamic quenching. This situation will be discussed later. Deviations from the linear Stern–Volmer relation can, in certain cases, be predicted (121). Positive deviations are due to the correlations among the are nondilute. When the fluorophores and quenchers are ions, the solution dielectric constant and ionic strength can also influence the deviation.

Typical representatives for sensors based exclusively on dynamic quenching are those for oxygen. The sensing material consists of a suitable indicator, such as an aromatic hydrocarbon with a long lifetime (e.g., pyrene). The indicator is dissolved in a polymer with good oxygen permeability and solubility. The observed linearity in the Stern–Volmer plots at oxygen pressures up to 250 mm indicates almost exclusive dynamic quenching.

Static quenching, in contrast, results from complexation between analyte and indicator in the ground state. The relation between quencher concentration [Q] and fluorescence intensity is described by a Stern–Volmer type of equation:

$$F^0/F = 1 + K_s[\mathrm{Q}] \tag{3.32}$$

with K_s (the binding constant) replacing K_d of Eq. (3.29). K_s is defined as the ratio of the concentrations of the complex between reagent or indicator R to the product of [R] and [Q] [see also Eq. (3.20)].

Despite the formal analogy between the equations describing dynamic and static quenching, there are several aspects which demonstrate the entirely different nature of the two processes and may serve to differentiate between these:

1. The temperature coefficients are opposite. Dynamic quenching becomes more efficiently with increasing temperature [Eq. (3.31)], whereas in static quenching the reverse is usually the case, since the stability of most complexes decreases with temperature. Since temperature also lowers the viscosity of the solvent, an additional increase in dynamic quenching efficiency with increasing temperature will be observed. According to Eq.(3.31), the increase is proportional to T/η.
2. Since collisional quenching only affects the excited states of the fluorophores, it will have no effect on the absorption spectra. Static quenching, which is a ground-state process, will lead to a perturbation of the absorption spectrum of the fluorophore. The changes are mostly small but can be recognized by subtraction of the spectra obtained in the presence and absence of a quencher.
3. The lifetime τ of a fluorophore is not changed after addition of a static quencher. In the case of dynamic quenching, however, $F^0/F = \tau_0/\tau$. Lifetime measurements are therefore a most definite method to distinguish between the two kinds of quenching processes.

A fluorosensor for mercury ion, based on efficient static quenching of indole immobilized on quartz, may be considered as a typical representative for a sensor based on static quenching (96). That the process is static is demonstrated by the practical invariance of τ_0 of indoles with mercury(II) concentration (122).

3.5.4. Multiple Quenching

It may be anticipated that to a certain extent, all quenching processes proceed both statically and dynamically. Frequently, however, one of the two so much prevails over the other that one can speak of exclusive static or dynamic quenching. On the other hand, a number of indicators are known to be quenched via both mechanisms at a comparable rate, a fact that can result in positive deviations of otherwise linear Stern–Volmer plots. The following describes how to obtain the kinetic data for the two processes.

The dynamic portion of the observed quenching can be determined by lifetime measurements, since

$$\tau_0/\tau = 1 + K_d[\mathrm{Q}] \tag{3.33}$$

The quenching constant for the static process cannot be extracted from a single measurement, since this will provide an apparent (composed) quenching constant K_{app}, defined as

$$K_{\mathrm{app}} = K_d + K_s(1 + K_d[\mathrm{Q}]) \tag{3.34}$$

where K_d and K_s are the quenching constants for the dynamic and static processes, respectively. The relation between fluorescence intensity F_0, K_{app}, and analyte conventration [Q] can be described (123) by

$$F^0/F = 1 + K_{\mathrm{app}}[\mathrm{A}] = 1 + (K_d + K_s)[\mathrm{Q}] + K_dK_s[\mathrm{Q}]^2 \tag{3.35}$$

with

$$K_{\mathrm{app}} = K_d + K_s + C_dK_s[\mathrm{Q}] = (F^0/F - 1)/[\mathrm{Q}] \tag{3.36}$$

Plotting K_{app} [as calculated at various quencher concentrations with the help of Eq. (3.36)] against [Q] results in a straight line with an intercept I of $(K_s + K_d)$ and a slope S of K_sK_d. Therefore, K_s and K_d can be calculated from

$$K_s^2 - K_sI + S = 0 \tag{3.37}$$

$$K_d^2 - K_dI + S = 0 \tag{3.38}$$

There are two solutions for each of these quadratic equations. The correct value for K_d can be obtained independently from lifetime measurements.

An excellent example of a quenching process that is composed of static and dynamic contribution is shown in Fig. 3.26. The Stern–Volmer plot (F^0/F versus [Q]) displays an upward curvature as the concentration of the

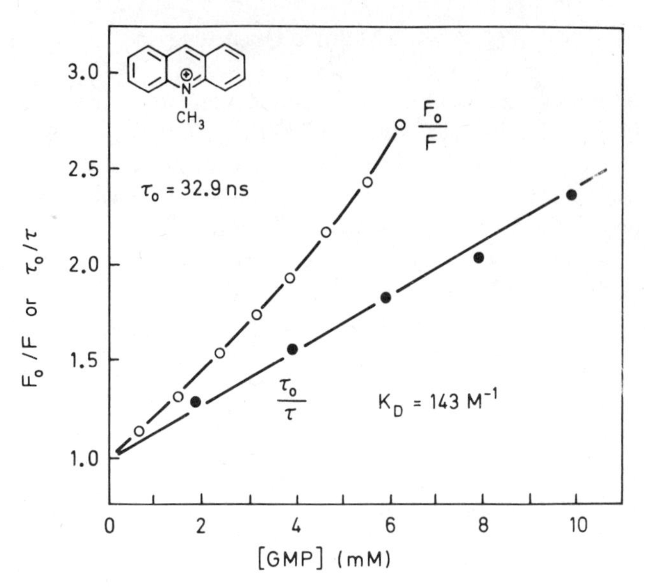

Fig. 3.27. Separation of the static and dynamic quenching constants in the quenching of 10-methylacridinium ion by guanosine monophosphate according to Eqs. (3.34) to (3.38). (Redrawn from ref. 123.)

quencher guanosine monophosphate (GMP) increases (124). The τ_0/τ plot, which refers to the dynamic portion of the overall process only, is linear and yields, via Eq. (3.33), a quenching constant of 143 M^{-1}. The lifetime in the absence of a quencher is 32.9 ns, which allows the calculation of the bimolecular quenching constant K_q via Eq. (3.29) and gives a value of 4.3 × 10^9 M^{-1}·s. Given the large size of both the quencher and fluorophore, this rate constant indicates efficient collisional quenching.

A plot according to Eq. (3.36) for the same system is shown in Fig. 3.27. Slope (= $K_d K_s$) and intercept ($K_d + K_s$) are found to be 9220 M^{-2} and 208 M^{-1}, respectively. Recalling Eq. (3.37) one can calculate with these data that K_s is 144 or 64 M^{-1}. K_d is known from lifetime measurements to be 143 M^{-1}, and hence the static quenching constant must be 64 M^{-1}.

Occasionally it is observed that there is some contribution from static quenching in addition to dynamic quenching because of apparent complex formation between analyte and indicator, despite an evident lack of complexation sites on the fluorophore. This phenomenon is observed at high quencher concentrations only and is interpreted by assuming a "sphere of action," within which the probability of quenching is unity since the quenching molecules are quite close to the fluorophore at the moment of excitation.

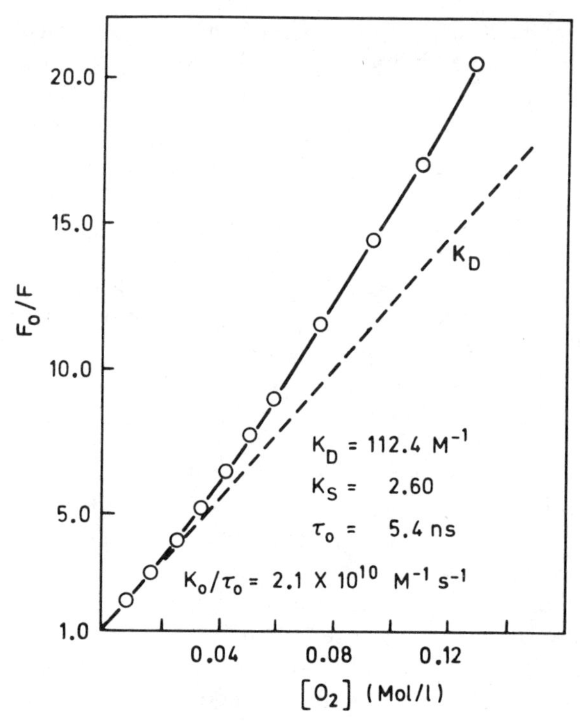

Fig. 3.28. Positive deviation in the Stern–Volmer curvature of the quenching of perylene in dodecane by molecular oxygen. (Redrawn from ref. 123.)

The modified Stern–Volmer equation that describes this situation is

$$F^0/F = (1 + K_d[\mathrm{Q}])e^{[\mathrm{Q}]\nu N/1000} \tag{3.39}$$

where ν is the volume of the sphere within which the probability of immediate quenching is 100%. This equation (in slightly rearranged form, with $\nu N/1000$ replaced by V) is given incorrectly in ref. 125.

Quenching of the perylene fluorescence by molecular oxygen in hydrocarbon solvent, shown in Fig. 3.28, is a typical example. Of course, the static quenching constant is rather small, since an oxygen concentration of 0.38 M ($1/K_s$) is required to "complex" 50% of the perylene molecules. At this concentration, the oxygen molecules are only about 16 Å apart on the average and so close to the fluorophore at the moment of excitation that quenching is 100% efficient when a light quantum impinges on a fluorophore molecule.

Numerous fluorescent indicators are known to be quenched by various interesting analytes, and all of them are therefore of potential interest in fluorescence optical sensing. However, most indicators lack specifity, giving rise to interferences from other quenchers. As long as the number of interfering quenchers can be kept small (typically, lower than 3), interferences can be eliminated by simple mathematical methods in combination with a multiple-sensor technique (126,127). It will now be discussed in some detail.

If the fluorescence of an indicator is quenched by several quenching analytes present in concentrations $Q_1, Q_2, \ldots, Q_n$, their contributions to the overall quenching process can be taken into account by adding further terms to the Stern–Volmer equation:

$$F^0/F = 1 + {}^1K[Q_1] + {}^2K[Q_2] + {}^3K[Q_3] + \ldots \tag{3.40}$$

To calculate the concentration of n analytes, one needs n independent equations, obtained by measuring F^0/F with n indicators. All K's are constants, which have to be determined once from plots of F^0/F versus [Q] for each indicator–quencher combination. For two quenchers in concentrations $[Q_1]$ and $[Q_2]$, a simple mathematical expression can be given.

The following abbreviations and definitions are used: F_a^0, fluorescence intensity of indicator A in the absence of quenchers; F_a, fluorescence intensity of indicator A in the presence of quenchers; F_b^0, fluorescence intensity of indicator B in the absence of quenchers; F_b, fluorescence intensity of indicator B in the presence of quenchers; ${}^1K_a$, Stern–Volmer constant for the quenching of indicator A by the first quencher; ${}^nK_a$, SV constant for the quenching of indicator A by the nth quencher; and ${}^nK_b$, SV constant for the quenching of indicator B by the nth quencher.

For further simplification we set

$$F_a^0/F_a - 1 = \alpha \quad \text{and} \quad F_b^0/F_b - 1 = \beta \tag{3.41}$$

For two indicators and two quenchers we get, by combining Eqs. (3.40) and (3.41),

$$\alpha = {}^1K_a[Q_1] + {}^2K_a[Q_2] \tag{3.42}$$

$$\beta = {}^1K_b[Q_1] + {}^2K_b[Q_2] \tag{3.43}$$

Hence it follows that

$$[Q_1] = \frac{\alpha {}^2K_b - \beta {}^2K_a}{{}^1K_a {}^2K_b - {}^2K_a {}^1K_b} \tag{3.44}$$

$$[Q_2] = \frac{\beta^1K_a - \alpha K_b}{^1K_a{}^2K_b - {}^2K_a{}^1K_b} \tag{3.45}$$

Note that to solve Eqs. (3.44) and (3.45) only α and β have to be measured.

Equation (3.44) is generally applicable for the quantitation of a variety of quenchers (ions as well as uncharged molecules), provided that they act independently. Specific indicators are no longer necessary. Among the analytically important quenchers that can be determined by this multiple sensor technique, mention should be made of oxygen, sulfur dioxide, dinitrogen monoxide, iodine, acrylic acid derivatives, and olefins, all of which quench the fluorescence of most polycyclic aromatic hydrocarbons. Anions such as chloride, bromide, iodide, cyanide, sulfite, isothiocyanate, and cyanate quench the fluorescence of various nitrogen heterocycles, among *which* 6-methoxyquinoline (and quinine) are the most thorougly studied.

There are several fluorophores known to be quenched specifically by one quencher only. Therefore, one of the constants in Eq. (3.44) or (3.45) becomes zero. In the most common case, ^1K_b will be set equal to zero, which simplifies Eqs. (3.44) and (3.45) to

$$[Q_1] = (\alpha^2K_b - \beta^2K_a)/{}^1K_a{}^2K_b \tag{3.46}$$

$$[Q_2] = \beta/{}^2K_b \tag{3.47}$$

Equation (3.47), of course, is a modification the classicial Stern–Volmer equation.

It is not necessary that there be two different indicators with their different excitation and emission maxima which complicate the optical system of a sensor. It is sufficient that the fluorescent indicator has two different quenching constants for the same analyte. This situation is simply achieved by dissolving the indicator in different solvents, for instance, in two different polymers. Provided that the quenching constants of the two interacting quenchers are different, the concentrations of the two analytes can be calculated using Eqs. (3.44) to (3.47).

Aside from a more simple optical system, the use of one indicator instead of two requires the determination of 3 (rather than 4) quenching constants, which improves the accuracy of the corresponding sensor.

In the most refined version, the sensing polymer membrane is covered with a polymer layer that is permeable for one quencher only. The second sensing membrane is absolutely identical with the first except that it is not covered with the partially impermeable layer. As a result, Eq. (3.44) can be greatly simplified: Since sensor 2 is no longer quenched by Q_2, ${}^2K_b$ becomes zero and Eq. (3.44) is again transformed into the Stern–Volmer equation

$$[Q_1] = \beta/{}^1K_b \tag{3.48}$$

More striking is the effect on Eq. (3.45). Since the quenching constants are the same in both sensor membranes, ${}^1K_a$ and ${}^1K_b$ become identical, so that Eq. (3.45) can be simplified to give

$$[Q_1] = (\alpha - \beta)/{}^2K_a \tag{3.49}$$

Equations (3.48) and (3.49) allow the determination of two quenching analytes by fluorimetry using a system of two sensing membranes, one of which is accessible to one analyte only. From a technical point of view, the manufacturing of such a sensor combination is very simple, since they differ only by a thin cover. The optical system can be kept simple because it is the same indicator whose fluorescence is measured in two sensing layers and precision can be expected to be better because two (instead of four) quenching constants with their inherent errors have to be determined. The method has successfully been applied to determine oxygen, halothane, or both, using a fiber catheter (see Section 3.7.7).

Stern–Volmer constants are frequently determined from plots of F^0/F or τ_0/τ versus quencher concentration. When both the static and dynamic quenching process is operative, the method of plotting K_{app} versus [Q] and graphical evaluation with the help of Eqs. (3.34) to (3.38) (as shown in Fig. 3.27) is suitable.

When two collisional quenchers interact with one fluorophore, the respective quenching constants can be obtained by investigating the action of each quencher separately [Eq. (3.29)], or by a simple graphic procedure.

The Stern–Volmer expression for the case of two collisional quenchers Q_1 and Q_2 can be written as

$$F^0/F = 1 + {}^1K[Q_1] + {}^2K[Q_2] \tag{3.50}$$

Keeping the concentration of the first quencher constant and varying that of the second, a graphic representation of the results gives a straight line with slope ${}^2K$ and an intercept of ${}^1K/[Q_1]$, provided that the indicators do not mutually interfere. A representative plot is shown in Fig. 3.29. This kind of plot allows a rapid evaluation of two quenching constants and can easily demonstrate whether or not the two quenchers act independently. In practice it shows whether a one-point calibration of a double sensor is necessary.

When a gaseous quencher is admixed to the other, the situation is slightly

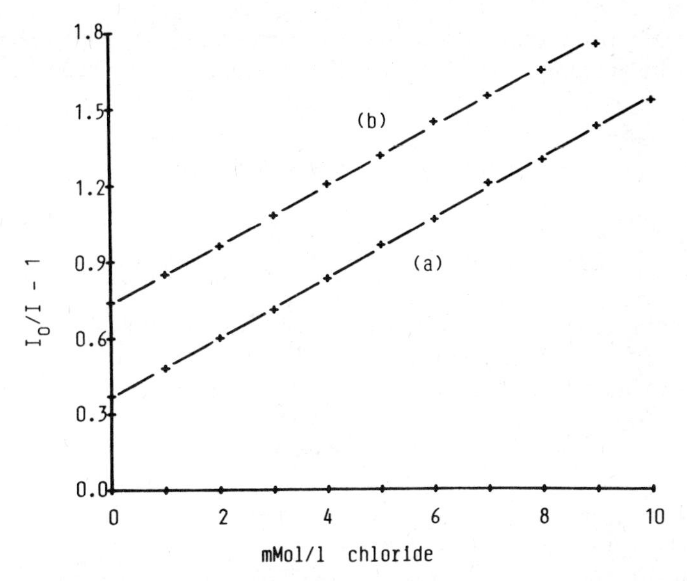

Fig. 3.29. Modified Stern–Volmer plots of the quenching of quinine sulfate by two quenchers simultaneously [Eq. (3.50)]: (*a*) quenching by chloride in the presence of 2 m*M* bromide; slope (= 2*K*) 115 M^{-1}, intercept (= ${}^1K$[Cl]) 0.37; (*b*) quenching by chloride in the presence of 4 *m*M bromide; slope 115 M^{-1}, intercept 0.74. Both graphs are linear and have the same slope, demonstrating the independent action of the two quenchers.

different. Assuming a clinically important case, namely the evaporation of the anesthetic halothane into a stream of breath air. The relation between the concentrations of oxygen ([O]) and halothane ([H]), both in vol %, is described by

$$[\mathrm{O}] = x(1 - [\mathrm{H}]/100) \tag{3.51}$$

with x being the initial concentration of oxygen in air (i.e., 21%). With Q_1 being oxygen and Q_2 being halothane, Eq. (3.50) can be written as

$$F^0/F = 1 + x^{\mathrm{O}}K(1 - [\mathrm{H}]/100) + {}^{\mathrm{H}}K[\mathrm{H}] \tag{3.52}$$

which can be rearranged to give

$$F^0/F = 1 + ({}^{\mathrm{H}}K - x^{\mathrm{O}}K/100)[\mathrm{H}] + x^{\mathrm{O}}K \tag{3.53}$$

Hence a plot of [H] versus $(F^0/F - 1)$ will result in a straight line with intercept $x^{\mathrm{O}}K$ and slope $({}^{\mathrm{H}}K - x^{\mathrm{O}}K/100)$. Again it is possible to estimate from these plots whether or not two quenchers act independently. Deviations of linearity are indicative of either additional quenching processes or mutual interaction of quenchers. Aside from the more complex interpretation that sloped Stern–Volmer plots require, a more sophisticated calibration process becomes necessary when sensors display sloped calibration plots.

3.6. SENSORS FOR CHEMICAL ANALYTES

This section describes in some detail the specific fluorosensors that have been developed for chemical analytes, many of which are of great interest to the clinical chemist as well.

3.6.1. Sensors for pH

The idea of measuring pH by colorimetry dates back to the early days when litmus was the indicator of choice. The concept has found its most widespread application in the form of pH indicator strips. Although fluorescence indicators were recognized as frequently being superior to absorption indicators, they have always been of limited practical interest. Only recently, with the advance of sophisticated instrumentation for fluorimetry, have fluorescent pH indicators received appropriate attention.

Several patents have so far been issued for inventions in the field of fiber optical pH sensors (93,107,128–131). The first fiber optical sensor ever described in detail is based on the changes in the absorption spectrum of phenol red with pH (62). Microspheres of polyacrylamide containing the indicator and smaller polystyrene microspheres (for light scattering) are packed in an envelope of a cuprophane tubing at the end of a pair of plastic fibers (Fig. 3.10). The dye, when copolymerized with acrylamide, was surprisingly found to be immobilized without previously introducing a chemical link (106).

The instrument is equipped with a tungsten halogen lamp as a light source and a photodiode light detector. It is calibrated by measuring the ratio R between intensities of dye-reflected green and red light, respectively, in two separate buffers after subtracting a background level. This ratio is related to pH by

$$\log(R/k) = -C/(10^{-\Delta} + 1) \tag{3.54}$$

Here k is the optical constant of the system, C the green optical density of the sensor when the dye is totally in the base form, and Δ the difference between the pH and the dye pK_a (7.57).

The sensor measures pH over the range 7.0–7.4 to the nearest 0.01 pH unit. It is of flexible construction and hence presents no risk of breakage into sharp bends. The dye is nontoxic, and the sensor is of potential utility for in-vivo measurement (see Section 3.7.1). The temperature coefficient, expressed as the change in pH indication per degree, was 0.017 between 20 and 40° (which is less than that of a pH electrode), and a change of 0.01 pH unit was observed per 11% change in ionic strength at pH 7.00. As would be expected for a diffusion-controlled situation, the response time for the signal to drop to $1/e$ of its initial value is 0.7 min.

A detailed description of the sensor with respect to mechanical, optical, and electronic compartments has been given (132). An exploded view of the detection module is shown in Fig. 3.30. In a more refined version (133,134) the same instrument was equipped with a highly sensitive photomultiplier tube with considerably improved signal-to-noise ratio. This permitted the use of much smaller fibers, although the precision is as good or even better than that of the initial version.

The advantage of this system lies in the ratio method, which can eliminate errors due to lamp fluctuations, certain detector sensitivity drifts, washout effects, local imhomogenities, and temperature changes. On the other hand, the measurement of light intensities at two analytical wavelengths makes the device more complicated because of the need for two optical systems along with choppers and triggers.

The concept of remotely monitoring the color of a pH-sensitive dye was also used by others. Suidan et al. (135) have briefly reported on a pH-sensing hypodermic needle containing a fiber light guide. Again, the ratio of green and red light, as returned by the indicator phenolsulfophthalein, was measured. The precision was said to be ± 0.013 pH units in the region pH 7.0–7.4 for both the in-vivo and in-vitro experiments.

In studies with various immobilized pH indicators covering the range pH 3–12, it was found (136) that cross-linked polymers of styrene and polystyrene firmly retain the dyes in the polymer matrix, after the beads had been soaked with a dye solution. Response versus pH and the effects of temperature (about -0.014 pH unit per °C increase) and ionic strength (very small) were studied in this work.

The findings led to the design of a pH optrode (38) based on absorption changes. The sensitive layer consists of bromthymol blue immobilized on polymer beads which are retained in position by a porous polytetrafluoroethylene membrane. It showed good response to pH in the range pH 8–10.5 which is hardly affected by the turbidity of the solution. There is a nonlinear temperature dependence between 19 and 50°C and—unlike with bromthymol blue in solution—a significant dependence of response with variations in the ionic strength of the solution. Probably because of the

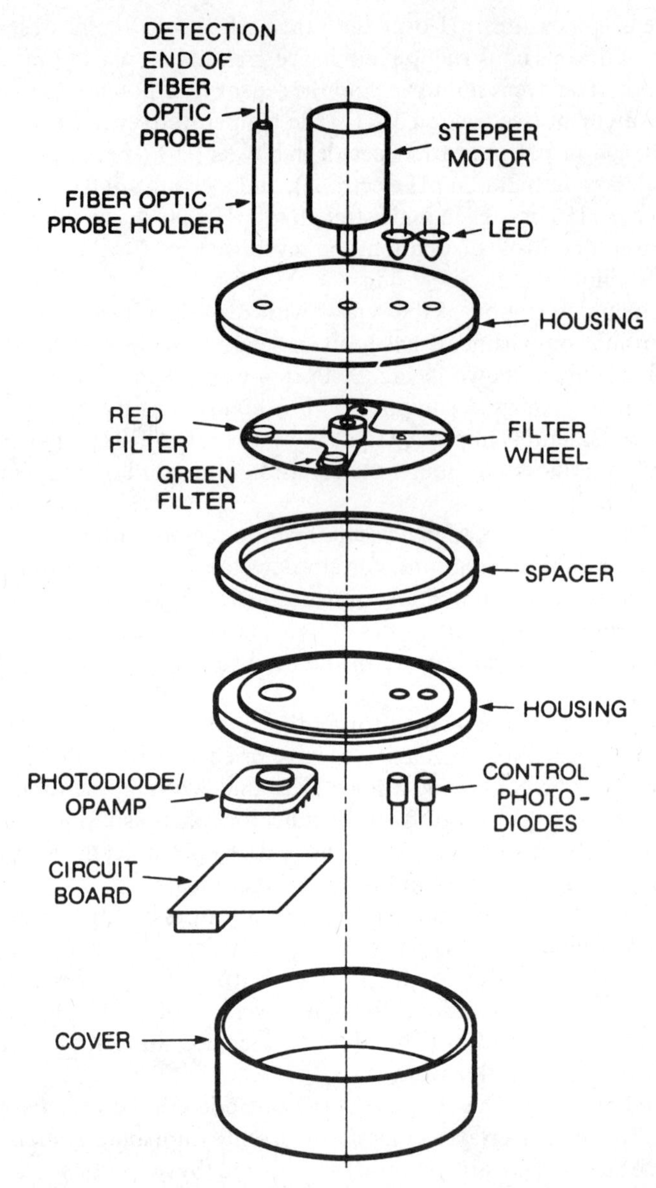

Fig. 3.30. Exploded view of the detection unit of a pH sensor. Light returning from the sensor head passes, respectively, a red or green filter and is measured with a photodiode (left). The filters are brought in position by a rotating filter wheel. Two control LEDs (right) serve as reference light sources to account for drifts in the photodiode dark current, as the unit warms up. (Reproduced from ref. 132 with permission of the author and the Am. Soc. for Mech. Engineering).

hydrophobic PTFE barrier, the response time is rather long, being about 5 min for 99% of the total signal change. An optrode that had been in daily use over a period of 5 months showed little change in response characteristics, thereby demonstrating the excellent long-term stability of this device.

Goldfinch and Lowe (65) have covalently immobilized bromcresolgreen and bromthymol blue on cellulose to obtain a pH-sensitive layer whose absorbance changes with pH. An LED served as a light source and a photodiode as a light detector. The bromthymol blue–dyed layer, containing approximately 10 μmol dye per gram dry weight, changes from yellow to blue over the pH range 6–10 and exhibits a pK_a of around 7.5. Both membrane types produce an output voltage change of 1.6–2.0 V per pH unit around their respective pK_a values and thus permit pH values within these limits to be determined to at least three decimal places. A pH glass electrode with Nerstian response yields a voltage change of 59 mV per pH unit only, albeit over a much wider range. This extreme sensitivity of the optoelectronic sensor to pH provides the basis for the enzyme sensors developed by the same group. They are described in Section 3.7.

Seitz and co-workers presented two types of pH sensors (94,95). The first one is based on immobilized fluorescein with its well-known strongly pH-dependent fluorescence. The solid support was controlled porous glass or powdered cellulose. Support plus immobilized indicator was attached to the end of a bifurcated fiber which was immersed into the sample whose pH was varied. Unfortunately, a relatively high background, in addition to some pH-independent fluorescence, limited the precision of the optrode.

Fluorescein has two overlapping pK_a values, which extend its analytically useful range (2 to 9) far over that of a coventional indicator. Thus although not as precise as other pH fluorosensors, the fluorescein-based sensor seems to be the only one at present that allows pH measurement over a large range. 5-Carboxyfluorescein seems to be an indicator that displays even more significant signal changes over this range. It seems attractive to immobilize this indicator on the end of a fiber to give a catheter suitable for invasive measurement of gastric pH.

The second sensor developed by Seitz and co-workers is suitable for quantifying pH in the range pH 6.5–8.5 (95). It is based on changes in the fluorescence of electrostatically immobilized 1-hydroxypyrene-3,6,8-trisulfonate (HPTS), which has been shown (96) to be an almost ideal pH indicator. HPTS fluoresces from the phenolate form between pH 1 and 11 irrespective of whether the phenol form (at 405 nm) or the phenolate form (at 470 nm) is excited. As in the case of the photometric pH sensor (62) the method of working at two analytical wavelengths was applied. The ratio of fluorescence intensities obtained under excitation at 470 and 405 nm was used to calculate pH.

With respect to practical application, the ratio method has the inherent disadvantage of requiring a considerably more complex instrumentation in order to (1) isolate two excitation wavelengths from one light source, and (2) discriminate, in the detection system, between the signals obtained under two different excitation wavelengths. On the other hand, the ratio method can eliminate a number of sources of error such as dye leaching, photobleaching, light source intensity fluctuations and detector drifts. Surprisingly enough, the pK_a of the dye was found not to be affected by changes in ionic strength, although HPTS has been presented as an example of an indicator whose pK_a is particularly sensitive to ionic strength in bulk solution because of its triple or quadruple negative charge (137,138).

Since in this type of sensor there is no concentration quenching evident, heavily loaded membranes may be applied which will allow prolonged use of the optrode even with slow losses or decomposition of HPTS. However, heavily loaded membranes have slow response and consume or release protons, a fact that can introduce errors in case of minute sample volumes.

The standard deviation of pH determinations evaluated with three serum samples of pH 7.32–7.40 was found to range from ± 0.08 to ± 0.02 pH unit. It should be noted that the pH values in this case were determined from a calibration curve rather than from ratio measurements, which may be expected to provide more precise results.

The sensor is slightly affected by temperature but not by oxygen. A variety of ions, including transition metal ions, was found to be inert, as was a 10% protein solution. Obviously, the protein is not able to penetrate into the small channels of the ion-changing membrane since in other sensor types proteins were found to cause a so-called protein error. The sensor changes its characteristics on prolonged standing. After several months the absorption spectra showed that 30–40% of HPTS remained in the acid form, independent of the pH of the solution in contact with the membrane.

Lübbers and co-workers have developed devices for the measurement of oxygen and carbon dioxide pressure and, to a lesser extent, pH values. Aside from several patents (128–130), they have published several papers on the application of fluorimetric methods to physiological problems. The authors call their device, which is shown in Fig. 3.31, by lingual analogy to electrodes, an optode. It was used to sense pH, pCO_2, and pO_2. As to the pH sensor, an interesting approach was made in that the pH indicator (4-methylumbelliferone, "4-MU") was incorporated into proton-permeable polyacrylamide capsules of 150–200 nm i.d. (105,139). The membrane protects the pH indicator from interferences resulting from effects of solvents, proteins and colloids, and fluorescent molecules.

The membrane does not protect against changes in ionic strength. Since membranes selectively permeable for protons are not available, a method has

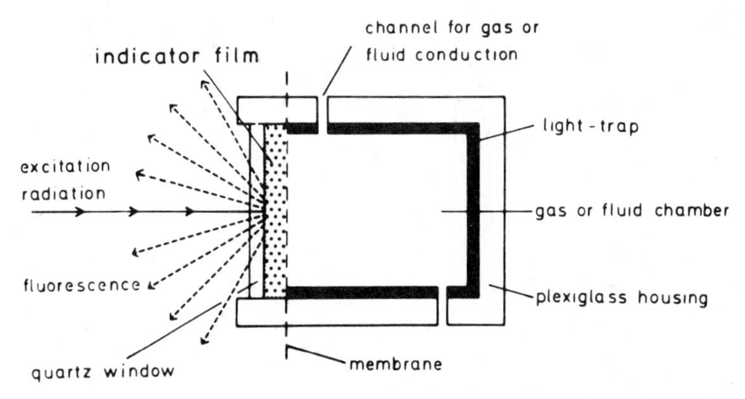

Fig. 3.31. Schematic of an optode. The indicator film is separated from the analyte by a membrane. It protects the indicator chemically and optically from disturbances and provides some selectivity since it acts as a barrier to most interfering analytes. The device has been used to sense pO_2 and pCO_2. (From ref. 16 with permission).

been worked out that accounts for pK_a changes of indicators as a result of changing ionic strength. The fluorimetric determination of true pH (as measured by an electrode) was shown to be possible (136) by applying two pH indicators of different charge. Consequently, the ionic strength dependence of the respective pK_a values is different. By using the indicator couple 4-MU (charge −1) and HPTS (charge −4) it was shown (140a) that the true pH of the solution can be obtained from an apparently indicated pH by the following equation:

$$\text{pH} = \text{pH}'_{\text{4-Mu}} - 0.17(\text{pH}'_{\text{HPTS}} - \text{pH}'_{\text{4-MU}}) \tag{3.55}$$

The method yielded pH values that were in excellent agreement with those measured with an electrode. The response times are between 10 and 20 s for 90% of the total signal change. This is faster than that of any other optical pH sensor.

It has been proposed to measure pH and oxygen simultaneously with one fluorescent indicator which suffers both shifts in its acid–base equilibrium by pH, and dynamic fluorescence quenching by oxygen (141). Various dyes with different pK_a values and sensitivities to oxygen have been presented. The oxygen partial pressure can be determined by measurement of the fluorescence intensity under excitation at the isosbestic wavelength. pH can be determined by measuring fluorescence under excitation into the anion band and taking into account the contribution by quenching of oxygen as determined by the first measurement.

In the author's laboratory, two kinds of pH sensors have been developed

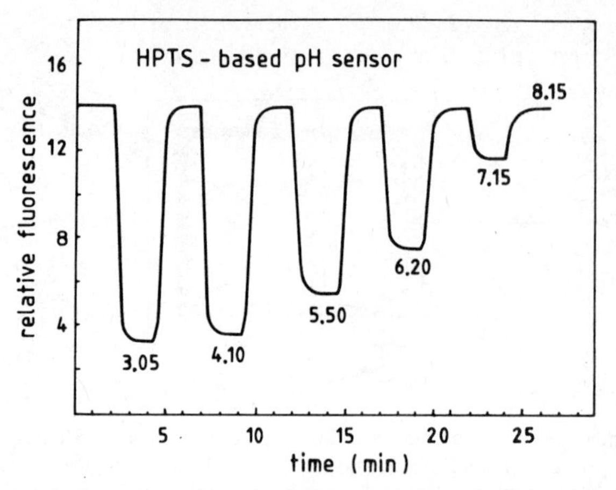

Fig. 3.32. Changes in fluorescence intensity and response time of cellulose-immobilized HPTS at analytical wavelengths 470 nm (excitation) and 530 nm (emission). Note that the response time for a change from acidic to basic milieu is different from a reverse change.

which have meanwhile found application in commercially available instrumentation. The first sensor type (142) is based on covalently immobilized HPTS. It has been linked to cellulose by various methods, including those involving interpenetrating networks (107). Cellulose is an ideal material with respect to response times, since it is easily penetrated by the sample, chemically modified, and handled, but it lacks the mechanical stability of glass supports.

The spectra of immobilized HPTS are only slightly different from those in fluid solution. Its pK_a value is considerably lowered by immobilization (from 7.3 to 6.7). This seems disadvantageous with respect to measuring physiological pH values, but the signal-to-noise ratio in the range pH 7.0–7.5 is still good enough (better than 200) to allow sufficient accuracy ($\pm$0.02 unit) in pH determination. The sensor was specifically designed for following, during operations, relative changes in pH, rather than determining absolute values.

The response of the HPTS-based sensor is shown in Fig. 3.32. It can be seen that the shapes of the response curves are not symmetric. Rather, the response time for a change from high pH to low pH is longer by about 20% than vice versa.

The second sensor type (143)—designed for in-vitro use—has a sensing layer that consists of 7-hydroxycoumarin-3-carboxylic acid immobilized on a porous glass surface which provides both a large specific surface and mechanical stability. When attached to the tip of a fiber, it allows the precise ($\pm$0.005) fluorimetric determination of pH values in the range 6.5–7.6. Effects of changing ionic strength were eliminated by chemical treatment of the glass surface with an aminosilane reagent which renders the indicator's environment highly charged. As a result, the changes introduced by electrolytes from the sample remain very small. The coumarin, which in contrast to HPTS has only a single negative charge in its pH-indicating form, is much less influenced by electrolytes than the fourfold negatively charged HPTS phenolate form. The differences in the apparent pH displayed by sensors of different surface modification have been utilized to determine pH and ionic strength in the near-neutral pH range (144).

Interferences from proteins and blood particles are eliminated by covering the sensor end with a very thin layer of an ion-permeable polymer and a 10-μm layer of cuprophane. This, however, prolongs the response time to around 1.5 min for 90% of the total signal change. The sensor is suitable for measuring pH in about 25 μL of blood.

It has frequently been stated that pH optrodes had been developed (2,4,6,25,146), but experimental details have never been given. Thus a pH optrode has been constructed by immobilization of a fluorescein-type indicator (probably 5-carboxyfluorescein) on collodion (6,25). The response time is said to be better than 3 min and it was expected that 0.01–0.02 pH unit should be resolvable.

Hirschfeld and co-workers (146,147) have reported on the coating of fiber tips with various pH-dependent fluorescers to give proton sensors. Immobilization has been accomplished by cross-linking reagents or using functionalized glass surfaces. pH is measured via the fluorescence intensity, internally referenced to the fiber Raman line, a pH-independent ("isosbestic") point in the dye spectrum, or a reference fluorescer simultaneously coated on the fiber.

Larger signals are possible by using a thicker dye layer that may be produced by physical dye immobilization via entrapment in cross-linked acrylamide beads, adsorption on non-cross-linked polyacrylamide layers, chemical binding to various aqueous gels, or binding to porous or fitted glass layers. This results in slower response times, which can be minimized by choosing the shape of the gel volume.

Specific optodes for industrial use of several 100 m length, 0.02 pH unit resolution, and 1 s response time were also described (147). Single dyes with near-IR fluorescence and He-Ne laser excitation having a limited response range as well as dye combinations were also tested (146). The best results were obtained with the aquo-complex series of fluorescent rare earth ions. Improved sensors with wider dynamic range use the pH dependence of the uranyl ion emission.

A fiberoptic pH sensor containing a fluorophore (eosin) and an absorber (phenol red), co-immobilized on the distal end of a fiber, has been developed. In this so-called energy transfer sensor, eosin is selectively excited and fluoresces in a region where the basic form of phenol red absorbs. Consequently, a pH-dependent energy transfer can occur from eosin to phenol red, leading to a pH-dependent fluorescence quenching. The sensor exhibits a precision of at least 0.02 pH units (197a). The fluorescence signal was considerably enhanced by co-immobilizing the fluorophore with the acrylamide directly onto the modified fiber surface (197b). Fluorescein may also be immobilized directly onto the chemically modified end of the fiber (197c).

3.6.2. Sensors for Oxygen

The major advantages of optical oxygen sensors over electrochemical ones are small size, lack of oxygen consumption, and inertness against sample flow rates, stirring, or bubbles. The first optode ever used for determination of oxygen was described in 1931, when it was observed (148) that the room-temperature phosphorescence of certain dyes adsorbed on silica gel was efficiently quenched by molecular oxygen under formation of singlet oxygen (149).

The effect was used to detect traces of oxygen (43,150) produced in the photosynthetic process (44,151) or to measure the diffusion coefficient of oxygen in acrylics (152). Trypaflavine, benzoflavine, euchrysine 3R, rheonine 3A, rhoduline yellow, safranine, chlorophyll, and hematoporphyrin adsorbed on silica gel or aluminum oxide were the dyes that were most efficiently quenched (148). Unfortunately, most dyes are photolabile, and traces of water or ammonia strongly interfere. Hydrogen, nitrogen, methane, ethylene, and carbon dioxide are without influence.

The method is suitable for determination of oxygen at partial pressures between 0.004 and 0.0005 mm Hg. The method is characterized by a fast response time but requires several seconds for regeneration. Other phosphorescence-based oxygen sensors are presented in Section 3.2.4.

Low oxygen concentrations are frequently encountered in waters heavily loaded with biodegradable waste. Zakharov and Grishaeva (47) have utilized the phosphorescence quenching effect to devise an optosensor for very low oxygen levels in water. Trypaflavine or acriflavine, adsorbed on silica gel, cellulose, or ion exchangers, were immersed into the sample and the phosphorescence monitored. Although the sensor layer has not been combined with a fiber optic, it may easily be so. Obviously, the interfering effect of humidity observed by others plays no role in an underwater instrument. Quenching usually does not obey the Stern–Volmer relation [Eq. (3.29), except within a small oxygen concentration range. Using silica gel as a support, the device is able to determine oxygen in the range between 0.06 and 1.00 $\mu g \cdot L^{-1}$.

Both sensitivity and selectivity can be improved by using a hydrophobic silica gel with average pore radius 20–60 nm as a solid support. Silica gel was made hydrophobic by treatment with derivatization reagents such as dichlorodimethylsilane. The phosphorescence is sensitive to oxygen concentrations below 0.05 $\mu g \cdot L^{-1}$, but a large persistence was observed. A stationary value after a change in oxygen concentration was sometimes established only after 3–5 min. Stern–Volmer plots showed a distinct positive deviation from linearity at oxygen levels above 0.01 $\mu g \cdot L^{-1}$. The shape of the plot seems to depend on irradiation intensity. The effect was traced back to photoabsorption of oxygen by adsorbates.

Freeman and Seitz (66b) reported on a chemiluminescence-based sensor for oxygen. The device consists of a reagent chamber at the fiber end, containing a solution of a basic amine that reacts with oxygen to yield chemiluminescence. The detection limit of this method is estimated to be as low as 1 ppm. A thermoluminescence assay for oxyen, as developed by Hendricks (145), is worth mentioning in this context.

Practically all other optical sensors for oxygen described so far are based on dynamic fluorescence quenching of a suitable indicator. Unlike most other molecules, oxygen has a triplet multiplicity in its electronic ground state, resulting in more or less expressed quenching of all fluorophores.

Dynamic quenching of the fluorescence of polycyclic aromatic hydrocarbons has been known for a long time, but the effect was not utilized for optical sensors until 1968, when Bergman described the first oxygen fluorosensor (39,154). Its arrangement is shown in Fig. 3.4. The oxygen-sensitive dye fluoranthene in the porous glass disk is excited by the UV glow lamp, and the intensity of the resulting emission is measured in a photocell. This intensity varies according to the oxygen tension. The Vycor-glass-adsorbed indicator is favored over an indicator solution in a 25- to 50-μm-thick polyethylene or silicone film because of the lower quenching efficiency of oxygen for dissolved dyes. Thus 63.5% of the fluorescence are quenched when fluoranthene is adsorbed on Vycor glass, whereas 9.9% are quenched only when the dye is dissolved in polyethylene.

Among the polycyclic aromatic hydrocarbons, pyrene and, less so, pyrenebutyric acid (PBA) are probably most efficiently quenched by molecular oxygen by virtue of their long fluorescence lifetimes. PBA in dimethylformamide or silicone rubber solution has been applied to measure oxygen (8,16,155) in an arrangement shown in Fig. 3.31. The indicator has an excited-state lifetime of around 100 ns and is therefore subject to strong quenching by oxygen. The Nernst distribution favors the presence of PBA in the polymer phase rather than in the aqueous sample phase, but it was noted that it is slowly washed out, resulting in a signal drift of the sensor.

This problem was overcome by immobilizing PBA on controlled porous glass that was previously sintered onto a glass slide to give a mechanically

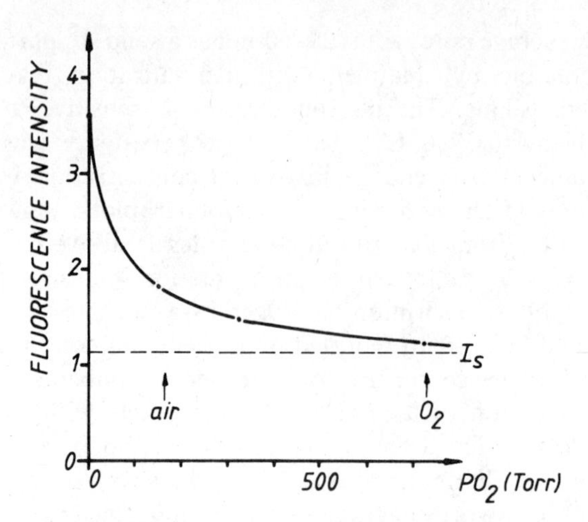

Fig. 3.33. Stern–Volmer plot of the quenching of glass-immobilized pyrenebutyric acid by oxygen. The effect is strongest in the range 0–200 torr (i.e., in the physiological range). I_s is the the contribution from stray light and background fluorescence from glass. (From ref. 40.)

stable oxygen sensor with a very fast response of less than 20 ms for the 90% value (40). Since stray light could not be separated quantitatively from fluorescence, its contribution (I_s) had to be taken into account by calculating the oxygen partial pressure (pO_2) according to

$$(I_0 - I_s)/(I - I_s) - 1 = K_d pO_2 \tag{3.56}$$

A plot of experimental data demonstrating the contribution of straylight I_s is shown in Fig. 3.33. The sensor is highly specific for oxygen. No interferences were observed with nitrogen, carbon monoxide, dinitrogen oxide, methane, carbon dioxide, and noble gases. Covering the sensitive layer with silicone rubber resulted in a 60% reduction of the quenching constant.

In an alternative approach (166), polyurethane microcapsules containing a solution of PBA in dioctyl phthalate were prepared and embedded in silicone rubber or suspended in water or dioctyl phthalate. The quenching efficiency was found to be slightly smaller than the one for the indicator in homogeneous solution, an effect that was attributed to boundary layer phenomena at the polyurethane/solvent interface.

Notwithstanding the stability and long decay time of pyrene and PBA, its shortwave excitation and emission maxima are disadvantageous with respect to suitable light sources, fiber material, and background from biological matter and fibers. It therefore became desirable to look for suitable indicators possessing more ideal spectral properties.

Berlman (156) has cataloged the oxygen quenching sensitivities of organic molecules of interest in scintillation counting, but none of them meets all requirements imposed on an ideal indicator: namely, visible excitation, photostability, nontoxicity, and efficient quenching. Among the other longwave-excitable indicators which are known to be strongly quenched, mention should be made of the ruthenium(II)–tris(bipyridyl) complex. It displays promising spectral properties with an excitation maximum at 460 nm and a far-red-shifted emission centered at 600 nm. The dye is strongly quenched by oxygen (157). Other LED-excitable oxygen indicators have also been described (158).

After screening approximately 70 dyes, Peterson et al. (37) found perylene–dibutyrate (color index no. 59,075) when adsorbed on amberlite XAD4 resin beads to be a most useful probe. It has exciation and emission maxima of, respectively, 468 and 514 nm, is stable, and is efficiently quenched by oxygen, thereby allowing a resolution of ± 1 torr up to 150 torr oxygen partial pressure. The resulting optrode, shown schematically in Fig. 3.10, consists of two small fibers for guiding the exciting and fluorescing light with fine particles bearing the fluorescent dye in a tube at the common end. To protect it from contamination it is covered with porous polyethylene of 25 μm thickness.

The sensor measures the ratio of scattered blue excitation light (I_0) and green fluorescence (I). An electronic circuit processes the blue and green signal in accord with the following relation:

$$pO_2 = (\text{gain})(I_0/I - 1)^m \tag{3.57}$$

Which is the Stern–Volmer equation rearranged with an exponent m added for curvature since Stern–Volmer plots are not linear. This ratio method can eliminate temperature effects and, partially, drifts resulting from photobleaching.

While hydrophilic quenchers are unlikely to interfere because of an impermeable polymer envelope, the sensor is affected by hydrophobic quenchers. Hence nitrous oxide has 2% of the effect of an equal partial pressure of oxygen. A more serious problem is caused by bromine-atom-containing anesthetics and narcotics which interfere heavily. The interference is cumulative, related to both concentration and time of exposure and changes both the sensitivity and zero adjustment of the sensor. If exposure is not too severe, the effect is reversible.

A new technique for surface flow pattern visualization exploits the same quenching principle. It uses visible-light excitation and streams of oxygen or nitrogen together with fluorescent yellow 1205781 adsorbed on silica gel and embedded in polyvinylpyrrolidone (159). The plate is illuminated with blue light and photographed. Streams of oxygen appear dark, those of nitrogen bright.

A steam-sterilizable oxygen sensor developed for use in bioreactors utilizes the same effect (160). It consists of a rigid light guide with an oxygen-sensing layer, a filter, and a photodetector at the tip. The sensor layer is covered with black gas-permeable material to avoid interferences from fluorescence from the sample. The rod, when immersed into a bioreactor, gives a response to oxygen that compares favorably with respect to response time (9–65 s), drift (−0.01 to −0.09% signal loss per hour), and effects of stirring (not observed). Unlike the polarographic electrode, it is not affected at all by carbon dioxide.

Cox and Dunn (161) investigated the oxygen quenching of 5 PAHs, including 9,10-diphenylanthracene, decacyclene, and rubrene, in a silicone rubber matrix. This material appears to be most ideal for oxygen-sensing layers due to its excellent solubilizing properties for oxygen. Excitation and emission spectra in addition to fluorescence intensities as a function of fluorophore concentration in silicones of different viscosities were reported. Figure 3.34 shows that the increase in fluorescence with increasing viscosity reaches a limiting value when the viscosity exceeds 10 poise.

Diphenylanthracene was considered as the most useful indicator because it is highly fluorescent in viscous solvents, stable, displays a good Stokes shift, and presents no health hazard. Its relative fluorescence intensity in poly(dimethyl siloxane) and silicone fluids as a function of temperature and oxygen partial pressure above the membrane is plotted in Fig. 3.35. Most noteworthy, the quenching process becomes more efficient with decreasing temperature, which is in contrast to the behavior in fluid solution or in the gas phase. This is possibly due to the strong increase in the oxygen solubility when going to lower temperatures.

The almost linear relations between I_0/I and oxygen pressure indicate that when the fluorophore is entangled or trapped in silicone, the only bimolecular deactivation process is oxygen quenching. From the temperature dependence of quenching the solution enthalpy ΔH was calcuated to be -3.0 $\mathrm{kcal \cdot mol^{-1}}$.

Silicone rubber is certainly the most suitable material for use in oxygen sensors because of its very good solubility and permeability for this gas. All other polymers seem to have less favorable properties. However, as mentioned briefly in Section 3.4.2, a mixture of two different polymers does not necessarily display the properties that would be expected from the properties of the pure components. It was found, for instance, that the oxygen-quenching efficiency of PBA dissolved in mixtures of silicone and poly(vinyl chloride) (PVC) decreases almost linearly as the percentage of PVC increases (162). If, however, the percentage of PVC exceeds 25%, the quenching

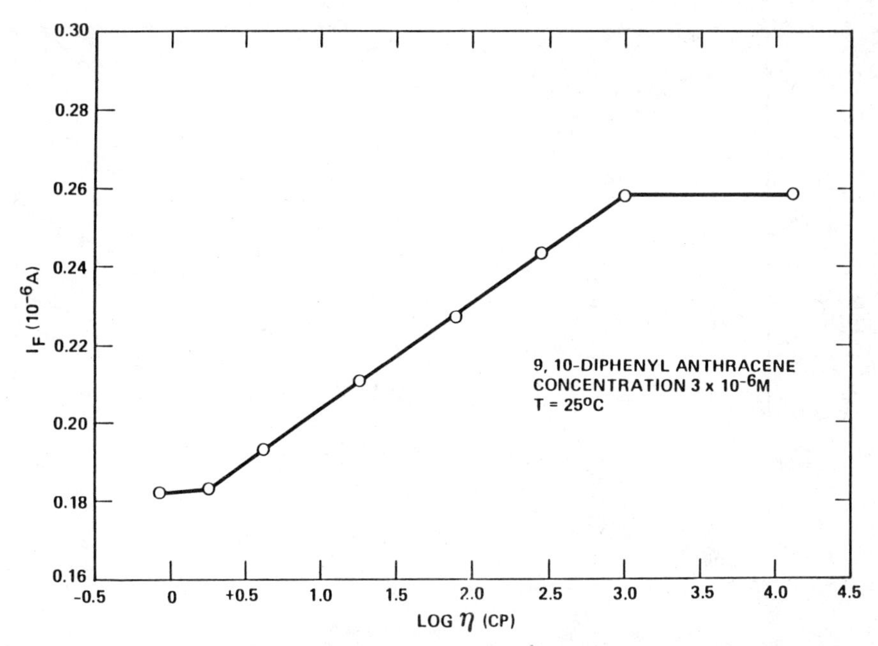

Fig. 3.34. Effect of viscosity of a silicone rubber matrix upon the fluorescence intensity of the oxygen-sensitive fluorophore 9,10-diphenylanthracene. The fluorescence is practically not affected by viscosity changes outside the range 1–1000 cP. (From ref. 161 with permission of the Optical Society of America.)

efficiency becomes very high again, being almost the same at 35% PVC as at 5% PVC.

We had developed (127) a fiber sensor for determination of oxygen, or halothane, or both, on the basis of the observation that (1) both analytes act as strong quenchers and (2) PTFE acts as a barrier to halothane. This sensor appears to be the first one that allows oxygen sensing in the presence of halothane. It is comprised of two sensing layers attached to the end of two fibers. One sensor layer is covered with PTFE, giving a signal α, defined as $({}^aI_0/{}^aI - 1)$. The second layer is uncovered and gives a signal β, defined as $({}^bI_0/{}^bI - 1)$. Concentrations of oxygen and halothane can be calculated with the help of Eqs. (3.48) and (3.49).

This two-sensor technique allows the determination of oxygen in the range 0–200 torr with a precision of ± 1–2 torr, even in the presence of halothane. Outside this range it is poorer, but still better than that of a Clark electrode. The response time is 10–15 s for 90% of the total signal change. It is useful for in-vivo catheters for continuous sensing in blood vessels, for in-vitro determination of oxygen in samples containing halothane, and for breath gas analysis during inhalation narcosis.

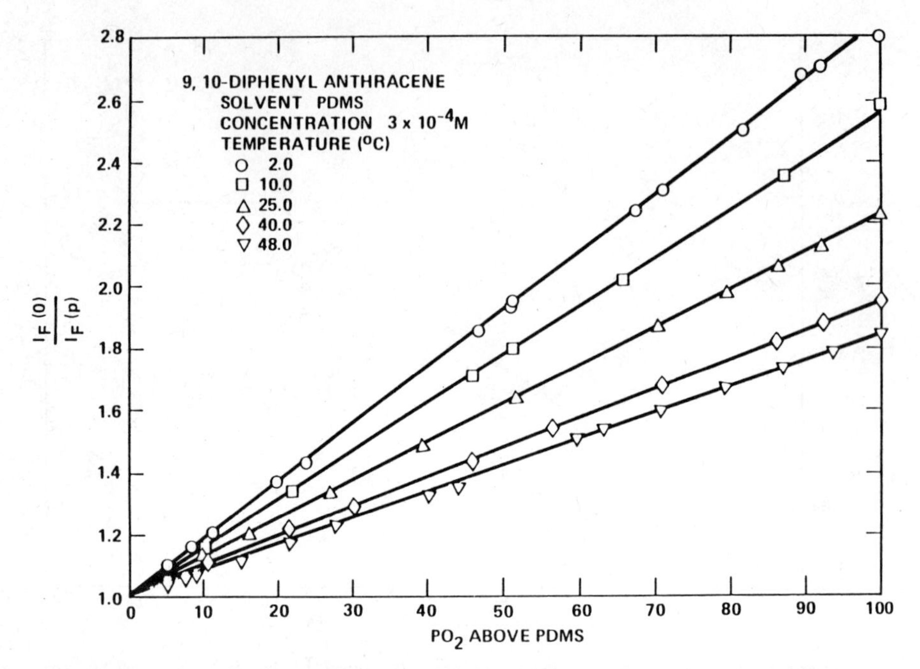

Fig. 3.35. Relative fluorescence intensity of diphenylanthracene versus oxygen partial pressure as a function of temperature. Quenching becomes more efficient with decreasing temperature. Solvent: poly(dimethyl siloxane). (From ref. 161 with permission of the Optical Society of America.)

In view of the advantages of LEDs as light sources, long-wave-excitable fluorosensors appear rather attractive. A LED-excitable sensor for oxygen is obtained by immobilizing perylene--tetracarboxylic acid on controlled porous glass using the established silyl reagent methods (50). The dye, when excited at 480–500 nm through one arm of a bifurcated fiber optic is rapidly and efficiently quenched by molecular oxygen, with a response time of less than 1 s (Fig. 3.36). The immobilized dye exhibits two emissions centered at 540 and 575 nm, whose respective quenching constants are different (0.022 and 0.028$\%^{-1}$ oxygen). Unless carefully covered with silicone, or operated continuously under water, it is also sensitive to humidity.

A new type of fiberoptic oxygen sensor exploits the fluorescence decay time (rather than its intensity) as the information carrier (128). A high frequency-modulated LED acts as a light source, and the phase shift between excitation the long-lived fluorescence (ca. 200 ns) is measured as a function of oxygen pressure. This type of sensor has advantages over the former type because of an internal referencing system which can compensate for photobleaching, indicator leach-out, light source and detector fluctuations,

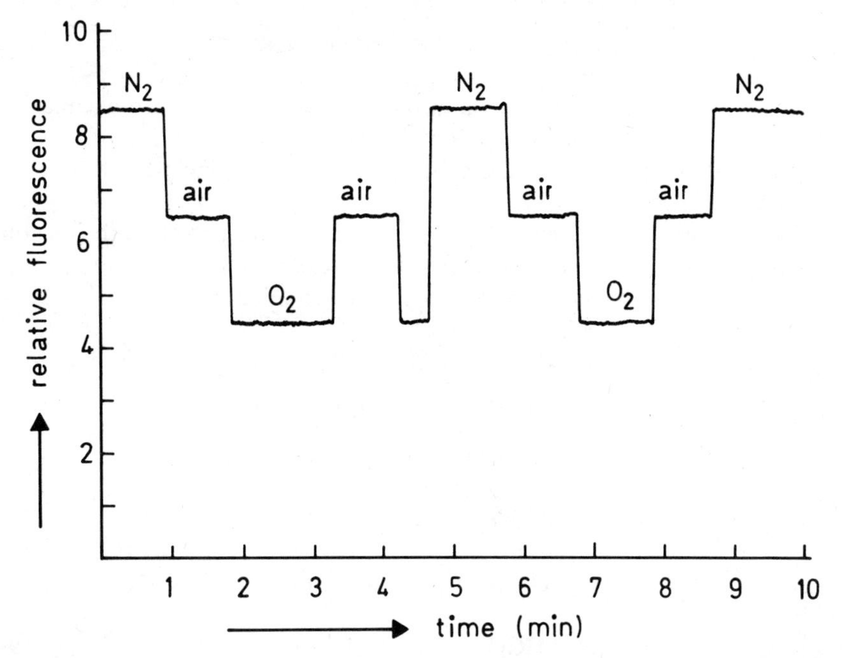

Fig. 3.36. Response time and reproducibility of the quenching of glass-immobilized perylene–tetracarboxylic acid diimide by molecular oxygen. Excitation at 500 nm using an LED, emission taken at 570 nm. The sensor has a fast response (ca. 2 s) and is not photobleached because of the great photostability of the indicator.

and therefore, is likely to have an excellent long-term stability.

Not unexpectedly for a matter of tremendous practical utility, a number of patents and patent disclosures covers the subject (36,39,45,46,129–130, 163–165,167–170). In addition to the patents that have already been discussed in some detail, mention should be made of an oxygen sensor consisting of a 2- to 80-μm-thick film of a solution of an indicator such as pyrene, coronene, *p*-terphenyl, or ovalene, in mineral oil, glycerol, or decaline (165). The layers are sensitive to oxygen, sulfur dioxide, nitrogen monoxide, and nitrogen dioxide. Aside from conventional excitation it has also been proposed to use a radioactive material as an excitation source (164) or to produce electroluminescence (165).

Kautsky's method can be improved upon by covering the support with a thin layer of a water-repelling polymer (46). A respective sensor was applied to determine oxygen in water (153,167). Poor quenching of dyes in rigid polymer solution can be improved by the addition of plasticizers (169). The effect of photobleaching can be compensated for by making use of a reference indicator (39,129).

3.6.3 Sensors for Carbon Dioxide

Carbon dioxide can be assayed via infrared absorption, or electrochemically by measuring the changes in the pH of a buffer solution as a result of varying CO_2 partial pressure above the solution. The latter principle can also be applied to optical sensors.

The response of a pH sensor to CO_2 occurs as a direct result of the proton concentration in the sensitive layer, which is related to the concentration of CO_2 through the following series of chemical equilibria:

(1) $CO_2(aq) + H_2O \leftrightharpoons H_2CO_3$ (hydration)

(2) $H_2CO_3 \leftrightharpoons H^+ + HCO_3^-$ (dissociation, step 1)

(3) $HCO_3^- \leftrightharpoons H^+ + CO_3^{2-}$ (dissociation, step 2)

These are governed by the following equilibrium constants:

$$K_h = [H_2CO_3]/[CO_2]_{aq} = 0.0026 \tag{3.58}$$

$$K_1 = [H^+][HCO_3^-]/[H_2CO_3] = 1.72 \times 10^{-4} \tag{3.59}$$

$$K_2 = [H^+][CO_3^{2-}]/[HCO_3^{2-}] = 5.59 \times 10^{-11} \tag{3.60}$$

In most studies on CO_2 sensors, the total analytical concentration of carbon dioxide (i.e., $[CO_2]_{aq} + [H_2CO_3]$) has been related to the response.

Lübbers and Opitz (16) followed the changes in fluorescence of a solution of 4-methylumbelliferone in 1 μM bicarbonate as a function of CO_2 partial pressure in the arrangement shown in Fig. 3.31. The indicator solution was covered with a 6-μm PTFE membrane and the ratio of fluorescence intensity at 445 nm measured under excitation at 318 nm and 357 nm related to pressure.

The optode showed good reproducibility in the CO_2 range 1–70 torr, with a response time of 3–4 s for 90% of the final value. It can be accelerated by addition of the enzyme carbonic anhydrase, which catalyzes the establishment of equilibrium (1). The short response is in striking contrast to the 0.5–2.0 min value of the CO_2 electrode. When coupled to a fiber optical system it allows the transcutaneous measurement of CO_2 pressure (171).

The same principle was applied to construct a compact (5 × 6 × 14 cm) fluorosensor for CO_2 consisting of a blue LED as a light source, a long-wave-absorbing indicator (HPTS) dissolved in bicarbonate and covered with a PTFE layer, two optical filters, and two photodiodes for light detection

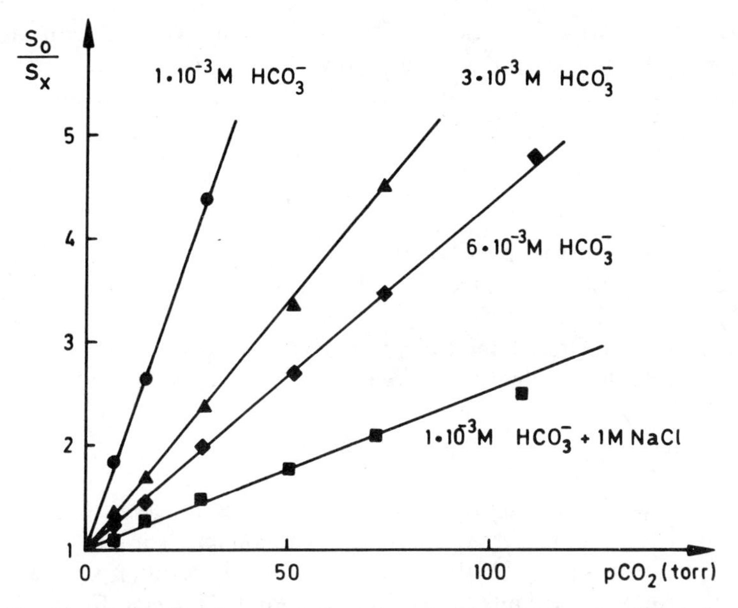

Fig. 3.37. Linearized calibration curves of a carbon dioxide sensor based on CO_2-induced pH changes of a buffer solution containing various amounts of bicarbonate and added electrolyte (NaCl). S_0, fluorescence intensity under zero CO_2 pressure; S_x, fluorescence under varying CO_2 pressure. (Redrawn from ref. 172.)

(172). Linearized calibration curves of this sensor type are shown in Fig. 3.37.

The slopes of the graphs can be governed by the pK_a of the indicator, ionic strength, and buffer concentration. Unfortunately, high buffer concentrations result in very long response times. These are further prolonged with increasing thickness of the sensor layer (typically, 20 μm), polytetrafluoroethylene covers (6 μm), and slow kinetics of the hydration of CO_2. For the system described above, the response for a change from 0 to 5% CO_2 was 15 s for 90% of the final value. The reverse change required about 34 s.

A fiber optical device for measuring CO_2 pressure in tissue and blood of humans and experimental animals has been described in a patent (173). A solution of phenol red in bicarbonate buffer is covered with a gas-permeable but ion-impermeable barrier such as silicone rubber. This allows equilibration of CO_2 between indicator solution and body fluid, while protons remain outside the barrier. The apparatus uses two plastic fiber optics, one carrying input light to the sensing tip and the other carrying the transmitted output signal, which is later converted to an electrical signal.

Zhujun and Seitz (174) used HPTS in bicarbonate, covered with silicone, to sense CO_2. Fluorescence is measured with a bifurcated fiber system. The equation that was used to relate the CO_2 partial pressure to hydrogen ion concentration is (175)

$$[H^+]^3 + N[H^+]^2 - (K_1C + K_w)[H^+] - K_1K_2C = 0 \qquad (3.61)$$

where N is the internal bicarbonate concentration; K_1, K_2, and K_w are the dissociation constants of carbonic acid [steps 1 and 2; see Eqs. (3.59) and (3.60)] and water, respectively; and C is the analytical CO_2 concentration, including both hydrated and unhydrated CO_2. It was shown that within a limited range, there is linearity between CO_2 pressure and $[H^+]$ according to

$$[H^+] = K_1/CN \qquad (3.62)$$

Figure 3.38 shows fluorescence intensity as a function of CO_2 concentration when the internal filling solution contains various amounts of hydrogen carbonate. Evidently, the choice of concentration is particularly critical because of the limited range of the internal pH sensor. In practice, the internal HCO_3^- concentration should be such that the CO_2 concentrations of interest yield pH changes between 6.5 and 8.0, where the pH sensor based on HPTS with its pK_a of 7.3 is most sensitive.

Aside from CO_2, the sensor also responds to sulfide and sulfite. Like the electrochemical CO_2 sensor, it may be affected by certain other ionic constituents (176). Recent progress (177) in the preparation of polymer membranes for use in CO_2 electrodes may also have beneficial effects CO_2 optrodes.

Heitzmann and Kroneis (178b) as well as Marsoner et al. (178a) have prepared CO_2-sensitive fluorescent membranes by soaking cross-linked polyacrylamide beads with a solution of HPTS in bicarbonate, and embedding them in silicone rubber. The response to CO_2 was varied by adding different amounts of bicarbonate, carbonate, and HPTS, all of which act as buffers. The polyacrylamide beads may be omitted, so that an emulsion of the HPTS/carbonate solution in silicone rubber is obtained.

A computer program was developed (178c) to calculate the response curve of a CO_2-sensing composition from (1) the initial carbonate concentration, (2) the initial bicarbonate concentration, (3) the amount of indicator added, (4) the CO_2 partial pressure and (5), the indicator pK_a. It also considers ionic strength and temperature effects and is useful for simulating both continuous flow and standing samples. The program calculates the pH of the solution under certain CO_2 pressure in addition to the amount of free carbonic acid, bicarbonate, carbonate, and indicator enion present. Plots of CO_2 pressure

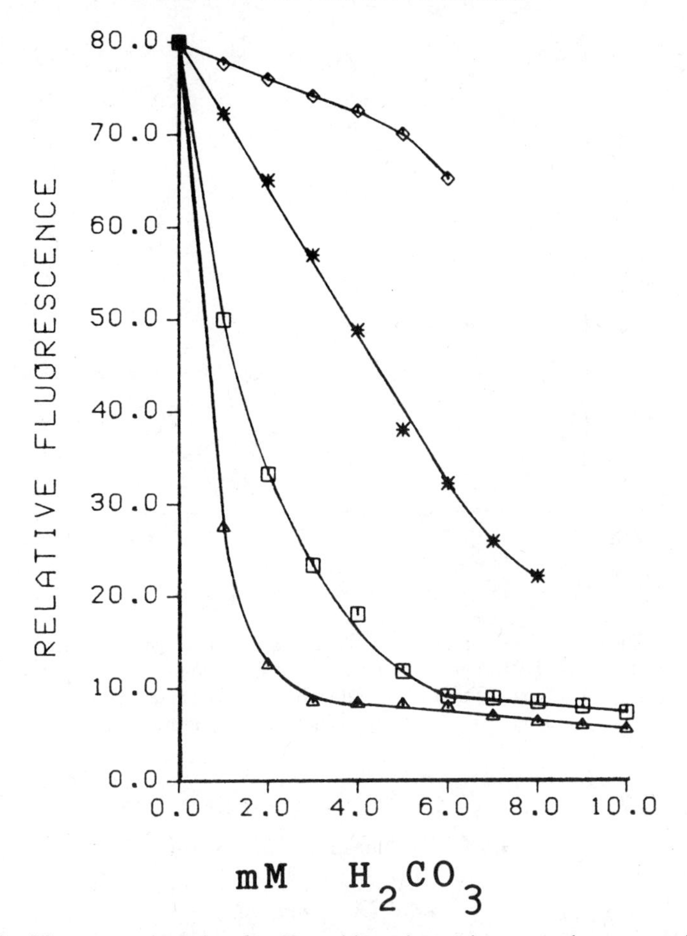

Fig. 3.38. Fluorescence intensity of a pH-sensitive anion-exchange membrane versus [H_2CO_3] at various bicarbonate concentrations of the filling solution. (△) 0.1 m*M*; (□) 1 m*M*; (*) 10 m*M*; (◇) 100 m*M*. [From ref. 174 with permission of Elsevier Publ. B.V. (Amsterdam).]

versus fluorescence intensity, pH, indicator anion concentration, carbonate, and bicarbonate may also be obtained.

The theoretically predicted performance of a typical solution versus CO_2 partial pressure is shown in Fig. 3.39. Obviously, there is a large change in intensity with increasing CO_2 pressure and almost linearity between CO_2 and fluorescence. In the CO_2 range above 110 mm, the signal change per mm CO_2 is rather small, so that precision is less expressed. Since, however, signal-to-noise ratios of 2000 are achieved, a sufficient precision is provided even under relatively high CO_2 partial pressure.

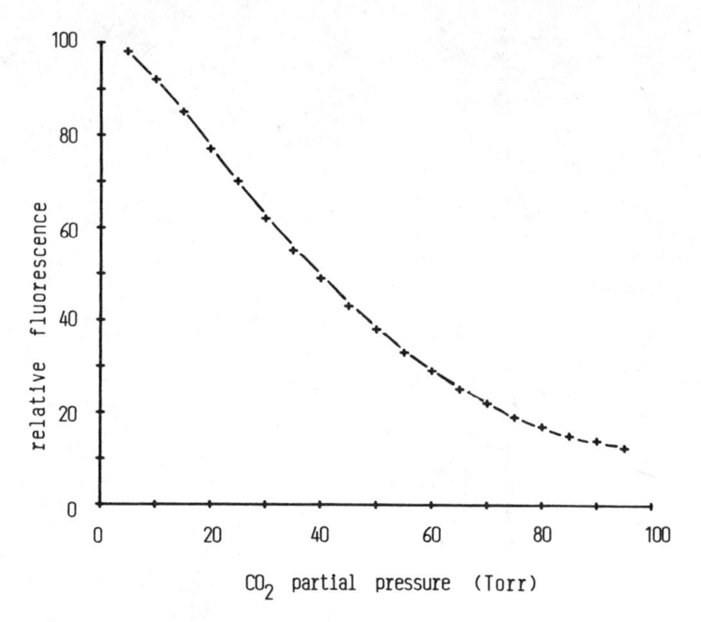

Fig. 3.39. Calculated fluorescence response of a solution containing 4.4 m*M* Na_2CO_3, 1.0 m*M* $NaHCO_3$, and 4.0 m*M* HPTS against variation in the partial pressure of carbon dioxide after full equilibration. This composition is favorable in that it responds strongly to pCO_2 in the physiological range (40–80 torr). The experimental results are practically identical with this curve.

3.6.4. Sensors for Other Gases

3.6.4.1. Ammonia

The well-known reaction between ninhydrin and ammonia producing a purple coloration was utilized to develop an ammonia probe. The basis for the analytical technique is the partial attenuation of light signal traversing a cylindrical optical waveguide by successive total internal reflections. A fiber rod was covered with a solution of ninhydrin in polyvinylpyrrolidone and the changes in the absorption of the evanescent wave were followed (179). Concentration and relative humidity both determine the slope of the transmission curves. The proble works irreversibly and is able to detect ammonia down to the 50-ppb range. Another ammonia sensor specifically designed for use in bioliquids is described in Section 3.7.1.

A reversible optical waveguide sensor for ammonia vapors was introduced more recently (180) consisting of a small capillary glass tube fitted with an LED and a phototransistor detector to form a multiple reflecting optical device. When the capillary was coated with a thin solid film composed of a pH-sensitive oxazine dye, the gauge was capable of reversibly sensing ammonia. Vapor concentrations from 100 ppm to below 60 ppm were easily and reproducibly detected. A preliminary qualitative kinetic model was proposed to describe the vapor–film interactions.

An ammonia fluorosensor based on the same principle as electrochemical ammonia sensors (i.e., the change in pH of an alkaline buffer solution caused by ammonia) has been reported by two groups: Arnold and Ostler (181a) followed the changes in the absorption of an internal buffer solution, to which *p*-nitrophenol was added. As ammonia passes by, it penetrates through the membrane and gives rise to an increase in pH. Wolfbeis and Posch (181b) used a fine emulsion of an aqueous solution of fluorescent indicators in silicone rubber. Both ammonium chloride and the indicators themselves were found to be useful buffers. The fluorescent dye again sees the change in pH as ammonia passes into the filling solution. Both sensors have a very slow response.

3.6.4.2. Sulfur Dioxide

Continuous determination of SO_2 in air can be performed by fluorimetry owing to its strong intrinsic fluorescence at about 330 nm (182a). In complex samples the method is not very specific. Advantage can be taken of the observation that the fluorescences of ovalen and related PAHs are quenched by SO_2. Interferences by oxygen are eliminated by a second sensor (the same indicator in a different polymer). SO_2 concentrations can be calculated according to Eqs. (3.44) and (3.45). 0.01% SO_2 in exhaust gases is easily detectable. The quenching of PAHs and rhodamines by SO_2 has been studied in some detail (182b).

3.6.4.3. Hydrogen Sulfide

An optrode for hydrogen sulfide (183a) is based on the following chemical reaction:

$$Pb^{2+} + 2OH^- + EDTA + H_2S \rightarrow PbS + 2H_2O + EDTA$$

Lead hydroxide, which is kept in solution by addition of ethylenediaminetetraacetic acid (EDTA) is neutralized by the weak acid H_2S to give an insoluble precipitate of lead(II) sulfide, thus rendering the reaction irreversible. The decrease in pH due to consumption of hydroxyl ion is monitored via the decrease in fluorescence intensity of added fluorescein. 2.3 ppm H_2S caused a linear 30% signal reduction over a period of 3 h.

Alternatively (183b), quaternized acridinium ion may be immobilized on a support such as cellulose or an ion-exchanging membrane. In slightly alkaline medium, hydrogen sulfide adds to the strongly fluorescent dye and renders it non-fluorescent. Interferences by ionic quenchers may be eliminated by covering the sensor with a thin layer of silicone which is permeable to H_2S, but not to ions.

3.6.4.4. Humidity

Humidity sensors are of interest in continuous monitoring of air moisture in airports and related locations to allow the timely prediction of fogging and icing. A simple version for a humidity sensor makes use of the color changes of cobalt dichloride when exposed to humidity. The optical sensor measures the reflectivity changes of a crystal and relates it to the intensity of scattered light of a certain wavelength (184).

Another type of humidity sensor (185) utilizes the effect of water vapor on the fluorescence quantum yield of adsorbed fluorophores. Thus silica gel–adsorbed or glass-immobilized perylene–tetracarboxylic acid diimide, when excited with an LED at 500 nm, shows intense fluorescence under dry air. Under moist air, the signal drops according to the degree of humidity. Although the response is slow, it is sufficiently fast to be used in meteorology.

A recent patent (186) describes a fluorescence-based moisture gauge comprised of a luminescent material whose fluorescence intensity varies in relation to the amount of moisture adsorbed. The method used to correlate moisture with the analytical signal is also discussed.

3.6.5. Sensors for Anions

3.6.5.1. Halides and Pseudohalides

Chloride has been determined (6) over a distance of 100 m with a fiber possessing a chloride-sensitive tip, which contains collodion-entrapped silver fluoresceinate. The salt is insoluble and nonfluorescent. When chloride (and probably other halides and pseudohalides as well) react with the salt, AgCl is precipitated. A corresponding amount of fluorescein is released and can be measured. A semipermeable membrane is used to prevent silver ion loss and to assure reversibility.

The chloride optrode works best in the chloride concentration range 10^{-3}–10^{-4} *M*, but the calibration curve of molarity versus fluorescence is not linear. Since it was expected that at lower concentrations the plot should tend to linearity, an optrode sensitive to lower concentrations (10^{-5} to 10^{-6} *M*) was also designed. It consisted of a cross-linked polymer substrate such as poly(hydroxyethyl methacrylate) in whose intersticies the fluoresceinate is retained. These coatings, however, could not be made uniform and the data were inconsistent.

Another halide sensor type exploits the effect of dynamic fluorescence quenching (187). The emission of quaternized heterocyclic fluorophores such as acridinium and quinolinium is quenched by chloride, bromide, or iodide with incresing efficiency. When immobilized onto glass the sensors are able to indicate the halide concentration in solution by virtue of the decrease in fluorescence intensity as a result of the quenching process. The sensitivity toward different halides can, to some extent, be varied by the choice of the indicator. However, quenching of the immobilized dyes is much less efficient than quenching of the same dyes in bulk solution.

A more convenient way to immobilize the aforementioned cationic heterocyclic indicators is provided by electrostatic binding to a cation exchanging membrane such as the Raipore R 1010 membrane (from RAI Res. Corp., NY 11787). Thus *N*-methylacridinium chloride is firmly bound to this membrane when it is immersed into its methanol solution. The resulting polymer sheet, when drawn onto the end of a fiber optic, can serve as an optosensor for halides. Its sensitivity characteristics is similar to the one based on covalently immobilized indicators.

A typical response of covalently immobilized acridinium ion toward halide concentration is shown in Fig. 3.40. Immobilized quinolinium, in contrast, is much more sensitive to chloride, although still not as sensitive as toward iodide. Detection limits are 0.15 m*M* for iodide, 0.40 m*M* for bromide, and 10.0 m*M* for chloride. Sulfite, isothiocyanate, cyanide, and cyanate interfere (or may be assayed by the same sensor type), while sulfate, phosphate, perchlorate, and nitrate are without effect.

Cyanide ion can be detected by using the first optical probe of the evanescent wave type ever described (188). The surface of a fiber rod was coated with a poly(vinyl alcohol) solution of sodium picrate, which is known to yield a reddish-brown product with cyanide. The reaction products change the refractive index and absorption coefficient of the coating. The resulting change in light transmission through the guide is proportional to the concentration of the cyanide species. The probe acts irreversibly.

3.6.5.2. Free Halogens

Although not anions in a chemical sense, free halogens shall be included in this section, owing to the similarity to halides in their quenching properties. Unlike halides, however, free halogens can penetrate hydrophilic polymers. Sensors for free iodine and bromine are of interest for use in the endpoint determination in bromometry and iodometry. Sensors for chlorine are of

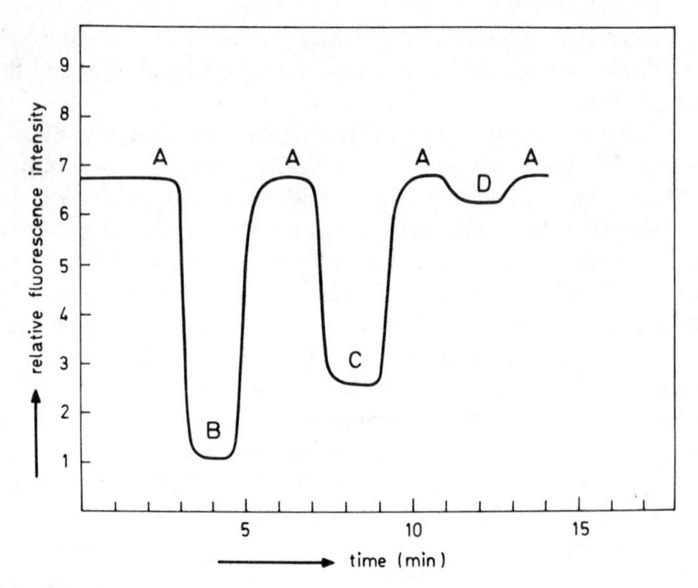

Fig. 3.40. Changes in fluorescence, response time, and reproducibility of a halide sensor based on glass-immobilized acridinium ion. *A*, Pure water; *B*, 0.1 *M* potassium iodide; *C*, 0.1 *M* potassium bromide; *D*, 0.1 *M* potassium chloride. Note that the response curves are not symmetric. (From ref. 187.)

potential utility in continuous monitoring of chlorinated bath- and wastewaters.

The fluorescence of rubrene (a PAH) is quenched by traces of iodine. When a fiber optic rod with a layer of a 20-ppm rubrene solution in polystyrene attached to its end is immersed into an aqueous solution containing 10 ppm iodine, iodine diffuses into the polymer membrane and causes a reduction in the rubrene emission intensity to around one-third of the initial intensity (189a). The sensor responds to bromine as well.

α-Naphthoflavone is another indicator whose fluorescence is quenched by bromine or iodine. Unlike rubrene, it is much more photostable and practically not quenched by oxygen. On the other hand, it requires UV excitation at around 350–365 nm. Its solution in silicone-type materials or PVC can serve as a sensing layer for free halogenides (50).

3.6.5.3. Other Anions

Probes probe for H_2S (and therefore probably also for sulfide) have been presented in Section 3.6.4.3. Alternatively, a sulfide-sensitive dye such as 2,6-dichloro-indophenol (189b) or *N*,*N*-dimethyl-*p*-phenylenediamine (189c), in immobilized form, may be used to sense sulfide by reflectometry. A sulfate sensor has been proposed (17), consisting of an optrode with immobilized barium chloranilate (the barium salt of chloranilic acid). The salt is nonfluorescent by itself. When sulfate is present, barium sulfate will precipitate and an equivalent amount of fluorescent chloranilic acid be released. Although preliminary work indicated that this technique will work, there is some concern about the ability to reach sensitivities of 1 ppm or below. One reason is that the "insoluble" barium sulfate is really soluble at 27 ppm, and this drives back the reaction equilibrium and reduces sensitivity.

A nitrate sensor is urgently needed for groundwater monitoring but has not been reported so far. A concept for a nitrate sensor is based on a two-step reaction (17). Nitrate is first chemically reduced to nitrite with hydrazine sulfate. The resulting nitrite, in the presence of acid, would quench the fluorescence of an aromatic amine as it is converted to a diazonium salt. A reservoir sensor can be used so that both reactions occur immediately in sequence. Nitrite would, of course, be determinable by an even simpler reaction sequence, since it would react with amines without previous chemical reduction.

The optical determination of surface potentials, which is likely to be applicable to the quantitation of anions as well, will be described in the next section.

3.6.6. Sensors for Cations

Opitz and Lübbers (190) have described an arrangement for the optical determination of differences in electrical potentials at the interface between an insulator and an electrolyte solution. By analogy to the principle of ion-selective electrodes, the device should allow the determination of various electrolytes (cations as well as anions), but no experimental results have been presented so far. A device is claimed in the patent that is suitable for the determination of potential differences, with the help of a potential-sensitive indicator such as a a cyanine dye, whose changes in fluorescence intensity are observed.

It consists of two parts: One is a very thin or even monomolecular layer of an indicator (eventually embedded in a lipid membrane) attached to the surface of a boundary to the electrolyte (fig. 3.41). The other is a potential-forming electrode of the usual type. The two parts are connected by an electrical lead, and the response of the fluorescent indicator toward changes in the potential are measured. Numerous other modifications of the instrument are described in the patent, which has 63 claims.

A different approach, that no longer requires a reference electrode, has been made by directly placing a potential-sensitive dye plus an ion carrier

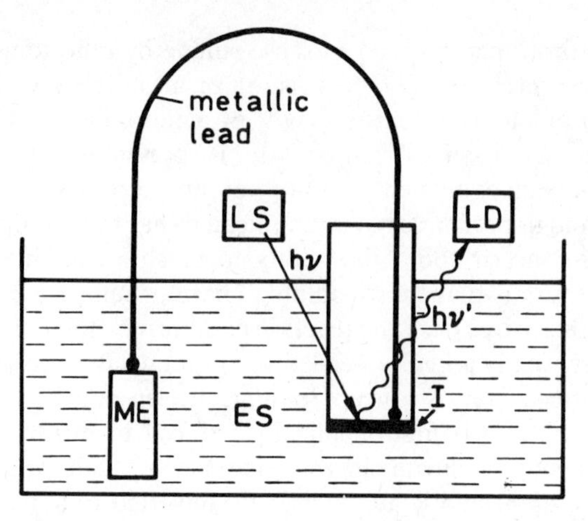

Fig. 3.41. Concept of an optical sensor for measurement of electrochemical potentials. ES, electrolyte solution; ME, metal reference electrode; I, indicator layer; LS, light source; LD, light detector. The classical electrochemical element creates a potential which is indicated by an potential-sensitive cyanine dye whose fluorescence varies with the field applied at the surface. (Redrawn from ref. 190.)

such as valinomycin into a thin lipid membrane (191). Dyes more stable than the carbocyanines have been used, and the Langmuir-Blodgett technology proved to be a most suitable method for preparing well-defined sensing layers (191b).

Zhujun and Seitz (192c) report on a fluorosensor for sodium based on the principle of ion-pair extraction. The components of the sodium-sensitive phase are (i) 8-anilino-1-naphthalenesulfonate (ANS), (ii) copper(II)-polyethyleneimine, and (iii) a sodium-selective ionophore. In the absence of sodium, the ANS binds to the copper(II) polyelectrolyte and fluorescence is quenched. When sodium (20–200 mM) is added, it forms a cationic complex with some of the ANS, causing it to fluoresce.

Various chelating agents are known which are essentially nonfluorescent by themselves but form fluorescent metal chelates. This can be the basis for metal ion sensors, provided that binding of the metal ion is not too strong or even irreversible. Alternatively, the chelating agent may exhibit fluorescence which is quenched by binding of transition metal ions.

Morin is known to form a fluorescent complex with aluminum(III). When immobilized on cellulose, attached to the common end of a fiber optic, and immersed into a solution of Al(III) ion, its fluorescence varies linearly from 1 to 100 μM Al(III) at pH 4.8 (116). At this pH the binding constant is 1.7×10^4 M^{-1} for the immobilized complex. The response time is 2 min. Above

100 μ*M* Al(III) there is a change in slope and the response levels off due to saturation of morin with Al(III). See also Fig. 3.25*a*.

The sensor is fairly specific at pH 4.8. Major interferences come from beryllium(II), which increases the signal, and iron(III) and copper(II), which quench fluorescence to an appreciable extent.

The response of immobilized morin to Be(II) has subsequently been characterized in more detail (115). The response is linear from 1 to 10 μ*M* Be(II). Above this concentration, there is a decrease in slope and the response gradually levels off, owing to saturation of the morin. The best response were observed at pH 5.3 and 11.2, whereas at pH 7.0 there was no fluorescence observed at all.

The excitation spectra of the immobilized Be(II)–morin complex differ considerably from the spectra of the complex in solution, in that they show an additional excitation maximum at around 440 nm. It was suggested that this band be used to distinguish Be(II) from Al(III) in solutions containing both ions. Interferences by Al(III) may alternatively be eliminated by complexing it with EDTA. Again, other major interferences are copper (II) and iron (III).

Calcein was immobilized on cellulose and tested in a fiber optic system for its utility in metal ion sensing (192a). It is not suitable for calcium analysis because the linkage used for immobilization hydrolyzes at high pH. When the sensor is immersed in a solution of a metal ion such as Cu(II), Co(II), or Ni(II) at pH 5 or 7, fluorescence decreases as a function of added metal ion. However, because the binding is so strong, the response is not reversible. The reagent completely extracts the metal ion from solution at these pH values.

The sensor based on calcein can be used to determine endpoints of complexometric titrations (192a). For example, Cu(II) was titrated with EDTA at pH 7.0. At the point of equivalence a large increase in calcein fluorescence is observed, since a very slight excess of EDTA is sufficient to pull the Cu(II) away from the immobilized indicator.

A fluorescence sensor for Al(III), Mg(II), Zn(II), and Cd(II) which is based on electrostatically immobilized oxin-5-sulfonate (8-hydroxyquinoline-5-sulfonic acid) has been presented (192b). Uncomplexed reagent displays only very little fluorescence, but when brought into contact with one of the metal ions above, a strong increase in fluorescence is observed. The response to metal concentrations was found to be nonlinear, and detection limits are below 1 μ*M*.

Dissolved and immobilized oxinesulfonate behave similarly with respect to pH and interferences, three types of which were reported: Metal ions that form fluorescent complexes interfere positively in the determination of other metal ions. Metal ions that form nonfluorescent complexes interfere negatively by combining with immobilized ligand and reducing the amount that is available to form fluorescent complexes. The third type of interference is due to the presence of ligands, which combine with the analyte and render it unavailable to immobilized ligand.

The recent observation (192d) that the fluorescence polarization of the ligand is a function of complex composition may be exploited to improve the selectivity of oxine-based sensors for Al(III), Mg(II), Zn(II), Be(II), Ca(II), Cd(II), and Pd(II). Copper(II) may be determined in plating baths by remote absorptiometry at 820 nm, using an infrared LED as a light source. 50–500 mM copper(II) in sulfuric acid are detectable over long distances with good accuracy (193).

Mercury is known to statically quench the fluorescence of indoles (122). When indole-3-acetic acid is covalently linked via long spacer groups to quartz beads welded to the tip of a quartz fiber or onto a quartz slide, its fluorescence (occurring at 345 nm) is quenched by mercury(II). Thus the emitted light provides a continuous and reversible signal for the Hg(II) concentration of the solution in contact with the fiber end (121). Hg(II) concentrations down to 1 μ*M* are detectable, but the strong background from both fiber and solution under UV excitation (290 nm) frequently limits sensitivity.

A great deal of work was dedicated to the development of fluorosensors for actinide ions such as UO_2(II). These are encountered in groundwater around nuclear fuel processing facilities and in soil after underground tests. Sensing of radioactive analytes now becomes possible without hazard by applying remote fluorosensing using fiber optics.

UO_2(II) ion shows an intrinsic fluorescence with maxima at 518 and 546 nm, the intensity of which is strongest in acidic solution at pH 1.6. In 8% phosphoric acid emission is particularly intense, being 150 times stronger than in water, probably by virtue of the combined effect of lowering the pH and dehydration by formation of the phosphate complex (194a). The lifetime is 197 μs in 8% phosphoric acid, but is strongly reduced by external quenchers such as iron(III). Unfortunately, the quantum yields are strongly affected by other factors, too: Lowering the temperature causes an exponential increase, added salts a decrease.

In a continuous reservoir sensor for uranyl(II) ion a 1% aqueous solution of phosphoric acid is injected into an interactive compartment into which the analyte sample can pass through a membrane (see Fig. 3.6). The concentration of the uranyl phosphate thus formed can be interrogated by a 415-nm laser flash (194b). The advantage of such a membrane optrode is that the environment of the fluorescing analyte can be controlled and potential interferences can more easily be eliminated. 10 μm uranyl ion is detectable with an exceptionally good signal-to-noise ratio. The use of 2% phosphoric acid in place of the 1% solution reduces the signal. There is some contradiction in the data given in ref. (194a) and 194b).

The fluorimetric detection limits can be enhanced dramatically by making use of the coprecipitation technique: When calcium fluoride is precipitated from solution, uranyl ion is reproducibly coprecipitated. Air-calcining at 800°C followed by quantitative fluorimetry offers a most sensitive method for UO_2(II) quantitation in the 50-ppb range (195). This technique has been applied to remote sensing of uranyl ion over long distances using laser excitation (27). Routine detection limits of 0.0002 ppb were reported, with an extrapolated sensitivity of 10^{-14} *M* (!). This, however, is probably not usable because of the ca. 10^{-13} *M* uranyl background in normal water.

The extension of the coprecipitation technique to remotely sense other fluorescers such as plutonium(III) is obvious. Detection of 10^{-14} *M* Pu(III), which in solution is nonfluorescent, was said to be a realistic expectation (27).

Remote fiber sensing of Pu(III) has obvious safety advantages, but the absence of a fluorescent plutonium species in *solution* is a difficulty in remote fiber fluorimetry. Pu(III) can be determined (1983) by virtue of its quenching effect. It has an energy level nearly coincident with one of the highly fluorescent terbium(III) ions, allowing the measurement of Pu(III) by the quenching of a known amount of added Tb(III). To heighten the effect, the lifetime of the Tb(III) fluorescence is lengthened by using D_2O as a solvent and adding complexing agents such as acetate or trifluoroacetate.

Gaseous uranium hexa fluoride can be determined by the remote fiber optical laser-induced fluorescence technique as well. An input near-UV laser pulse (2 mJ, 5 ns) is transported through a 50-m single fiber to the sample. The fluorescence of UF_6 is collected by a fiber bundle and conveyed to a photomultiplier. Aspects of laser beam profiles, fiber attenuation, fiber breakdown thresholds, fiber sizes, optical spatial resolution and sensitivity, coupling and monitoring techniques, and other pertinent experimental details were discussed (196).

3.6.7. Sensors for Titrimetry

Titrimetry is another field of application, where fiber optics can replace electrodes. Although fluorescence titrations have been known for a long time, their application has been limited in practice to titration in which the endpoint can be recognized visually. This seems to be a result of the formerly limited availability of fluorimeters and fiber optics. Moreover, the laborious process of transfering a sample from the titration beaker to the fluorimeter cuvette, doing a measurment, returning the sample to the beaker, and continuing titration has prevented analysts from using the method more often. Visual endpoint determination is more convenient but is operator-

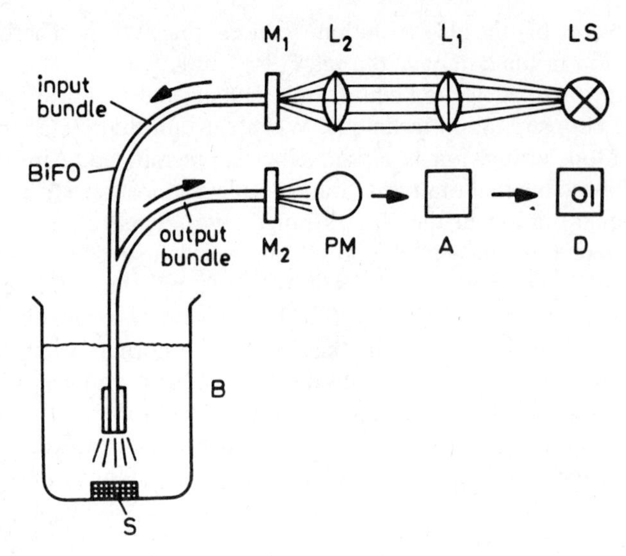

Fig. 3.42. Schematic of a fiber optical titrator. (From ref. 199.)

dependent and limited to strongly fluorescent indicators having visible emission.

The remedy for this situation is the fiber optic light guide that can be dipped into the solution to be titrated. Aside from a considerably simplified experimental setup and procedure, the fiber method allows the use of UV indicators, measurement of small spectral changes, and an operator-independent endpoint determination.

Fluorescence titrations provide a more expanded range of analytical possibilities than absorptiometric titrations. Thus acid–base indicators, redox indicators, chelating reagents, and adsorption indicators are usually suitable for both absorptiometry and fluorimetry, but quenching titrations are confined to the latter method. Therefore, the known fluorescent indicators which are dynamically quenched by titrant or titrand offer a considerbly enlarged field of application.

The experimental setup for performing fiber optical titration is shown in Fig. 3.42. Light from a light source (LS) passes lenses L_1 and L_2. After passing an interference monochromator M_1 it is coupled into the end of the input bundle of a bifurcated fiber optic (BiFO). The common end is immersed into the titration beaker containing analyte plus indicator (if necessary). Fluorescence is collected by the strands of the ouput bundle and guided to a photomultiplier after it has spectrally been isolated from scattered light by the emission monochromatior M_2. The signal is amplified in *A* and displayed in *D*.

A fiber optical titration system can be relatively inexpensive. The light source can be a halogen tungsten lamp or, even cheaper, an LED (198a). The latter are available for wavelengths above 440 nm only. Both cutoff filters and plastic fibers are also inexpensive. Costly photomultipliers can be replaced by cheap photodiodes.

Since fibers can be made smaller than any electrodes, the use of 0.1-mm-thick fibers in a bundle enables the titration of very small sample volumes. As a result, the low-cost optoelectronic equipment required for fiber optical titrations will probably result in reasonably priced fiber titration equipment.

In order to restrict the adverse effect of ambient light entering the output fiber bundle, titration should be performed in a brown or black Teflon-covered beaker. But as long as the noise level (i.e., ambient light) is not too high, it will not interfere, since it is the relative change in intensity that is observed during a titration, not an absolute value.

3.6.7.1. Acid-Base Titrations

Both strong acids and strong bases can be titrated with a precision of better than 1% using fiber optics in place of electrodes (197). Figure 3.43 shows typical plots obtained when HPTS is used as an indicator. The precision is not affected when methylene blue (an intensely colored and redox-active dye) is present or when the solution was made turbid by adding silver nitrate plus potassium chloride.

Aside from strong acids and bases, acetic acid as well as mono-and dibasic phosphate (Fig. 3.44) were titrated successfully using either fluorescein or HPTS as indicators. In each case the precision was better than 1%, provided that the ratio of the concentrations of acid or base to indicator was greater than 100. Otherwise, the titration of the indicator which itself is a weak acid will cause interferences.

Phosphoric acid with its two pK_a values of 2.16 and 7.21 can be titrated using an indicator having pK_a values in the same region. Both steps are easily recognizable in the titration plot. Hydrofluoric acid is a challenging candidate for fiber optical titration, since it cannot be titrated with the help of the glass electrode. Using a plastic fiber or a parafilm-covered glass fiber, this acid has been titrated in concentrations down to 0.001 M with excellent precision using HPTS as an indicator (198b).

Rather than adding a fluorescent indicator to the solution, one of the fiber sensors described in Section 3.6.1 may as well be utilized for this purpose. These, however, have certain limitations. The immobilized indicator may be stripped off by strong acid or base, the response time is usually longer than 10 s, and the pK_a of known pH sensors is not suitable for titration of, for example, ammonia or dibasic phosphates. The latter fact appears to be only a matter of time.

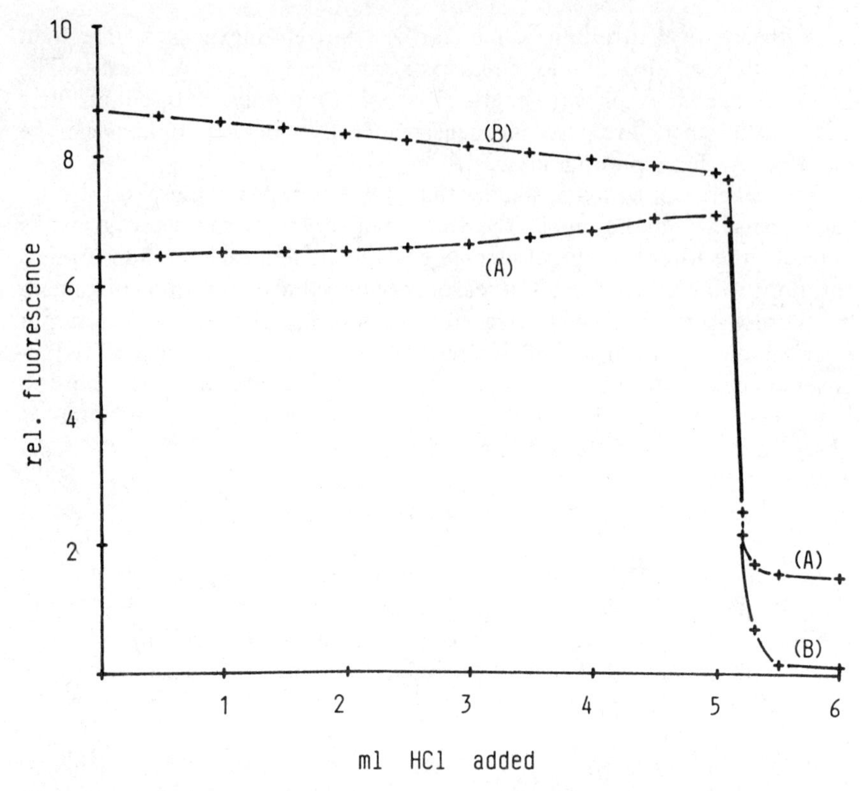

Fig. 3.43. Typical curve as obtained in the fiber optical pH titration of 5 ml 0.1 *N* sodium hydroxide with 0.1 *N* hydrochloric acid in the presence of 0.2 mg of methylene blue (curve A) and 20 mg of silver chloride suspension (curve B). Both the strongly colored dye and the white suspension do not affect the sharpness of the endpoint. Indicator: 2 μ*M* 1-hydroxypyren-3,6,8-trisulfonate (HPTS). (From ref. 198a).

3.6.7.2. Argentometry

Sulfide has been determined in a precipitation titration using silver nitrate and a sulfide sensitive fluorescent indicator (198b). The analytically useful range is 1–10 mM, the average error 0.5%. Another example is the titration of chloride or related halides with silver nitrate at pH 2.3 in the presence of acridine (199). Other indicators, such as quinine and phenanthridine, may also be used. The analytical principle in this case is dynamic fluorescence quenching of the indicator by both halide and silver ion.

Figure 3.45 shows the resulting plot obtained by following with a plain two-armed fiber bundle the fluorescence intensity changes of acridinium

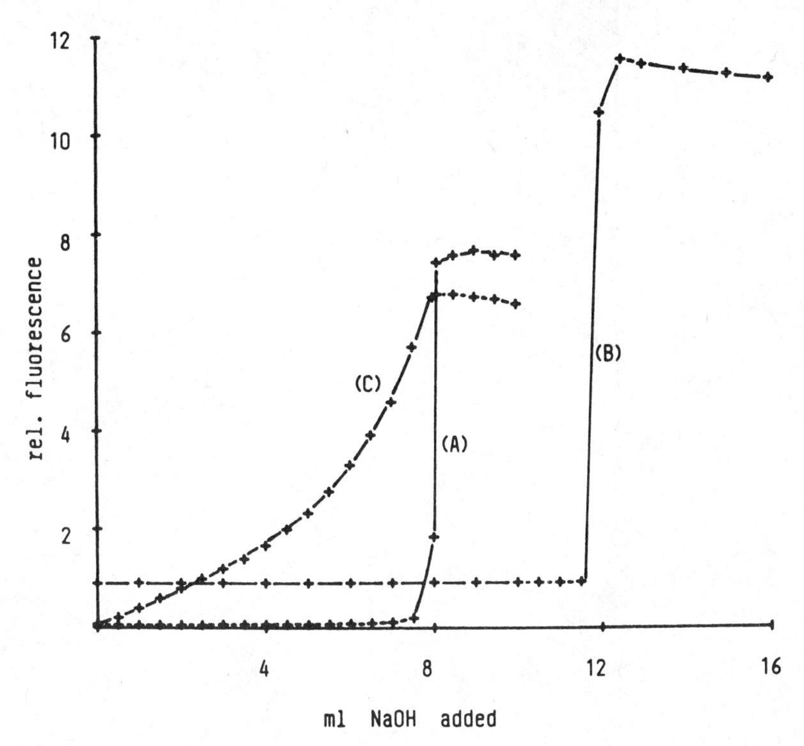

Fig. 3.44. Fiber optical titration: curve A, 5 ml of acetic acid (0.16 *M*); curve B, 5 ml of hydrofluoric acid (0.24 *M*); curve C, 5 ml of monobasic phosphate (0.16 *M*) with 0.1 *N* sodium hydroxide. Indicator: 2 μ*M* 1-hydroxypyren-3,6,8-trisulfonate (HPTS).

cation, when 0.1 *N* $AgNO_3$ is added to 0.01 KCl. At the beginning fluorescence is low due to dynamic quenching by chloride. As the titration proceeds, the concentration of chloride becomes smaller and reaches a minimum at the endpoint. At this stage practically all chloride is precipitated as AgCl and fluorescence quenching is not observed. Addition of excess silver(I), which acts as a quencher, results in a decrease in fluorescence again.

The endpoint is clearly evident in the plot, even in 2 *N* acid. In contrast to the popular Fajans method, where a *color change* is observed, there is an *intensity change* evident in this method. The standard deviation in the titration of 0.01 *N* chloride was $\pm 1.8\%$, and 0.001 *M* halides were titrated with 0.01 *N* $AgNO_3$ with practically the same precision. Other ions that have successfully been titrated by this method involve bromide, iodide, isothiocyanate, sulfide, and silver(I) ion. The endpoint is not affected by the turbidity of the solution, and surface potential effects as observed with Ag/AgCl electrodes were not found.

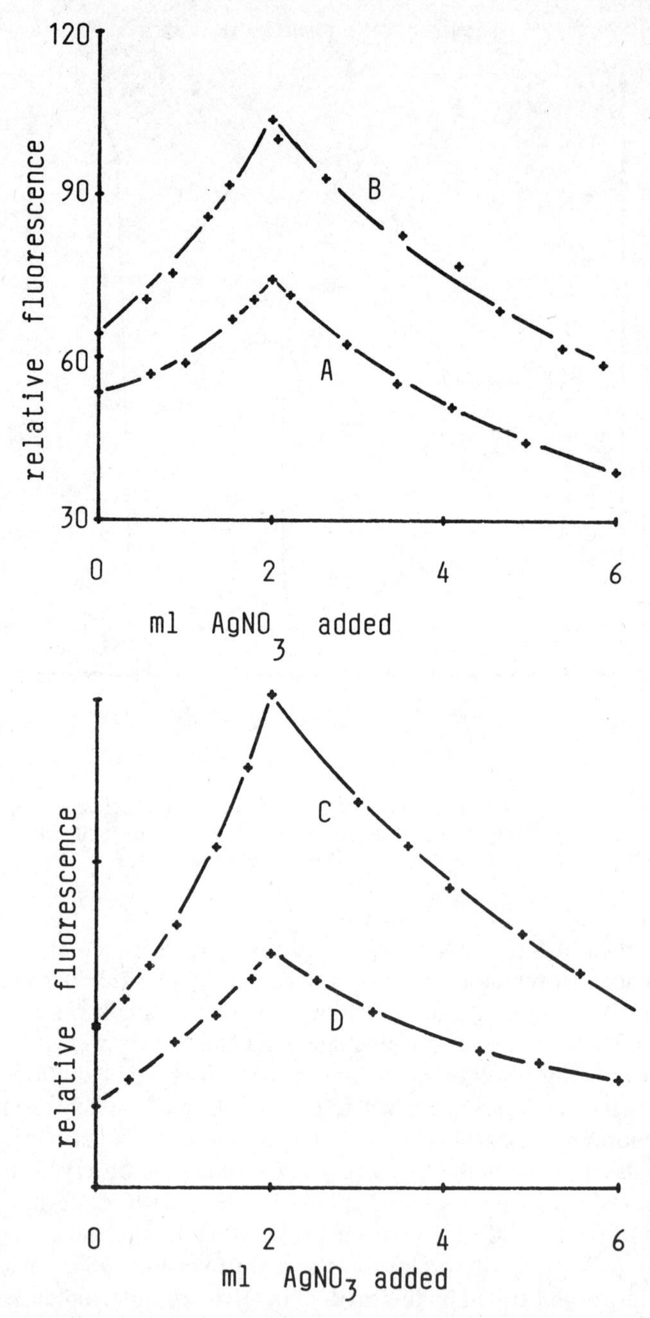

Fig. 3.45. Fiber optical titration of 20 ml 0.01 *N* potassium chloride with 0.1 *N* silver nitrate in aqueous solution of pH ca. 2.3 using acridine as an indicator. Excitation at 410 nm, emission taken at 480 nm. Acridine content: curve A, 1.6 mg; curve B, 2.9 mg; curve C, 5.4 mg; curve D, 9.9 mg. (From ref. 199).

Instead of adding the halide-sensitive indicator to the solution and following its intensity changes with a plain fiber, the fiber optical halide sensor described in Section 3.6.5 may be utilized to follow the course of argentometric titration. Although this may save the addition of indicator to each sample, the less efficient quenching of immobilized dye makes the shape of the titration curve less characteristic and, consequently, the endpoint less sharp.

3.6.7.3. Complexometry

The old problem of direct complexometric determination of aluminum(III) with ethylenediaminetetraacetic acid (EDTA) has successfully been solved by fiber optical fluorotitration using morin as an endpoint indicator (200). The fluorescent Al(III)–morin complex is slowly destroyed by addition of EDTA titer solution and reaches a low and constant value at the endpoint (Fig. 3.46). Aluminum at concentrations above 20 ppm may be determined by the method with good precision, even in colored or turbid solution, although the slow complexation kinetics may be annoying. Interestingly, the point of equivalence predicted by stoichiometry is different from the experimental value, resulting in an empirical but fully reproducible relation between EDTA consumption and Al(III) concentration.

Saari and Seitz (192a) have applied a sensor based on immobilized calcein to titrate Cu(II) with EDTA. Initially, fluorescence is very low because of total fluorescence quenching by Cu(II). At the endpoint, there is a sharp increase in fluorescence when Cu(II) is removed from calcein because it binds more strongly to EDTA.

Matsuo et al. (201) described the use of membrane photosensors which can be immersed into a test solution during acid-base or complexometric titrations. They were shown to be able to indicate the endpoint of the phototitration of copper(II) and zinc(II) with EDTA, and of strong acids with strong bases. Although not specifically coupled to fiber optics, they easily could be.

3.6.7.4. Redox Titrations

Cerimetry has the potential of being followed by fluorescence sensors, since cerium(III) is fluorescent, whereas cerium(IV) is not. The rather broad excitation and emission bands have peaks at 275 and 375 nm, respectively. Glass light guides with their UV absorption shifts the broad excitation maximum to 350 nm, and fluorescence is best observed at 400 nm. Since cerimetry is performed mostly in strong acid and elevated temperature, the

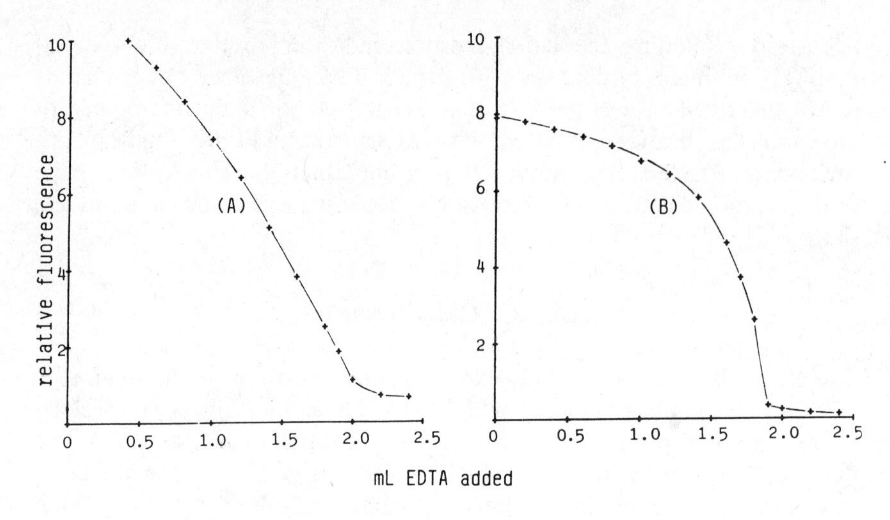

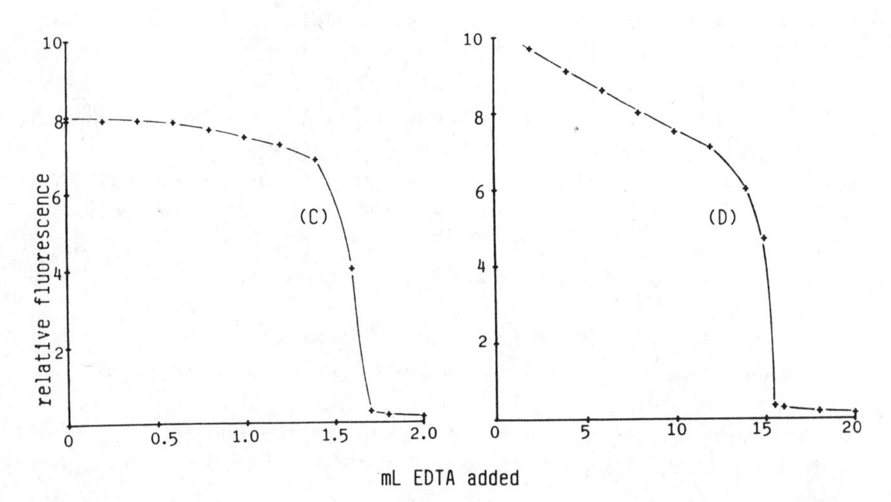

Fig. 3.46. Titration of aluminum(III) with EDTA using morin (10 μ*M*) as an indicator: (*a*) 0.96 ppm Al(III) with 0.001 *M* EDTA; (*b*) 9.6 ppm Al(III) with 0.01 *M* EDTA; (*c*) 96 ppm Al(III) with 0.1 *M* EDTA; (*d*) 960 ppm Al(III) with 0.1 *M* EDTA. Excitation at 420 nm, emission at 490 nm. The concentration range over with titration can be performed is very large, but at less than 20 ppm Al(III) the titration kinetics is very slow. (From ref. 200).

use of fibers offers particular advantages in following the course of the reaction, because they are resistant to both acid and heat.

The course of bromometric and iodometric titrations may be followed using the fluorosensors known for elemental bromine and iodine, or those for bromide or iodide (Section 3.6.5).

3.6.8. Sensors for Ionic Strength

Ionic strength (IS) is a measure for the total electrolyte concentration of a solution. Its knowledge is of interest in monitoring chromatographic eluents, since it provides a continuous record of all changes in buffer composition. Changes in IS of blood are observed as a result of perturbances in electrolyte homeostasis after saline infusion and myocardial shock. Generally, IS is known to affect the response of ion-selective electrodes (ISEs) and makes the major contribution to the differences between measured and true ion concentration.

According to Lewis and Randell, ionic strength I is defined as

$$\mathrm{I} = 0.5\ \Sigma\ C_i z_i^2 \tag{3.63}$$

with C_i being the concentration of each ion in solution, and z_i being its charge. I is known to affect the thermodynamic pK_a value of a pH indicator according to

$$\mathrm{p}K^{\mathrm{I}} = \mathrm{p}K_a^{\mathrm{th}} \pm 0.512 z^2 I^{1/2}/(1 + 1.6 I^{1/2}) \tag{3.64}$$

with pK^{I} being the pK_a of the indicator at ionic strength I. This equation shows that, in principle, all pH determinations by either photometry or fluorimetry are IS dependent. Unfortunately, its magnitude can frequently not be calculated from Eq. (3.63) since the composition of the sample solution is unknown.

The feasibility of fluorimetric IS determination has been discussed (137,138). A two-pH-sensor system was proposed with different IS–dependent pH response for each sensor unit. The two sensors display different pH values for the same solution, except when IS is zero. The difference (ΔpH) may be used for determination of IS.

A system consisting of two different indicators would put considerable constraints on both preparation of the two sensors and the optical layout. It has therefore been proposed (138) to measure the true pH by an electrode and to relate the apparent pH, as measured by the optode, to IS.

A more simple sensor was developed by utilizing the effect of varying microenvironment upon the IS dependence of an immobilized pH indicator (144). Thus when 7-hydroxycoumarin-3-carboxylic acid (7-HCC) is immobilized on porous glass, its pK_a (7.1) can be made practically IS independent by treating the glass surface with the silylating agent aminopropyl triethoxysilane (APTS). This is probably due to the fact that the immobilized indicator becomes surrounded by a large number of protonated aminopropyl groups, so that additional changes in electrolyte composition have only little effect (S_1 in Fig. 3.47).

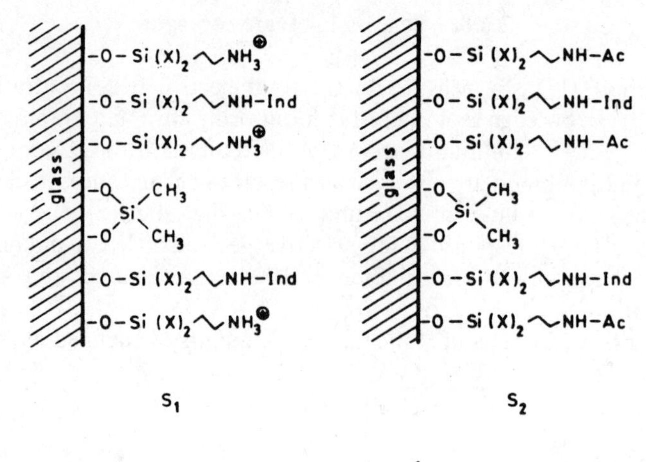

Fig. 3.47. Schematic of the surface of the two sensors used for determination of ionic strength via differences in their pH display. Sensor 1 (S_1) has a surface in which the indicator is surrounded by ammonium groups so that changes in the ionic strength of the analyte solution in contact with the surface have little effect on the pK_a value of the immobilized indicator. In sensor 2 (S_2) the ammonium groups are removed by acetylation. The indicator is fully exposed to changes in the ionic strength of the solution. (From ref. 144.)

If this sensor is treated with acetic anhydride, the protonated amino groups will become acetylated, and no local charge will be present any more (S_2 in Fig. 3.47). Consequently, the pH sensor becomes highly sensitive toward changes in IS. The apparent difference in the pH of the same solution as displayed by the two sensors can be related to IS. Figure 3.48 shows the fluorimetric titration curve of sensor 1 and sensor 2 at various IS values. The response of sensor 1 is independent of IS. The differences, shown at the right, are most expressed near the pK_a of the indicator, but vanishingly small outside the pH range 6–8.

Empirically it was found (202) that the pK of sensor 2 at IS 0.1 M was pK^{th} − 0.1075. Hence, at unknown IS, ΔpH is equal to 0.1075 − log γ. Since log γ is related to IS by

$$\log \gamma = -0.512 z^2 I^{1/2}/(1 + 1.6 I^{1/2}) \tag{3.65}$$

one gets

$$\Delta pH = 0.1075 - 0.512 I^{1/2}/(1 + 1.6 I^{1/2}) \tag{3.66}$$

In practice, a correction factor k has to be added to the right term of Eq. (3.65) to account for surface and neighbor group effects.

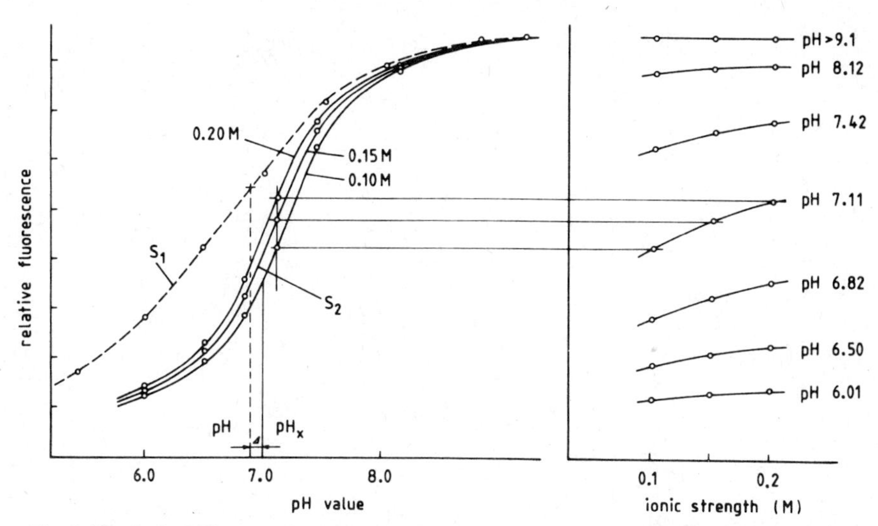

Fig. 3.48. Left: Differences in the response of the ionic strength-independent pH sensor (S_1) and the ionic strength-dependent sensor (S_2) toward pH at ionic strength 0.10 M, 0.15 M, and 0.20 M NaCl. The difference ($\triangle$) displayed by the two sensors can be related to ionic strength via Eq. (3.66). Right: Effect of ionic strength on the apparent pH display of a highly ionic strength-dependent pH sensor (S_2 in Fig. 3.46). The differences are most expressed near the pK_a of the indicator. (From ref. 144.)

Figure 3.49 shows a plot of ΔpH of two sensors with different microenvironments. The IS-dependent sensor was calibrated at I = 100 mM. Curve B was obtained from Eq. (3.66). The correction factor was chosen so as to fit the calculated curve to the experimental curve.

This sensor combination for determination of IS is simple in that (1) the preparation of the two sensor layers differ only in the treatment with acetic anhydride, and (2) the optical system can be the same for both. It allows a simultaneous determination of IS and pH at pH values within ± 1.2 units of the pK_a of the indicator.

3.6.9. Monitoring Water Flow and Quality

3.6.9.1. Water Flow

The use of fluorescent tracers in groundwater studies is important for following the route of the water flow from specific injection sources. In recent work (6) tracers such as rhodamine 6G have been shown to be detectable after laser excitation over a distance of 100 m in 10-ppb concentrations. By increasing the laser power to a few milliwatts, tracers can be detected at around 1 ppm over distances as long as around 1 km.

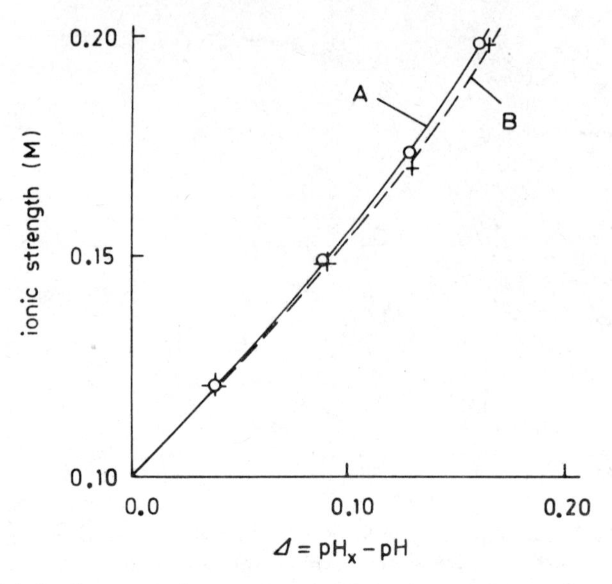

Fig. 3.49. Relation between ionic strength and difference in pH display of an ionic strength sensor based on immobilized 7-hydroxycoumarin which is embedded in two different microenvironments (see Figs. 3.46 and 3.47): curve A, experimental values; curve B, calculated with Eq. (3.66). (From ref. 144.)

3.6.9.2. Mineral Oils and PAHs

Polycyclic aromatic hydrocarbons and oil spills can now be detected with airborne fluorosensors using UV excitation, preferentially from a laser source. The field performance of a laser fluorosensor for the detection of oil and dye spills was described (18). It consists of two main subsystems, a spectrometer and a lidar altimeter. It is capable for operation in full daylight and has a 20-nm resolution in the range 380–700 nm. A correlation technique clearly differentiates between two crude oils, dye spills, and background of ocean water.

In an effort to remotely monitor potentially hazardous effluents from developmental coal processing operations (77), two major considerations arose in the design of aerial UV fluorosensors. The rapid increase of solar irradiance with wavelength indicates that for daylight operation, the fluorosensor detection bandpass should be at as short a wavelength as is consistent with the emission of the material to be detected. Atmospheric absorption, however, increases sharply at wavelengths below around 300 nm, causing attenuation of both the exciting light and emission. These factors led to the selection of fluorosensors using either nitrogen (337 nm) or krypton fluoride lasers (249 nm).

The nitrogen laser system exhibited high selectivity during nighttime operation in flight tests conducted at altitudes from 105 to 308 m, although, as anticipated, the detectivity decreased with increasing solar background (202). Areal testing of KrF laser sensors showed this type to be more promising (77). Its two-spectral-channel system (295 and 395 nm) successfully deteted a variety of pollutant-type materials on the ground at night and full daylight.

The probably most useful fiber sensor for oil pollution monitoring (203) consists of a fiber without cladding, inserted through a stainless steel capillary, and coated with an organophilic material (a long-chain alkyl group). The input radiation is at 632.8 nm from a low-power laser. The normal total internal reflection is changed, as the pollutant is absorbed by the organophilic cladding, because of a change of its refraction index. The sensor can detect diesel oil (17 $mg{\cdot}L^{-1}$) and crude oil (3 $mg{\cdot}L^{-1}$) in water. Of course, it is not very selective.

A cumulative probe for detection of reactive hydrocarbons in the environment is based on a porous catalytic matrix that generates opaque tars after absorption which cause light attenuation. (204). It exploits an irreversible chemical reaction, but may possibly be modified for level monitoring by the use of photobleaching.

An optical sensor for observation of exhaust gases has been described in a patent (205). Deposited particles such as soot on the windows of optical sensors are catalytically oxidized by metal oxide particles deposited on the window surface.

3.6.9.3. Phytoplankton and Chlorophyll

Airborne laser fluorosensing of chlorophyll (Chl) and phycoerythrine promises to be a valuable technique for quantitation of phytoplankton abundance and distribution in water (19,206). To enhance and control photosynthesis, most free-drifting algae contain a number of so-called primary absorbing pigments. The absorbed excitation energy is transferred to Chl A, whose emission centered at 685 nm is predominatly observed. Alternatively, the emission of the strongly fluorescent photoreceptor phycoerythrine may be measured.

Compared with passive radiometric techniques, the fluorosensor methods for phytoplankton have several attractive features which are unique. These include relative insensitivity to daytime or cloudiness, specifity of response to Chl, and the potential for depth profiling using range-gating techniques. On the other hand, they are sensitive to other factors, such as energy transfer, water composition, algae species, and sediments (206,207).

A method that has evolved to circumvent the lack of understanding physical processes in remote Chl sensing is the normalization of fluorescence signals by the OH strech water Raman signal occurring at a frequency shift of 3418 cm^{-1} (207). Nearly linear relationships between normalized fluorescence and in-situ Chl concentration (0.2–20 $\mu g \cdot L^{-1}$) were observed by Brislow et al. (208) under 470-nm excitation and by Hoge and Swift under 532-nm excitation (209).

Linear vertical gradients in Chl concentration may be evaluated by the same method by applying a combination of stochastic and analytical techniques. This makes it possible to model lidar systems with realistic geometric constraints and reduced computer resources (210). In this approach the media are represented by combinations of layers with constant Chl concentrations within each layer, but varying Chl concentration in different layers. Statistically significant differences can be seen under certain conditions in the water-Raman-normalized fluorescence signals between homogeneous and nonhomogeneous cases when the Chl concentrations vary linearly with depth in the water medium.

The green and brown algae fractions of phytoplankton populations in ocean waters were estimated by an index which is the ratio of Chl A fluorescence excited by narrow band light at 539 nm, to that excited at 454 nm (211). The fluorescence at 685 nm was detected by a remote airborne fluorosensor.

A simple and sensitive algae fluorosensor based on fiber optics was described by Lund (32). It was used for in situ studies of natural sweet water. Excitation spectroscopy of four algae species revealed differences in their excitation spectra, which therefore may yield more specific and more accurate information than data obtained under single wavelength excitation.

3.6.9.4. Industrial Waste

Natural ground and creek water is known to exhibit blue emissions due to the presence of fluorescent humic and fulvic acids (13). Lignin sulfonate pollution manifests itself in an increase in the fluorescence of wastewater. In many cases it is strong enough to be of use in continuous fluorimetric sensing of pollutants. Lakshman (212) has analyzed a number of soil and runoff samples for the nature and stability of the intrinsic fluorescence. A strong correlation exists between intensity and water quality parameters such as total carbon, total organic carbon, and total inorganic carbon. It can be used in the quantitative measurement of agricultural pollution.

Chudyk et al. (213) report on the results obtained in a test of remote fluorescence analysis of groundwater contaminants using UV laser light sources along with fiber optics. Several priority pollutants, such as phenols, toluene, and xylenes, as well as naturally occurring humic acid were readily detected over a distance of about 5 m. Detection limits were at or below the ppb level.

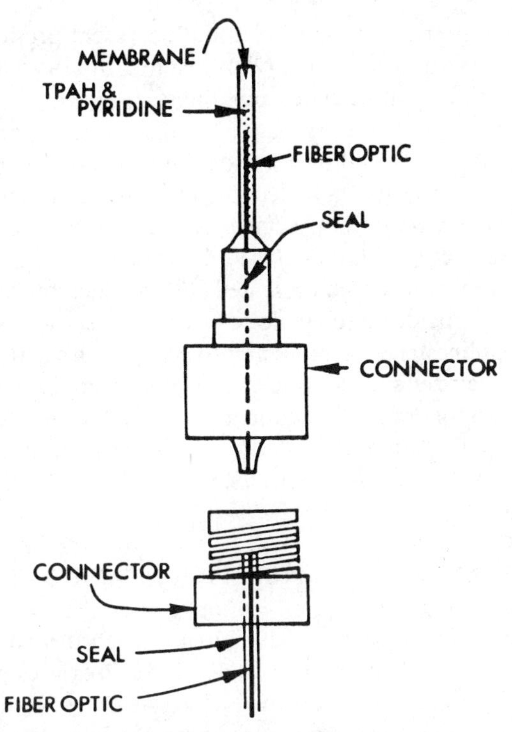

Fig. 3.50. Configuration of the organic chloride optrode. Volatile organic chlorides such as chloroform diffuse into the pyridine-containing compartment. In the presence of a little tetrapropylammonium hydroxide (TPAH) the chlorides undergo a Fujiwara reaction with pyridine. This yields a product with red fluorescence which is viewed by the glass core fiber. (Reprinted from ref. 17 with permission.)

3.6.9.5. Groundwater Quality

Chlorinated hydrocarbons are frequently encountered in groundwater, drinking water, and wastewater. A method for the detection of organic chloride has been developed (214). It makes use of the Fujiwara color reaction. This is a general assay for organic compounds that have two or more chlorides (215). Reaction of warm alkaline pyridine with polychlorinated organic compounds yields a red fluorescent chromophore.

The sensor consists of a single 200-μm glass core fiber to transmit exciting light (the argon laser 488-nm line) to the optode and to carry the fluorescence back to the spectral sorter. The sensing unit (Fig. 3.50) is fixed to the tip of the fiber with a 400-μm glass capillary. Pyridine is kept in position by the use of a Mylar membrane which keeps the pyridine and water in but passes the organic chloride. A 10 *M* hydroxide solution keeps the pyridine basic, since the Fujiwara reaction yields a yellow, nonfluorescent product when it turns acidic. Preliminary measurements (214) resulted in sensitivities of less than 1 ppm for chloroform and related pollutants.

Although many pesticides (organophosphates) are fluorescent by themselves, the spectra are spread over a wide range. A number of measurements would therefore be required to detect the fluorescent ones with a plain fiber. A more general way to detect pesticides is to group them by their biological effect, since they act as cholinesterase inhibitors.

A reservoir sensor has been presented (17) in which the enzyme-inhibiting properties of pesticides are exploited. In this sensor type a layer of immobilized cholinesterase is brought into contact with a fluorogenic enzyme substrate and the resultant fluorescence level measured through a fiber. The presence of inhibitors retards reaction and reduces fluorescence. The signal is referenced to the Raman backscattering due to the fiber itself, thus normalizing out the source, detector, spectral sorter, and system fluctuations.

3.7. SENSORS FOR CLINICAL ANALYTES

Numerous fiber optical sensors have already been described that measure physical parameters of the human body (216). Pressure, physiologic flow, strain, motion, displacement, or flow velocity can be monitored by optical methods such as variable reflection, laser Doppler velocimetry, optical holography, or diffraction. In this section the application of fluorosensing methods for the determination of *molecular species* encountered in clinical and biomedical analysis will be described.

The sensors for pH, pO_2, pCO_2, and chloride as discussed in Sections 3.6.1 to 3.6.4 have been designed mainly for biomedical application. Some of them have found their first commercialization in blood pH and blood gas analyzers. A serious problem for many optrodes is a limited long-term stability. This, however, plays no role in the case of disposable optrodes which are in use only for the duration of an operation.

Disposable sensor heads for clinical analytes seem to be the most promising candidates for practical use at present. Current problems of photobleaching and signal drifts do not severely limit the performance of these sensors. Actually, they are not intended to provide very precise absolute analytical data. Rather, it is the relative change of a parameter that has to be recognized as soon as possible during the course of operations on the critically ill.

An interesting feature of fiber optical sensing is the possibility of combining several minute fiber sensors (typically 100 μm in diameter) to form a bundle of fibers applicable for simultaneous sensing of, for example, physiological pH, oxygen, electrolytes, anesthetics, glucose, creatinine, pressure, and flow rate. The sensors will not necessarily rely on fluorimetric methods exclusively, since other optical methods may allow better monitoring in certain instances.

3.7.1. Sensors for pH, Oxygen, Carbon Dioxide, and Electrolytes

3.7.1.1. Sensing pH Values

Most pH sensors described in Section 3.6.1 are suitable for measuring physiological pH values. In fact, they have been developed mainly for this purpose. They exhibit sufficient precision to allow a resolution of ± 0.02 pH unit or better in the near-neutral pH range. Some have-already been tested with respect to the in-vivo or ex-vivo performance.

An evaluation of the Peterson et al. pH sensor (62) in sheep demonstrated its aptitude for in-vivo blood pH measurement. Although not identical with the values obtained with an invasive microelectrode and with data obtained on withdrawn blood, the agreement between numerical data and trends is very good. It was not possible to say which of the measurements was most correct. This demonstrates that the fiber optic method is generally useful for blood pH measurements in vivo, and gives as good an indication of pH levels as electrode methods. The blood pH of a ewe sheep as determined by an electrode, a blood gas analyzer, and a fiber optic device is shown in Fig. 3.51.

An even more precise version of this sensor type was specially designed for in-vivo application consisting of a 25-gauge hypodermic needle with an ion-permeable side window and 75-μm fibers. It had a response time of 30 s. Along with computerized signal processing and three-point calibration, a precision of 0.001 pH unit was achieved. The sensor was used in studies of the transmural pH gradient in myocardial ischemia (133,217a) and for continuous measurement of intravascular pH (217b).

3.7.1.2. Sensing Oxygen

The oxygen sensors described in Section 3.6.2 have been shown to exhibit excellent in-vivo performance. Thus, an in-vivo evaluation of the fiber optical oxygen sensor based on dynamic fluorescence quenching (37) was done in the blood stream of a 31-kg ewe sheep after the sensor had been inserted through a Teflon catheter. Figure 3.52 shows the continuous oxygen record of the fiber sensor during a 3-h experiment compared with the blood values.

Oximetry has been performed with plain fibers utilizing the difference in the absorption spectra of oxyhemoglobin and hemoglobin. Because of the high absorbance of blood at analytical wavelengths below 600 nm practi-

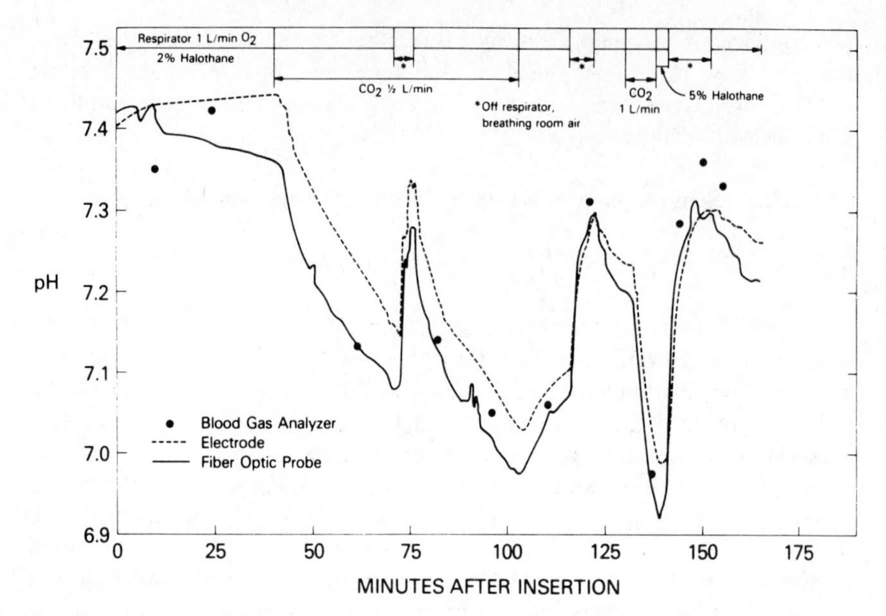

Fig. 3.51. In-vivo evaluation of a pH sensor in sheep undergoing varying respiration conditions. The blood pH analyzer data were obtained ex-vivo, the electrode data with an inserted microelectrode. [Reprinted from ref. 62 with permission. Copyright (1980) Americal Chemical Society.]

cally, all work has been performed at wavelengths close to or beyond the red end of the visible spectrum. The advent of red and infrared LEDs along with red-sensitive photodiodes has considerably contributed to stimulate these experiments.

Generally, the analytical wavelength is set to between 600 and 750 nm. Internal standardization is achieved by measuring reflection at around 930 nm, which is one of the isosbestic wavelengths in the absorption spectra of hemoglobin and oxyhemoglobin (218).

Aside from oxygen partial pressure, cardiac output is another important parameter in the diagnosis of heart diseases. Polyani and Hehir 67 have developed a method for measuring cardiac output which seems to be the first application of fiber optics in bioengineering. A dye was injected into the bloodstream and its appearance measured at different locations by inserting a small fiber catheter. Cardiac output and vascular oxygen saturation were later measured via catheters by various other groups (219).

The theory, design, and evaluation of an optical sensor for the measurement of blood oxygen saturation and hematocrit was presented (220). It is said to meet the requirements of a large dynamic range even in the deep-lying

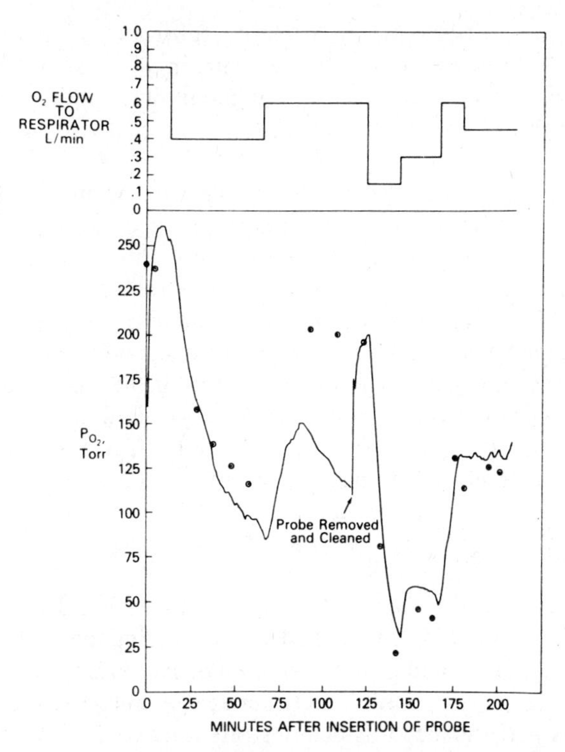

Fig. 3.52. In-vivo test of an oxygen fluorosensor in an ewe. The breath gas composition was varied (top) and the change in oxygen pressure in blood followed with an invasive catheter. The data points are values obtained ex-vivo with a blood gas analyzer. [Reprinted from ref. 37 with permission. Copyright (1984) Americal Chemical Society.)

vessels, a high reliability and stability over extended periods of tifme, and biocompatibility. It is based on the measurement of diffuse reflectance by a ratio method using 3 LEDs (one red, two infrared) as light sources placed on an implantable $6 \times 1.6\ mm^2$ area.

A solid-state linear diode array spectrometer has been applied for the simultaneous determination of relative amounts of hemoglobin, carboxyhemoglobin, and oxyhemoglobin in whole blood (221). The fiber is incorporated in a hypodermic syringe needle and the analytical wavelength is set to between 520 and 620 nm. Derivative spectra, signal averaging, and least-squares calculations were applied to extract the maximum possible information from the spectra.

A fluorosensor for monitoring blood gases and pH in an extracorporeal loop is commercially available (222a). Essentially, it is based on fluorosensors described in Sections 3.6.1 to 3.6.3. Arterial or venous oxygen and carbon dioxide pressure, pH, and temperature can continuously be determined during cardiopulmonary bypass surgery. The system consists of a microprocessor-based instrument, bifurcated fiber optical cables, and a disposable sensor head with fluorescent spots sensitive to the respective analytes.

The same company has recently published first results obtained with an invasive catheter for continuous monitoring of oxygen, CO_2, and pH. The sensor head is only 1 mm thick and consists of three single fibers with the appropriate chemistry at each end. The results compare excellently with the data obtained with conventional blood gas analyzers (222b,c).

Pyrenebutyric acid in fluid solution and entrapped in a 6-μm-thick PTFE layer was used in a sensing unit for monitoring the surface oxygen partial pressure of the isolated guinea pig heart (223c) via fluorescence quenching. The system had a response time of 2–3 s only. It is a modification of the oxygen sensor developed in the same group (see Section 3.6.2).

3.7.1.3. *Others*

The principle of the fiber optical pH sensor (62) led Vurek and colleagues to devise a CO_2 sensor (173,223a,b). Instead of coupling the pH indicator dye to an insoluble polymer, and isotonic solution of salt, bicarbonate, and dye was used which was covered with a CO_2-permeable silicone rubber membrane. The sensor's performance was demonstrated invivo.

The carbon dioxide fluorosensor based on spectral changes of 4-methyl umbelliferone (16) was exploited to monitor the changes in the CO_2 pressure of the guinea pig heart (223c). The device is similar in design to the one shown in Fig. 3.31. The PTFE membrane is brought into tight contact with tissue (area about 28 mm^2) to permit the gas to enter the reaction chamber easily. The filling solution consists of a 0.01 M indicator solution in 1 μM bicarbonate plus 10% dimethylformamide. Agarose was also added for stabilization. The response time is 3–4 s for the 90% value.

Two kinds of fiber optical sensors for ammonia have been described in Section 3.6.4. In addition, an evanescent wave technique was applied for vapor-phase determination of ammonia in blood and serum (224). It utilizes the ninhydrin reaction occurring in the polymer coating of the fiber. The probe is applicable to clinical determinations normally carried out in the vapor phase, but works irreversibly. A linear relationship exists between absorbance and ammonia concentration in the clinicially useful range 0–4.0 $\mu g \cdot mL^{-1}$. Comparison with the reference method showed a correlation coefficient of 0.92.

A fiber optic reservoir sensor was developed for studying calcium(II) transients in the cerebral cortex of cats (225). The calcium-binding enzyme aequorin in introduced through a thin channel into the extracellular space. The fluorescence emitted by aequorin after interaction with Ca(II) is monitored with an optical fiber and is related to the actual Ca(II) activity. The probe was used to study epileptic seizures.

3.7.2. Sensors for Glucose, Metabolites, and Coenzymes

Glucose is probably the most frequently assayed low-molecular nonionic analyte in clinical chemistry, but only recently, reliable sensors have been developed for this species. At present, the lack of a suitable glucose sensor can be regarded as the rate-limiting step in the development of an artificial pancreas.

The first fluorosensor for glucose was described in 1980 and is based on the measurement of oxygen consumed during its enzymatic oxidation by glucose oxidase in the presence of catalase (155,226,227). The enzymes were covalently immobilized on cross-linked bovine serum albumin, spread on 12-μm cellophane, and covered with a 12-μm Teflon membrane. These reagent layers were covered with an oxygen-sensitive fluorescent layer [i.e., a solution of pyrene-butyric acid (PBA) in a viscous solvent].

When glucose solution is brought into contact with the sensing multilayer, an increase in fluorescence intensity is observed as a result of oxygen consumption. The response to glucose is proportional to its concentration, provided that (1) the oxygen supply for the first enzymatic reaction remains constant and (2) glucose does not saturate the enzyme. Since the system is based on diffusion processes, the response does also depend on factors such as thickness, viscosity, enzyme concentration, and diffusion characteristics of all layers. The performance is most sensitive to oxygen supply from outside the sensor. The effects of substrate saturation, glucose oxidase concentration, and layer thickness were reported and found to be in agreement with a model of slow diffusion that determines response.

A glucose sensor exploiting the phenomenon of competitive binding of substrates to an enzyme has also been described (228,229). Fluorescein-labeled dextran competes with glucose for binding to concanavalin A which is immobilized on sepharose. Figure 3.53 shows a schematic of the device.

The end of a bifurcated fiber optic fits into a hollow fiber with a plug on the end. The substrate is fixed on the walls of the fiber and thus in a position out of its numerical aperture. Therefore, it cannot be seen by the fiber. Glucose can freely diffuse through the hollow fiber or a dialysis membrane, which is impermeable to the large dextrane molecules. Increasing glucose concentration displaces the labeled dextrane, causing it to diffuse into the illuminated solution volume. Thus fluorescence intensity as seen by the fiber follows the glucose concentration.

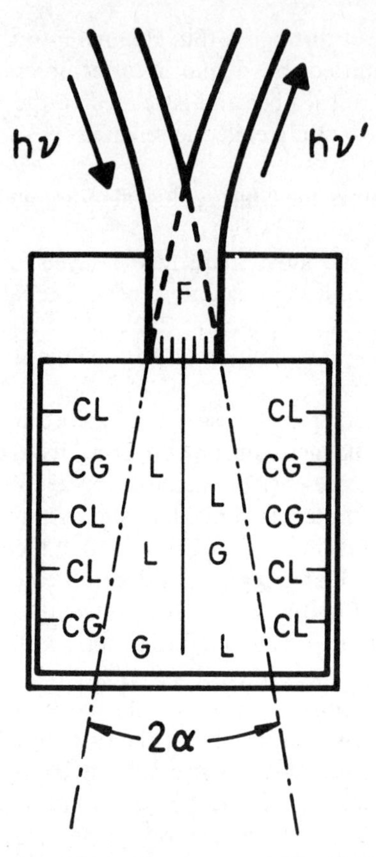

Fig. 3.53. Schematic of a glucose sensor based on competitive binding to concanavalin A. The geometry of the sensor is chosen in a way that the fiber "sees," within its numerical aperture 2α, only the labeled dextran in solution, not the bound one. G, Glucose; L, labeled dextran; C, concanavalin A immobilized on the walls of the hollow fiber; F, bifurcated fiber. (Redrawn from ref. 228.)

This sensor principle is of particular interest because the design idea has broad potential for application to any analytical problem for which a specific competitive binding system can be devised.

Co-immobilization of pH-sensitive dyes and proton-consuming or proton-generating enzymes to a thin transpoarent membrane sandwiched between a LED light source and a silicon photodiode-amplifier detector system permits the characteristic color change observed on contact with a transparent substrate to be measured. Goldfinch and Lowe (65) used bromcresol green immobilized on cellophane to measure the decrease in pH produced by the catalytic action of immobilized glucose oxidase on the oxidation of glucose. The glucose optoelectronic sensor responds linearly to glucose over the range 0–70 mM and displays a half-life of 7–8 days.

The same principle was applied to sense urea (65). Immobilized bromthymol blue changes its color because of an increase in pH that is produced by the activity of the enzyme urease. The sensor respons to urea concentration in the range 0–40 mM linearly up to 10 mM and with an output voltage change of approximately 125 mV·min^{-1} at 10 mM urea. The sensor is completely reversible and is regenerated with pH 7.0 buffer. The half-life is largely conditioned by the nature of the enzyme and the sensor preparation, being on the order of 3–4 weeks.

A device similar to that used by Uwira et al. (227) for glucose was used by Lübbers et al. to determine lactate in an enzyme optode (230). Immobilized lactate oxidase converts lactate into acetate (plus CO_2 and H_2O_2) and consumes 1 mol of oxygen. Two oxygen-sensing layers were used to measure the difference in the oxygen partial pressure that is enzymatically produced within the complex sensor membrane. Two oxygen-sensitive indicators with different spectral properties (PBA and perylene) were applied. Their excitation and emission maxima (342/395 and 436/490 nm, respectively) are sufficiently separated so that both sensor layers can be submitted to simultaneous fiber fluorimetry. An almost linear ratio between lactate concentration and normalized fluorescence intensity was found in the concentration range 0–1 mg·mL^{-1}.

Oxygen is consumed when alcohol dehydrogenase oxidizes alcohol. This enables the continuous assay of ethanol by measurement of the oxygen gradient across a sensing membrane (231). Layers of cellulose, immobilized alcohol oxidase, Teflon, and a viscous solution of PBA (in that order) are spread on glass. Alcohol enters the sensing membrane through the cellulose layer and is oxidized to produce an oxygen pressure gradient across the enzyme layer.

As oxygen is consumed during the reaction, quenching of the indicator by oxygen becomes less efficient. The resulting increase in fluorescence intensity can be related to the ethanol concentration ([EtOH]) by the following modified Stern–Volmer equation:

$$F^0/F = 1 + k_d[\mathrm{Q}] - K'[\mathrm{EtOH}] \tag{3.67}$$

where F^0, F, K_d, and [Q] (equal to $\alpha p O_2$) have the same meaning as in Eq. (3.29). K is a parameter specific for the enzymatic layer. The relation does, however, hold only if

1. Oxygen is continuously available from the substrate solution side.
2. The enzyme is not overloaded.

3. A diffusion equilibrium is established.
4. Acetaldehyde and hydrogen peroxide (formed during reaction) do not retard the enzyme activity.

In a short note it was also stated that the oxygen optode allows determination of xanthine and cholesterol by immobilizing the appropriate oxidase (232). Unfortunately, no experimental details were given.

Plain fibers were used to measure the intrinsic fluorescence of various low-molecular-weight biomolecules. The use of light guides in direct contact between a fluorimeter and the animal brain has made possible experiments with the conscious mobile animal (233). A fiberoptical multiwavelength time-sharing apparatus afforded flexibility and versatility in measuring absorption, reflection, or fluorescence at four wavelength channels (234). The simultaneous measurement of the surface fluorescence of NADH and oxidized flavoproteins provides a promising approach to estimate the redox state and oxygen supply of perfused organs. In later work the construction of much smaller trifurcated lightguides containing one or more strands of 25- or 80-μm-diameter optical fibers in each channel has been reported. They were used to measure the NADH fluorescence in various biological materials (235).

Coleman et al. (63) determined bilirubin in serum with a fiber optical system terminated in a 19-gauge hypodermic needle. The sensor was expected to be applicable to the photometric determination of other important clinical analytes such as drugs, toxins, and biomolecules as well, but the limited selectivity and sensitivity of spectrophotometry will possibly limit the scope of the method.

3.7.3. Sensors for Proteins, Antigens, and Antibodies

A solid-phase optoelectronic sensor for serum albumin has been described (64). Bromcresol green, covalently attached to a cellophane membrane, is sandwiched between a red LED (λ_{max} 630 nm) and a silicon photodiode. Adsorption of serum albumin to the membrane at pH 3.8 causes a characteristic yellow to blue-green color change and can be monitored as a fall in the output voltage of the detector system. The response is reproducible and linear over the range 5–35 $mg \cdot mL^{-1}$ albumin with precision $\pm 1.4\%$.

A simple fiber can be used to monitor the adsorption of chemical species such as proteins onto an optical fiber core (58). Fluorescence is induced by the evanescent wave field of the light propagating in the core. For example, the intrinsic fluorescence of the amino acid tryptophan in immuno-γ-globulin (IgG) as well as the fluorescence of fluorescein-labeled IgG was detected and used to calculate the amount of adsorbed protein on hydrophilic glass and quartz surfaces. Fluorescent, nonadsorbing dextrane came to use as a calibration molecule which approximates the diffusion properties of the protein. Remote spectroscopic sensing of adsorption of rhodamin B-labeled IgG at the tip of a 600-μm fiber optic was also described (236,237).

Fluorescence detection of antibody–antigen reactions has been demonstrated by using the evanescent wave sensing technique. The emission of fluorescein-labeled antibody bound to a hapten–protein conjugate adsorptively immobilized on a quartz plate in contact with the antibody solution was monitored (238). The quartz plate acts as a waveguide. The presence of any free hapten in solution reduces the amount of antibody free to bind to the surface and thus reduces the fluorescence signal. Following the decrease in fluorescence gives a measure of the concentration of free hapten being present. Because of its low penetration depth the evanescent wave is interacting mainly with the immobilized material rather than with the bulk solution. It therefore reports the binding processes occurring at the interface. The technique is simple, fast, and has high intrinsic sensitivity and specifity. Free morphine at a concentration of 2.10^{-7} M is readily detected.

Recently, an examination of the binding kinetics of rhodamine-labeled antibodies to dinitrophenol, also immobilized to a quartz light guide, was presented (239). In a similar experiment rhodamine-labeled IgG and insulin were shown to adsorb nonspecifically to serum albumin-coated fused silica with both reversible and irreversible components. The characteristic time of the most rapidly reversible component measured was around 5 ms and is limited by the rate of bulk diffusion. The binding was followed by fluorescence which, collected by a microscope from a surface area of about 5 μm^2, spontaneously fluctuates as the solute molecules randomly bind to, unbind from, and/or diffuse along the surface in chemical equilibrium.

The general characteristics of these immunoassay systems includes kinetic monitoring of the immunological reactions without major interferences from the bulk solution and without a formal spearation step of antibody bound from free antigen. The approaches were extended by attaching the specific antiserum to the surface of the waveguide and monitoring the sequestration of antigen by the labeled antibody (57).

Theory, instrumentation, and preliminary application of the evanescent wave probing technique specifically for the case of human IgG were described in some detail (240a). The antibody is covalently immobilized onto the surface of either a planar microscope slide or cylindrical waveguide made of fused quartz. The reaction of immobilized fluorescein-labeled antibody with an antigen in solution is detected by the evanescent wave component of the light beam.

Specificially, methotrexate (MTX, a cancer drug) was detected by the reduction in light transmission as the molecule was bound to anti-MTX

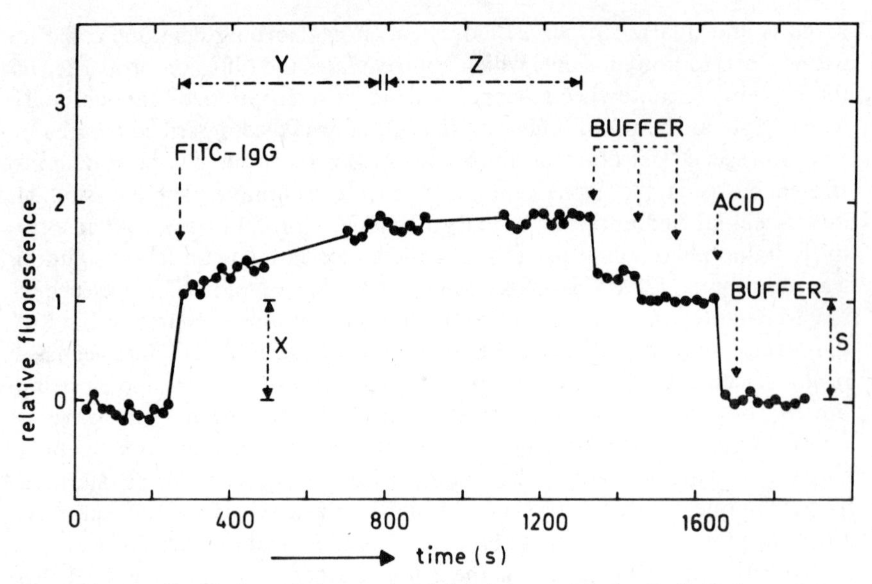

Fig. 3.54. Specific binding of fluorescein-labeled immunoglobulin G (IgG) by antiserum immobilized at a glass–liquid interface of a quartz slide. The binding process is monitored by the increase in fluorescence, which is excited by the evanescent wave of the beam propagating in the slide. When labeled IgG is added, there is an increase in fluorescence (X) due to free molecules fluorescing close to the waveguide surface. Over distance Y a plateau level Z is reached. After removing supernatant IgG with buffer, an intensity S remains which is an absolute measure of specific binding. Treatment with strong acid (0.01 M hydrochloric acid) removes bound IgG and makes the probe reusable. (Redrawn from ref. 55.)

serum immobilized on the liquid waveguide interface. Flushing the slide with 0.01 N hydrochloric acid disrupts the antibody– antigen binding and makes the probe reuseable.

IgG was measured by a sandwich immunofluoroassay. First, immobilized sheep antiserum was incubated with antigen to form one half of the sandwich. After washing out excess IgG the solid-phase complex was reacted with FITC-labeled antiserum and the reaction monitored for 15 min.

Figure 3.54 shows the detection of the binding process on the surface of a slide. Following the injection of FITC-labeled anti-IgG, a rapid increase in fluorescence intensity is observed (X). This is due mainly to free molecules fuorescing within the penetration depth of the evanescent wave. Second, a slower increase in binding over the next 500 s is evident, reaching a plateau after approximately 600 s. The curve obtained when no antigen was present shows fluorescence because of the presence of free molecules, but lacks the rapid binding event.

A no-label, homogeneous optical immunoassay for human IgG was

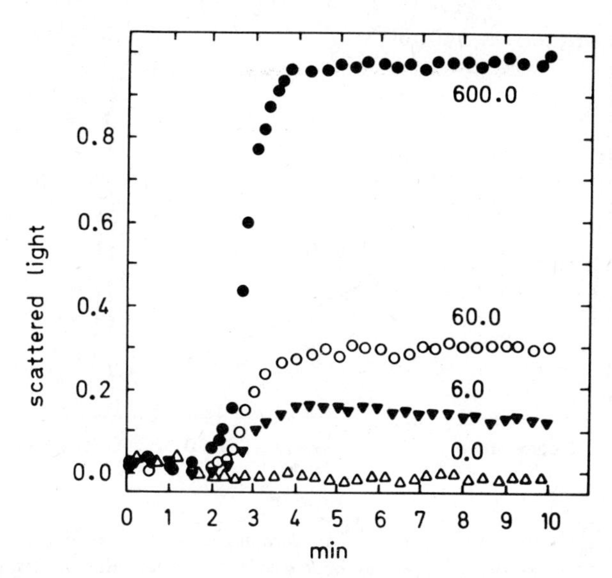

Fig. 3.55. Generation of light-scattering signal at the surface of an optical waveguide bearing immoblized antibody. IgG was added after 2 min in concentrations ranging from 0.0 to 60.0 μg·mL^{-1}, and scattering monitored at 440 nm. (Redrawn from ref. 241.)

developed and tested (240b) (Fig. 3.55). Following adsorption of antiserum to the surface of an optical waveguide, the immobilized antibody was then reacted with a solution containing antigen. The reaction was detected utilizing the evanescent wave component of a light beam totally internally reflected within the waveguide. The growing antigen–antibody layer resulted in an increase of scattered light which was monitored kinetically.

Optical waveguide techniques have a great potential for studying immunoreactions and performing immunoassays (241). Several patents covering the subject have been filed (242–244). The use of an energy-transfer fluorescence system may reduce background signal since fluorescence occurs only when the two antibodies involved in the immunoassay are in close proximity (i.e., bound to the same antigen molecule). Tromberg et al. (250) describe a fiberoptic chemical sensor based on competitive-binding fluorescence immunoassay (250). The binding of labeled rabbit IgG to unlabeled anti-IgG was monitored via the fiber. Alternatively (251), the fiberoptic approach may be coupled to an enzyme-linked immunosorbent assay (ELISA), or the evanescent wave technique may be exploited to monitor binding processes occurring at the quartz-sample interface (252,253). Lanthanide chelates with their extremely long fluorescence decay time and large Stokes shift have led to the design of a new generation of nonisotopic

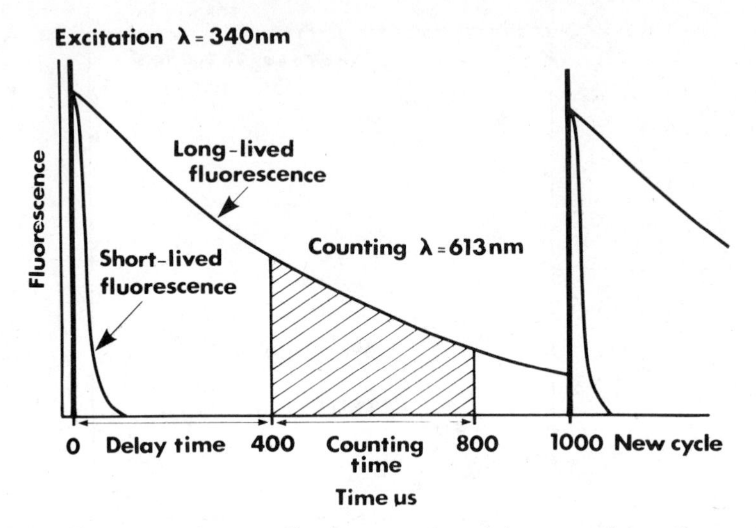

Fig. 3.56. Fluorescence decay profile of an europium chelate as used in a time-resolved fluorescence immunoassay. Background fluorescence disappears after a few nanoseconds, whereas the chelate decays in the millisecond time regime. (Permission for reproduction by courtesy of LKB Produkter, Bromma, Sweden).

immunoassays (245). Measuring fluorescence after a lag phase, during which the emission intensity of practically all other molecules has dropped to zero, allows an immunoassay to be performed with great sensitivity (Fig. 3.56). In this assay, an antibody is labeled with an europium chelate of the EDTA type which is only weakly fluorescent by itself. After the immunoreaction has occurred, the europium ion is coupled to another chelating agent by adding a reagent of the β-diketone type plus a surfactant. This results in a dramatic increase in fluorescence intensity that can be measured with unique selectivity and sensitivity.

Arnold (246) has immobilized alkaline phosphatase together with light-scattering material at the end of a fiber optic. When immersed into a substrate solution such as *p*-nitrophenyl phosphate, the second fiber bundle "sees" the formation of yellow *p*-nitrophenolate. Although not specifically used in combination with fluorogenic substrates, the method could easily be so.

A fiberoptic probe for the kinetic determination of enzyme activities has been described (247). It was designed for in-vivo determination of hydrolases such as esterases, phosphatases, sulfatases, and amylases. The synthetic substrates used in this method do not present a health risk when applied in vivo since it is the acid or sugar component that is released. Effects of pH may be taken into account by a second sensor that measures pH. Effects of interfering isoenzymes can, to some extent, be eliminated with additional sensor with immobilized substrates having different K_m and/or v_{max} values.

3.7.4. Sensors for Drugs and Anesthetics

Plain fibers have been used to make in-vivo measurements in minute volumes of body fluid using conventional fluorimetry, two-photon-excited fluorimetry, and sequentially excited fluorimetry (28,248). Detection limits of approximately 0.5 μM were determined for adriamycin (doxorubicin) in cerobrospinal fluid and whole blood and are comparable for the three techniques. The sensor consisted of a fiber that was threaded into a blunt-ended 26-gauge needle of a syringe. When the fiber was withdrawn a known distance, a defined sampling chamber was created within the needle. The utilization of this fluorosensor for pharmacokinetic studies has also been reported.

The pH-sensitive membrane described by Goldfinch and Lowe (65) was applied to determine penicillin and related antibiotics by virtue of a proton-producing reaction catalyzed by penicillase. An optoelectronic probe consists of a cellulose membrane containing the enzyme and coimmobilized bromcresol green, which responds to penicillin G in the concentration range 0–20 mM. The change in output voltage of the silicone diode light detector was approximately linear within the range 0.5–5.0 mM, but displayed inceasing desaturation at coincentrations above 10.0 mM. Response of the probe to ampicillin and cephaloridine was also observed. Subsequent to each determination, the membrane was regenerated by flushing the cell with 5 mM phosphate buffer of pH 7.0.

A fiber optical sensor for the widely used inhalation narcotic halothane in the presence of varying levels of oxygen was presented (127). It is based on dynamic fluorescence quenching of decacyclene by halothane in a silicone rubber matrix. Interferences by molecular oxygen are taken into account by a second sensor layer covered with PTFE. The latter is halothane—impermeable, so that the concentration of halothane can be calculated with Eq. (3.49).

Figure 3.57 shows the response of a sensor membrane to halothane in nitrogen, air, and oxygen. The signal changes are large enough to allow a quantitation of halothane, or oxygen, or both with a precision of $\pm 0.1\%$ for halothane, and ± 1 torr for oxygen in the physiological concentration range. The sensor is practically specific for the two analytes since other anesthetics have little or no effect.

Penicillin can be assayed using a pH fiber sensor onto which penicillinase is immobilized (249). The enzyme catalyzes the hydrolysis of penicillin to

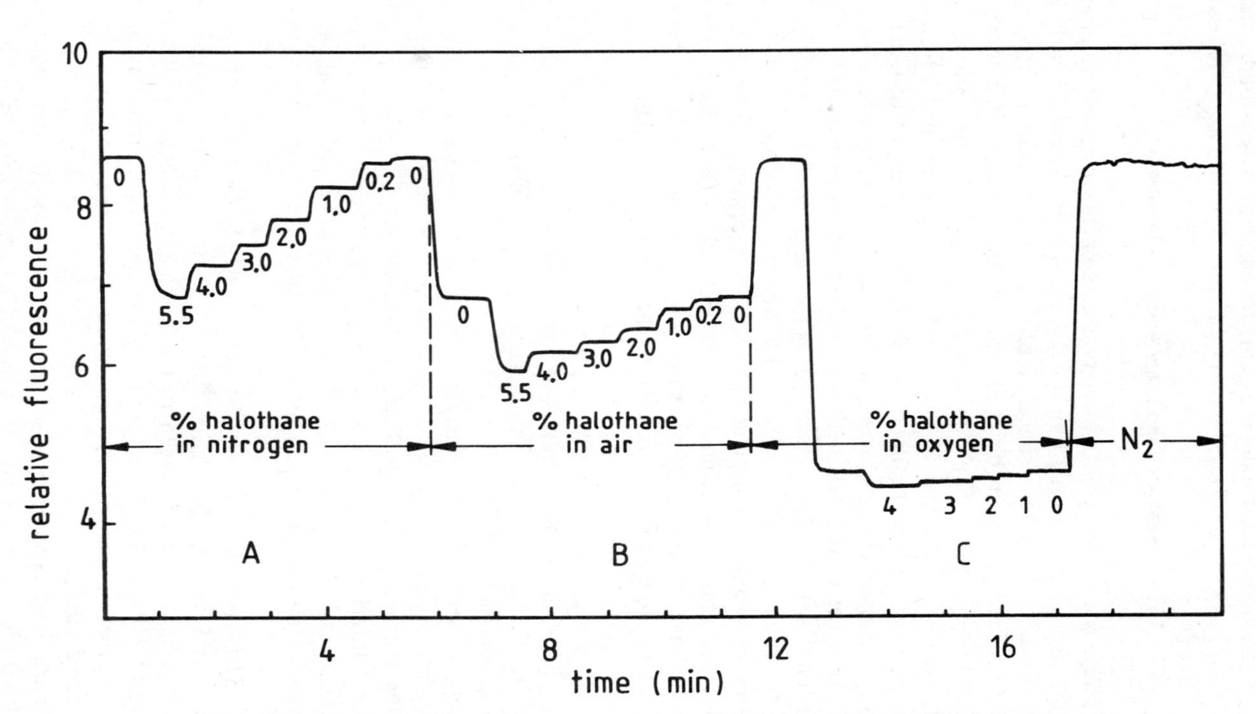

Fig. 3.57. Response time and reversibility of the fluorescence quenching of decacyclene (0.05%) in silicone rubber. Both oxygen and halothane act as quenchers. The sensing layer has a thickness of 50 µm. Excitation wavelength 400 nm, emission taken at 510 nm. (From ref. 127.)

penicilloic acid, and the resulting increase in the hydrogen ion concentration can be followed. As may be expected, the relative signal change depends on the buffer capacity of the sample solution.

REFERENCES

1. N. N., *Anal. Chem.*, **58**, 766A (1986).
2. J. F. Alder, *Fresenius' Z. Anal. Chem.* **324**, 372 (1986).
3. I. Chabay, *Anal. Chem.*, **54**, 1071A (1982).
4. T. H. Maugh, *Science*, **218**, 875 (1982).
5. T. G. Giallorenzi, J. A. Bucarro, A. Dandridge, G. H. Sigel, J. H. Cole, S. C. Rashleigh, and R. G. Priest, *IEEE J. Quant. Electron.*, **QE-18**, 626 (1982).
6. T. Hirschfeld, T. Dayton, F. Milanovich, and S. Klainer, *Opt. Eng.*, **22**, 527 (1983).
7. W. R. Seitz, *Anal. Chem.*, **56**, 16A (1984).
8. D. W. Lübbers and N. Opitz, *Sens. Actuators*, **4**, 641 (1983).
9. C. R. Lowe, *Trends Biotechnol.*, **2**, 59 (1984).
10. J. I. Peterson and G. G. Vurek, *Science*, **224**, 123 (1984).
11. T. Hirschfeld, J. B. Callis, and B. R. Kowalski, *Science*, **226**, 312 (1984).
12. O. S. Wolfbeis, *Pure & Appl. Chem.*, **59**, 663 (1987).
13. G. G. Guilbault, *Practical Fluorescence: Theory, Methods, and Techniques*, Marcel Dekker, New York, 1973.
14. O. S. Wolfbeis, "*Fluorescence of Organic Natural Products*," in S. G. Schulman, Ed., *Molecular Luminescence Spectroscopy: Methods and Applications*, Part 1, Wiley, New York, 1985, Chap. 3.
15. A. Fernandez-Gutierrez and A. Munoz de la Pena, "*Determination of Inorganic Substances by Luminescence Methods*," in S. G. Schulman, Ed., *Molecular Luminescence Spectroscopy: Methods and Applications*, Part 1, Wiley, New York, 1985, Chap. 4.

16a. N. Optiz and D. W. Lübbers, *Eur. J. Physiol.*, **335**, R120 (1975).

16b. D. W. Lübbers and N. Opitz, *Z. Naturforsch.*, **30C**, 532 (1975).

17. T. Hirschfeld, T. Deaton, F. Milanovich, and S. Klainer, *The Feasibility of Using Fiber Optics for Monitoring Groundwater Contaminants*, EPA Report AD-89-F-2A074 (1983).
18. R. A. O'Neil, L. Buja-Bijunas, and D. N. Rayner, *Appl. Opt.*, **19**, 863 (1980), and refs. 1 to 8 cited therein.
19. H. H. Kim, *Appl. Opt.*, **12**, 1454 (1973).
20. F. Milanovich, T. Hirschfeld, J. Roe, and N. Aziz, *Air Pollution Sensing Optrodes for Optical Fibers*, paper presented at the Pittsburgh Conference, Atlantic City, N.J., 1984.
21. F. P. Milanovich and T. Hirschfeld, *Adv. Instrum.*, **38**, 407 (1983).

22a. J. C. Fjeldsted, B. E. Richter, W. P. Jackson, and M. L. Lee, *J. Chromatogr.*, **279**, 423 (1983).

22b. J. Sepaniak, *Clin. Chem.*, **31**, 671 (1985); review.

22c. J. Gluckman, D. Shelly, and M. Novotny, *J. Chromatogr.*, **317**, 443 (1984).
23. G. Kychakoff, M. A. Kimball-Linnne, and R. K. Hanson, *Appl. Opt.*, **22**, 1426 (1983).
24. W. E. Howard, A. Greenquist, B. Walter, and F. Wogoman, *Anal. Chem.*, **55**, 878 (1983).
25. F. P. Milanovich, T. B. Hirschfeld, F. T. Wang, S. M. Klainer, and D. Walt, *SPIE Proc.*, **494**, 18 (1984).
26. J. R. North, *Biomedical Sensors*, lecture given at the Biomedical Business International Symposium on Emerging Medical Technologies in a Cost-Conscious Environment, Zürich, October 1984.
27. S. Klainer, T. Hirschfeld, H. Bowman, F. Milanovich, D. Perry, and D. Johnson, *A Monitor for Detecting Nuclear Waste Leakage in a Sub-Surface Repository*, Lawrence Berkeley Lab. Ann. Report LBL 11981 (1981).
28. B. J. Tromberg, J. F. Eastham, and M. J. Sepaniak, *Appl. Spectrosc.*, **38**, 38 (1984).
29. O. S. Wolfbeis and M. Leiner, *Anal. Chim. Acta*, **167**, 203 (1985).
30. J. N. Demas and R. A. Keller, *Anal. Chem.*, **57**, 538 (1985).
31. T. V. Samulski, P. T. Chopping, and T. Haas, *Phys. Med. Biol.*, **27**, 107 (1982).
32. T. Lund, *IEEE Proc.*, **131H**, 49 (1984), and references cited therein.
33. D. R. Demers and C. D. Allemand, U.S. Pat. 4,432,644 (Feb. 21, 1984), to Baird Corp.; *Chem. Abstr.*, **100**, 167,383 (1984).
34. H. H. Köhler, D. L. Redhead, and M. A. Nelson, *SPIE Proc.*, **404**, 119 (1983).
35a. J. C. Mills and R. J. Hodges, *Appl. Spectrosc.*, **38**, 413 (1984).
35b. L. M. Faires, T. M. Bieniewski, C. T. Apel, and T. M. Niemczyk, *Appl. Spectrosc.*, **39**, 9 (1985);
35c. K. Fujiwara and K. Fuwa, *Anal. Chem.*, **57**, 1012 (1985).
36. H. C. Christoph, East-Ger. Pat. 106,086 (May 20, 1974).
37. J. I. Peterson, R. V. Fitzgerald, and D. K. Buckhold, *Anal. Chem.*, **56**, 62 (1984).
38. G. F. Kirkbright, R. Narayanaswami, and N. A. Welti, *Analyst*, **109**, 1025 (1984).
39. I. Bergman, Brit. Pat. 1,190,583 (May 6, 1970).
40. O. S. Wolfbeis, H. Offenbacher, H. Kroneis, and H. Marsoner, *Mikrochim. Acta (Vienna)*, **I**, 153. (1984).
41. K. A. Wickersheim, U.S. Pat. 4,075,493 (Feb. 21, 1978).
42. K. V. Alves, J. Christol, M. Sun, and K. A. Wickersheim, *Adv. Instrum.*, **38**, 925 (1983).
43. H. Kautsky and A. Hirsch, *Z. Anorg. Allg. Chem.*, **222**, 126 (1935); a brief summary on this and related work was given in H. Kautsky, *Trans. Faraday Soc.*, **35**, 216 (1939).
44. M. Pollak, P. Pringsheim, and D. Terwoord, *J. Chem. Phys.*, **12**, 295 (1944).
45. J. Marcoll and M. Schmidt, Ger. Auslegeschrift 2,823,318 (Nov. 15, 1979), to Draegerwerk AG.; *Chem. Abstr.*, **92**, 69,036 (1980).
46. W. Barnikol and O. Burkhard, Ger. Off. 3,148,830 (June 23, 1983); *Chem. Abstr.*, **99**, 84,752 (1983).

47. I. A. Zakharov and T. I. Grishaeva, *Zh. Anal. Khim.*, **35**, 481 (1980); *ibid.*, **36**, 112 (1981); *Zh. Prikl. Spektrosk.*, **31**, 703 (1979).
48. H. Kautsky and G. O. Müller, *Naturwissenschaften*, **29**, 150 (1941); *Z. Naturforsch.* **2A**, 167 (1947).
49. T. Vo-Dinh, *Room Temperature Phosphorescence for Chemical Analysis*, Wiley, New York, 1984.
50. H. E. Posch, Ph.D. thesis, University of Graz, 1985.
51. S. Scypinski and L. J. Cline-Love, *Int. Lab.*, 60 (April 1984).
52. T. Hirschfeld, U.S. Pat. 3,604,927 (1971).
53. G. J. Müller, "Spectroscopy with the Evanescent Wave in the Visible Region of the Spectrum," *Am. Chem. Soc. Symp. Ser.*, **102**, 239 (1979).
54. J. H. W. Cramp and R. F. Reid, Eur. Pat. Appl. 61,884 (Oct. 6, 1982); *Chem. Abstr.*, **98**, 83,017 (1983).
55. R. M. Sutherland, C. Dähne, J. F. Place, and A. S. Ringrose, *Clin. Chem.*, **30**, 1533 (1984); see also Refs. 240 and 241.
56. E. H. Lee, R. E. Benner, J. B. Fenn, and R. K. Chang, *Appl. Opt.*, **18**, 862 (1979).
57. K. Newby, W. M. Reichert, J. D. Andrade, and R. E. Benner, *Appl. Opt.*, **23**, 1812 (1984).
58. S. A. Rockhold, R. D. Quinn, R. A. van Wagenen, J. D. Andrade, and W. M. Reichert, *J. Electroanal. Chem.*, **150**, 261 (1983).
59. P. O'Connor and J. Tanc, *Opt. Comm.*, **24**, 135 (1978).
60. J. F. Rappold, R. Santo, and J. D. Swalen, *Appl. Spectrosc.*, **34**, 517 (1980).
61a. B. W. Dodson, *SPIE Proc.*, **497**, 91 (1984).
61b. G. B. Harper, *Anal. Chem.*, **47**, 348 (1975).
62. J. I. Peterson, S. R. Goldstein, R. V. Fitzgerald, and D. K. Buckhold, *Anal. Chem.*, **52**, 864 (1980).
63. J. T. Coleman, J. F. Eastham, and M. J. Sepaniak, *Anal. Chem.*, **56**, 2246 (1984).
64. M. J. Goldfinch and C. R. Lowe, *Anal. Biochem.*, **109**, 216 (1980).
65. M. J. Goldfinch and C. R. Lowe, *Anal. Biochem.*, **138**, 430 (1984).
66a. J. Ruzicka and E. H. Hansen, *Anal. Chim. Acta*, **173**, 3 (1985).
66b. T. M. Freeman and W. R. Seitz, *Anal. Chem.*, **53**, 98 (1981).
67. M. L. Polanyi and R. M. Hehir, *Rev. Sci. Instrum.*, **33**, 1050 (1962).
68. M. L. Polanyi, *Dye Curves*, University Park Press, Baltimore, 1974, pp. 267, 284.
69. M. Stenberg and H. Nygren, *Anal. Biochem.*, **127**, 183 (1982).
70. G. R. Trott and T. E. Furtak, *Rev. Sci. Instrum.*, **51**, 1493 (1980).
71. G. E. Walfren and J. Stone, *Appl. Spectrosc.*, **26**, 585 (1972).
72. J. C. Schaefer and I. Chabay, *Opt. Lett.*, **4**, 227 (1979).
73. A. C. Eckbreth, *Appl. Opt.*, **18**, 3215 (1980).
74. H. Yamada and Y. Yamamoto, *J. Raman Spectrosc.*, **9**, 401 (1980).
75. A. D. Schwab and R. L. McCreery, *Anal. Chem.*, **56**, 2199 (1984).
76a. M. Seifert, K. Tiefenthaler, K. Heuberger, W. Lukosz, and K. Mosbach, *Anal. Lett.*, **19**, 205 (1986).

76b. T. N. K. Radju and D. Vidyasagar, *Med. Instrum.*, **16**, 154 (1982), and references cited therein.

77. G. A. Capelle, L. A. Franks, and D. A. Jessup, *Appl. Optics*, **22**, 3382 (1983).

78. K. O. White and W. R. Watkins, *Appl. Optics*, **14**, 2812 (1975).

79. K. O. White and S. A. Schleusener, *Appl. Phys. Lett.*, **21**, 419 (1972).

80. T. Broogardh and Ch. Ovren, U.S. Pat. 4,473,747 (Sept. 25, 1984), to ASEA AB (Sweden).

81. J. P. Dakin and A. J. King, *IEEE Proc.*, **131H**, 273 (1984).

82. N. Uesugi, K. Noguchi, N. Hata, N. Shibata, and Y. Negishi, *Electron. Lett.*, **20**, 1068 (1984).

83. S. E. Miller and A. G. Chynoweth, *Optical Fiber Telecommunications*, Academic Press, New York, 1979.

84. G. E. Walrafen, *Appl. Spectrosc.*, **29**, 179 (1975).

85. F. R. Aussenegg, M. E. Lippitsch, E. Schiefer, U. Deserno, and D. Rosenberger, *Appl. Spectrosc.*, **32**, 587 (1978).

86. W. Carvalho, P. Dumas, J. Corset, and V. Neumann, *SPIE Proc.*, **404**, 82 (1983).

87. GEO Centers, Inc., *Report 1983*, GC-TR-82-219; *Chem. Abstr.*, **101**, 77,530 (1984).

88. J. E. Jones and R. C. Spooncer, *J. Phys. E*, **16**, 1124 (1983).

89. (a) R. P. Haugland, "Covalent Fluorescent Probes," in R. F. Steiner, Ed., *Excited States of Biopolymers*, Plenum Press, New York, 1983, p. 29; (b) A. Hülshoff and H. Lingeman, in S. G. Schulman, Ed., Molecular Luminescence Spectroscopy: Methods and Applications, Part 1, Wiley, New York, 1985, Chap. 7.

90. H. Yasuda and V. Stannett, "Permeability of Polymers," in J. Brandrup and E. H. Immergut, Eds., *Polymer Handbook*, Wiley, New York, 1975, p. III-232ff.

91. M. E. Cox and B. Dunn, *J. Polymer Sci.*, **24A**, 621 and 2395 (1986).

92. H. Marsoner, H. Kroneis, and O. S. Wolfbeis, *Eur. Pat. Appl.* 109,959 (May 30, 1984).

93. O. S. Wolfbeis, H. Kroneis, and H. Offenbacher, Ger. Off. 3,343,636 and 3,343,637 (June 7, 1984), to AVL AG (Switzerland).

94. L. A. Saari and W. R. Seitz, *Anal. Chem.*, **54**, 821 (1982).

95. Zh. Zhujun and W. R. Seitz, *Anal. Chim. Acta*, **160**, 47 (1984).

96. P. Hochmuth, H. E. Posch, and O. S. Wolfbeis, unpublished results (1984).

97. *The Pierce Handbook and General Catalog*, Pierce Chem. Co., Rockford, Ill., 1985.

98. L. Goldstein and C. Manecke, "The Chemistry of Enzyme Immobilization," in *Appl. Biochem. Bioeng.*, **1**, 23 (1976).

99. K. Mosbach, Ed., *Immobilized Enzymes*, Vol. 44 in *Methods in Enzymology*, Academic Press, New York, 1976.

100. W. Wegscheider and G. Knapp, "Preparation of Chemically Modified Cellulose Exchangers and Their Use for the Preconcentration of Trace Elements," in *CRC Critical Reviews in Analytical Chemistry*, CRC Press, Cleveland, Ohio, 1981, p. 79ff.

101. D. E. Leyden and W. T. Collins, *Silylated Surfaces*, Goldon and Breach, New York, 1980.
102. E. P. Plueddemann, *Silane Coupling Agents*, Plenum Press, New York, 1982.
103. P. W. Carr and L. D. Bowers, *Immobilized Enzymes in Analytical and Clinical Chemistry*, Wiley, New York, 1980.
104. J. Woodward, Ed., *Immobilised Cells and Enzymes*, IRL Press, Oxford, 1985.
105. D. W. Lübbers, N. Opitz, P. P. Speiser, and H. J. Bisson, *Z. Naturforsch.*, **32C**, 133 (1977).
106. J. I. Peterson, U.S. Pat. Appl. 855,397 (June 23, 1978), to U.S. Dept. of Health, Education, and Welfare.
107. O. S. Wolfbeis, H. Kroneis, and H. Offenbacher, Ger. Off. 3,343,637 (June 17, 1984); *Chem. Abstr.*, **101**, 126,317 (1984).
108. C. A. Parker, *Photoluminescence of Solutions*, Elsevier, New York, 1968.
109. O. Stern and M. Volmer, *Phys. Z.*, **20**, 183 (1919).
110. C. A. Parker and W. J. Barnes, *Analyst*, **82**, 606 (1957).
111. J. F. Holland, R. E. Teets, P. M. Kelly, and A. Timnick, *Anal. Chem.*, **49**, 706 (1977).
112. H.-P. Lutz and P. L. Luisi, *Helv. Chim. Acta*, **66**, 1929 (1983).
113. J. Eisinger and J. Flores, *Anal. Biochem.*, **94**, 15 (1979).
114. E. H. Ratzlaff, R. G. Harfmann, and S. R. Crouch, *Anal. Chem.*, **56**, 342 (1984).
115. L. A. Saari and W. R. Seitz, *Analyst*, **109**, 655 (1984).
116. L. A. Saari and W. R. Seitz, *Anal. Chem.*, **55**, 667 (1983).
117. J. N. Miller, *Pure Appl. Chem.*, **57**, 515 (1985).
118. G. Kortüm, *Reflectance Spectroscopy*, Springer-Verlag, New York, 1969.
119. J. Goldman, *J. Chromatogr.*, **78**, 7 (1973).
120. V. Pollak, *Opt. Acta*, **21**, 51 (1974).
121a. J. Keizer, *J. Am. Chem. Soc.*, **105**, 1494 (1983).
121b. R. I. Cukier, *J. Am. Chem. Soc.*, **107**, 4115 (1985).
122. R. F. Chen, *Arch. Biochem. Biophys.*, **142**, 552 (1971).
123. J. R. Lakowicz, *Principles of Fluorescence Spectroscopy*, Plenum Press, New York, 1983.
124. J. R. Lakowicz, *Principles of Fluorescence Spectroscopy*, Plenum Press, New York, 1983, Chap. 9.
125. T. Lim and R. M. Dowben, "Nanosecond Pulse Fluorimetry of Proteins," in R. F. Steiner, Ed., *Excited States of Biopolymers*, Plenum Press, New York, 1983, p. 80.
126. O. S. Wolfbeis and E. Urbano, *Anal. Chem.*, **55**, 1904 (1983); US-Pat. 4,580,059.
127. O. S. Wolfbeis, H. E. Posch, and H. Kroneis, *Anal. Chem.*, **57**, 2556 (1985).
128. M. E. Lippitsch, J. Pusterhofer, M. J. P. Leiner, and O. S. Wolfbeis, *Anal. Chim. Acta*, in press (1988).
129a. D. W. Lübbers and N. Opitz, Ger. Offen. 2,508,637 (Sept. 9, 1976); *Chem. Abstr.*, **85**, 173,867 (1976).
129b. D. W. Lübbers and N. Opitz, Ger. Offen. 2,720,370 (Nov. 16, 1977); *Chem. Abstr.*, **90**, 214,720 (1979).

130. D. W. Lübbers, K. P. Völkl, and N. Opitz, Ger. Offen. 3,001,669 (Aug. 6, 1981); *Chem. Abstr.*, **96**, 14814 (1982).
131. J. I, Peterson and R. S. Goldstein, U.S. Pat. Appl. 855,384 (July 21, 1978); *Chem. Abstr.*, **90**, 99677 (1979).
132. S. R. Goldstein, J. I. Peterson, and R. V. Fitzgerald, *J. Biomech. Eng.*, **102**, 141 (1980).
133. R. M. Watson, D. R. Markle, Y. M. Ko, S. R. Goldstein, D. A. McGuire, J. I. Peterson, and R. E. Patterson, *Am. J. Physiol. Heart Circ. Physiol.*, **15**, H232 (1984).
134. G. A. Taite, R. B. Young, G. J. Wilson, D. J. Steward, and D. C. MacGregor, *Am. J. Physiol. Heart Circ. Physiol.*, **12**, H1027 (1982).
135. J. S. Suidan, B. K. Young, F. W. Hetzel, and H. R. Seal, *Clin. Chem.*, **29**, 1566 (1983).
136. G. F. Kirkbright, R. Narayanaswany, and N. A. Welti, *Analyst*, **109**, 15 (1984).
137. N. Opitz and D. W. Lübbers, *Sens. Actuators*, **4**, 473 (1983).
138. D. W. Lübbers and N. Opitz, Ger. Offen. 3,222,325 (Dec. 15, 1983).
139. D. W. Lübbers, *Arzneim. Forsch. (Drug Res.)*, **28**, 705 (1978).
140a. N. Opitz and D. W. Lübbers, *Adv. Exp. Med. Biol.*, **169**, 907 (1984).
140b. O. S. Wolfbeis and H. Offenbacher, Ger. Pat. 3,430.935.
141. H. Kroneis and O. S. Wolfbeis, Austrian Pat. 377,364 (Mar. 11, 1985).
142. O. S. Wolfbeis, E. Fürlinger, H. Kroneis, H. Marsoner, H. Offenbacher, and B. Trathnigg, unpublished results (1979–1982).
143. H. Offenbacher, O. S. Wolfbeis, and E. Fürlinger, *Sens. Actuators*, **9**, 73 (1986).
144. O. S. Wolfbeis and H. Offenbacher, *Sens. Actuators*, **9**, 85 (1986).
145. H. D. Hendricks, US-Pat. 3,709,663 (1973).
146. T. Hirschfeld, F. Wong, and T. Deaton, *Optimization and Performance of pH Optrodes*, abstract 382, paper presented at the Pittsburgh Conference, Atlantic City, N.J. 1983.
147. T. Hirschfeld, T. Deaton, R. Malstrom, and F. Wong, *Specific Optrode Development for Industrial Applications*, paper presented at the Federation of Analytical Chemists Society Meeting, Philadelphia, September 1982.
148. H. Kautsky and A. Hirsch, *Ber. Dtsch. Chem. Ges.*, **64**, 2677 (1931); a short account of Kautsky's work has been given in *Trans. Faraday Soc.*, **35**, 216 (1939).
149. H. Kautsky, H. de Bruijn, R. Neuwirth, and W. Baumeister, *Ber. Dtsch. Chem. Ges.*, **66**, 1588 (1933).
150. H. Kautsky and G. O. Müller, *Z. Naturforsch.*, **2A**, 167 (1947).
151. H. Kautsky, A. Hirsch, and F. Davidshöfer, *Ber. Dtsch. Chem. Ges.*, **65**, 1762 (1932).
152. G. Shaw, *Trans. Faraday Soc.*, **63**, 2181 (1967).
153. I. A. Zakharov and T. I. Grishaeva, in (a) Zh. Prikl. Spektrosk., **36**, 980 (1982); engl. ed., p. 697; (b) *Zh. Anal. Khim.*, **37**, 1753 (1982); (c) *Zh. Prikl. Khim.*, **57**, 1240 (1984); (d) *Zh. Anal. Khim.*, 36, 112 (1981).
154. I. Bergman, *Nature*, **218**, 396 (1968).
155a. D. W. Lübbers and N. Opitz, *Anal. Chem. Symp. Ser.*, **17**, 609 (1983).
155b. D. W. Lübbers and N. Opitz, *Adv. Exp. Med. Biol.*, **75**, 65 (1976).

156. I. B. Berlman, *Handbook of Fluorescence Spectra of Aromatic Molecules*, Academic Press, New York, 1971, p. 443.
157. R. Brady, W. V. Miller, and L. Vaska, *Chem. Comm.*, 393 (1974).
158. O. S. Wolfbeis and F. M. Carlini, *Anal. Chim. Acta*, **160**, 301 (1984).
159. J. I. Peterson and L. V. Fitzgerald, *Rev. Sci. Instrum.*, **51**, 670 (1980).
160. H. W. Kroneis and H. J. Marsoner, *Sens.* Actuators, **4**, 587 (1983).
161. M. E. Cox and B. Dunn, *Appl. Opt.* **24**, 2114, (1985).
162. N. Opitz and D. W. Lübbers, in D. Bruley, H. I. Bicher, and D. Reneau, Eds., *Oxygen Transport to Tissue*, Vol. IV, Plenum Press, New York, 1984, p. 261.
163. J. I. Peterson and R. V. Fitzgerald, U.S. Pat. 4,476,870 (1984).
164. B. Stevens, U.S. Pat. 3,612,866 (Oct. 12, 1971); *Chem. Abstr.*, **76**, 20945 (1972).
165. C. C. Stanley and J. L. Kropp, U.S. Pat. 3,725,658 (Apr. 6, 1973), to TRW Inc.
166. N. Opitz and D. W. Lübbers, *Adv. Exp. Med. Biol.*, **169**, 899 (1984).
167. I. A. Zakharov and T. I. Grishaeva, USSR Pat. Appl. 893,853 (Dec. 30, 1981); *Chem. Abstr.*, **96**, 205,174 (1982).
168. R. V. Fitzgerald, U.S. Pat. Appl. 363,425 (Nov. 5, 1982); *Chem. Abstr.*, **98**, 85,777 (1983).
169. H. Marsoner and H. Kroneis, Eur. Pat. Appl. 109,958 (May 30, 1984); *Chem. Abstr.*, **101**, 65,204 (1984); US-Pat. 4,587,101.
170. H. Marsoner, H. Kroneis, and O. S. Wolfbeis, Eur. Pat. Appl. 109,959 (May 30, 1984); *Chem. Abstr.*, **101**, 65,205 (1984); US-Pat. 4,587,101.
171. D.W. Lübbers, F. Hannebauer, and N. Opitz, *Birth Def. Orig. Art. Ser.*, **15**, 123 (1979).
172. N. Opitz and D. W. Lübbers, in D. Bruley, H. I. Bicher, and D. Reneau, Eds., *Proceedings of the International Symposium on Oxygen Transport Tissue*, Vol. VI, Plenum Press, New York, 1984, p. 757.
173. G. G. Vurek, U.S. Pat. Appl. 470,920 (July 22, 1983); *Chem. Abstr.*, **99**, 118,844 (1983).
174. Z. Zhujun and W. R. Seitz, *Anal. Chim. Acta*, **160**, 305 (1984).
175. M. A. Jensen and G. A. Rechnitz, *Anal. Chem.*, **51**, 1972 (1979).
176. M. E. Lopez, *Anal. Chem.*, **56**, 2360 (1984).
177. W. N. Opdycke, S. J. Parks, and M. E. Meyerhoff, *Anal. Chim. Acta*, **155**, 11 (1983).
178a. H. Marsoner, H. Kroneis, and O. S. Wolfbeis, Eur. Pat. 105.870.
178b. H. A. Heitzmann and H. Kroneis, U.S. Pat. 4,557,900 (1985).
178c. M. J. P. Leiner, unpublished results (1985).
179. D. J. David, M. C. Wilson, and D. S. Ruffin, *Anal. Lett.*, **9**, 389 (1976).
180. J. F. Giuliani, H. Wohltjen, and N. L. Jarvis, *Opt. Lett.*, **8**, 54 (1983).
181a. M. A. Arnold and T. J. Ostler, *Anal. Chem.*, **58**, 1137 (1986).
181b. O. S. Wolfbeis and H. E. Posch, *Anal. Chim. Acta*, **185**, 321 (1986).
182a. D. A. Helm and W. J. Zolner, U.S. Pat. 3,845,309 (Oct. 29, 1974), to Thermo Electron Corp.; *Chem. Abstr.*, **82**, 126,949 (1975).
182b. A. Sharma and O. S. Wolfbeis, *Spectrochim. Acta, Part A*, in press.
183a. T. Hirschfeld, *Remote Analysis Using Optical Fibers*, paper presented at the 13th congress of the International Committee for Optics, Sapporo (Japan), August 20–24, 1984.

183b. O. S. Wolfbeis, unpublished results (1986).
184a. Jpn. Kokai JP 81,112,636 (Sept. 5, 1981), to Tokyo Shibaura Electric Co.; *Chem. Abstr.*, **96**, 87,570 (1982).
184b. A. P. Russell and K. S. Fletcher, *Anal. Chim. Acta*, **170**, 209 (1985).
185. H. E. Posch and O. S. Wolfbeis, *Sens. Actuators, in press* (1988).
186. Jpn. Kokai JP 59,19843 [84,19,843] (Feb. 1, 1984), to Tateishi Electronics Co.; *Chem. Abstr.*, **101**, 83,210 (1984).
187. E. Urbano. E. Offenbacher, and O. S. Wolfbeis, *Anal. Chem.*, **58**, 427 (1984).
188. E. E. Hardy, D. J. David, N. S. Kapany, and F. C. Unterleitner, *Nature*, **257**, 666 (1975).
189a. T. Hirschfeld and T. Deaton, *Remote Fiber Fluorimetry: Specific Analyte Optrodes*, lecture presented at the Pittsburgh Conference, Atlantic City, N.J., March 1982.
189b. R. Narayanaswami and F. Sevilla, *Analyst*, **111**, 1085 (1986).
189c. A. Martinez, M. Moreno, and C. Camara, *Anal. Chem.*, **58**, 1877 (1986).
190. N. Opitz and D. W. Lübbers, Ger. Pat. Appl. 33,019 (Jan. 21, 1983).
191a. O. S. Wolfbeis and P. Hochmuth, Eur. Pat. Appl. 198,815 (1986).
191b. O. S. Wolfbeis and B. P. H. Schaffar, *Anal. Chim. Acta*, 1987, in press.
192a. L. A. Saari and W. R. Seitz, *Anal. Chem.*, **56**, 810 (1984).
192b. Z. Zhujun and W. R. Seitz, *Anal. Chim. Acta*, **171**, 251 (1985).
192c. Z. Zhujun, J. L. Mullin, and W. R. Seitz, *Aanal. Chim. Acta*, **184**, 251 (1986).
192d. S. D. Dowling and W. R. Seitz, *Spectrochim. Acta*, **40A**, 991 (1984).
193. J. E. Freman, A. G. Childers, A. W. Steele and G. M. Hieftje, *Anal. Chim. Acta*, **177**, 121 (1985).
194a. R. A. Malstrom and T. Hirschfeld, *Anal. Chem. Symp. Ser.*, **19**, 25 (1984); in this article the legends of Figs. 4 and 5 should apparently interchanged.
194b. R. Malstrom, T. Hirschfeld, and T. Deaton, Abstracts of papers presented at the 14th Natl. Meeting of the Am. Chem. Soc., Kansas City, Sep. 1982.
195. W. Campen and K. Bachmann, *Mikrochim. Acta (Vienna)*, **II**, 159 (1979).
196. S. W. Allison, D. W. Magnuson, and N. R. Cates, *SPIE Proc.*, **380**, 369 (1983).
197a. D. M. Jordan, D. R. Walt, and F. P. Milanovich, *Anal. Chem.*, **59**, 437 (1987).
197b. C. Munkholf, D. R. Walt, F. P. Milanovich, and S. M. Klainer, *Anal. Chem.*, **58**, 1427 (1986).
197c. Y. Kawabata, K. Tsuchida, T. Imasaka and N. Ishibashi, *Anal. Sci.*, **3**, 7 (1987).
198a. O. S. Wolfbeis, B. P. H. Schaffar, and B. Kaschnitz, *Analyst*, **111**, 1331 (1986).
198b. W. Trettnak and O. S. Woifbeis, *Fresenius' Z. Anal. Chem.*, **326**, 547 (1987).
199. O. S. Wolfbeis and P. Hochmuth, *Mikrochim. Acta (Vienna)*, **III**, 129 (1984).
200. O. S. Wolfbeis, B. P. Schaffar, and R. A. Chalmers, *Talanta*, **33**, 867 (1986).
201. T. Matsuo, Y. Masuda, and E. Sekido, Talanta, **33**, 665 (1986).
202. L. A. Franks, G. A. Capelle, and D. A. Jessup, *Appl. Opt.*, **22**, 1717 (1983).
203. F. K. Kawahara, R. A. Fiutem, H. S. Silvus, F. M. Newman, and J. H. Frazar, *Anal. Chim. Acta*, **151**, 315 (1983).
204. F. Milanovich, T. Hirschfeld, J. Roe, and N. Aziz, *Air Pollution Sensing Optrodes for Optical Fibers*, paper presented at the Pittsburgh Conference, Atlantic City, N.J., 1984.

205. W. Sarholz, Ger. Offen. 3,143,480 (May 11, 1983), to R. Bosch GmbH.; *Chem. Abstr.*, **99**, 76,079 (1983).
206. F. E. Hoge and R. N. Swift, "Airborne Mapping of Laser-Induced Fluorescence of Chlorophyll and Phycoerythrin in a Gulf Stream Warm Core Ring," in A. Zirino, Ed., *Mapping Strategies in Chemical Oceanography, Adv. Chem. Ser.*, **209**, 353 (1985).
207. L. R. Poole and W. E. Esaias, *Appl. Opt.*, **22**, 380 (1983); *ibid.*, **21**, 3756 (1982).
208. M. Brislow, D. Nielson, D. Bundy, and R. Furtek, *Appl. Opt.*, **20**, 2889 (1981).
209. F. E. Hoge and R. N. Swift, *NASA Conf. Publ. 2188*, U.S. Government Printing Office, Washington, D.C., 1981, p. 349.
210. D. D. Venable, A. R. Punjabi, and L. R. Poole, *Appl. Opt.*, **23**, 970 (1984), and references cited therein.
211. F. H. Farmer, *NASA Conf. Publ. 2188*, U.S. Government Printing Office, Washington, D.C., 1981, p. 349.
212. G. Lakshman, *Water Resour. Res.*, **11**, 705 (1975).
213. W. A. Chudyk, M. M. Carrabba, and J. E. Kenny, *Anal. Chem.*, **57**, 1237 (1985).
214a. F. Milanovich, T. Hirschfeld, H. Miller, W. Anderson, F. Miller, S. M. Klainer, and R. Gaver, *Groundwater Monitoring by Spectroscopy with Optical Fibers and Optrodes*, Technical Report prepared for the US EPA, Las Vegas, Nev. (July 1984).
214b. F. P. Milanovich, *Environ. Sci. Technol.*, **20**, 441 (1986).
215. G. A. Lugg, *Anal. Chem.*, **38**, 1532 (1966), and references cited therein.
216. M. R. Neuman, D. G. Fleming, P. W. Cheung, and W. H. Ko, *Physical Sensors for Biomedical Applications*, CRC Press, Boca Raton, Fla., 1980.
217a. D. R. Markle, D. A. McGuire, S. R. Goldstein, R. E. Patterson, and R. M. Watson, in D. C. Viano, Ed., *1981 Advances in Bioengineering*, American Society of Mechanical Engineers, New York, 1981, p. 123.
217b. E. Abraham, D. R. Markle, S. Fink, H. Ehrlich, M. Tsang, M. Smith, and A. Meyer, *Anesth. Analg.*, **64**, 731 (1985).
218. M. L. J. Landsman, N. Knop, G. Kwant, G. A. Mook, and W. G. Zijlstra, *Pflueger's Arch. (Eur. J. Physiol.)*, **373**, 273 (1978), and refs. 1, 4, 5, 8, 15 therein.
219. Refs. 2–5 in: J. I. Peterson and G. G. Vurek, *Science*, **224**, 123 (1984).
220. J. M. Schmitt, J. D. Meinl, and F. G. Mihn, *IEEE Trans. Biomed. Eng.*, **33**, 98 (1986).
221. M. J. Milano and K.–Y. Kim, *Anal. Chem.*, **49**, 555 (1977).
222a. D. W. Lübbers, J. Gehrich, and N. Opitz, *Life Support Systems*, **4**, 94 (1986).
222b. J. L. Gehrich, D. W. Lübbers, N. Optiz, D. R. Hansmann, W. W. Miller, J. K. Tusa, and M. Yafuso, *IEEE Trans. Biomed. Eng.*, **33**, 117 (1986).
222c. W. W. Miller, M. Yafuso, C. F. Yan, H. Hui, and S. Arick, *Clin. Chem.*, **33**, 1538 (1987).
223a. G. G. Vurek, J. I. Peterson, S. W. Goldstein, and J. W. Severinghaus, *Fed. Proc. Fed. Am. Soc. Exp. Biol.*, **41**, 1484 (1982);
223b. G. G. Vurek, P. J. Feustel, and J. W. Severinghaus, *Ann. Biomed. Eng.*, in press.

223c. N. Opitz, H. Weigelt, T. Barankay, and D. W. Lübbers, in I. A. Silver, M. Erecinska, and H. I. Bicher, Eds., *Oxygen Transport to Tissue, Vol. III*, Plenum Press, New York, 1978, p. 99.

224. P. L. Smock, T. A. Orofino, G. W. Wooten, and W. S. Spencer, *Anal. Chem.*, **51**, 505 (1979).

225. E. Labeyrie and Y. Koechlin, *J. Neurosci. Methods*, **1**, 35 (1979).

226. K. P. Völkl, Ph.D. thesis, University of Bochum (FRG), 1980.

227. N. Uwira, N. Opitz, and D. W. Lübbers, *Adv. Exp. Med. Biol.*, **169**, 913 (1984).

228. J. S. Schultz, U.S. Pat. 4, 344, 438 (1982).

229. J. S. Schultz, S. Mansouri, and I. J. Goldstein, *Diabetes Care*, **5**, 245 (1982).

230. D. W. Lübbers, K. P. Völkl, U. Grossmann, and N. Opitz, in D. W. Lübbers, H. Acker, R. P. Buck, E. Eisenmann, M. Kessler, and W. Simon, Eds., *Progress in Enzyme and Ion Selective Electrodes*, Springer-Verlag, Berlin, 1981, p. 67.

231. K. P. Völkl, N. Opitz, and D. W. Lübbers, *Fresenius Z. Anal. Chem.*, **301**, 162 (1980).

232. K. P. Völkl, U. Grossmann, N. Opitz, and D. W. Lübbers, *Adv. Physiol. Sci.*, **25**, 99 (1984).

233. S. Kobayashi, K. Kaede, K. Nishiki, and E. Ogata, *J. Appl. Physiol.*, **31**, 693 (1971).

234. B. Chance, V. Legallais, J. Sorge, and N. Graham, *Anal. Biochem.*, **66**, 498 (1975).

235. A. Majevsky and B. Chance, *Science*, **217**, 537 (1982).

236. K. Newby, W. M. Reichert, J. D. Andrade, and R. E. Benner, *Appl. Opt.*, **23**, 1812 (1984).

237. R. A. Van Wagenen, S. Rockhold, and J. D. Andrade, "Probing Protein Adsorption: Total Internal Reflection Intrinsic Fluorescence," in S. L. Cooper and N. A. Pappas, Eds., *Interfacial Phenomena and Applications*, Wiley, New York, 1982, p. 351.

238. M. N. Kronick and W. A. Little, *J. Immunol. Methods*, **8**, 235 (1975).

239. N. L. Thompson and D. Axelrod, *Biophys. J.*, **43**, 103 (1983).

240a. R. M. Sutherland, C. Dähne, J. F. Place, and A. R. Ringrose, *J. Immunol. Methods*, **74**, 253 (1984).

240b. R. Sutherland, C. Dähne, and J. Place, *Anal. Lett.*, **17**, 43 (1984).

241. J. F. Place, R. M. Sutherland and C. Dähne, *Biosensors*, **1**, 321 (1985).

242. T. Hirschfeld, U.S. Pat. 4,447,546 (1984).

243. M. J. Block and T. B. Hirschfeld, U.S. Pat. Appl. 406,324 (Aug. 9, 1982).

244. H. J. Nöller, Ger. Pat. Appl. 3,028,591 (June 28, 1980).

245. E. Soini and H. Kojola, *Clin. Chem.*, **29**, 65 (1983).

246. M. A. Arnold, *Anal. Chem.*, **57**, 565 (1985).

247. O. S. Wolfbeis, *Anal. Chem.*, **58**, 2874 (1986).

248. M. J. Sepaniak, B. J. Tromberg, and J. F. Eastham, *Clin. Chem.*, **29**, 1678 (1983).

249. M. S. Fuh, L. W. Burgess and G. D. Christian, Pittsburgh Conf. Paper nr. 601, Atlantic City, N.J., 1987.

250. B. M. Tromberg, M. J. Sepaniak, T. Vo-Dinh, and G. D. Griffin, *Anal. Chem.*, **59**, 1226 (1987).
251. T. Vo-Dinh, G. D. Griffin, and K. R. Ambrose, *Appl. Spectrosc.*, **40**, 696 (1986).
252. G. K. Iwamoto, L. C. Winterton, R. S. Stoker, R. A. Van Wagenem, J. D. Andrade, and D. F. Mosher, *J. Coll. Interface Sci.*, **106**, 459 (1985).
253. J. D. Andrade, R. A. Van Wagenem, D. E. Gregonis, K. Newby, and J. N. Lin, *IEEE Trans. Electr. Dev.*, **32**, 1175 (1985).

CHAPTER

4

HIGHLY RESOLVED MOLECULAR LUMINESCENCE SPECTROSCOPY

J. W. HOFSTRAAT, C. GOOIJER, AND N. H. VELTHORST

Department of General and Analytical Chemistry
Free University
Amsterdam, The Netherlands

4.1. INTRODUCTION

Molecular fluorescence and phosphorescence spectrometry are well-known methods of analysis for organic compounds. Fluorimetry is obviously the most frequently employed of the two because it can be applied straightforwardly to fluid samples. On the contrary, phosphorimetry in general requires rigid frozen systems in order to prevent collisional deactivation of the long-living triplet state. Recently, however, interesting applications of the latter technique to fluid samples have been reported (1.1,1.2).*

In comparison to UV/Vis absorption spectrometry, luminescence techniques are less universally applicable, but, in principle, obviously more sensitive. Whereas absorption of near UV and visible radiation is a general

* References in this chapter are numbered within each section of the chapter; ref. (1.1) refers to the first reference in Section 1 of Chapter 4.

property of analytes with chromophoric groups, many of these compounds are poor luminophores. On the other hand, this makes the emission methods more selective. Additional selectivity is obtained in fluorimetry and phosphorimetry because two wavelengths, instead of one as in absorption measurements, can be chosen for the determination of a compound, namely, the excitation and emission wavelengths.

Nevertheless, fluorimetry has been little used in qualitative analysis of fluid samples. It is almost exclusively utilized for quantitation purposes. The reason is that both the fluorescence (phosphorescence) emission and excitation spectra of fluid samples exhibit bands that are broad and relatively featureless, like those in absorption spectra. In general the same holds for spectra of rigid samples, and even the bands of not too small molecules (i.e., larger than benzene) in the gas phase are relatively broad (1.3).

Of course, the potential of the emission techniques would be extended considerably if an increase in resolution of the spectra could be realized. In fact, investigations with this purpose in mind have been going on since the 1950s along different lines. An impression of the gain in information that can be realized by improving the spectral resolution is given in Fig. 4.1. It is the aim of the present chapter to give an outline of the techniques that yield narrowing of the bands in molecular emission spectra and to discuss their potential in analytical chemistry. This can only be done on the basis of a treatment of some fundamental aspects. Primarily, the origins of line broadening in electronic spectroscopy have to be elucidated. Only if the various contributions to the spectral bandwidth are known may it be possible to find experimental procedures to reduce or eliminate them.

4.1.1. Linewidths in Electronic Spectroscopy

Before discussing the various contributions to the linewidths in molecular luminescence spectra, it is appropriate to discriminate between two fundamentally different types of broadening, homogeneous and inhomogeneous broadening. *Homogeneous line broadening* is composed of contributions that are the same for every molecule in the sample. The shape of spectral bands that are dominated by homogeneous broadening processes is typically Lorentzian (1.4). The Lorentzian line shape function $g(\nu)$ is in its most general form given by

$$g(\nu - \nu_0) = \frac{a}{(\nu - \nu_0)^2 + b^2} \qquad (1.1)^*$$

in which a and b are constants. The resulting lineform is characteristic for

* Equations in this chapter are numbered within sections.

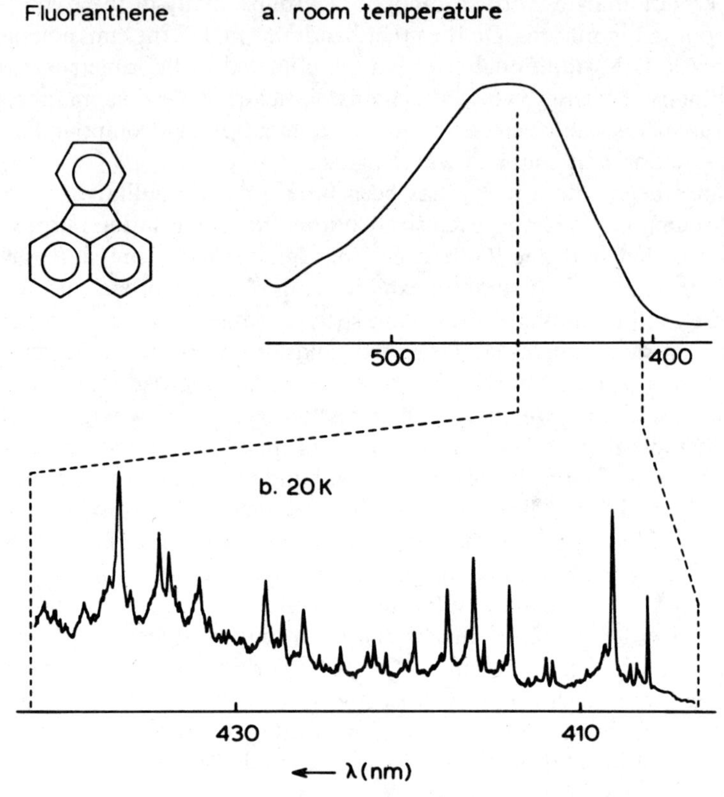

Fig. 4.1. Fluorescence spectra of fluoranthene: (*a*) in liquid solution (methanol/water 80:20, v/v) at room temperature, $\lambda_{exc} = 280$ nm; (*b*) in *n*-hexane at $T = 20$ K, $\lambda_{exc} = 284$ nm.

damped oscillatory motion and shows a sharp maximum with magnitude a/b^2 at frequency $\nu = \nu_0$; the full width at half maximum (FWHM) $\Delta\nu = 2b$.

Inhomogeneous or *heterogeneous broadening* results from the statistical average of a property that is not the same for all particles in the sample. The overall line shape is formed by an ensemble of homogeneously broadened Lorentzian lines centered at different transition frequencies. The spectral bands that are dominated by inhomogeneous broadening processes are described by a Gaussian function with general form (c and d are constants) (1.5)

$$g(\nu - \nu_0) = ce^{-d(\nu - \nu_0)^2} \tag{1.2}$$

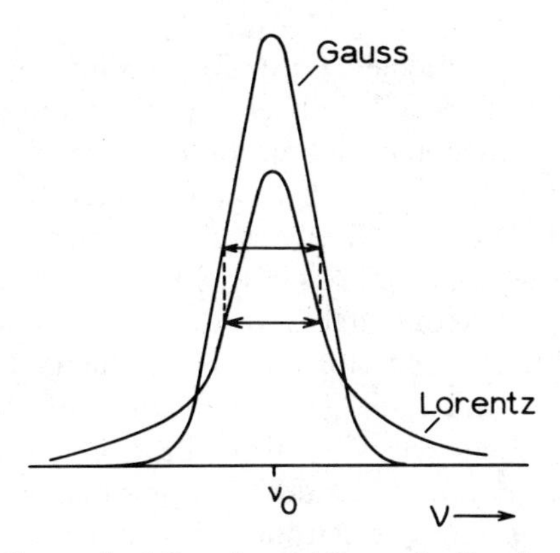

Fig. 4.2. Comparison of normalized Gaussian and Lorentzian line shapes with equal FWHMs.

The Gaussian line shape also has a maximum for $\nu = \nu_0$; the FWHM is $2\sqrt{\ln 2}/d$. As is displayed in Fig. 4.2, the intensity decreases much faster for a Gaussian profile than for a Lorentzian profile if both have the same FWHM.

It will be shown later that the high-resolution techniques to be discussed are mainly inclined to reduce the inhomogeneous broadening terms. To date, this has been achieved in gaseous and solid matrices but not in fluid systems.

4.1.1.1. Gaseous Samples

Let us first consider a gaseous sample of identical atoms with a low vapor pressure. In the spectral region under concern here only electronic transitions take place. Even under the most favorable conditions such transitions are not observed as infinitely sharp lines. There is a definite reproducible line shape with FWHM $\Delta\nu_N$ due to natural broadening, which can readily be conceived on the basis of the Heisenberg uncertainty principle. According to the Heisenberg principle a finite lifetime τ of an atomic or molecular state leads to an uncertainty of at least $\Delta E = h/2\pi\tau$ joule in the energy of the state. Therefore, for a transition between two states with energies E_i and E_f and lifetimes τ_i and τ_f, the corresponding spectral band will have a FWHM of (see, e.g., ref. 1.6):

$$\Delta\nu_N = \frac{\Delta E_i + \Delta E_f}{h} = \frac{1}{2\pi}\left(\frac{1}{\tau_i} + \frac{1}{\tau_f}\right) \tag{1.3}$$

If one of the two states involved in the electronic transition is the ground state of the atom (or molecule), Eq. (1.3) simplifies since this state has an infinite lifetime. Equation (1.3) implies that the natural line broadening is inversely proportional to the lifetime of the excited state: For a typical fluorescence lifetime of 10 ns, ΔE is at least 1.1×10^{-26} J and $\Delta\nu_N =$ 16 MHz (5.5×10^{-4} cm^{-1}). In molecular spectroscopy, for the higher excited electronic states, lifetimes as low as 10^{-13} s are encountered, yielding natural linewidths of about 200 GHz (50 cm^{-1}). Furthermore, vibronic fluorescence transitions occur which involve as the final state a vibrationally excited electronic ground state. Typical lifetimes of such states are in the order of 1 ps, so that the vibronic transitions will acquire an extra contribution to their natural linewidth of approximately 5 cm^{-1} with respect to the corresponding, purely electronic, 0–0 transition. Of course, natural broadening belongs to the homogeneous type; therefore, the form of the spectral line of a naturally broadened transition is given by the (normalized) Lorentzian line-shape function;

$$g(\nu - \nu_0) = \frac{\Delta\nu_N}{4\pi^2[(\nu - \nu_0)^2 + (\Delta\nu_N/2)^2]} \tag{1.4}$$

In practice, however, the linewidths in the electronic spectra of atoms in gases exceed the natural broadening significantly (even if the limited resolution of the spectrometer does not induce any apparent broadening). This is caused by Doppler broadening and pressure broadening.

The Doppler broadening is of inhomogeneous nature. It can be attributed to the fact that the atoms move very fast in different directions relative to the source of radiation. Let us assume that the actual spectral line has a frequency ν_0. If an atom is moving with a velocity v away from the light source for an observer the apparent frequency is not ν_0 but ν, given by

$$\nu = \nu_0/(1 - v/c) \tag{1.5}$$

where c is the speed of light. Hence every atom has its own apparent resonance frequency, depending on its velocity. The velocity distribution of a sample of atoms with mass m in the gas phase, in thermal equilibrium at absolute temperature T, has been derived by Maxwell (see, e.g. ref. 1.7). The density $n_i(v_z)$of atoms in electronic state i with velocity component v_z in the interval v_z to $v_z + dv_z$ is given by

$$n_i v_z dv_z = N_i \left(\frac{m}{2\pi kT}\right)^{1/2} e^{-(1/2\,m/kT)v_z^2}\, dv_z \tag{1.6}$$

where N_i is the total number of atoms in state i per unit of volume and k is the

Boltzmann constant. Combination of Eqs. (1.5) and (1.6) yields a Gaussian intensity profile of the Doppler-broadened spectral line,

$$I(\nu) = I_0 \exp\left\{-\frac{1}{2}\frac{m}{kT}\left[\frac{c(\nu - \nu_0)^2}{\nu_0}\right]\right\} \tag{1.7}$$

corresponding to a FWHM $\Delta\nu_D$ of

$$\Delta\nu_D = \frac{2\nu}{c}\left(\frac{2kT\ln 2}{m}\right)^{1/2} \tag{1.8}$$

In contrast with the Doppler broadening, the so-called pressure (collision) broadening is of homogeneous character. This type of broadening is caused by collisions of the particles in the sample with each other or with the walls of the vessel in which they are contained. The collisions induce energy transfer, thus shortening the lifetime of the excited state. Of course, this broadening term strongly depends on the pressure of the gas and the dimensions of the sample cell (1.8).

The foregoing implies that in conventional atomic (and certainly also molecular) gas-phase spectrometry it is not easy, or even impossible, to obtain spectra in which only transitions with the natural linewidth are seen. However, as will be discussed in a separate section, the Doppler line broadening can be at least partially removed by introducing the sample in an atomic (or molecular) beam, in which all particles are moving in the same direction within a limited range of velocities. This technique is commonly denoted as supersonic jet spectroscopy.

The foregoing considerations apply also for molecular samples. The spectra of molecules, however, are far more complex than those of atoms. The reason is that (in a simplified description), simultaneous with the electronic transitions, changes occur in the molecular vibrational and rotational states. Especially the latter are so close together that they are usually barely resolved. In addition to this strong increase in transition possibilities, the spectra reflect the number of populated vibrational and rotational levels. At room temperature the number of accessible vibrational and especially rotational levels can be very high, so that the spectra may consist of hundreds or even thousands of lines. Such spectra are too congested for individual lines to be resolved. In this case the introduction of the molecule in a supersonic jet has an additional advantage: Rotational temperatures as low as 0.2 K and vibrational temperatures of 10–100 K have been achieved, so that in most cases only the rotational and vibrational ground states are populated. Thus in supersonic jet spectroscopy not only is the width of the spectral lines decreased significantly, but also, the number of lines (and the congestion) is impressively reduced.

4.1.1.2. Rigid Samples

Analytes present as dilute solutes in frozen samples have, of course, no freedom of translation and rotation, so that Doppler and pressure (collision) broadening play no role. Furthermore, at sufficiently low temperatures usually only the lowest molecular vibrational mode is populated. Hence, it might be expected that under these conditions the molecular spectra are much less complicated than for gaseous samples. However, other effects on the line shape have to be taken into account, due to the interactions between the luminescent molecule and its directly surrounding matrix atoms or molecules.

For convenience, we assume that the fluorescent and/or phosphorescent molecules are matrix isolated; in other words, the molecules are trapped in the rigid cage of a chemically inert substance, the matrix, at low temperature. This implies that they are not able to interact with each other chemically and physically, that chemical reactions with the matrix do not occur, and that the excited electronic states of the solute molecules are, to a good approximation, not coupled to those of the matrix. Nevertheless, interaction of the sample molecules with the surrounding matrix atoms or molecules ensures vibrational and phonon relaxation (see below) to be so rapid that radiative emissions start exclusively from the lowest vibrational and phonon state of S_1 (for fluorescence) and T_1 (for phosphorescence), the lowest excited electronic states with singlet and triplet character, respectively.

The influence the solid matrix exercises on the spectral bandwidths of the dissolved molecules is of two kinds. First, there is an interaction between the guest molecules and the lattice vibrations (phonons) of the host. The strongly temperature-dependent, periodic movements of the matrix molecules or atoms result in a time-dependent electric field experienced by the guest molecules. Since the lattice vibrations are very fast compared to the lifetimes of the excited states, this interaction leads to a temperature-dependent homogeneous broadening of the spectral bands. At temperatures below 30 K this broadening leads to a broadening of not more than a few cm^{-1}, which does not impede the vibrational resolution of the low-temperature spectra (1.9,1.10). The coupling between the electronic transitions of the dissolved molecules and the matrix phonons also has a marked effect on the general shape of the spectral bands. These appear to consist of a narrow line, the "zero-phonon line," accompanied by a broad band, the "phonon wing," at its low-energy side in the emission spectrum. The phonon wing is the direct result of the electron–phonon coupling, which will discussed extensively in the next section.

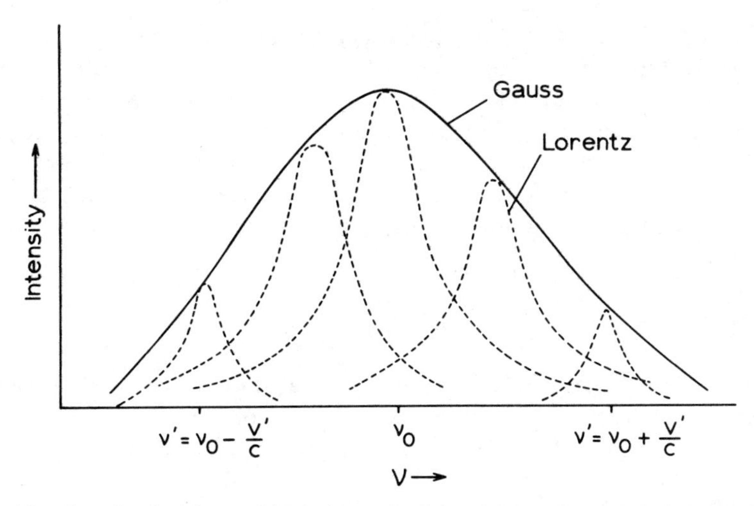

Fig. 4.3. Gaussian line shape which is the result of the addition of a statistical distribution of Lorentzian line shapes centered at different frequencies. This example demonstrates the Guassian Doppler profile, but is also exemplary for the inhomogeneous broadening in solid matrices.

Second, the energies of the electronic states are shifted under the influence of static interactions with the lattice (e.g., mechanical stress, electrostatic interactions, or the action of dispersion forces). If the host lattice is perfect, the interaction is the same for all guest molecules. Unfortunately, the shift generally depends strongly on the immediate environments (the "sites") of the solute molecules, which may differ very much. Spectra of molecules dissolved in amorphous solids (such as glasses), for instance, usually exhibit bandwidths of hundreds of cm^{-1}. This type of broadening, which is of course inhomogeneous in nature, is the main impediment to the observance of highly resolved spectra for low-temperature solid solutions. As illustrated in Fig. 4.3, the spectral bands observed consist of a large number of Lorentzian lines centered at different frequencies, which add up to a broad Gaussian profile.

All techniques aimed at improving the resolution of solid-state spectra strive for reduction of the inhomogeneous broadening in low-temperature solutions. This can be achieved in two ways. The first is by incorporating the molecules in a more uniform matrix. For polynuclear aromatic hydrocarbons (PAHs), line narrowing can be accomplished by dissolving the analyte either in a single crystal of another hydrocarbon with approximately the same molecular dimensions (in the so-called mixted-crystal technique) or in a suitable *n*-alkane (in the Shpol'skii technique). In this way linewidths of a few cm^{-1} can be achieved, sufficient for the recording of highly characteristic vibronically resolved spectra.

The second manner in which line narrowing can be achieved is more universally applicable. It appears that the homogeneous linewidth of bands in inhomogeneously broadened luminescence spectra can be extracted by making use of selective, narrow-line excitation. The laser is, of course, preeminently suited as an excitation source in this so-called fluorescence (phosphorescence) line-narrowing or optical site-selection technique.

Both types of techniques will be discussed in more detail in separate sections.

4.1.1.3. Fluid Samples

In fluid samples the linewidths of the spectral transitions of the dissolved molecules are greatly broadened by inelastic collisions with the solvent molecules, which induce radiationless transitions and thus shorten the lifetimes of the excited states. Since in liquids the mean time between successive inelastic collisions is of the order of 0.1–10 ps, this collisional broadening is of the homogeneous type. Furthermore, the solute molecules experience an extra contribution to the homogeneous broadening as a consequence of the strongly fluctuating interactions with their rapidly changing cage of solvent molecules. Since both types of broadening in the case of molecular samples are added to many closely spaced rotational–vibrational lines per electronic transition, absorption and emission spectra of fluid samples generally yield a broad continuum of signals with hardly any fine structure. The techniques employed to induce line narrowing in low-temperature amorphous systems will not be successful in the fluid state due to the nonstatic character of the guest–host interactions in the latter case. One way to obtain somewhat more resolved spectra in fluid samples is the application of double resonance excitation methods (1.11,1.12). In this case the analyte is brought first into a vibrationally excited state by selective laser excitation; then, before vibrational relaxation can take place, another laser excites the molecule to an upper electronic state, which may fluoresce. Since the vibrational energies are not strongly influenced by environmental effects, in this way better resolved emission spectra may also be obtained in fluid samples. This technique will not be considered in this chapter.

4.1.2. Outline of This Chapter

After a short introduction to processes that determine the linewidths of molecular fluorescence and phosphorescence spectra in the different aggregation states, in the following sections attention will be paid to three aspects of highly resolved molecular luminescence spectroscopy.

First, in Section 4.2 a theoretical account will be presented of factors that influence such spectra. Attention will be paid to both electronic and vibrational transitions in isolated molecules and to the (dynamic and static) influence of the environment on these molecular transitions.

Second, general instrumental aspects that are involved in the realization of vibrationally resolved luminescence spectra will be discussed in Section 4.3. The major part of this section will be devoted to the application of laser sources for excitation and to several types of cryogenic apparatus used to cool the samples. Also, the most commonly employed detection techniques will be considered.

The third and main part of this chapter is taken up by the major techniques that are employed to obtain highly resolved luminescence spectra. Most of these techniques make use of low-temperature solid samples. Successively, matrix isolation spectroscopy (Section 4.4), Shpol'skii spectroscopy (Section 4.5), and fluorescence and phosphorescence line-narrowing spectroscopy (Section 4.6) will be elaborately discussed. Section 4.7 is devoted to a gas-phase technique, supersonic jet spectroscopy. All sections describing the different experimental methods are modular in character: they can be studied separately. As a starting point for these sections the material presented in the general sections 4.1 to 4.3 suffices. In Sections 4.4 to 4.7, first, some characteristic features of the technique concerned will be commented upon. Both theoretical and experimental characteristics that have not been touched in the first three sections will be discussed. Furthermore, some general applications will be mentioned; a major part of the section, however, will be taken up by a discussion of the analytical applications and a presentation of the analytical state of the art.

Finally, in Section 4.8 a résumé of the accomplishments and pitfalls of the various techniques will be given. Also, in this section we will try to indicate some future tends in high-resolution molecular luminescence spectroscopy as a detection method in analytical chemistry.

References to Section 4.1

1.1. R. J. Hurtubise, *Anal. Chem.*, **55**, 669A (1983).
1.2. C. Gooijer, N. H. Velthorst, and R. W. Frei, *Trends Anal. Chem.*, **3**, 259 (1984).
1.3. J. W. Birks, Ed., *Organic Molecular Photophysics*, Vol. 1, Wiley, London, 1973.
1.4. R. Loudon, *The Quantum Theory of Light*, Clarendon Press, Oxford, 1973.
1.5. A. M. Stoneham, *Rev. Mod. Phys.*, **41**, 82 (1969).
1.6. A. Messiah, *Quantum Mechanics*, Vol. 1, North-Holland, Amsterdam, 1961.
1.7. W. J. Moore, *Physical Chemistry*, Longman, London, 1972.
1.8. P. R. Berman, *Appl. Phys.*, **6**, 283 (1975).
1.9. F. P. Burke and G. J. Small, *Chem. Phys.*, **5**, 198 (1974).
1.10. F. P. Burke and G. J. Small, *J. Chem. Phys.*, **61**, 4588 (1974).

1.11. A. Seilmeier, W. Kaiser and A. Laubereau, *Opt. Comm.*, **26**, 441 (1978).
1.12. J. C. Wright, *Appl. Spectrosc.*, **34**, 151 (1980).

4.2. THEORETICAL BACKGROUND

When a molecule is embedded in a more or less ideal crystal lattice (as in a mixed crystal or as in Shopl'skii matrix), or when a molecule is selectively excited by "monochromatic" laser light, inhomogeneous broadening of the luminescence spectra is removed to a large extent. The appearance of the vibrationally resolved spectra that are observed then is determined by two factors: first, the vibronic structure of the molecule involved, that is, the relative positions of its electronic and vibrational energy levels and their mutual interaction (vibronic coupling), and second, the interaction between the electronic transitions of the molecules and their surroundings. The latter interaction, of course, only plays a role in the solid-state high-resolution techniques. In supersonic jet spectroscopy the molecules are virtually isolated and thus the latter term is absent.

The Hamiltonian which describes the optically active molecule in an optically inert crystalline solid, the so-called impurity center, can be expressed as.

$$\mathscr{H}_{\text{impurity center}} = \mathscr{H}_{\text{isolated molecule}} + \mathscr{H}_{\text{molecule-matrix}} \tag{2.1}$$

In the latter term all interactions between the molecule and the matrix are incorporated. It is appropriate in this context to distinguish static and dynamic interactions. Static interactions lead to a shift and a possible splitting of the molecular transitions under the influence of the crystal field. The dynamic interactions lead to the typical appearance of the bands in solid-state high-resolution spectra, as is shown in Fig. 4.8. Due to the coupling of the molecular electronic transitions with the low-energy matrix lattice vibrations (the phonons), apart from narrow "quasi-lines," broad features are also observed in the spectra. The broad bands, generally referred to as phonon wings, are situated at the low-energy side of the narrow lines in emission spectra and are attributed to the creation or annihilation of matrix phonons during the electronic transition. The narrow zero-phonon lines correspond to transitions that occur in the molecule without a change in the number of matrix phonons. The ratio of the intensities of the sharp zero-phonon line and the broad phonon wing, which is a measure for the quality of the highly resolved spectrum, is expressed as the Debye–Waller factor:

$$\alpha = \frac{I_{\text{ZPL}}}{I_{\text{ZPL}} + I_{\text{PW}}} \tag{2.2}$$

The Debye–Waller factor depends strongly on the temperature and the strength of the electron–phonon coupling, experimental parameters that determine whether or not a high-resolution spectrum is obtained for a particular guest–host combination.

In highly resolved luminescence spectra, vibronic transitions can be observed. Therefore, in the theoretical treatment, first the vibronic spectra of isolated molecules will be discussed. Only the lowest excited electronic state needs to be considered because luminescence from higher excited electronic states, in general, is precluded by very rapid radiationless processes (2.1). Subsequently, extensive attention will be paid to electron–phonon coupling, which is most often the decisive factor for the observation of highly resolved spectra in the solid-state techniques. Discussion of the electron–phonon coupling will be limited to impurity centers in ideal host lattices, so that only homogeneous line-broadening effects play a role. Inhomogeneous broadening due to static interactions of the guests with different solvent cages will be dealt with qualitatively.

4.2.1. Spectra of Isolated Molecules

The vibronic spectrum for an isolated molecule can be obtained theoretically by solving the Schrödinger equation:

$$[\mathscr{H}_{\text{isolated molecule}} - E_{\text{vibronic}}]\psi_{\text{vibronic}}(q, Q) = 0 \tag{2.3}$$

The relevant Hamilton operator for an isolated molecule, neglecting rotational motions, depends on both electronic (q) and nuclear (Q) coordinates and is given by

$$\mathscr{H}_{\text{isolated molecule}} = T_N(Q) + T_e(q) + V(q, Q) \tag{2.4}$$

where $T_N(Q)$ and $T_e(q)$ represent the kinetic energies of nuclei and electrons, respectively, and $V(q, Q)$ is the potential energy of the system. Unfortunately, the associated wave function $\psi_{\text{vibronic}}(q, Q)$ cannot be calculated from the Schrödinger equation (2.3) because the potential energy $V(q, Q)$ depends in a complex way on nuclear and electronic coordinates. For this reason one generally assumes that the electrons adapt themselves more or less instantaneously to the nuclear configuration (the Born–Oppenheimer approximation (2.2–2.4). $\psi_{\text{vibronic}}(q, Q)$ can then be written as a product of an electronic wave function $\varphi_i(q, Q)$ and a vibrational wave function $\chi_{iv}(Q)$, that is,

$$\psi_{\text{vibronic}}(q, Q) \approx \psi_{i,v}(q, Q) = \varphi_i(q, Q)\chi_{iv}(Q) \tag{2.5}$$

The subscripts i and v denote an electronic state and a vibrational mode, respectively.

In the simplest approach to the solution of Eq. (2.3) it is assumed that the electronic wave function is independent of the nuclear motions (the crude Born–Oppenheimer, or Franck–Condon approximation). Then this function can be calculated for a fixed (equilibrium) nuclear configuration Q_0. The electronic Hamiltonian is set equal to

$$\mathscr{H}_e^\circ = T_e(Q) + V(q, Q_0) \tag{2.6}$$

The total vibronic wave function $\psi_{i,v}(q, Q)$ is approximated as

$$\psi_{i,v}^\circ(q, Q) = \varphi_i^\circ(q, Q_0)\chi_{iv}^\circ(Q) \tag{2.7}$$

$\chi_{iv}^\circ(Q)$ is only an approximation of $\chi_{iv}(Q)$ as the vibrational function in the crude Born–Oppenheimer approximation corresponds to the motion of the nuclei in the averaged stationary field of the electrons.

In luminescence spectra generally, the transition from $iv = 10$ to $iv = 0j$ is observed. The transition moment $\mathbf{M}_{10,0j}^\circ$ on the basis of Eq. (2.7) is given by

$$\mathbf{M}_{10,0j}^\circ = \langle\psi_{10}^\circ|\mu|\psi_{0j}^\circ\rangle = \langle\varphi_1^\circ|\mu|\varphi_0^\circ\rangle\langle\chi_{10}^\circ|\chi_{0j}^\circ\rangle \tag{2.8}$$

in which μ is the electric dipole operator that works only on the electronic wave functions. The intensity of the transition is proportional to $|\mathbf{M}_{10,0j}^\circ|^2$. The overall intensity of the transition is governed by the first (electronic) term in Eq. (2.8). The second term, the Franck–Condon overlap integral, determines the intensity distribution over fundamentals and overtones. Only those vibrations in the electronic ground state will, in principle, be visible that have a symmetry equal to that of χ_{10}, which is totally symmetric. Thus these transitions will have the same polarization as the purely electronic transition ($j = 0$), commonly denoted as the 0–0 transition (2.5,2.6).

Of course the treatment above is rather crude, because the electronic wave functions will actually depend on the nuclear motions. This dependence can be taken into account to some extent by means of perturbation theory applying the term $\Delta V(q, Q) = V(q, Q) - V(q, Q_0)$ as a perturbation operator. In this so-called Herzberg–Teller coupling scheme the electronic crude Born–Oppenheimer functions [which solve the Schrödinger equation with Eq. (2.6) as Hamiltonian] are regarded as a complete set from which the "adiabatic electronic wave functions," denoted as $\varphi_i'(q, Q)$, are constructed:

$$\varphi_i'(q, Q) = \sum_k c_{ik}(Q)\varphi_k^\circ(q, Q_0) \tag{2.9}$$

Of course, the coefficient c_{ii} will have the largest value since φ'_i will be much like φ°_i. Nevertheless, other crude Born–Oppenheimer functions φ°_k also play a role, especially those with an energy $E^\circ_{e,k}$ close to $E^\circ_{e,i}$. This can be readily shown. Expansion of $\Delta V(q, Q)$ as a Taylor series in the nuclear cordinates gives

$$\Delta V(q, Q) = \sum_n \left(\frac{\partial V}{\partial Q_n}\right)_{Q_0} Q_n + \frac{1}{2}\sum_n\sum_m \left(\frac{\partial^2 V}{\partial Q_m \partial Q_n}\right)_{Q_0} Q_m Q_n + \ldots \quad (2.10)$$

If only the linear terms are considered, this leads to

$$c_{ik}(Q) = \sum_n \frac{\langle \varphi^\circ_k | (\partial V/\partial Q_n) Q_0 | \varphi^\circ_i \rangle}{E^\circ_{e,k} - E^\circ_{e,i}} Q_n \quad (2.11)$$

Hence Eq. (2.9) can be rewritten as

$$\varphi'_i(q, Q) = \varphi^\circ_i + \sum_{k \neq i} \sum_n a^n_{ik} Q_n \varphi^\circ_k \quad (2.12)$$

Of course, the vibrational functions associated with $\varphi'_i(q, Q)$ will differ a little from χ_{iv}. If, for convenience, this difference is ignored, the transition moment can be expressed simply as

$$\mathbf{M}'_{10,0j} = \mathbf{M}^\circ_{10,0j} + \sum_{k \neq i} \sum_n a^n_{ik} \langle \varphi^\circ_1 | \mu | \varphi^\circ_k \rangle \langle \chi^\circ_{10} | Q_n | \chi^\circ_{0j} \rangle \quad (2.13)$$

Equation (2.13) implies that nontotally symmetric modes (for which $\mathbf{M}^\circ_{10,0j} = 0$) may also be observed in the spectra because there will be normal coordinates Q_n with the same symmetry as χ_{0j}. They "borrow" their intensities from higher electronic transitions that also determine the polarization direction of the vibronic band (see Fig. 4.4).

Since in the adiabatic approximation sketched above $\mathbf{M}'_{1j,00} = \mathbf{M}'_{10,0j}$, absorption and emission spectra are expected to be symmetric as far as intensities are concerned (2.7,2.8). However, severe deviations from absorption–emission symmetry are observed, especially if the excited electronic states are separated by a relatively small energy gap. Such effects can be explained to some extent by taking into account nonadiabatic coupling of the adiabatic wave functions with $T_N(Q)$ as a perturbation operator (2.7–2.12). When the excited electronic states are (nearly) degenerate, perturbation theory is no longer appropriate. (Pseudo-) Jahn–Teller effects will then play a role (2.2,2.13–2.15). Discussion of nonadiabatic coupling and Jahn–Teller effects on vibronic spectra falls outside the scope of this chapter. Very thorough and recent discussions on the theory and practice of vibronic coupling effects are given in refs. 2.16 and 2.17.

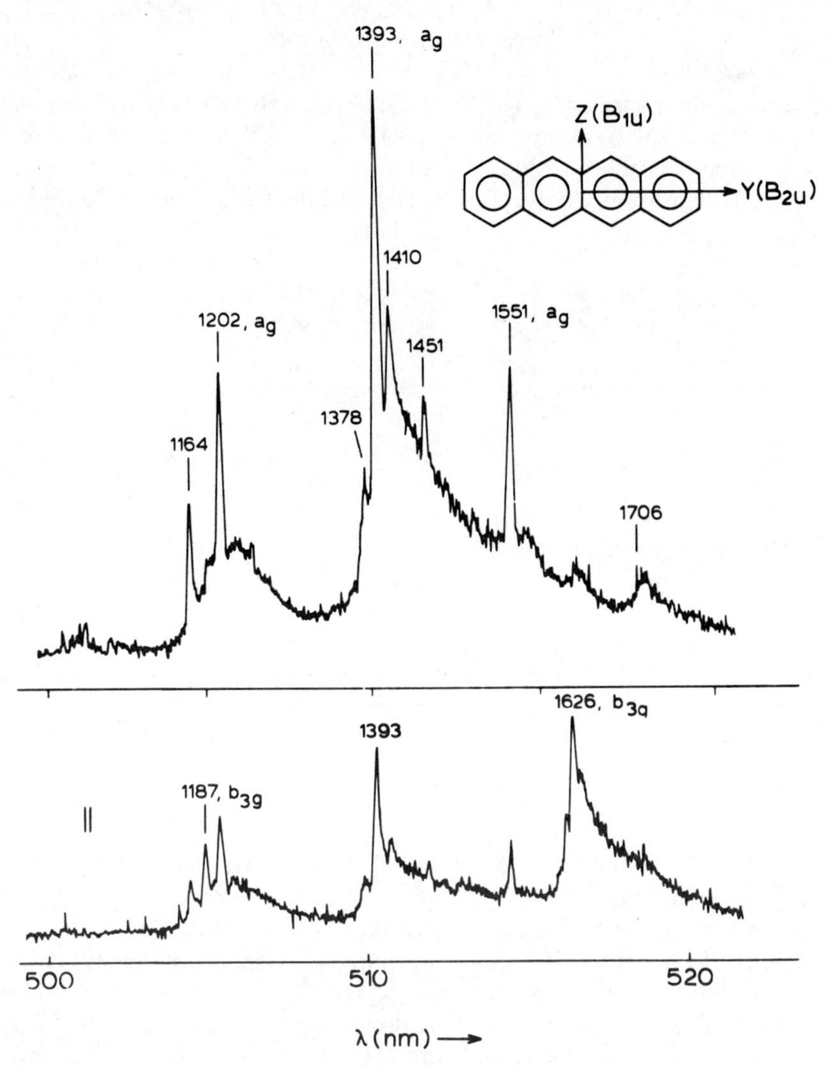

Fig. 4.4. Fluorescence line-narrowing spectra of tetracene in stretched polyethylene polymer film (T = 5 K, λ_{exc} = 476.5 nm). By stretching the polymer film the tetracene molecules are oriented so that their long axes are preferentially directed in the stretching direction. When polarized laser excitation is used, followed by polarized detection of the fluorescence emission, straightforward information on the symmetry of the individual vibronic lines is obtained. This information is vital for the clarification of vibronic coupling phenomena. The upper part of the figure shows the polarized fluorescence spectrum measured perpendicular to the stretching direction i.e., the direction of polarization of the $S_1(B_{1u})$–$S_0(A_{1g})$ 0–0 transition. The lower part shows the spectrum obtained for emission detected parallel to the stretching direction. Totally symmetric modes (like the 1164-, 1202-, 1393-, and 1551-cm^{-1}a_g modes) are clearly polarized in the same direction as the 0–0 transition; nontotally symmetric modes, the 1187- and 1626-cm^{-1}b_{3g} modes, which are Franck–Condon forbidden, are polarized perpendicular to the 0–0 transition. They become allowed through vibronic coupling between the $S_1(B_{1u})$ state and the S_2 and S_3 states, which both have B_{2u} symmetry (namely, the direct product of B_{1u}, B_{3g}, and B_{2u} in point group D_{2h} yields A_{1g}).

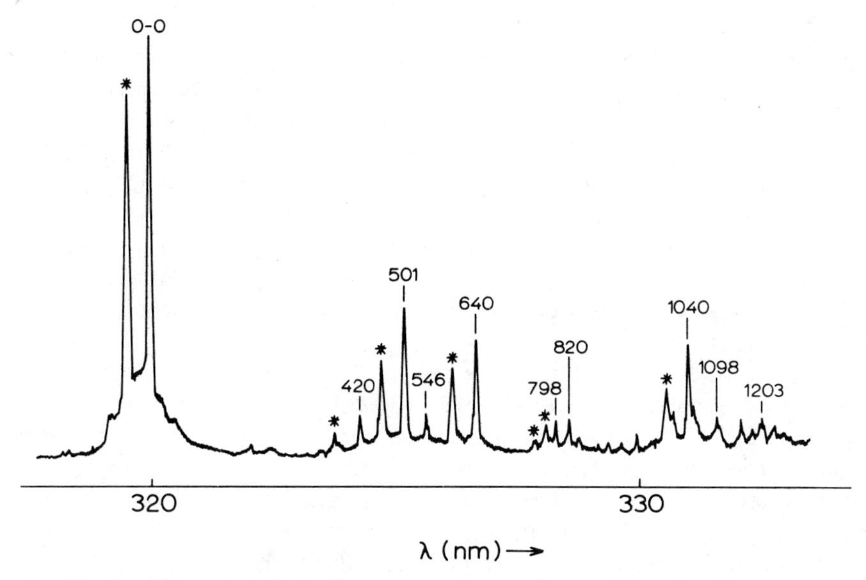

Fig. 4.5. Doublet structure in the fluorescence spectrum of acenaphthene in *n*-pentane at T = 20 K. The vibrational bands have been indicated by the wavenumbers; the peaks marked by an asterisk represent lines belonging to the high-energy site.

4.2.2. Interactions between Impurity Molecules and Host Lattice

For a discussion of the interactions between impurity molecules and the host lattice surrounding them, it is appropriate to divide the problem into three more or less independent parts. Hence the interaction Hamiltonian, $\mathscr{H}_{\text{molecule-matrix}}$, is written as

$$\mathscr{H}_{\text{molecule-matrix}} = \mathscr{H}_{\text{molecule-static matrix}} + \mathscr{H}_{\text{dynamic matrix}} + \mathscr{H}_{\text{molecule-dynamic matrix}} \quad (2.14)$$

The first term in Eq. (2.14), $\mathscr{H}_{\text{molecule-static matrix}}$, describes the influence of the averaged static environment of the molecule on, mainly, the electronic energy levels. Mostly, this influence results in a small red shift of the 0–0 transition, which increases with increasing polarizability of the host (2.18).

Static guest–host interactions may also lead to splitting of the peaks in the vibronic spectrum. In most cases multiplet structure in luminescence spectra is caused by multiple site effects. In this case, molecules in different types of solvent cages (sites) undergo slightly different shifts of their 0–0 transitions, leading to a multiplet structure in the spectra. An example of multiplet structure in the spectrum of acenaphthene in *n*-pentane at 20 K is given in Fig. 4.5. Another reason for matrix splitting may be the lifting of degeneracy as a result of a lowering of the symmetry of the dissolved molecule through the interaction with the solvent (2.19,2.20).

Finally, the static interaction between molecule and lattice is the predominant cause of inhomogeneous broadening. Variations in the crystal field due to strains in the crystal are a major source of broadening, but defects in the lattice also contribute (2.21). In amorphous matrix materials, such as glasses, linewidths of hundreds of cm^{-1} may be observed due to the strongly varying environment of the impurity centers in such inherently disordered media. The inhomogeneously broadened electronic band shape found for disordered solids is adequately described by a model using a statistical distribution of static site excitation energies (2.22). The band shape is thus given by a Gaussian function, as expected for an inhomogeneously broadened band (see Section 4.1.1).

The second term in Eq. (2.14), $\mathscr{H}_{\text{dynamic matrix}}$, represents the vibrational energy of the solid-state matrix, which is described in terms of lattice phonon modes (2.23,2.24). As the phonon transitions usually lie in the low-energy infrared region (they form a "continuum" of very close states with energies up to about 150 cm^{-1}), they do not interfere directly with the electronic spectra. The strength of the interaction between the electronic transitions of the impurity molecule and the phonon transitions of the host matrix, however, is crucial for the observation of highly resolved luminescence spectra in the solid state.

The so-called electron–phonon coupling, which is represented in Eq. (2.14) by the term $\mathscr{H}_{\text{molecule-dynamic matrix}}$, will be discussed separately in Section 4.2.3.

Finally, it is remarked that no interaction of the electronic transitions of the impurity centers with electronic transitions of the matrix molecules is taken into account. Since the electronically excited states of the host, giving the exciton band, lie in general at considerably higher energies than the relevant energy levels of the guest molecules, such interactions can be neglected in good approximation.

4.2.3. Electron–Phonon Coupling

The theoretical background and practical implications of electron–phonon coupling have been discussed extensively in the literature by Rebane (2.25) and, more recently, by Sapozhikov (2.26) and O'sadko (2.27, 2.28). In this section only a brief summary of the most relevant features will be given.

Before attempting to give a quantitative description of electron–phonon coupling, a simple qualitative impression will be presented. To a first approximation, the origin and shape of the vibronic lines of the molecular electronic transitions can be elucidated by use of Fig. 4.6. In this figure the electronic ground state (energy E_0) and first excited state (energy E_1) of an impurity molecule are depicted in interaction with a matrix phonon i (energy

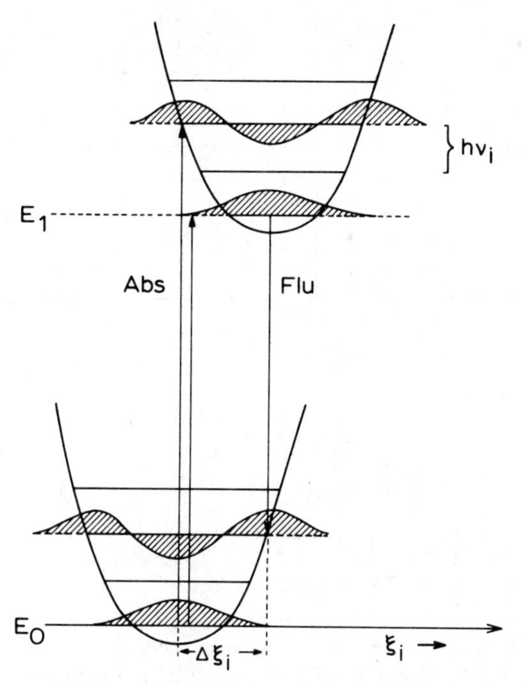

Fig. 4.6. Model to explain the origin of the phonon wing in solid-state electronic spectra. Both electronic (E_0,E_1) and phonon ($h\nu_i$) energy levels are represented. As phonons are collective vibrations of the host molecules, the coordinate ξ_i represents a normal coordinate of the host lattice.

$h\nu_i$). A number of phonon quanta, which all are taken to be described by the harmonic oscillator model, are incorporated in the figure. This implies that their wave functions can be represented by the striped shapes. Along the horizontal axis the lattice normal coordinate ξ_i, belonging to phonon mode i, is plotted. Because the interaction with the matrix is influenced by the electronic distribution of the guest, the minima in the curves for S_0 and S_1 generally lie at different values of ξ_i (the difference amounting to $\Delta\xi_i$). As for intramolecular vibronic transitions,the Franck–Condon principle applies, (i.e., the electronic transitions take place so fast that the nuclear positions will remain fixed). This means in Fig. 4.6 that transitions may be represented by vertical lines. As illustrated in the figure, a number of transitions can occur with, in most cases, simultaneous change of the vibrational state of the host and the electronic state of the guest. The relative probabilities of the transitions are given in good approximation by the Franck–Condon overlap integrals of the harmonic phonon wave functions (see Section 4.2.1 for the intramolecular analog). Depending on the magnitude of the displacement

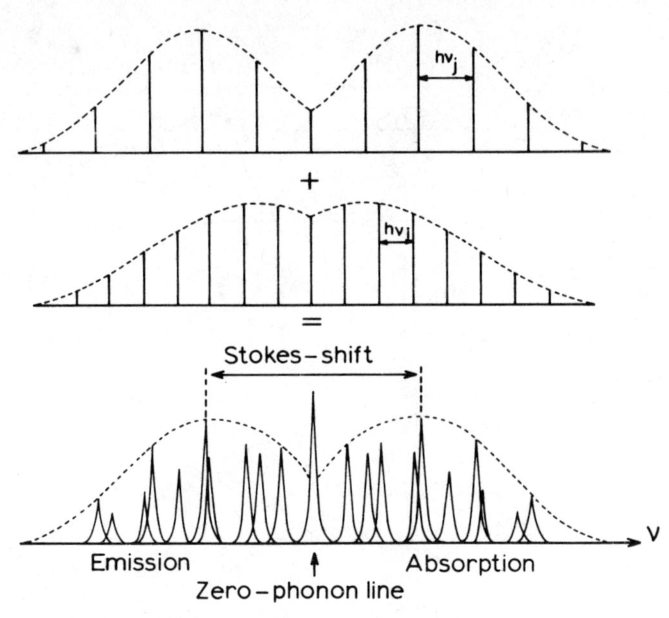

Fig. 4.7. Transition probabilities for two phonon modes, *i* and *j*. Note that only in the zero-phonon transition the contribution of the two phonon modes add. Upon addition of a large number of phonon modes, a sharp zero-phonon line and a broad, structureless phonon wing is observed.

$\Delta\xi_i$, one of the electron–phonon transitions will be the most probable. In Fig. 4.6 the two-phonon transition will be the most intense one. The transition probabilities of the different phonon transitions for mode *i* are depicted in Fig. 4.7. In the harmonic oscillator approximation a number of equidistant lines is observed; the distance between the intensity maxima of the electron-phonon transitions in absorption (at the high-frequency side of the zero-phonon transition) and emission (at its low-frequency side) is denoted as the Stokes shift. It is obvious that the larger the displacement $\Delta\xi_i$, the larger the Stokes shift and the smaller the intensity of the zero-phonon transition.

Evidently, a very large number of normal coordinates ξ_i, each with its characteristic frequency ν_i, is needed to describe the vibrational movements of the matrix. One therefore has to sum over a large number of electron–phonon spectra, as is illustrated in Fig. 4.7 for two modes, *i* and *j*, In practice there are many phonons, which are so close together that their states can be considered as a quasi-continuum. The summation over all these phonon states leads to the typical line form shown in Fig. 4.8 with its barely structured phonon wing. The sharp zero-phonon line can become intense for

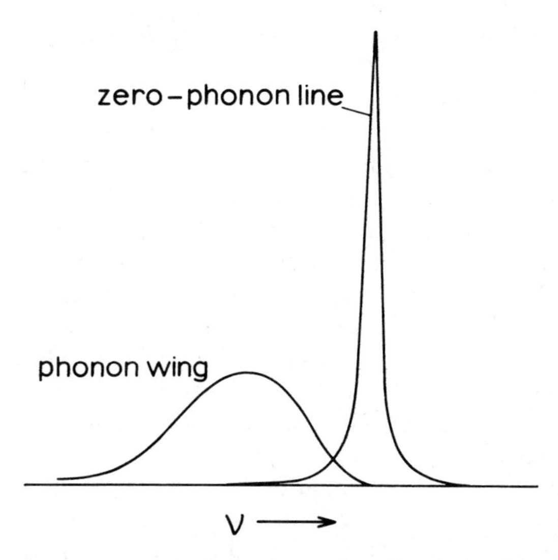

Fig. 4.8. Schematic representation of the shape of a vibronic band in the luminescence spectrum of a guest molecule in a solid matrix.

two reasons. First, the spectral transitions which are not accompanied by a change in phonon state have exactly the same energy for all ξ_i. Second, the natural broadening of these lines is relatively small because only the initial states involved in these transitions have a limited lifetime, while the final states live indefinitely. If the final state is a higher phonon state, its energy is not sharply defined since its lifetime is limited because of phonon relaxation [for a comparison, see Eq. (1.3)]. Of course, this point also supports the lack of structure in the phonon wing.

The theoretical description of electron–phonon coupling shows a strong resemblance with the theory outlined in Section 4.2.1. This can readily be conceived: in both cases one is concerned with vibronic interactions since phonons can be considered as intermolecular vibrations.

Fortunately, the simplest approximation outlined in Section 4.2.1 appears to suffice in most cases [i.e., an expression similar to Eq. (2.8) can be applied]. This means that the intensity distribution of the fundamentals and overtones of the phonon modes is governed by vibrational overlap integrals. For convenience we confine ourselves to electronic transitions in which the intramolecular vibrations do not change. For fluorescence, the probability of the transition from phonon mode p with quantum number $n = 0$ in the first excited electronic state to phonon mode p in the electronic ground state can be written as

$$P_{10,0p}(n) \sim |\langle\chi_{\mathrm{ph},1p}(0)|\chi_{\mathrm{ph},0p}(n)\rangle|^2|\langle\varphi_i^\circ|\mu|\varphi_0^\circ\rangle|^2 g(E - E_1 + nh\nu_i) \qquad (2.15)$$

The function $g(E - E_1 + nh\nu_i)$ is a line form function, which ensures sharp spectral features with distance $h\nu_i$ in the fluorescence spectrum. $\chi_{\mathrm{ph},1p}$ and $\chi_{\mathrm{ph},0p}$ represent the phonon wave functions for mode i in the electronic excited state and ground state, respectively, φ_1° and φ_2° are the electronic wavefunctions of the excited and ground states. The total intensity per phonon quantum follows from a summation over all phonon modes.

The crucial point in the evaluation of $P_{10,0p}(n)$ is the calculation of the overlap integrals $\langle\chi_{\mathrm{ph},1p}(0)|\chi_{\mathrm{ph},0p}(n)\rangle$. To facilitate the calculation, in general, use is made of the harmonic approximation, which applies well for small displacements of the nuclei. In this approximation the potential in the vibrational Schrödinger equation is taken to depend exclusively on the squared displacements of the nuclear positions. The phonon wave functions that solve this Schrödinger equation appear to be harmonic oscillator functions. Thus the problem of calculating the vibrational overlap integrals is reduced to the calculation of the overlap between two harmonic oscillator functions (one in the excited and one in the ground state of the molecule). This problem can be solved elegantly by expanding the change of adiabatic potential upon electronic excitation, $\Delta V(Q) = V_1(Q) - V_0(Q)$, in the normal coordinates of the ground-state crystal:

$$\Delta V(Q) = \sum_i U_i \xi_i + \frac{1}{2}\sum_{i,j} W_{ij}\xi_i\xi_j \qquad (2.16)$$

In this way the two harmonic oscillator functions are defined in the space spanned by the same set of normal coordinates, which greatly facilitates the calculations.

The implications of Eq. (2.16) are twofold. If only linear coupling, represented by the first term, is taken into account, the situation sketched in the qualitative example above is encountered. The excited-state potential has the same shape as the ground-state potential, but is shifted in the space spanned by the crystal normal coordinates. The overlap between two displaced harmonic oscillators has been derived by Keil (2.29). The linear electron–phonon coupling appears adequate to describe the Debye–Waller factor [Eq. (2.2)] and its temperature dependence. If the total intensity of the band shape is normalized to 1, the Debye–Waller factor is given by

$$\alpha = \frac{I_{\mathrm{ZPL}}}{I_{\mathrm{ZPL}} + I_{\mathrm{PW}}} = \sum_i |\langle\chi_{\mathrm{ph},1p}(0)|\chi_{\mathrm{ph},0p}(0)\rangle|^2 \qquad (2.17)$$

This factor can easily be calculated for $T \to 0$ (2.30):

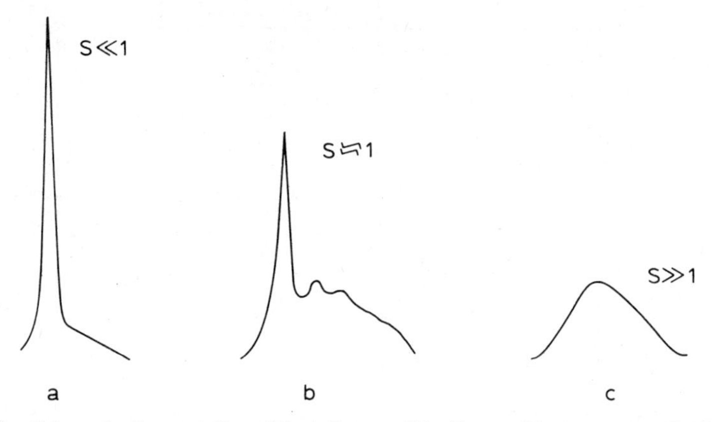

Fig. 4.9. Schematic representation of the influence of the Huang–Rhys parameter S, which is a measure for the strength of the electron-phonon coupling, on the shape of the vibronic bands in solid-state electronic spectra.

$$\alpha(T = 0) = \exp(-S) \tag{2.18}$$

where

$$S(T = 0) = \frac{1}{2}\sum_i \Delta\xi_i^2 \frac{\mu_i \nu_i}{h} \tag{2.19}$$

μ_i and ν_i are reduced mass and frequency, respectively, of modi i. The Huang–Rhys parameter S is a dimensionless quantity that indicates the strength of the electron–phonon coupling for a particular guest–host combination. The influence of S on the band shape is depicted schematically in Fig. 4.9. For strong electron–phonon coupling ($S \gg 1$) no narrow zero-phonon line will be observed in the spectra, not even at very low temperature.

The temperature dependence of the zero-phonon line intensity can be expressed as

$$I_{\mathrm{ZPL}} \sim \exp\left[-S + \frac{4S}{\nu_D{}^2}\int_0^{\nu_0} \frac{\nu}{\exp(h\nu/kT) - 1}\,d\nu\right] \tag{2.20}$$

In this equation it is assumed that the Debye distribution gives a good description of the density of states of the phonons. ν_D is the Debye frequency for the matrix under consideration. In the limit of low temperature,

$$I_{\mathrm{ZPL}} \sim \exp(-S)\exp[-8S(kT/h\nu_D)^2] \tag{2.21}$$

It follows that at very low temperatures the zero-phonon line intensity

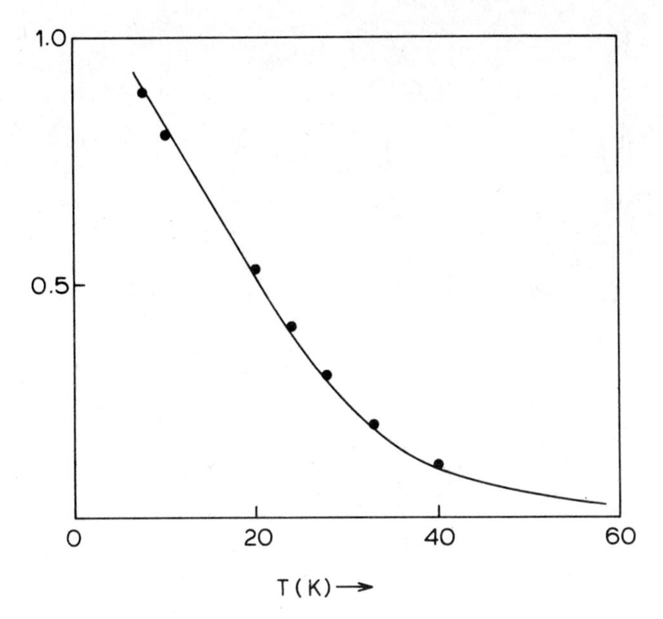

Fig. 4.10. Temperature dependence of the Debije–Waller factor α for perylene in *n*-heptane.

decreases approximately quadratically with increasing temperature; this decrease is accompanied by a broadening and increase in intensity of the phonon wing. An impression of the temperature dependence of the zero-phonon line intensity reduction for perylene in *n*-heptane is given in Fig. 4.10.

The second term of Eq. (2.16) represents the quadratic electron–phonon coupling. It incorporates the change in shape of the potential of the excited state with respect to that of the ground state. The quadratic coupling coefficients W_{ij} describe the change in energy of the excited state resulting from the change in the normal oscillator frequencies during excitation ($i = j$) and from the mixing of the normal coordinates in the excited state [($i \neq j$, Duschinsky effect (2.30)]. A detailed description of quadratic electron–phonon coupling effects will not be presented here. These effects are manifested in the thermal broadening and shift of the impurity zero-phonon lines. It appears that in general, the temperature effects on position and width of the zero-phonon lines are small under 40 K (2.26,2.31,2.32).

The theory can be further refined by introducing Herzberg–Teller coupling (2.27,2.28). The effect of this type of coupling on the electron–phonon band shape is a departure from mirror symmetry of conjugate phonon wings in absorption and emission. In contrast to the quadratic electron–phonon coupling, the Herzberg–Teller coupling may also influence the temperature dependence of the Debye–Waller factor (2.33).

The electron–phonon coupling theory as sketched above in principle applies to perfect crystals, as it makes use of the Debye theory to describe the phonon density of states. It also gives only the homogeneous band shapes. Thus it is not surprising that the theory gives a good description of the features of mixed crystal and Shpol'skii spectra [although the spectral band shapes in such spectra also contain a significant inhomogeneous contribution (2.34)], but is less accurate for fluorescence (phosphorescence) line-narrowed spectra. The main deviations from the theory concern the widths of the homogeneous zero-phonon lines in amorphous matrices which are typically one to two orders of magnitude larger than those observed in mixed molecular crystals (2.35,2.36), and the much stronger temperature dependence of the Debye–Waller factor. The discrepancies can be accounted for in two ways. First, by realizing that apart from the delocalized acoustic phonon modes amorphous materials also possess many very low-frequency quasi-localized modes (2.37). The density of states of these modes is, of course, characterized by local peaks and thus much different from the Debye density of state. Second, the anomalies may be explained by modeling the amorphous solid as a broad distribution of two-level systems (2.38,2.39), which may be coupled electrostatically to the optical impurity levels (2.40). Those theoretical aspects will be discussed elaborately in the section on fluorescence (phosphorescence) line-narrowing spectroscopy.

References to Section 4.2

2.1. M. Kasha, *Discuss. Faraday Soc.*, **9**, 14 (1950).
2.2. H.C. Longuet-Higgins, in *Advances in Spectroscopy*, Vol. 2, Wiley, New York, 1969, p. 429.
2.3. T. Azumi and K. Matsuzaki, *Photochem. Photobiol.*, **25**, 315 (1977).
2.4. G. Herzberg, *Electronic Spectra of Polyatomic Molecules*, D. Van Nostrand, Princeton, N.J., 1966.
2.5. W. Siebrand, *J. Chem. Phys.*, **46**, 440 (1967).
2.6. R. M. Hochstrasser, *Molecular Aspects of Symmetry*, W. A. Benjamin, New York, 1966.
2.7. P. A. Geldof, R. P. H. Rettschnik, and G. J. Hoytink, *Chem. Phys. Lett.*, **10**, 549 (1971).
2.8. A. R. Gregory, W. H. Henneker, W. Siebrand, and M. Z. Zgierski, *J. Chem. Phys.*, **65**, 2071 (1976).
2.9. R. M. Hochstrasser and C. A. Marzzacco, in E. C. Lim, Ed., *Molecular Luminescence*, W. A. Benjamin, New York, 1969.
2.10. N. Kanamaru and E. C. Lim, *Chem. Phys. Lett.*, **35**, 303 (1975).
2.11. N. Kanamaru and E. C. Lim, *Chem. Phys.*, **10**, 141 (1975).
2.12. G. Orlandi and W. Siebrand, *J. Chem. Phys.*, **58**, 4513 (1973).
2.13. H. C. Longuet-Higgins, U. Öpik, M. H. L. Pryce, and R. A. Sack, *Proc. R. Soc. A*, **244**, 1 (1958).

2.14. C. S. Sloane and R. Silbey, *J. Chem. Phys.*, **56**, 6031 (1972).
2.15. R. Englman, *The Jahn-Teller Effect in Molecules and Crystals*, Wiley, New York, 1972.
2.16. I. B. Bersuker, *The Jahn-Teller Effect and Vibronic Interactions in Modern Chemistry*, Plenum Press, New York, 1984.
2.17. H. Köppel, W. Domcke, and L. S. Cederbaum, in I. Prigogine and S. A. Rice, Eds., *Advances in Chemical Physics*, Vol. LVII, Wiley, New York, 1984, p. 59.
2.18. B. Meyer, *Low Temperature Spectroscopy*, Elsevier, New York, 1971.
2.19. W. M. Pitts, A. M. Merle, and M. A. El-Sayed, *Chem. Phys.*, **36**, 437 (1979).
2.20. G. Jansen, M. Noort, N. van Dijk, and J. H. van der Waals, *Mol. Phys.*, **39**, 865 (1980).
2.21. A. M. Stoneham, *Rev. Mod. Phys.*, **41**, 82 (1969).
2.22. J. Klafter and J. Jortner, *Chem. Phys.*, **26**, 421 (1977).
2.23. M. Born and K. Huang, *The Dynamical Theory of Crystal Lattices*, Clarendon Press, Oxford, 1954.
2.24. S. Califano, V. Schettino, and N. Neto, *Lattice Dynamics of Molecular Crystals*, Lecture Notes in Chemistry, Vol. 26, Springer-Verlag, New York, 1981.
2.25. K. K. Rebane, *Impurity Spectra of Solids*, Plenum Press, New York, 1970.
2.26. M. N. Sapozhnikov, *Phys. Status Solidi (b)*, **75**, 11 (1976).
2.27. I. S. O'sadko, *Sov. Phys. Usp.*, **22**, 311 (1979).
2.28. I. S. O'sadko, in V. M. Agranovich and R. M. Hochstrasser, Eds., *Spectroscopy and Excitation Dynamics of Condensed Molecular Systems*, North-Holland, Amsterdam, 1983.
2.29. T. Keil, *Phys. Rev.*, **140**, A601 (1965).
2.30. F. Duschinsky, *Acta Physicochem.*, **7**, 551 (1937).
2.31. F. P. Burke and G. J. Small, *Chem. Phys.*, **5**, 198 (1974).
2.32. F. P. Burke and G. J. Small, *J. Chem. Phys.*, **61**, 4588 (1974).
2.33. O. N. Korotaev and M. Yu. Kaliteevskii, *Sov. Phys. JETP*, **52**, 220 (1980).
2.34. A. P. Marchetti, W. C. McColgin, and J. H. Eberly, *Phys. Rev. Lett.*, **35**, 387 (1975).
2.35. P. M. Selzer, D. L. Huber, D. S. Hamilton, W. M. Yen, and M. J. Weber, *Phys. Rev. Lett.*, **36**, 813 (1976).
2.36. S. Völker, R. M. Macfarlane, and J. H. van der Waals, *Chem. Phys. Lett.*, **53**, 8 (1978).
2.37. G. Winterling, *Phys. Rev. B*, **12**, 2432 (1975).
2.38. P. W. Anderson, B. I. Halperin, and C. M. Varma, *Philos. Mag.*, **25**, 1 (1972).
2.39. W. A. Phillips, *J. Low. Temp. Phys.*, **7**, 351 (1972).
2.40. H. Morawitz and P. Reineker, *Solid State Commun.*, **42**, 609 (1982).

4.3. INSTRUMENTAL ASPECTS

In this section some instrumental aspects which are relevant for the achievement of highly resolved luminescence spectra will be discussed. A rough division has been made into three parts: excitation system, sample compartment, and detection system. Attention will be paid only to those features that are of general importance for all high-resolution techniques; specific aspects will be discussed in the sections devoted to the technique concerned.

4.3.1. Excitation

Under certain conditions (i.e., when the inhomogeneous broadening of the spectral lines is sufficiently reduced by the employment of a suitable matrix material or in a supersonic jet) highly resolved *emission* spectra can be obtained via conventional broad-banded light sources. Xenon, mercury, tungsten,and deuterium lamps cover the entire spectral range relevant for the detection of all kinds of molecules. The properties of such lamp sources are elaborately described in Parker's standard work (3.1).

High-resolution *excitation* spectra require the use of a monochromator with sufficient spectral resolving power (bandwidths down to 0.2 nm must be attained to gain satisfactory information) in combination with a lamp that produces a continuous wavelength output. Unfortunately, this condition requires extremely narrow monochromator slits, so that only very low light intensities are obtained, which makes high-resolution excitation spectra of less analytical use. Nevertheless, selectivity can be enhanced by choosing optimal excitation wavelengths if a number of emitting compounds are present in the sample.

The development of the laser or, more precisely, the availability of laser systems which are easy to operate has had a strong impact on high-resolution luminescence spectroscopy. For fluorescence and phosphorescence line-narrowing spectroscopy the laser is a basic requirement. Shpol'skii, matrix isolation, and supersonic jet spectroscopy can, in principle,be done using classical excitation sources, but in these fields the introduction of the laser has also created tremendous possibilities.

Lasers have several important advantages over classical radiation sources, such as the mercury and xenon lamp. Of particular importance for applications in analytical spectroscopy are their high degree of monochromaticity and spatial coherence, which makes the photon density that is available for excitation of the sample extremely high. The latter property is usually denoted as "laser intensity," a subject that is discussed below. Other important features, depending on the particular type of laser under consideration, are the relatively low source noise and the ability to produce pulses which are short on the fluorescence time scale. As disadvantages of laser systems their relative complexity and high cost must be mentioned. The most serious disadvantage, however, is that especially at short wavelengths ($\lambda < 320$ nm), where many compounds have their highest absorptions, it is difficult to obtain wavelength-tunable laser output (use can be made of nonlinear frequency mixing techniques; see below). At present the laser tenchnology is still a rapid developing field.

In this chapter we limit ourselves to some general remarks with regard to the applicability of lasers in analytical spectroscopy. For a more detailed discussion of lasers and their applications, one is referred to the literature (3.2–3.5).

The point of laser intensity deserves some comment. It is noted that in a luminescence experiment the power of the excitation source is not a good indication of the number of sample molecules that can be excited. Essential is the number of photons with a defined wavelength that in a certain period of time "collides" with the sample molecules. This number is determined on the one hand by the brightness of the light source and on the other hand by the divergence of the radiation. Brightness is equivalent to radiance, that is, the power of the electromagnetic radiation (joule/second = watt) per unit area of the source (m^{-2}) per unit solid angle ($steradian^{-1}$). In comparison to other light sources, lasers in general do not have extremely high powers, but both the emittive surface of the source and the solid angle are very small. As a consequence of the small solid angle (a divergence of less than 0.5 mrad can be achieved) the collimation of the laser beam is very good, which implies that almost 100% of the available radiation can be used in a luminescence experiment. For comparison, classical sources emit equally in all directions (solid angle 4π) so that in practice no more than 15% of the available radiation can be employed unless one resorts to complex light collection systems.

Furthermore, it is emphasized that power is energy of radiation per second, so that in a pulsed source, during the pulse the power can be very high (e.g., 1 joule in 10^{-6} second corresponds to a power of 1 megawatt during the pulse); the mean power, however, is determined by the pulse width and the repetition frequency (the number of pulses per second).

Finally, it is important to realize that under many circumstances it is the *spectral* radiant power that determines the experimental possibilities [i.e., the power per wavelength (W/nm)]. Lasers are highly monochromatic compared to classical sources. Hence, the spectral radiant power of laser sources is very high, so that even with spectral bandwidths of 0.1 nm and lower, luminescence experiments can be performed with reasonable sensitivities. For classical sources spectral bandwidths lower than 2 nm are not appropriate.

Two comments have to be made. First, from a fundamental point of view it is obvious that the luminescence intensity does not increase unlimitedly upon enhancing the laser excitation power, because of saturation effects. These are, however, much more probable in atomic than in molecular spectroscopy, because atomic absorption cross sections are four to five orders of magnitude higher than molecular ones (3.5). Second, it is not always detrimental to the detection limits to employ conventional lamp sources with low spectral radiant power. When compounds are studied with unresolved spectra, effective use can be made of broad-banded excitation. Van Geel and Winefordner, for instance, reported comparable detection limits when studying lamp and laser excited fluorescence of several organic molecules in liquid solutions (3.7). For compounds with highly resolved absorption spectra, such as found in Shpol'skii matrices and in supersonic jets, the availability of an excitation source with high spectral radiant power is, of course, favorable.

Within this context it is appropriate to discuss in short the most important characteristics of a number of laser systems that have been or are expected to become applied in (high resolution) molecular luminescence spectroscopy: the N_2 laser, the He-Cd laser, the frequency tripled or quadrupled Nd-YAG (or Nd-glass) laser, the Ar- and Kr-ion gas lasers, the excimer lasers, and finally, the various dye laser types that can be used in combination with suitable pump lasers or with flash lamps.

In Table 4.1 the main characteristics of the pump lasers are collected (see also Fig. 4.11). Details of the principles of operation can be found in the *Laser Handbooks* (3.8,3.9). A disadvantage of these lasers is that their output is discontinuous in wavelength. This severely limits the applicability of such systems in fluorescence and phosphorescence line-narrowing spectroscopy, since then the laser excitation wavelength sould be energetically not too far from the S_1–S_0 0–0 transition. Continuously tunable radiation can be obtained by employing a dye laser. In such a laser an organic dye is excited by a laser source or a flash lamp, the fluorescence form the dye being the active medium in the dye laser. By incorporating a wavelength-selective element in the laser cavity the dye laser output can be continuously varied over the fluorescence envelope of the dye. For one dye thus a wavelength-scannable region of 40–60 nm can be achieved, so that several dyes are needed to obtain tunability over the whole spectral range; that range is determined mainly by the pumping source. Also, the other characteristics [pulsed or continuous-wave (CW) power] strongly depend on the manner of excitation. Typical values for the most important characteristics of dye lasers in several pumping modes are summarized in Table 4.2 (see also Fig. 4.12). The construction of the dye laser in pulsed and CW systems is generally different (3.10): In the ion-laser pumped dye lasers the active medium is usually formed by a free jet of dye solution formed through a polished nozzle; in the pulsed laser systems (nitrogen, excimer, and Nd-YAG) the dye solution is most often contained in some sort of cuvette. The latter configuration is, of course, profitable when the laser dyes have to be changed regularly. Flash lamp pumped dye lasers have the advantage that they do not need an expensive pump laser; they also

Table 4.1. Characteristics of Relevant Laser Sources[a]

Type	Wavelength (nm)	Output Power[b] (W)	Pulse Repetition Rate (Hz)	Pulse Length (ns)
Excimer				
ArF	193	Up to 10	0.1–200	10
KrF	249	Up to 25	0.1–125	20
XeCl	308	Up to 8	0.1–200	20
XeF	351	Up to 7	0.1–100	20
Nitrogen	337	0.001–0.3	Up to 1000	0.3–10
Ion				
Argon	several lines, 330–530	0.005–20	CW[c,d]	—
Krypton	several lines, 330–800	0.005–6	CW[d]	—
He-Cd	325	0.002–0.01	CW	—
	442	0.002–0.04	CW	—
He-Ne	633	0.0001–0.05	CW	—
Nd-YAG	1064	Up to 600	CW[e]	—
	532[f]	Up to 10		
	355[g]	Up to 3		
	266[h]	Up to 1		

[a] For details on commercial systems, see *Laser Focus* Buyers' Guide, 20th ed. 1985, or *Lasers and Applications* 1985 Designers' Handbook and Product Directory.
[b] For pulsed laser systems the average power is indicated.
[c] CW, continuous wave.
[d] Can also be operated in the mode-locked mode; then typical picosecond pulses are obtained at a repetition rate of 76 MHz (for a 1.80-m-long ion laser).
[e] Can also be driven by a flash lamp, yielding pulsed output with a repetition rate of 0.05–300 Hz and an average power up to 400 W.
[f] Frequency-doubled 1064-nm line.
[g] Frequency-tripled 1064-nm line.
[h] Frequency-quadrupled 1064-nm line.

have a typical construction, different from the laser pumped systems (3.10). The spectral properties of the flash lamp pumped dye lasers are generally worse than those of the laser pumped ones. Especially, they supply long pulse lengths (typically hundreds of nanoseconds long) and when a high degree of monochromaticity is required, provide relatively low powers.

A very important feature of laser application in analytical spectroscopy is the attainable wavelength region. To be able to detect all types of compounds it is necessary to have laser output available in the UV. Excimer lasers

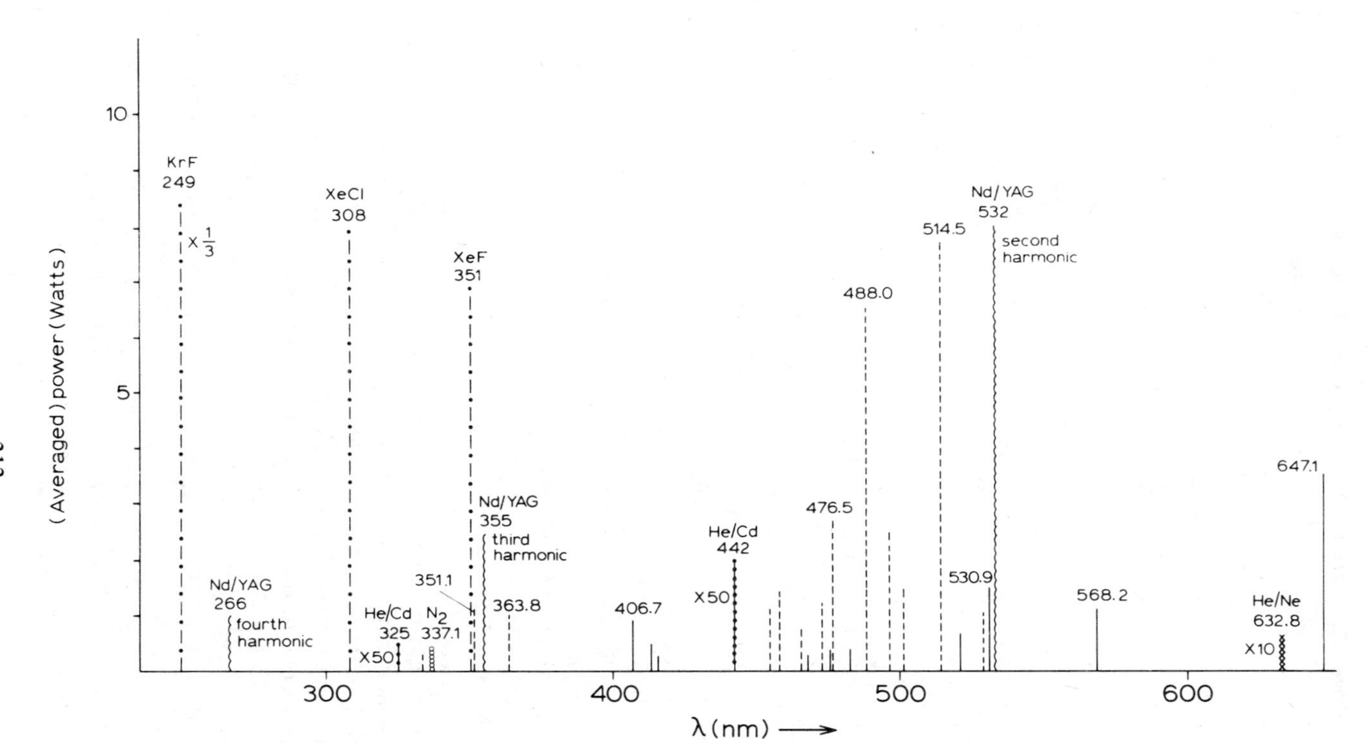

Fig. 4.11. Overview of the accessible wavelength region for pump laser sources (see Table 4.1). The different pump laser types are indicated. Most laser sources provide only one emission wavelength. Exceptions are the argon-ion and krypton-ion lasers, which have a number of emitting lines in the UV and visible region. Argon-ion laser lines are indicated by dashes lines, krypton-ion laser lines with solid lines.

Table 4.2. Characteristics of Dye Laser Performance[a]

Pumping Mode	Wavelength (nm)	Output Power[b] (W)	Pulse Repetition Rate (Hz)	Pulse Length (ns)
Excimer	320–1000	Up to 5	0.1–200	3–20
Flash lamp[c]	335–850	0.0001–50	0.03–50	200–1000
Nitrogen	360–1000	0.0002–0.2	Up to 1000	0.5–10
Nd-YAG (frequency tripled)	380–950	Up to 2	0.05–40	0.05–12
Ion	400–1000	0.02–1	CW[d]	—

[a] For details on commercial systems see *Laser Focus* Buyers' Guide, 20th ed. 1985, or *Lasers and Applications* 1985 Designers' Handbook and Product Directory.
[b] For pulsed laser systems the average power is indicated. Of course, the output power also depends on the dye imployed.
[c] The bandwidth of the most powerful flash lamp pumped dye lasers is too large to be applied for high-resolution spectroscopic purposes; then the upper limit becomes 1 W.
[d] Continuous wave; if synchronously pumped by a mode-locked ion laser, picosecond pulses can be attained at a typical repetition rate of 76 MHz (when a 1.80-m-long ion laser is used as a pump source).

provide intense output in the UV (apart from the types mentioned in Table 4.1, a number of other excimers can be used, such as Ar_2 at 126 nm, Kr_2 at 146 nm, Xe_2 at 170 nm, KrCl at 222 nm, and XeBr at 282 nm), but unfortunately they operate only in a very narrow wavelength region. Another way to obtain laser operation in the UV is based on the nonlinear interaction of intense radiation with atoms or molecules in crystals or in the liquid or gaseous phase. Second (or even third and fourth, as mentioned for Nd-YAG in Table 4.1), harmonic generation (3.11), sum frequency mixing (3.12–3.14) and stimulated Raman scattering (3.15) can be used to reach far into the UV. When nonlinear techniques are employed in combination with a dye laser, wavelength-tunable UV laser light can also be obtained. Since the efficiency of nonlinear processes is strongly dependent on the intensity of the fundamental light, the highest outputs are generally realized with pulsed laser sources which have high peak powers. For CW lasers the efficiency is low; the highest output is obtained when the nonlinear element is incorporated in the laser cavity (3.16,3.17). An overview of the accessible wavelength region for different types of laser sources is presented in Figs. 4.11 and 4.12.

An alternative to nonlinear optical techniques for probing molecules which absorb in the UV can be the application of two-photon excitation (3.18). Due to the very high intensity of the fundamental laser sources, two-photon excited luminescence can become practical even though the transition probabilities are very low.

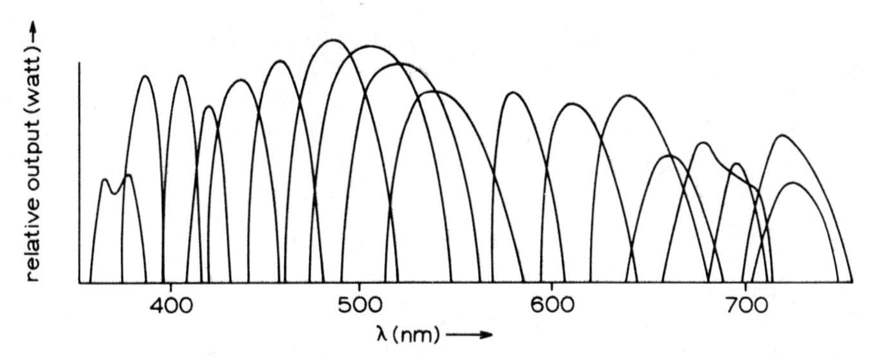

Fig. 4.12. Overview of the accessible wavelength region for dye laser sources. the curves in this figure are typical for a N_2-laser pumped dye laser. Dyes that are frequently used are polyphenyls (blue-violet), coumarins (green), and rhodamines (red) (see also Table 4.2).

4.3.2. Sample

For all high-resolution luminescence techniques described in this chapter the sample needs to be measured at low temperatures. Except in the case of supersonic jet spectroscopy, which will be discussed separately, cooling is achieved by employing cryogenic liquids. With Shpol'skii spectroscopy, temperatures of 77 K can be sufficient to observe reasonably highly resolved spectra. Temperatures down to 63 K are obtainable with nitrogen under reduced pressure as a cooling substance. Better resolution in Shpol'skii spectroscopy is created at lower temperatures; for fluorescence and phosphorescence line-narrowing spectroscopy and matrix isolation spectroscopy, very low temperatures are indispensable. Such temperatures are reached with helium as a refrigerant.

Two major types of helium cryostats can be discerned. In the simplest construction the sample is immersed in liquid helium. A sample temperature between 4.2 and 1.2 K (at reduced pressure) can be achieved in this way. A more sophisticated design involves a sample chamber separated from the liquid helium reservoir by a valve through which liquid or vapor can pass. In combination with a heater, temperatures between 2 and 300K can be maintained in such a system. Both types of apparatus described above involve cooling by direct thermal contact between sample and (liquid or gaseous) refrigerant. Thus optimal cooling is assured. A disadvantage of this type of cooling is the presence of several sets of optical windows on excitation and detection side, which is not only expensive but also leads to significant losses of radiation, and the possible disturbance of the measurement by the presence of the cryogenic liquid or gas in the optical pathways (e.g., bubbles in the case of boiling helium at 4.2 K).

In the second type of cryostat the sample, placed in a vacuum space, is attached to a cold finger which is in contact with the cryogenic medium on the other side. In this setup cooling is accomplished via thermal conductance. Crucial is a good thermal contact between the sample holder and the cold tip, which can be improved by employing a high-thermal-conductivity varnish or indium sheet as intermediate. Nevertheless, temperature gradients can occur in the sample, especially when a material with poor thermal conductance is studied. Local heating effects may play an important role when high-intensity laser excitation is used. Heating as a result of 300-K radiation can be minimized by attaching a gilded copper radiation shield to the cold tip, surrounding the sample as completely as possible. An advantage of this type of cryostat is that only one set of optical windows is present on the excitation and emission sides and that no cryogenic liquid or gas comes into the light-path.

Two type of conductance cryostats may be discerned. In the first type, liquid helium is sprayed against the cold tip; in combination with a heater, temperatures between 3.5 and 300 K can be achieved in such a continuous-flow cryostat. The second type makes use of expansion of helium gas near the cold finger for cooling. In the closed-cycle helium refrigerator the flow is cyclic, so that no expensive liquid helium is consumed. Depending on the number of cold stages, temperatures of 10–20 or 5–10K can be attained. When a heater is employed in combination with such a system, the whole temperature range up to 300 K is available.

Temperatures can be measured by employing resistor thermometers (e.g., carbon, carbon/glass, germanium, platinum, rhodium/iron), thermocouples (e.g., copper/constantan, chromel/alumel, gold/iron/chromel), or diodes (silicon, gallium arsenide). Of these, the carbon, carbon/glass, and to a lesser extent germanium resistors can be applied only in the lower temperature range (up to 50 K). The platinum resistor becomes impractical below 40 K. The other temperature sensors can be employed over the whole temperature range of interest. For measurement of temperatures below 4.2 K in a helium bath cryostat one can use an accurate manometer to establish the reduced pressure, which is a good measure of the temperature.

4.3.3. Detection

The manner of detection is strongly dependent on the excitation light source used. When laser excitation is employed, strong scattering effects and severe problems with "ghosts" in the emission monochromator are to be expected, due to the high degree of directionality and coherence of the laser light. So laser excitation imposes special requirements on the detection side, particularly in fluorescence and phosphorescence line-narrowing spectroscopy, where the difference between excitation and emission wavelengths is small. There are several ways to reduce the effects of laser scattering, depending on whether the laser operates in the continuous or the pulsed mode. When time-continuous excitation is applied the laser scattering can be removed by a high-quality emission monochromator. A double monochromator (consisting of two monochromators in tandem) is especially suitable for this purpose since it has excellent stray-light rejection properties. In another approach, modulated excitation of the sample followed by phase-sensitive detection is employed (3.19). For a pulsed laser, gated detection is the solution to scattering problems. Since scattering phenomena occur on a much faster time scale than do fluorescence and, most certainly, phosphorescence, the laser stray light can be totally removed via time-resolved methods. In a boxcar integrator a window of 100 ps to 2ms width can be obtained with a variable delay time after the laser pulse. Repetition rates up to 5 MHz are feasible, which is sufficient for pulsed lasers as N_2 and excimer lasers but too slow for mode-locked CW lasers and synchronously pumped dye lasers (the repetition frequency for a 1.80-m-long Ar-ion system, for instance, is approximately 76 MHz). Pulse picking—with as a disadvantage a reduction of laser intensity—or cavity dumping will also render the latter systems compatible with boxcar integrators. Apart from boxcar integrators, gated photon counters or sampling scopes are also suited to achieve temporal resolution. For detection of long-living phosphorescence, scattering can be removed by simple mechanical shuttering (rotating can). Time resolution is not appropriate only to reduce scattering. For pulsed laser sources, as a N_2 or excimer system, there is an additional advantage since such lasers have very low duty cycles (e.g., 10^{-7} for a N_2 laser at 10 Hz). Also, temporal resolution can be employed as an extra means to distinguish between compounds on the basis of differences in luminescence lifetimes.

Several principles of detection are equally valid for conventional and laser excitation. In order to observe the high-resolution spectra, high-quality monochromators are necessary to separate the emitted light. To discriminate among structural isomers a resolution of 0.2 nm is generally sufficient. This resolution can readily be achieved by employing a relatively simple 0.3-m monochromator (3.20). In this configuration the bandwidths observed are, in general, instrumentally determined. The limiting bandwidth that can be obtained with a monochromator lies in the 0.01 nm range. To measure such small bandwidths very narrow slits have to be used, which impedes sensitive detection. Higher resolution can be achieved with an interferometer. The resolution obtained via such an apparatus is, at present, of no analytical use.

In the simplest configuration, detection can be realized via dc integration of the voltage generated across a load resistor on the output of a (preferably cooled) photomultiplier tube (PMT). Under certain conditions use of a lock-in amplifier after the PMT may be profitable. Then the excitation light is chopped at, for instance, 1 kHz and the emitted light is detected in phase. Thus low-frequency noise may be eliminated. This method is not feasible for combination with a low-repetition-rate pulsed laser. Low-level light intensities can be detected more effectively with a photon counting system coupled to a fast PMT (rise time in the low-nanosecond range). In such an instrument the photoelectrons are counted one-by-one over a certain period of time instead of measuring the averaged photocurrent. PMT dark-current effects can be suppressed by setting a discriminator threshold which eliminates low-voltage pulses. Additionally, coupling to a computer affording sophisticated data handling is facilitated since the photon counting instrument transforms the irregular photoelectron pulses into standard TTL pulses. Extremely low limits of detection can be measured via the time-correlated photon counting technique (3.21). Photon counting instruments cannot be used when a low-repetition-rate pulsed laser is applied as excitation source.

Instead of a PMT, a photodiode can be employed. Photodiodes have a much faster response than PMTs, but generally a lower amplification factor. A large number of photodiodes can be mounted in an array which may be placed in the monochromator at the place of the exit slit. In this way a big part of the luminescence spectrum can be detected simultaneously. Such photodiode array detectors may offer similar sensitivity as PMTs, especially if "sensitivity enhanced" avalanche diodes are applied, and give a tremendous gain in time. The apparatus needed to digest the impressive stream of data from the array, however, is costly. Other instruments that can act as "optical multichannel analyzers" are the silicon-intensified target (SIT), intensified SIT (ISIT), or vidicon cameras, also coupled to a microprocessor for data storage. OMAs of this kind contain a target consisting of a microscopic array of photodiodes which are illuminated on one side by the signal to be detected and are scanned on the other side by a focused electron beam (diameter about 20 μm) to retrieve the signal (3.22).

References to Section 4.3

3.1. C.A. Parker, *Photoluminescence of Solutions*, Elsevier, Amsterdam, 1968.
3.2. O. Svelto, *Principles of Lasers*, Heyden, London, 1976.
3.3. W. Demtröder, *Laser Spectroscopy*, Springer-Verleg, New York, 1982.
3.4. N. Omenetto, Ed., *Analytical Laser Spectroscopy*, Wiley, New York, 1979.
3.5. T. R. Evans, Ed., *Applications of Lasers to Chemical Problems*, Wiley, New York, 1982.
3.6. J. C. Wright, in T.R. Evans, Ed., *Applications of Lasers to Chemical Problems*, Wiley, New York, 1982, Chap. 2.
3.7. T. F. van Geel and J. D. Winefordner, *Anal. Chem.*, **48**, 335 (1976).

3.8. F. T. Arecchi and E. O. Schultz-Dubois, Eds., *Laser Handbook*, Vols. 1 and 2, North-Holland, Amsterdam, 1972.
3.9. M. L. Stitch, *Laser Handbook*, Vol. 3, North-Holland, Amsterdam, 1979.
3.10. F. P. Schäfer, Ed., *Dye Lasers*, Springer-Verlag, New York, 1973.
3.11. H. Dewey, *IEEE J. Quantum Electron.*, **QE-12**, 303 (1976).
3.12. G. A. Massey and J. C. Johnson, *IEEE J. Quantum Electron.*, **QE-12**, 721 (1976).
3.13. S. Blit, E. G. Weaver, F. B. Dunnings, and F. K. Tittel, *Opt. Lett.*, **1**, 58, (1977).
3.14. F. B. Dunnings, *Laser Focus*, **14**, (5), 72 (1978).
3.15. V. Wilke and W. Schmidt, *Appl. Phys.*, **18**, 177 (1979).
3.16. M. Brieger, H. Büsener, A. Hese, F. van Moers, and A. Renn, *Opt. Commun.*, **38**, 423 (1981).
3.17. B. Couillaud, Ph. Dabkiewicz, L. A. Bloomfield, and T. W. Hänsch, *Opt. Lett.*, **7**, 265 (1982).
3.18. M. J. Wirth and F. E. Lytle, *ACS Symp. Ser.*, **85**, 24 (1978).
3.19. C. Th. J. Alkemade, *Appl. Spectrosc.*, **35**, 1 (1981).
3.20. J. C. Brown, J. A. Duncanson, Jr., and G. J. Small, *Anal. Chem.*, **52**, 1711 (1980).
3.21. V. J. Koester and R. M. Dowben, *Rev. Sci. Instrum.*, **49**, 1186 (1978).
3.22. L. Perko, J. Haas, and D. Osten, *Proc. SPIE*, **116**, 56 (1977).

4.4. MATRIX ISOLATION FLUORESCENCE SPECTROSCOPY

One of the older methods successfully used to obtain better resolution in fluorescence spectra is the matrix isolation (MI) technique. An example of a MI fluorescence spectrum obtained for a PAH mixture using mercury lamp excitation is shown in Fig. 4.13. Especially if MI sample preparation techniques are combined with laser-induced fluorescence to reduce residual inhomogeneous broadening, highly resolved fluorescence spectra are obtained which are appropriate for both qualitative and quantitative analysis. For inorganic samples, one of the first applications of such a combination was published in 1969 by Shirk and Bass (4.1). Nevertheless, data on organic compounds have been reported only quite recently (i.e., from the late 1970s onward) (4.2). In the foregoing period the spectroscopic techniques applied in combination with MI were almost exclusively infrared (IR) and electron spin resonance (ESR) and, in some cases, UV-Vis absorption spectroscopy. Even now these techniques dominate the field.

MI is a cryogenic sample preparation technique originally developed in the 1950s to study transient (especially inorganic) species (4.3,4). In the simplest approach the sample is vaporized and mixed with an inert gas, the matrix gas. Then the mixture is deposited on a cryogenic surface held at a temperature of 15 K or less, which can easily be obtained with closed-cycle helium cryostats or liquid-helium cryostats. As matrix material, originally

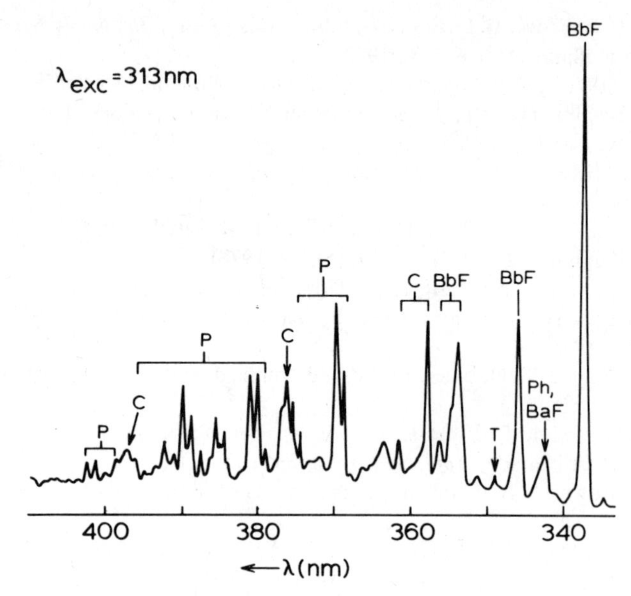

Fig. 4.13. MI fluorescence spectra for a six-component PAH mixture in nitrogen at 16 K. The sample contained 0.25 μg of benz[*b*]fluorene (B*b*F), 0.38 μg of benz[*a*]fluorene (B*a*F), 0.40 μg of chrysene (C), 0.50 μg of pyrene (P), 0.50 μg of phenanthrene (Ph), and 1.37 μg of triphenylene (T). For excitation a 2500-W mercury-xenon lamp was used (λ_{exc} = 313 nm, bandpass 7 nm). (Reprinted with permission from R. C. Stroupe, P. Tokousbalides, R. B. Dickinson, Jr., E. L. Wehry, and G. Mamantov, *Anal. Chem.*, **49**, 701. Copyright 1977 American Chemical Society.)

the noble gases Ne, Ar, and Xe were most frequently used, but for organic compunds N_2 and even *n*-alkanes and *n*-perfluoroalkanes are also applied.

The purpose of this sample handling technique is to prepare a solid matrix in which the "solute" species are "matrix isolated." This means that the following criteria hold:

1. No solute aggregates have been formed in the matrix; any solute molecule interacts solely with matrix atoms or molecules as near neighbors. The matrix material is chosen so as to minimize these interactions. The isolation of the solute molecules is achieved by using a large excess of the diluent gas (in general, a ratio of 10^4–10^8 on a molar basis is employed).
2. Diffusion of the solute molecules and rearrangement of the matrix lattice are both inhibited because of the low temperature of the sample.
3. Chemical reactions of the solute with the matrix or with other solvated species are excluded.

Considering these experimental conditions, it is obvious that the MI technique is very suitable to attain highly resolved luminescence spectra of organic compounds for quantitative and qualitative analysis. In the first place, the technique is well established. Detailed descriptions have been presented in the literature, for instance in the book of Cradock and Hinchcliffe (4.5) and the monograph edited by Barnes et al. (4.6). The MI technique is quite simple to apply, especially to stable and readily vaporized compounds. Second, aggregation of solutes, a phenomenon that provides a serious impediment to the acquisition of high-resolution spectra of polar compounds in frozen solutions, can easily be prevented. Third, circumstances can be chosen so that intermolecular energy transfer (also long-range transfer according to the Förster mechanism, which can be effective over distances longer than 5 nm) can be excluded. This is particularly important if quantitative analysis of mixtures composed of a number of fluorescent compounds is performed. Intermolecular interactions between these solutes make quantitative analysis inaccurate. Finally, the matrices applied in MI are inert so that the interaction between the species to be measured and the matrix are expected to be very weak, which will be favorable in terms of the Debye–Waller factor. Hence it is anticipated that sharp spectral lines are attainable, provided that inhomogeneous line broadening can be sufficiently reduced.

If lamp excitation is performed, the spectral resolution observed for organic compounds in conventional matrices as nitrogen and argon appears to be slightly better than that obtained in glassy frozen solutions. However, linewidths are definitely poorer than those that may be achieved in Shpol'skii matrices (4.2,4.7). Apparently, inhomogeneous broadening is far from negligible in MI fluorimetry. Therefore, the laser has been introduced in the MI fluorescence technique. In this way the inhomogeneous broadening is reduced by utilizing the fluorescence line-narrowing effect (see Section 4.6) (4.8). On the other hand, "Shpol'skii-like" matrix gases such as *n*-alkanes and *n*-perfluoroalkanes have been applied with the aim of obtaining narrower, Shpol'skii-like spectra, even upon conventional continuum source excitation (see Section 4.5) (4.9). As illustrated in Fig. 4.14, when an annealing procedure is employed, the same highly resolved spectrum is obtained in the MI sample (Fig. 4.14c) as in the sample prepared by freezing the liquid *n*-alkane solution (Fig. 4.14a).

4.4.1. Experimental Aspects

4.4.1.1. Temperature Control

In the conventional MI technique temperature plays a central role since there

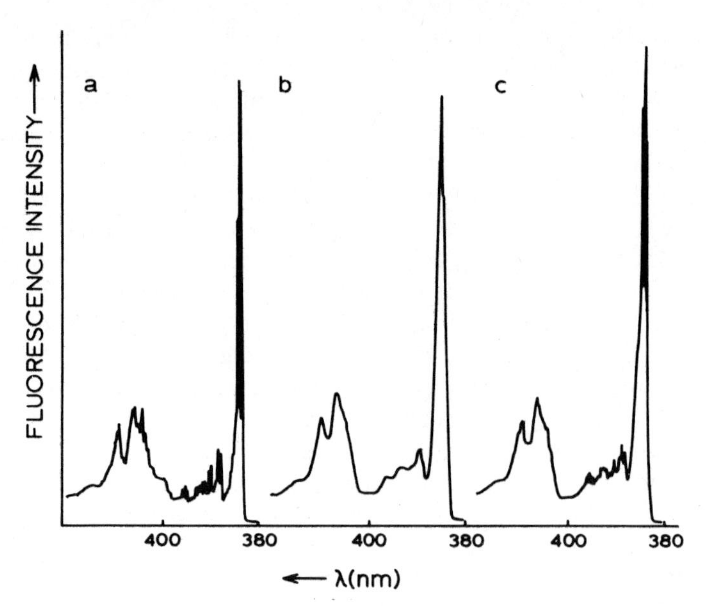

Fig. 4.14. Fluorescence spectra of benz[*a*]anthracene in *n*-heptane at 16 K (λ_{exc} = 278 nm): (*a*) in frozen solution (c = 1.4×10^{-5} *M*); (*b*) matrix isolated (benz[*a*]anthracene/heptane mole ratio = 1:200,000) before annealing; (*c*) after annealing B at 140 K. (Reprinted with permission from P. Tokousbalides, E. L. Wehry, and G. Mamantov, *J. Phys. Chem.*, **81**, 1769. Copyright 1977 American Chemical Society.)

is a need not only for low (cryogenic) temperatures but also for precise temperature control and scanning. It should be realized that during the matrix deposition process heat is released. Consequently, local annealing and (especially for small species) diffusion can take place, although the temperature of the surface seems to be low enough to prevent these processes. Within this context it is illustrative to mention the rules of thumb for inorganic MI: If T_m is the matrix melting temperature in K, below $0.3T_m$ the matrix is considered absolutely rigid, from $0.3T_m$ to $0.5T_m$ annealing may occur, and above $0.5T_m$ diffusion of the trapped species must also be taken into account (4.10). As a result, for reactive species only very slow deposition rates can be applied, so that sample preparation may require hours (4.11). Fortunately, for bulky organic molecules diffusion is expected only near the melting point, so that rapid sample preparation is allowed. In some cases, for instance with *n*-alkanes or perfluoroalkanes as matrix materials, annealing of the sample after deposition is necessary to obtain better resolution of the spectra.

Furthermore it should be realized that local changes in temperature may be destructive in fluorescence line-narrowing spectroscopy. This aspect is currently studied in our laboratory. It is expected that local heating of the microenvironment of the species concerned is especially important if they have low fluorescence quantum yields, so that a large fraction of the absorbed (intense) laser light is transferred into heat. Similar effects are well known to occur in experiments where reactive species are formed by photolysis of trapped compounds ("photolysis in situ").

4.4.1.2 Sample Preparation

In MI, samples are vaporized, mixed with the matrix gas, and the mixture is finally deposited on the cold surface. Hence the technique can be applied straightforwardly to samples which are readily vaporized and composed of thermally stable compounds. Unfortunately, it is not easily applicable to compounds that cannot be vaporized without decomposition, as are frequently encountered in clinical and biochemical analysis. The necessary vapor pressure of the solute must be 10^{-2}–10^{-4} torr; to reach such a pressure, ovens are frequently utilized. In MI one has to be alert for aggregation problems which may be observed for molecules which are strongly associated in the gas phase. As stressed before, however, aggregation effects will play a much smaller role in MI than in the other low-temperature techniques, which, in general, are based on liquid solutions for sample preparation.

Furthermore, it should be realized that all kinds of impurities may be condensed on the extremely cold surface, because the matrix is formed under high-vacuum conditions (about 10^{-6} torr). The more volatile impurities present in the sample will play an especially important role.

For MI fluorescence in organic analysis the matrix gas is present in large excess, mostly at least 10^4 on a molar basis. It is generally assumed that the matrix gases can be freed from fluorescent contaminants relatively easily, a favorable point as compared to Shpol'skii and fluorescence line-narrowing spectroscopy performed with organic solvents, which are harder to purify. Nevertheless, in MI fluorescence spectroscopy detection limits are also background limited. If *n*-alkanes and perfluoroalkanes are applied as matrix materials, backgrounds are much higher than with the conventional matrix gases. Nitrogen gas has proven to be a very suitable matrix for general-purpose spectroscopic measurements: for instance, for MI Fourier transform infrared (FT/IR) spectroscopy.

Apart from the classical continuous, slow, spray-on deposition, more recently, pulsed deposition (4.12–4.14) and pulse expansion deposition (4.15) have also been employed. Pulsed deposition can be much faster than spray-on deposition and will thus have fewer interferences from impurities introduced via small leaks in the vacuum system. In pulse expansion deposition the sample is prepared via expansion through a small orifice with a high backing pressure. During the expansion the sample is cooled before the cold surface is reached (see Section 4.7), resulting in very clear matrices with minimal aggregation.

The experimental procedures used in MI are elaborately described in refs. 4.5, 4.10 and 4.16.

4.4.2. Applications

Currently, MI sample preparation techniques are used mainly in combination with vibrational spectroscopy, in particular FT/IR. Fluorescence and UV/Vis absorption studies of organic compounds in recent years have been devoted mainly to unstable reactive species (4.17). Andrews' group has paid special attention to the study of UV/Vis absorption spectra of matrix-isolated organic cations (see, e.g., refs. 4.18–4.20). Bondybey's group has been concerned with laser-induced fluorescence emission and excitation spectra of halobenzene radical cations (4.21,4.22) and is interested mainly in the study of molecular relaxation processes. Extensive overviews on applications of MI on organic and inorganic substances have appeared in the Specialist Periodical Reports of the Chemical Society in London by Downs and Peake (4.23) and Chadwick (4.24,4.25). Applications in analytical chemistry have been studied almost exclusively by Wehry's group and were devoted mainly to polynuclear aromatic hydrocarbons (PAHs).

4.4.3. Analytical Aspects

A thorough discussion of the analytical aspects of organic MI luminescence spectroscopy has been presented by Wehry and Mamantov (4.26) and Wehry et al. (4.27). The obvious goal of the method is to acquire reliable qualitative and quantitative results for mixtures of considerable complexity.

The fundamental requirement is that the characteristic spectral features, that is, the wavelengths of maximum excitation and emission and the luminescence quantum yield are totally independent of the composition of the sample. In other words, the effects of quenching, intermolecular interactions, and long-range energy transfer must be negligible. These conditions appear to be simply fulfilled in MI fluorimetry for the compounds and samples investigated thus far.

With broad-band lamp excitation for intense fluorescent compounds, detection limits down to 10 pg have been reported as well as linear dynamic ranges from the detection limit up to 0.1–1 μg (i.e. more than five decades) (4.28). If a laser is used for excitation, so that fluorescence line narrowing is induced, depending on the matrix material (either conventional matrix gas or Shpol'skii solvent) a more or less impressive improvement in spectral

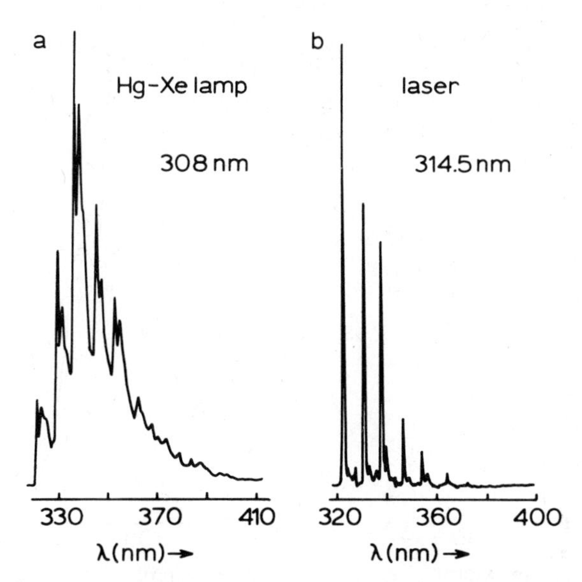

Fig. 4.15. MI fluorescence spectra of 600 ng of 2,7-dihydroxynaphthalene in perfluoro-*n*-hexane with (*a*) lamp excitation (λ_{exc} = 308 nm); (*b*) dye laser excitation (λ_{exc} = 314.5 nm). (Reprinted with permission from J. R. Maple and E. L. Wehry, *Anal. Chem.*, **53**, 266. Copyright 1981 American Chemical Society.)

resolution is obtained (see Fig. 4.15). For complex samples the qualitative analysis is thus greatly facilitated. Unfortunately, spectra measured via laser excitation exhibit an excitation wavelength-dependent site structure in addition to the high spectral resolution, which may complicate quantitative analysis. Tunability of the laser over the whole spectral range of interest is therefore a necessary requirement in order to obtain relatively simple fluorescence spectra for compounds to be quantitated. Furthermore, it has been observed that laser excitation in general does not further improve detection limits compared to broad-band excitation, as detection limits are determined by the solvent blank fluorescence. In addition, laser excitation may cause problems due to the local heating of the sample, which are easily produced since the thermal conductivity of the matrix materials is relatively low. Such heating may lead to unwanted diffusion of the solutes or annealing of the matrix.

Probably, better detection limits will be attainable if more advanced techniques are applied, such as suppression of solvent Rayleigh scattering via time-resolved measurements (4.7) or by means of polarizers (4.30).

In favorable cases, time resolution can also be utilized to facilitate qualitative analysis (4.29). To this end, fluorescence lifetimes of the sample

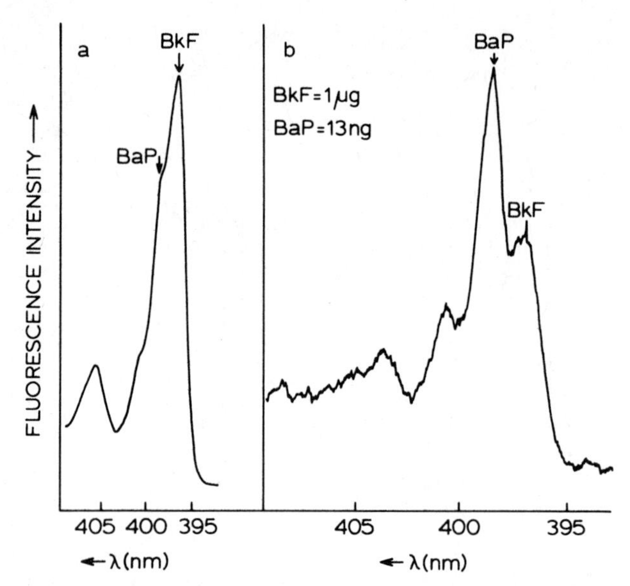

Fig. 4.16. Steady-state (*a*) and time-resolved, delay = 90 ns, (*b*) MI fluorescence spectra of a mixture of 1 μg of benz[*k*]fluoranthene and 15 ng of benz[*a*]pyrene. (Reprinted with permission from R. B. Dickinson, Jr., and E. L. Wehry, *Anal. Chem.*, **51**, 778. Copyright 1979 American Chemical Society.)

constituents must be significantly different. Unfortunately, this is, in general, not the case for compounds with similar structure. Application of a 90-ns delay in the fluorescence detection of a mixture of benz[*a*]pyrene and benz[*k*]fluoranthene, compounds with very similar spectra affords the sensitive detection of benz[*a*]pyrene even in the presence of a large excess of benz[*k*]fluoranthene (see Fig.4.16).

Finally, it is emphasized that internal standardization is necessary for quantitative analysis, as not all analyte molecules initially present in the sample will finally be deposited on the cold window. For most quantitative applications, however, a simple internal quantitation procedure appears to be sufficient (4.28).

4.4.4. Analytical State of the Art

MI high-resolution luminescence spectroscopy for analytical purposes has been developed almost exclusively by Wehry's group. PAHs have been measured in highly complex samples as chromatographic fractions of coking plant wastewater (4.31–4.33), untreated solvent refined coal (4.7,4.32–4.34), liquid chromatographic fractions of a shale oil (4.35), and Synthoil coal

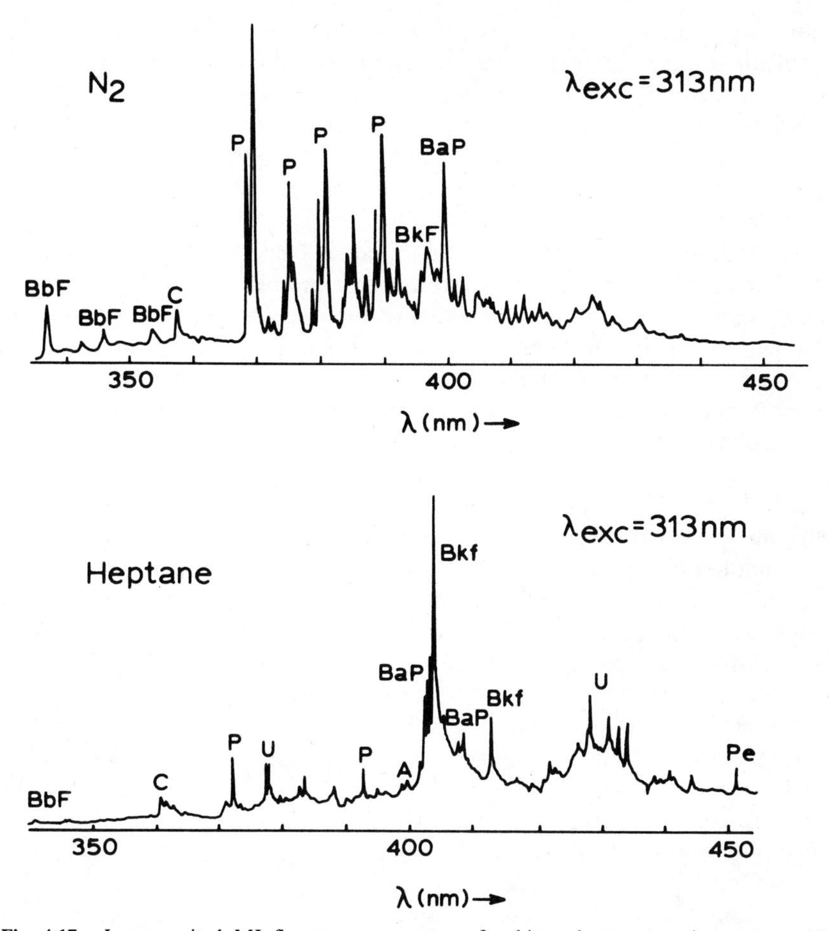

Fig. 4.17. Lamp-excited MI fluorescence spectra of coking plant water chromatographic fraction in nitrogen (top) and *n*-heptane (bottom) matrices at 15 K. Compounds: benzo-[*b*]fluorene (B[*b*]F), chrysene (C), pyrene (P), benzo[*k*]fluoranthene (B[*k*]F), benzo[*a*]pyrene (B[*a*]P), perylene (Pe). U, Unidentified. (Reprinted with permission from J. R. Maple, E. L. Wehry, and G. Mamantov, *Anal. Chem.*, **52**, 920. Copyright 1980 American Chemical Society.)

liquid (4.28). In the coking plant fraction, for instance, at least six PAHs could be identified simultaneously with mercury-xenon lamp excitation of both the N_2 and *n*-heptane matrix isolated sample (4.33) (see Fig. 4.17). The advantages of extremely selective laser excitation have been demonstrated for benz[*a*]pyrene determination in solvent-refined coal (4.7) and in coking plant water (4.33). In the former experiment the untreated coal-derived material was matrix isolated in *n*-octane, thus providing a Shpol'skii-like absorption spectrum so that selective excitation with a dye laser could be applied. In the latter experiment benz[*a*]pyrene was selectively excited in a chromatographic fraction of a coking plant sample. The selectivity of the MI fluorescence technique is very nicely illustrated by the monomethylchrysenes (4.36): The six isomeric compunds, difficult to identify with other methods, could be measured separately with MI fluorescence without any chemical or physical separation procedure.

Time resolution has been applied to reduce background scatter (4.7) and to discriminate between benz[*k*]fluoranthene and benzo[*a*]pyrene (4.29), two compounds with very similar, closelying fluorescence spectra (see Fig. 4.16). Under certain circumstances *n*-alkanes and perfluoroalkanes have shown to be useful as matrix materials (4.9,4.33,4.34). Recently, a hybrid of MI and mixed-crystal techniques has been reported (4.37): Chloronaphthalenes were vapor-deposited in a naphthalene matrix. It appeared that each of the chloronaphthalenes in the 10-component mixture could be selectively laser excited without interferences from the other compounds. For 2-chloronaphthalene linear calibration curves were obtained from 10 pg to 100 ng using two-photon excitation.

Thus far, few reports on polar compounds have been published, although for model systems the technique has shown to be applicable (4.38): Mono- and dihydroxynaphthalene isomers have been examined in Ar, N_2, *n*-heptane, and perfluorohexane matrices (see Fig. 4.15). The detection limits achieved were not lower than 5 ng, which is at least partly attributable to the low absorptivities of the compounds at the laser excitation wavelengths applied.

It is expected that in the near future the combination of gas chromatography (GC) and MI high-resolution fluorescence will become important. Combination of these techniques seems to be evident because similar conditions apply for the samples. Such a semi-on-line coupling (of course, the sample must be frozen on a cold surface) is expected to be very promising because the analytical speed is greatly increased and, at the same time, errors due to losses of sample or the presence of contaminants are avoided. The GC-MI coupling has already been realized for FT/IR detection by Reedy et al. (4.39) and Bourne et al. (4.40) and has resulted in the recent conception of a commercially available instrument (4.41). The application of fluorescence detection in combination with the stored matrix isolated gas chromatorgram is hampered by the long scan times (5–30 min) which are required for mechanical scanning of the diffraction grating in high-resolution monochromators. Much faster scan times can be realized with optical multichannel analyzers as photodiode array detectors or SIT vidicon tubes (4.42). The applicability of photodiode array detection in combination with highly resolved spectra is discussed in ref. 4.43. The GC-MI coupling with FT/IR and even more with fluorescence detection has an advantage over GC-MS coupling, as the spectroscopic methods afford simple distinction between isomers. In contrast with the GC-MI coupling, the combination of MI and HPLC seems to be much more troublesome and will probably not be introduced in the near future.

The MI technique is expected to be applied to a far wider variety of compounds than studied thus far. Much work has to be done to make the method applicable to nonfluorescent compounds. Within this context the suggestion of Wehry that even nonfluorescent compounds may be measurable via gas-phase fragmentation into molecular fragments with reasonable fluorescence quantum efficiencies may be important (4.31). Such decomposition techniques are well known from classical MI studies. For nonfluorescent compounds, furthermore, apart from straight UV-Vis absorption spectroscopy, photoacoustic detection may be useful (4.44).

References to Section 4.4

4.1. J. S. Shirk and A. M. Bass, *Anal. Chem.*, **41**, 103A (1969).
4.2. E. L. Wehry and G. Mamantov, *Anal. Chem.*, **51**, 643A (1979).
4.3. E. Whittle, D. A. Dows, and G. C. Pimentel, *J. Chem. Pys.*, **22**, 1943 (1954).
4.4. I. Norman and G. Porter, *Nature*, **174**, 508 (1954).
4.5. S. Cradock and A. J. Hinchcliffe, *Matrix Isolation: A Technique for the Study of Reactive Inorganic Species*, Cambridge University Press, Cambridge, 1975.
4.6. A. J. Barnes, W. J. Orville-Thomas, A. Müller, and R. Gaufrès, Eds., *Matrix Isolation Spectroscopy*, D. Reidel, Dordrecht, The Netherlands, 1981.
4.7. E. L. Wehry, V. B. Conrad, J. L. Hammons, J. R. Maple, and M. B. Perry, *Opt. Eng.*, **22**, 558 (1983).
4.8. B. Dellinger, D. S. King, R. M. Hochstrasser, and A. B. Smith III, *J. Am. Chem. Soc.*, **99**, 7138 (1977).
4.9. P. Tokousbalides, E. L. Wehry, and G. Mamantov, *J. Phys. Chem.*, **81**, 1769 (1977).
4.10. B. Meyer, *Low Temperature Spectroscopy*, American Elsevier, New York, 1971.
4.11. L. Andrews, *Appl. Spectrosc. Rev.*, **11**, 125 (1976).
4.12. M. M. Rochkind, *Anal. Chem.*, **40**, 762 (1968).
4.13. M. M. Rochkind, *Spectrochim. Acta*, **27A**, 547 (1971).
4.14. R. N. Perutz and J. J. Turner, *J. Chem. Soc. Faraday Trans.*, **69**, 452 (1973).
4.15. L. H. Jones, S. A. Ekberg,and B. I. Swanson, *J. Chem. Phys.*, **82**, 1055 (1985).
4.16. H. E. Hallam, Ed., *Vibrational Spectroscopy of Trapped Species*, Wiley, London, 1973.
4.17. I. R. Dunkin, *Chem. Soc. Rev.*, **9**, 1 (1980).
4.18. B. W. Keelan and L. Andrews, *J. Am. Chem. Soc.*, **103**, 822 (1981), and references therein.
4.19. B. J. Kelsall and L. Andrews, *J. Phys. Chem.*, **88**, 5893 (1984), and references therein.

4.20. L. Andrews, R. S. Friedman, and B. J. Kelsall, *J. Phys. Chem.*, **89**, 4016 (1985).
4.21. V. E. Bondybey and L. E. Brus, *Adv. Chem. Phys.*, **41**, 269 (1980).
4.22. V. E. Bondybey and T. A. Miller, in T. A. Miller and V. E. Bondybey, Eds., *Molecular Ions: Spectroscopy, Structure and Chemistry*, North-Holland, Amsterdam, 1983, p. 125.
4.23. A. J. Downs and S. C. Peake, in R. F. Barrow, D. A. Long, and D. J. Millen, Eds., *Molecular Spectroscopy*, Vol. 1, The Chemical Society, London, 1973, p. 523.
4.24. B. M. Chadwick, in R. F. Barrow, D. A. Long, and D. J. Millen, Eds., *Molecular Spectroscopy*, Vol. 3, London, 1975, p. 281.
4.25. B. M. Chadwick, in R. F. Barrow, D. A. Long, and J. Sheridan, Eds., *Molecular Spectroscopy*, Vol. 6, The Chemical Society, London, 1979, p. 72.
4.26. E. L. Wehry and G. Mamantov, in E. L. Wehry, Ed., *Molecular Fluorescence Spectroscopy*, Vol. 4, Plenum Press, New York, 1981, p. 193.
4.27. E. L. Wehry, R. R. Gore, and R. B. Dickinson, Jr., in G. M. Hieftje, J. C. Travis, and F. E. Lytle, Eds., *Lasers in Chemical Analysis*, Humana Press, Clifton, N. J., 1981, p. 201.
4.28. R. C. Stroupe, P. Tokousbalides, R. B. Dickinson, Jr., E. L. Wehry, and G. Mamantov, *Anal. Chem.*, **49**, 701 (1977).
4.29. R. B. Dickinson, Jr., and E. L. Wehry, *Anal. Chem.*, **51**, 778 (1979).
4.30. J. R. Maple and E. L. Wehry, *Anal. Chem.*, **53**, 1244 (1981).
4.31. E. L. Wehry, G. Mamantov, D. M. Hembree, and J. R. Maple, in A. Björseth and A. J. Dennis, Eds., *Polynuclear Aromatic Hydrocarbons: Chemistry and Biological Effects*, Battelle Press, Columbus, Ohio, 1979, p. 1005.
4.32. E. L. Wehry and G. Mamantov, *Am. Chem. Soc. Symp. Ser.*, **169**, 251 (1981).
4.33. J. R. Maple, E. L. Wehry, and G.Mamantov, *Anal. Chem.*, **52**, 920 (1980).
4.34. M. B. Perry, E. L. Wehry, nad G. Mamantov, *Anal. Chem.*, **55**, 1893 (1983).
4.35. E. L. Wehry, G. Mamantov, R. Kemmerer, R. C. Stroupe, P. T. Tokousbalides, E. R. Hinton, D. M. Hembree, R. B. Dickinson, Jr., A. A. Garrison, P. V. Bilotta, and R. R. Gore, in P. W. Jones and R. I. Freudenthal, Eds., *Carcinogenesis*, Vol. 3, *Polynuclear Aromatic Hydrocarbons*, Raven Press, New York, 1978, p. 193.
4.36. P. Tokousbalides, E. R. Hinton, Jr., R. B. Dickinson, Jr., P. V. Bilotta, E. L. Wehry, and G. Mamantov, *Anal. Chem.*, **50**, 1189 (1978).
4.37. C. F. Pace and J. R. Maple, *Anal. Chem.*, **57**, 940 (1985).
4.38. J. R. Maple and E. L. Wehry, *Anal. Chem.*, **53**, 266 (1981).
4.39. G. T. Reedy, S. Bourne, and P. T. Cunningham, *Anal. Chem.*, **51**, 1535 (1979).
4.40. S. Bourne, G. T. Reedy, and P. T. Cunningham, *J. Chromatogr. Sci.*, **17**, 460 (1979).
4.41. G. T. Reedy, D. G. Ettinger, J. F. Schneider, and S. Bourne, *Anal. Chem.*, **57**, 1602 (1985).
4.42. E. L. Wehry, G. Mamantov, D. M. Hembree, and J. R. Maple, in A. Björseth and A. J. Dennis, Eds., *Polynuclear Aromatic Hydrocarbons: Chemistry and Biological Effects*, Battelle Press, Columbus, Ohio, 1980.
4.43. J. W. Hofstraat, M. Engelsma, J. H. de Roo, C. Gooijer, and N. H. Velthorst, *Appl. Spectrosc.*, **41**, 625 (1987).

4.44. H. E. Howell, G. Mamantov, E.L. Wehry and R. W. Shaw, *Anal. Chem.*, **56**, 823 (1984).

4.5. SHPOL'SKII SPECTROSCOPY

In 1952 the Soviet scientist Shpol'skii reported the observation of narrow spectral lines in the emission spectra of aromatic molecules dissolved in *n*-alkanes, measured at low temperatures ($T \leq 77$ K) (5.1). From a comparison with Raman and infrared spectra, Shpol'skii and co-workers were able to show that the narrow bands, or "quasi-lines," correspond with transitions involving vibrationally excited states of the electronic ground state. Since then many vibrationally resolved spectra have been obtained following Shopl'skii's method.

The reduction of the spectral bandwidths in the Shpol'skii technique is a matrix-induced effect. It has been found that a large number of compounds, mainly polynuclear aromatic hydrocarbons (PAHs) but also polyenes and large biomolecules such as porphyrins and phthalocyanines, exhibit highly resolved spectra in frozen *n*-alkane solutions.

In principle, all solvents that yield crystalline matrices at low temperatures may be used to reduce inhomogeneous broadening of guest molecules. For instance, the mixed-crystal technique is based on the preparation of solid solutions of the analyte in a single crystal of another hydrocarbon with approximately the same dimensions, so that the guest molecules occupy nearly identical sites in the host lattice (5.2). However, analytical application of this technique seems limited as the compatibility of the molecular geometries of guest and host is critical and the preparation of the samples involves a tedious process of crystal growing. In contrast, preparation of the samples in Shpol'skii spectroscopy is easy and the demands on the fit in the *n*-alkane matrix are much less stringent, two reasons that have made this method at present the widest applied of the low-temperature high-resolution techniques.

It is interesting to note that *n*-alkanes may not be the only solvents that easily produce highly resolved spectra when cooled to cryogenic temperatures. Kirkbright and De Lima report the observation of quasi-linear fluorescence and phosphorescence spectra for a number of aromatic and heterocyclic compounds in tetrahydrofuran (5.3,5.4). Additionally, for some aromatic compounds in cyclohexane (5.5), methylcyclopentane (5.6,5.7), hexanol (5.8), isooctane and carbon tetrachloride (5.9), and toluene (5.10), well-resolved spectra were obtained. As only a limited number of compounds has been investigated, the potential of these solvents is not clear, in contrast to the "classical" Shpol'skii solvents, the *n*-alkanes, which have been proven to be appropriate for an impressive number of species.

The fact that the Shpol'skii effect is a matrix-induced phenomenon has several important implications. First, the technique does not require selective laser excitation (as fluorescence line narrowing) to invoke vibrationally resolved spectra. The selectivity of the Shpol'skii method can, however, be impressively improved by making use of narrow-band excitation in the S_1-vibronic region. Second, the S_1–S_0 *absorption* spectrum is also narrow banded. (Transitions to higher excited electronic states are, in general, significantly broadened by their extremely short lifetimes.) Unfortunately, recording of highly resolved absorption spectra is feasible only for very concentrated solutions, due to the opaqueness of the low-temperature matrix. Third, phosphorescence spectra are also vibrationally resolved, without the requirement of tuning the excitation source to the highly improbable T_1–S_0 transition as in phosphorescence line narrowing.

Since the early 1960s analytical applications of Shpol'skii spectroscopy have appeared in the literature. By now, hundreds of analyses have been performed with this technique, which is especially suitable for identification or quantitative analysis of a limited number of compounds in complex samples. No attempts will be made in this chapter to give a complete survey. The most recent complete review by Nurmukhametov dates from 1969 (5.8).

4.5.1. Principles

The Shpol'skii effect is a matrix-induced phenomenon. By freezing of the *n*-alkanes a polycrystalline rather than a glassy solid matrix is obtained. In the polycrystalline matrix a reduction in the inhomogeneity of the environment of the guest molecules is achieved so that narrow spectral bands may be observed, both in absorption and in emission. The inhomogeneous broadening, however, is by no means completely removed. Typical bandwidths in Shpol'skii spectroscopy are 1–10 cm^{-1}, which is sufficient to obtain vibrational resolution and discern among isomers but lies several orders above the limiting natural linewidth. Abram et al. have shown that the width of the 0–352 cm^{-1} vibronic fluorescence band of perylene in *n*-octane at 4.2 K could be reduced from 4 cm^{-1} to the instrumentally limited value of 0.4 cm^{-1} by selective laser excitation (5.11).

Apart from the lifetime-determined natural linewidth, the electron–phonon coupling also plays a role in Shpol'skii spectra. In general, however, the electron–phonon coupling is weak for organic guest–host systems as Shpol'skii matrices, so that even at higher temperatures, such as 77 K, relatively well resolved spectra can be measured. An example of the temperature dependence of the Shpol'skii fluorescence spectrum of tetracene in *n*-nonane is shown in Fig. 4.18.

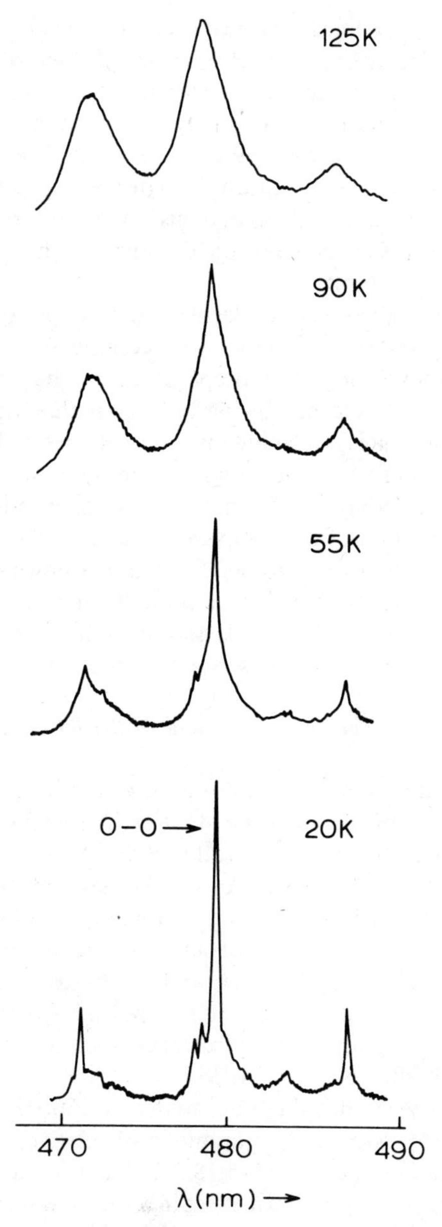

Fig. 4.18. Temperature dependence of the 0–0 transition region of the fluorescence spectrum of tetracene in *n*-nonane.

Frequently, the quasi-line Shpol'skii spectra show two additional characteristic features. First, the spectra may be complicated by the presence of a multiplet structure, as depicted in Fig. 4.5 for the fluorescence spectrum of acenaphthene in *n*-pentane. This spectrum is composed of two identical spectra which are displaced over 59 cm^{-1}. Second, apart from quasi-lines, relatively broad bands, due to solute aggregates or to guest molecules that do not fit well in the *n*-alkane matrix, may also be observed. Both features are extremely sensitive to the choice of the solvent and the thermal history and concentration of the sample.

The way in which the solute molecules are incorporated in the *n*-alkane host lattice is of primary importance for the reduction of inhomogeneous broading. The composition and preparation of the sample is therefore crucial. At present, it is generally agreed upon that the guest molecules occupy substitutional sites, replacing several *n*-alkanes in the host lattice. The requirement of a good fit of guest and host imposes some demands on the choice of solvent (although by far not as strict as in mixed-crystal techniques). In the preparation of the sample one has to reckon with the fact that the sample is at least a binary system. The manner in which such a system is cooled may have a profound influence on its final structure.

Solvent choice, multiplet structure, and the influences of cooling rate and concentration will be discussed in separate subsections.

4.5.1.1. Solvent Choice and Multiplet Structure

In general, the best spectral resolution is observed if there is a close match between the longest dimension of the guest molecule and the length of the *n*-alkane. Bolotnikova studied the optical spectra of naphthalene, anthracene, and naphthacene (tetracene) in various *n*-alkanes and found the narrowest bands in *n*-pentane, *n*-heptane, and *n*-nonane, respectively (5.12). The investigation was expanded by Shpol'skii who denoted these findings as the "lock-and-key principle" (5.13,5.14), and by Pfister, who formulated the "key-and-hole rule" for guest and host in Shpol'skii spectroscopy (5.15). The appearance of the naphthalene fluorescence spectrum in several *n*-alkane solvents at 20 K is shown in Fig. 4.19.

It appears, however, that there are many exceptions to this simple key-and-hole rule (5.16). Aromatic compounds of relatively large size, such as coronene (5.17–5.19), perylene (5.18,5.20,5.21), and 1,12-benzperylene (5.9,5.18,5.20), show the Shpol'skii effect in a rather wide range of *n*-alkanes.

The general validity of the key-and-hole rule suggests that the solute molecules occupy substitutional sites in the *n*-alkane lattice, which is polycrystalline when rapidly grown. This assumption is supported by several experiments. First, the exact position of coronene in slowly frozen *n*-octane

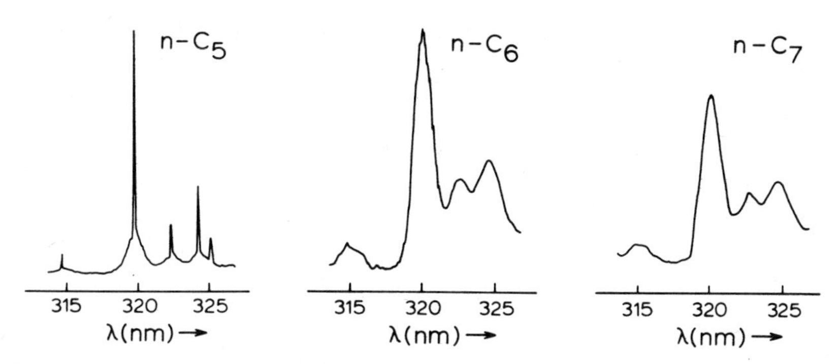

Fig. 4.19. Part of the fluorescence spectrum of naphthalene in n-C_5, n-C_6, and n-C_7 polycrystalline matrices.

single crystals could be established via ESR and x-ray diffraction methods (5.22) and microwave-induced delayed phosphorescence (5.23). Second, the orientation of several PAHs in *n*-alkane single crystals could be established through measurement of their polarized fluorescence spectra (5.24,5.25). Comparison with spectra obtained in polycrystalline solutions indicates that the guest molecules are trapped in similar substitutional sites in both types of matrices.

Often, Shpol'skii spectra are composed of several identical spectra which are displaced with respect to one another over a slight energetic distance (generally, not more than 100 cm^{-1}). This multiplet structure is extremely matrix dependent, as is clearly illustrated by Fig. 4.20, in which the 0–0 transition region of benz[*k*]fluoranthene in *n*-hexane and *n*-octane, both at 20 K, is shown.

Originally, the multiplet structure was ascribed to the presence of rotational isomers of the solvents (5.13,5.14). Nowadays, it is attributed primarily to molecules inserted in different substitutional sites in the host lattice, thus experiencing different microenvironments. Experimental evidence for the existence of distinct sites was presented by Vo-Dinh and Wild (5.26), who were able to selectively excite the molecules in one of the sites of the fluorescence spectrum of coronene in *n*-heptane with monochromatic laser radiation. Additionally, Jansen et al. (5.27) could determine the orientation of the two main sites observed in the phosphorescence spectrum of palladium porphin in a *n*-heptane single crystal from the dependence of the Zeeman effect on the direction of the magnetic field. They found a distinctly different orientation of the molecules giving rise to the two sets of spectral bands observed in the spectrum.

The multiplet structure of the spectra is not only matrix dependent but is also influenced by the thermal history of the sample.

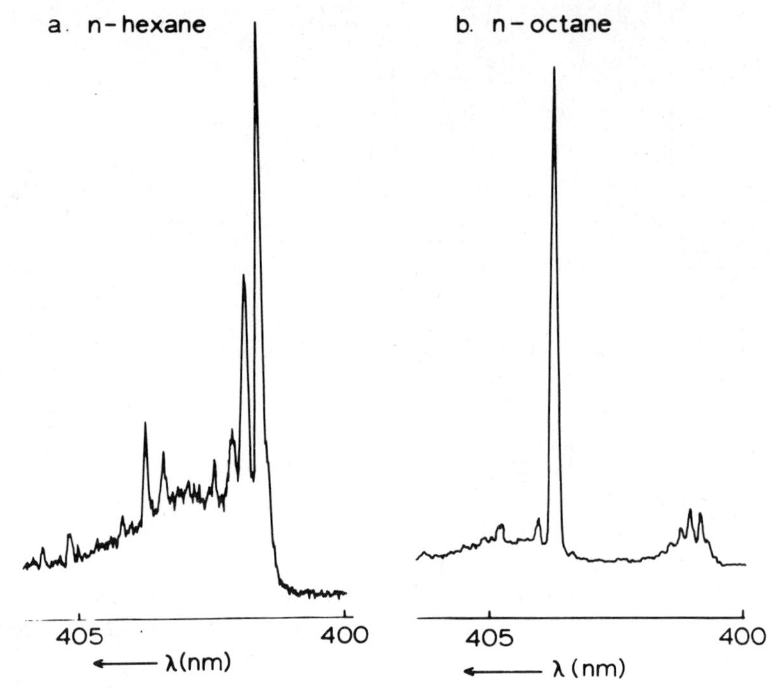

Fig. 4.20. Multiplet structure in the fluorescence spectrum of benz[*k*]fluoranthene in (*a*)*n*-hexane and (*b*)*n*-octane at 20 K. Only the 0–0 transition area is depicted.

4.5.1.2. Thermal History and Concentration

In Shpol'skii spectroscopy the samples are either cooled very fast by immersion in liquid nitrogen or helium, or more slowly in a closed-cycle or continuous-flow helium cryostat, where cooling takes place at 8–10°C/min. But even when the relatively slow cooling process is applied in general, metastable, polycrystalline solids are obtained, since the cooling rate is too fast to maintain thermodynamic equilibrium. Therefore, it is not surprising that the thermal history of the sample often has a profound influence on the appearance of the Shpol'skii spectra. Furthermore, the speed of cooling will strongly influence segregation processes, which are very likely to occur since a Shpol'skii sample is at least a binary system.

Within this context, according to Rima et al., two types of aromatic compounds can be discerned (5.28). The first consists of guest molecules (such as larger PAHs) whose dimensions require the substitution of two or more *n*-alkane molecules. For this group the Shpol'skii effect appears to be independent of cooling rate, at concentrations which are low enough to preclude solute aggregate formation. At concentrations above 10^{-5}–10^{-6} *M*, broad bands due to aggregates (5.29,5.30) and excimers (5.31) are observed which are, in general, red shifted with respect to the quasi-lines.

The second type of compounds consists of molecules whose molecular dimensions are not easily compatible with the host matrix. For such molecules, examples of which are the polyacenes and linearly ortho-condensed aromatics, segregation effects are expected to play an important role. Depending on cooling rate and concentration, these molecules give quasi-line spectra, broad-banded spectra, or both. Dekkers et al. suppose that only nonequilibrium solutions give rise to quasi-line spectra, while broad-banded spectra are correlated with thermodynamic equilibrium situations (5.32). This point of view is supported by Rima et al. (5.28). The dynamics of the crystallization process appears crucial as to which samples produce well-resolved spectra (5.16). Rima et al. (5.28) studied absorption spectra of phenanthrene and dibenzofuran, both type 2 molecules, in several *n*-alkanes using fast cooling of the samples by immersion in liquid nitrogen. They found pure quasi-linear spectra only at guest concentrations below 10^{-6} *M*. In contrast, Dekkers et al. in a study of the concentration dependence of the fluorescence spectrum of acenaphthene in several *n*-alkanes, observed that the best Shpol'skii spectra were obtained for the more concentrated sample (10^{-2}–10^{-3} *M*; for an example, see Fig. 4.21) (5.32). The latter authors, however, used a closed-cycle helium cryostat that took about 30 min to reach 20 K. They suggest that at the lower cooling rates, thermodynamic equilibrium is reached for low concentration samples, leading to segregation and broad-banded spectra. In the concentrated samples equilibrium is not achieved at the fixed cooling rate employed, so that nonequilibrium polycrystalline mixed crystals are formed which give rise to the quasi-line spectra. Clearly, to understand the mechanism of reduction of inhomogeneous broadening for type 2 molecules, knowledge of the thermodynamic aspects of the formation of the solid solution is necessary. More detailed studies on the complicated interplay of concentration and rate of cooling effects in Shpol'skii systems are required. We conclude from the experimental studies done so far that well-resolved Shpol'skii spectra can be obtained by fast cooling of samples with concentrations below 10^{-6} *M*.

The thermal history of the sample also has a pronounced effect on the multiplet structure of the Shpol'skii spectrum. For instance, slow cooling of pyrene in *n*-heptane and 3,4-benzpyrene in *n*-hexane leads to a reduction in the number of sites (5.33). Another thermal process which may seriously influence the site structure is the annealing procedure. This procedure involves a rise in temperature of the sample from its original low temperature to a temperature around its melting point, where it is kept for some time (in general, varying from several minutes to 1 h), followed by cooling to the original temperature. If this treatment is applied to samples prepared via fast cooling procedures, the number of sites is reduced. Pfister observed the disappearance of the site situated at the highest wavenumber in the Shpol'skii spectra of coronene, pyrene, and benz[*a*]pyrene upon annealing (5.15,5.34). She explained this behavior by assuming that at temperatures slightly below the melting point, hindered rotations of the *n*-alkane chains around their axis may occur. If the rotational barrier is not too high, the site will be changed into a thermodynamically more stable one, leading to the simultaneous disappearance of the corresponding quasi-lines. Annealing treatments have a particularly strong influence on the Shpol'skii spectra of unstable species, as radicals or biradicals, obtained by photolysis of precursors in the low-temperature *n*-alkane matrix (5.35,5.36). An example of this effect for *o*-durylene in *n*-hexane is shown in Fig. 4.22.

Other accounts concerning the principles of Shpol'skii spectroscopy have been presented in the reviews by Leach (5.16) and Shpol'skii and Bolotnikova (5.37). Although many studies on the influences of solvent, concentration, and thermal history have been undertaken, no consistent theory can yet be presented. To be able to predict the optimal circumstances for the production of quasi-lines, knowledge of the *spatial* configuration of the low-temperature Shpol'skii sample (distribution and surroundings of guest molecules) seems essential. Unfortunately, this information is lacking to date.

4.5.2. Experimental Aspects

As the Shpol'skii effect is matrix-induced, conventional lamps can be employed for excitation. To enhance the selectivity of the technique, use can be made of selective excitation, preferably to the narrow-line S_1 vibronic-region. This can be accomplished by employing a strong continuous source (e.g., a xenon lamp) in combination with a high-resolution monochromator or by utilizing a laser. Especially the laser appears a useful means to enhance the selectivity of Shpol'skii spectroscopy ["laser-excited Shpol'skii spectroscopy," LESS (5.38)]. Also x-ray sources have been used for excitation in Shpol'skii spectroscopy (5.39,5.40).

Even at rather high temperature (T = 77 K), reasonably well resolved Shpol'skii spectra are obtained due to the often weak electron–phonon coupling for the Shpol'skii guest–host system. So even with easily available liquid nitrogen as cooling medium, Shpol'skii spectra can be measured. For some species lower temperatures are required [e.g., *o*-durylene in *n*-hexane: only below 40 K is a quasi-line spectrum observed (5.35)]. Also, by lowering the temperature the width of the zero-phonon lines is somewhat reduced, which can be useful in some cases (see Section 4.2.3 for a description of the

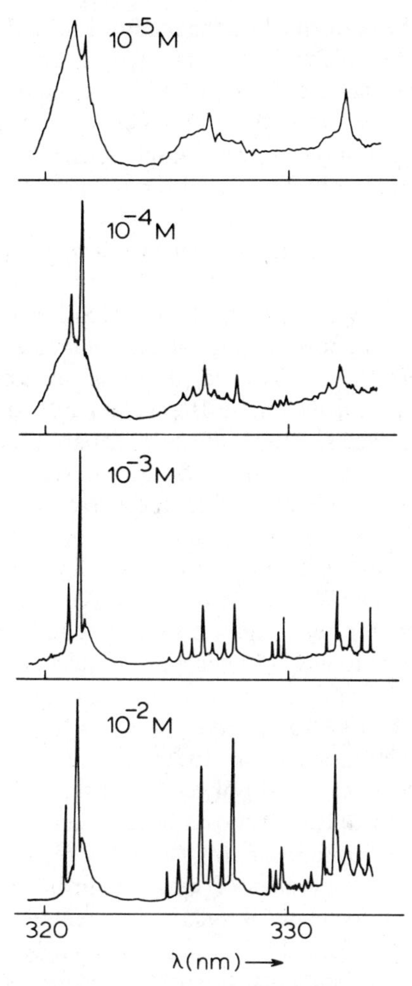

Fig. 4.21. Concentration dependence of a part of the fluorescence spectrum of acenaphthene in *n*-hexane at 20 K.

temperature dependence of the zero-phonon linewidth). Lower temperatures can be reached with helium-based cryogenic apparatus. In Section 4.5.1.2 the need for fast cooling of the samples was stressed; this is easily done in bath cryostats, but requires some attention in conductance cryostats. In such apparatus, fast cooling can be realized by first immersing the sample in liquid nitrogen followed by further cooling in the cryogenic system. Alternatively, a setup as described by Paturel et al. (5.41) can be employed.

Detection can be performed by the methods described in Section 4.3. The low-temperature Shpol'skii matrix is opaque and serves as an efficient scatterer. However, scattered light can easily be removed by optical filtering, as it is not necessary to excite the samples close to the S_1–S_0 0–0 transition. An evaluation of excitation sources, sample cells, and detection systems used for 77 K Shpol'skii spectrometry has been published by Causey et al. (5.42).

4.5.3. Applications

Shpol'skii spectroscopy, as all the high-resolution spectroscopic emission techniques, has been applied mainly for the detailed examination of vibronic spectra. Since both the S_1–S_0 absorption and emission spectra are highly resolved, the Shpol'skii technique is preeminently suited for the study of vibronic coupling effects. These effects are manifest in the appearance of nontotally symmetric modes with or without overtones and also frequently in absorption–emission asymmetry (i.e., the corresponding emission and absorption spectra show clear distinctions with respect to both intensity and position of the vibronic bands). As the technique has been applied to numerous compounds [according to Shpol'skii and Bolotnikova in 1974 quasi-linear spectra for about 500 compounds had been obtained (5.37)], no attempt will be made to give a full account of its accomplishments here. The state of the art at the end of the 1960s has been adequately summarized by Nurmukhametov in 1969 (5.8). In Table 4.3 a survey is presented of some more recent studies in an effort to give an impression of the types of compounds apt to yield well-resolved Shpol'skii spectra. Apart from substituted and nonsubstituted homo- and heterocyclic aromatic compounds more exotic species also appear to give highly resolved emission spectra in Shpol'skii matrices. A number of reactive unstable species (radicals and ions), generated photochemically in the low-temperature matrix, has been studied (5.36,5.54–5.57). Also, rather surprisingly, a number of bulky, nonplanar molecules such as hexahydrohelicene (5.67) and several naphthalene derivatives (5.55,5.56,5.67,5.68) have yielded well-resolved Shpol'skii spectra. An interesting phenomenon was reported by Platenkamp et al. (5.66). They tried in vain to obtain vibrationally resolved spectra of lumiflavin in *n*-alkane matrices, but observed that lumiflavin derivatives substituted with a long alkyl chain did give quasi-lines. Probably the long alkyl tail provides a sort of anchor which locks the molecule in the matrix in a well-defined way.

At present the bulk of the spectra has been measured at 77 K, although more and more spectra acquired at lower temperatures are being reported. A major barrier to the assessment of the state of the art in present-day Shpol'skii spectroscopy is the fact that most of the studies in this field are published in the Soviet literature.

4.5.4. Analytical Aspects

Inevitably, Shpol'skii spectra at an early date attracted the attention of analytical chemists since the combination of its high selectivity and the inherent sensitivity of luminescence is very interesting from an analytical point of view. In 1967, Shpol'skii himself published a paper concerning the application of Shpol'skii spectroscopy to trace analysis (5.69). The analytical potential of the method was underlined by its "inventor" by referring to the papers of Muel and Lacroix (5.70) and of Dikun et al. (5.71), who independently identified the carcinogenic benz[*a*]pyrene in cigarette smoke based on the coincidence of 60 lines within 0.1–0.2 nm. Furthermore, Shpol'skii showed that for compounds with a high fluorescence quantum yield, such as perylene and benz[*a*]pyrene, even at $5\ 10^{-11}\ M$ appreciable peaks were measured. Nevertheless, early Soviet investigators realized that for quantitative purposes the use of an internal standard or a standard addition procedure is inevitable (5.72).

In 1972, Winefordner and co-workers critically evaluated the analytical potential of Shpol'skii spectroscopy (5.73). They noted that, at that time, widespread applicability had been suggested by various authors, but very few publications concerned with basic analytical data, such as reproducibility, accuracy, and precision, had actually appeared in the literature. In the opinion of these authors the usefulness of the Shpol'skii method in analytical chemistry is limited due to several restrictions:

1. Suitable instrumentation is not commercially available and will be expensive;
2. Efforts required for the preparation of unknown samples, the choice of solvent, and the consideration of matrix effects must be unduly great;
3. The intensity of the quasi-lines often has an irregular and nonreproducible concentration dependence;
4. Intensity and width of quasi-lines are a function of the rate of cooling of the sample.

Obviously, these are important points. Nevertheless, since the early 1970s numerous analytical applications have been reported, presumably initiated by the thorough studies of Kirkbright and co-workers (5.3,5.42,5.74). At present, the technique has been applied both with and without prior chromatographic separation to a wide variety of real samples. Apparently, the limitations formulated above appear less restrictive now than anticipated in 1972. For instance, Fassel and co-workers (5.75) stressed that both nonreproducibility in the relative site populations (leading to nonreprodu-

Table 4.3. Survey of Compounds Studied with Shpol'skii Spectroscopy

Compound	Solvent	Temperature (K)	Fluorescence/ Phosphorescence (F1/Ph)	References
1. Homocyclic Aromatic Compounds and Derivatives				
Naphthalene	*n*-Pentane	20	Fl + Ph	5.43
Phenanthrene	*n*-Hexane	20	Fl + Ph	5.29
Methylphenanthrenes	*n*-Hexane	4.2	Fl	5.44
Pyrene-h_{10}	*n*-Octane	20	Fl	5.45
Pyrene-d_{10}	*n*-Octane	20	F1	5.45
Substituted pyrenes (halogeno-, phenyl-, amino-, methyl-, hydroxy-, and methoxy-)	*n*-Hexane	63	Fl	5.46
Coronene	*n*-Heptane	4.2	Fl + Ph	5.47

2. Heterocyclic Aromatic Compounds and Derivatives

Dibenzofuran	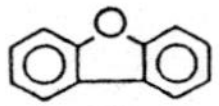	*n*-Hexane	4.2	Fl + Ph	5.48
Xanthone	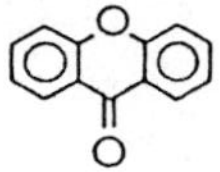	*n*-Hexane	77,4.2	Ph	5.49
Dibenzothiophen	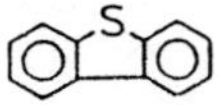	*n*-Hexane	4.2	Fl + Ph	5.50
10,11-Epithiobenzo[*a*]pyrene	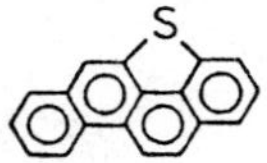	*n*-Hexane	63	Fl	5.51
Carbazole (and $-NH_3$ complex)	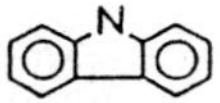	*n*-Heptane	4	Fl	5.52
Acridine derivatives (methyl-, benz-, and dibenz-)	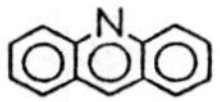	*n*-Octane	77,4	Fl + Ph	5.53

3. Radicals and Ions

Acenaphthenyl		*n*-Pentane	20	Fl	5.36
Phenalenyl		*n*-Pentane	20,4.2	Fl	5.54
1,3-Perinaphthadiyl		*n*-Pentane *n*-Hexane	20,4.2	Fl	5.55

Table 4.3. (*Continued*)

Compound	Solvent	Temperature (K)	Fluorescence/ Phosphorescence (F1/Ph)	References
1,4-Perinaphthadiyl	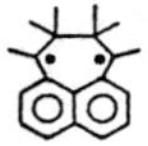 *n*-Pentane	20,4.2	Fl	5.56
m-Xylylene (and methyl derivatives)	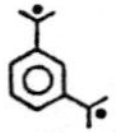 *n*-Pentane *n*-Hexane	5–10	Fl	5.57
1-Acenaphthenium cation	 *n*-Pentane	20	Fl	5.36
Benzocycloheptenium cation	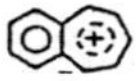 *n*-Hexane	20	Fl	5.36
4. Polyenes				
1,3,5,7-Octatetraene	*n*-Hexane	4.2	Fl	5.58,5,59
Diphenyloctatetraene	*n*-Tetradecane	4.2,1.8	Fl	5.60
2,12-Dimethyltridecahexaene	*n*-Undecane	4.2	Fl	5.61

5. Compounds of Biological Interest					
Porphyrins		*n*-Heptane/ *n*-dodecane	4.2	Fl + Ph	5.27,5.62, 5.63
Phthalocyanines		*n*-Octane	4.2	Fl + Ph	5.64,5.65
N_3-Undecyllumiflavin	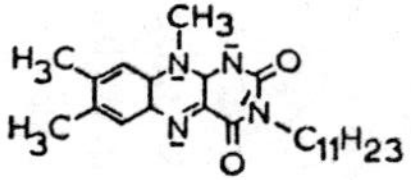	*n*-Decane	4.2	Fl	5.65
6. Nonplanar Compounds					
Hexahydrohexahelicene	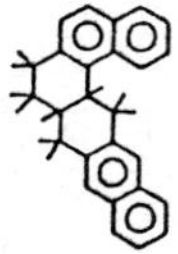	*n*-Pentane	77	Fl	5.67
Exo-naphtho[6:3,4]tricyclo [4.2.1.$0^{2,5}$]nonane	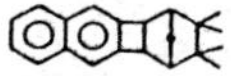	*n*-Hexane	5	Fl	5.68
6*b*,7*a*-Dihydro-7*H*-cyclo prop[*a*]acenaphthylene		*n*-Hexane	20	Fl	5.59
1,4-Dihydro-1,4-ethanonaphtho [1,8-*de*] [1,2]diazetine		*n*-Pentane	20	Fl	5.56

cible relative fluorescence intensities of site-specific lines) and aggregate formation (providing a restricted linear working range) can be prevented by applying dilute solutions (concentrations below 10^{-6} M) and reproducible cooling rates. Inner filter and enhancement effects and intermolecular interactions which complicate quantitative application of Shpol'skii spectroscopy (5.76) can be circumvented or reduced by utilizing standard addition and internal reference techniques (5.75). Moreover, it is noted that in comparison with conventional forms of fluorescence and phosphorescence spectroscopy, both inner filter effects and energy transfer processes are less important in Shpol'skii spectrometry. Of course, short-range (collision-induced) energy transfer is excluded in a frozen matrix. Moreover, both inner-filtering and (Förster) long-range energy transfer are governed by the amount of overlap of the absorption spectrum of the acceptor and the emission spectrum of the donor. This overlap is generally small because of the quasi-linear character of the Shpol'skii spectra.

The dependence of the spectral resolution on the host *n*-alkane necessitates the use of several solvents to determine all the compounds in a complex sample. It should be realized however, that for applications where mixtures of similar compounds, such as a group of isomers, are analyzed, this disadvantage hardly plays a role. Of course, the basic requirement in Shpol'skii spectroscopy is that the compound under consideration must be soluble in *n*-alkanes, a requirement which is not met for polar and ionic compounds. This puts a serious limitation to the scope of the Shpol'skii method. Solubility problems may be solved by dissolution of the analytes in a suitable solvent and subsequent dilution with the appropriate *n*-alkane (5.77) or by making use of, for instance, tetrahydrofuran as a matrix material (5.3,5.4). An alternative way to circumvent the solubility problem is to apply matrix isolation sample preparation utilizing an *n*-alkane as the matrix gas ["pseudo-Shpol'skii" spectroscopy (5.78); see also Section 4.4].

4.5.5. Analytical State of the Art

In view of the above it is not unexpected that by far the largest number of applications concerns *qualitative analysis* aimed at the identification of, at most, a few compounds in complex real samples, if necessary after partial chromatographic fractionation of the sample. For this purpose the quasi-line positions play the central role while their intensities are of minor importance.

Originally, only broad-band excitation at 77 K was applied; later, lower temperatures, down to 4.2 K, and more selective excitation were utilized. This extends the applicability of the method since various compounds are known which exhibit quasi-linear spectra only at temperatures (far) lower than 77 K. Additionally, upon further cooling linewidths are generally reduced which at first sight seems to be favorable with respect to selectivity. However, this reduction is only recorded if (more expensive) monochromators with higher resolving power are employed. Furthermore, a more pronounced site structure may become visible, so that the number of spectral lines is substantially enlargened. Therefore, decrease of temperature does not in all cases lead to an obvious selectivity enhancement. Site-selective excitation (preferably performed by means of a tunable dye laser) does, of course, provide an improvement of selectivity.

Selective excitation of a single compound in a mixture or even a single site in a spectrum exhibiting multiplet structure is based on the fact that the Shpol'skii effect is caused by the matrix, so that also in absorption (as far as the S_1–S_0 electronic transition is concerned) spectra are quasi-linear. In 1976, Farooq and Kirkbright showed that quasi-linear excitation spectra can be obtained with a modified conventional spectrometer (5.74). Of course, for this purpose narrow bandpass excitation is required. The reduction in light throughput is, however, not as dramatic, since the narrow-band excitation profile is absorbed most efficiently due to the narrowness of the absorption bands. This explains why selectively excited Shpol'skii luminescence spectra can be obtained with a Xe lamp, although evidently, a tunable laser source with its inherently high intensity monochromatic output is preferable.

Quantitative analysis on the basis of Shpol'skii spectrometry is far more difficult than qualitative analysis, and generally rather elaborate since appropriate internal standards must be found or a standard addition procedure has to be applied. Especially deuterated analogs are very suited as internal standards, as they interact in the same way with the *n*-alkane matrix as the analytes but are spectroscopically easily discerned (5.45); the spectra are virtually identical, apart from a small spectral shift (about 100 cm^{-1}). Optimally, a combination of standard addition and internal standardization is used, because at higher analyte concentrations problems with nonreproducibility of line intensities arise. Hence reference compounds at comparable concentrations should be added.

An interesting alternative has been suggested by Wittenberg et al. (5.80), who utilized the mercury excitation line scattered by the polycrystalline matrix as a standard intensity. Evidently, this method cannot be applied for Xe-lamp excitation.

For convenience, two main streams will be distinguished along which Shpol'skii spectroscopy in analytical chemistry has been developed. The first is directed to the utilization of (modified) conventional spectrometers and low-cost cooling instruments based on liquid nitrogen, boiling either under atmospheric pressure so that 77 K is reached, or under reduced pressure so that, for instance, measurements can be carried out at 63 K (5.79). The second makes use of the most advanced instrumental techniques available for excitation (tunable dye lasers, x-ray sources), sample cooling (closed-cycle helium refrigerators or helium bath cryostats), and light detection (e.g., temporal resolution). We shall only consider the excitation source as a criterion.

4.5.5.1. Conventional Shpol'skii Spectrometry

In the following, the main developments in conventional Shpol'skii spectrometry since 1970 will be discussed. A good impression of the state of the art at the end of the 1960s is given in the review paper by Shpol'skii (5.69). Up until then the number of compounds studied from an analytical point of view was restricted to eight PAHs.

First, attention will be paid to studies directed at the creation of a data base of reference spectra and the assembling of relevant analytical information. These investigations are generally performed on academic solutions with the eventual aim to utilize the data for identification purposes and analysis of real samples. Within this context developments along the instrumental line will also be considered. Second, the application of Shpol'skii spectrometry to the analysis of complex mixtures is treated. Usually, prior to the analysis some kind of chromatographic fractionation is employed.

It is emphasized that the light sources in conventional Shpol'skii spectroscopy are commonly stronger (e.g., operated at 500 or 1000 W) than those used in commercial spectrometers. Of course, the intensity of the lamp employed is crucial if results on selective excitation or sensitivities in publications by different groups are compared.

Reference Data of Academic Solutions. In 1974, Gaevaya and Keshina extended the number of compounds investigated in quantitative determinations from 8 to 15, utilizing both fluorescence and phosphorescence spectra (5.81). They reached detection limits ranging from 10 to 0.1 ng/mL and relative standard deviations lower than 10%. In 1978 an atlas of emission spectra at 77 K assembled by Tepletskaya et al. appeared (5.82). Also, Colmsjö and Östman composed an atlas of Shpol'skii spectra (5.83).

Winefordner's group in 1982 published relevant analytical data for 23 typical PAHs obtained at 77 K with a simple spectrometer equipped with a 300-W Xe lamp (5.84). The spectra in five solvents (i.e., the linear alkanes from *n*-pentane to *n*-nonane) were compared. Furthermore, optimum excitation and emission wavelengths were tabulated together with detection limits. The latter were generally on the order of 10 ng/mL, about three orders of magnitude higher than for room temperature fluorimetry in liquid solutions. The linear dynamic range for 1-methylpyrene appeared to cover three orders. An extension of this data base was given in a subsequent paper, in which phosphorescence characteristics were also incorporated (5.85).

Various attempts have been performed to realize unambiguous identification of compounds in complex mixtures without making use of prior chromatographic separation. Vershinin et al., for instance, have developed a computer-processed procedure for qualitative analysis based only on emission spectra (5.86). They investigated 85 spectra recorded in *n*-hexane at 77 K and calculated for each spectrum an identification factor which is equal to the total number of "characteristic lines" [i.e., quasi-lines without overlap with other lines, using data from the spectral atlas of Tepletskaya et al. (5.82)]. It appeared that 34 of the 85 investigated spectra are sufficiently characteristic to allow identification without the need of chromatographic separation or selective excitation. It is emphasized that possible errors remain if unknown compounds, whose spectra are not included in the basis set, are present. Furthermore, absence of a characteristic line is not a sufficient proof that the corresponding compound is not present in the mixture, since inner filter effects or inefficient excitation may play a role. Sogliero et al. considered the applicability of pattern recognition methods for the classification of low-temperature luminescence spectra (5.87).

As noted before, the detection limits in Shpol'skii spectrometry and the precision and accuracy of the measurements are strongly influenced by the experimental setup applied. In a thorough study, Causey et al. (5.42) evaluated three excitation sources (i.e., a 150-W Xe lamp, a medium-pressure Hg lamp, and a 3-mW He-Cd laser), two sample cells (i.e., a commercial Dewar flask system and a home-built copper cryostat cell), and three techniques for recording the luminescence signal received by the photomultiplier (i.e., dc integration, photon counting, and signal averaging). The detection limits obtained with the Xe and Hg lamps are similar; the laser gives a significant improvement (factor of 5) only if one of its fixed wavelength emissions occasionally exactly coincides with a quasi-line absorption. The copper cryostat cell provided a strongly improved reproducibility compared to the Dewar flask, principally because the sample can be positioned more reproducibly in the optical path. Furthermore, noise due to boiling nitrogen in the light path is prevented. For instance, for 2.5×10^{-6} M coronene in *n*-hexane (emission at 445.05 nm) the relative standard deviation was reduced from 12% to 2.5%. Finally, the detection limits were improved enormously by employing signal integration (charge measurement) instead of direct presentation of the voltage generated across a load resistor on the output of the photomultiplier, especially when the background luminescence is subtracted (for benz[*a*]pyrene in *n*-octane the LOD improved from 5×10^{-7} M to 1×10^{-9} M). Unfortunately, this procedure is not appropriate if mixtures are analyzed for which the emission spectrum must be scanned over a wide wavelength range.

The requirement of selective excitation for the analysis of a complex mixture hampers the applicability of Shpol'skii spectroscopy in routine analysis as for every analyte another excitation wavelength has to be used. From this point of view the development of constant-energy synchronous scanning spectroscopy by Winefordner and co-workers is interesting (5.88,5.89). Constant wavenumber differences from 1400 to 2000 cm^{-1} between the excitation and emission wavenumbers were applied and spectral separation of PAH isomers and PAH alkyl homologs was achieved (5.89).

Analysis of Real Samples. For the analysis of real samples, almost without exception, prior to the application of the Shpol'skii method some form of sample pretreatment was invoked. The combination with thin-layer chromatography (TLC) has been applied successfully to the analysis of automobile exhaust samples at 63 K by Colmsjö and Stenberg, (5.90). The PAH samples were obtained by vacuum sublimation from the TLC plates; benz[*a*]pyrene, pyrene, and benz[*ghi*]perylene could be identified.

Frequently, sample treatment is performed by reversed-phase high-performance liquid chromatography (HPLC); in this case the Shpol'skii method is applied to the different fractions (after removal of the solvent a *n*-alkane is added). For instance, Colmsjö and co-workers have employed this procedure for analysis of automobile exhaust gases (5.91). Also, sulfur-subsituted PAHs, as 10,11-epithiobenz[*a*]pyrene (see Table 4.3), have been identified by this group, even in the presence of large amounts of their parent PAHs (5.92,5.93). Similarly, Colmsjö et al. investigated the identification of partially hydrogenated PAHs (5.94). Ewald and co-workers have investigated petroleum fractions, extracts from marine sediments, and medicinal white oils, at both 15 and 4 K (5.95–5.105). They have shown that the method can be used successfully in oceanographic studies for the identification of geochemical markers: aromatic compounds with bulky saturated substituents (5.102). Furthermore, the Bordeaux group was able to identify the monomethylated isomers of pyrene, phenanthrene, and chrysene, and to establish their relative distribution. The distribution of isomers in a sample is important because it determines the overall toxic activity; for instance, 5-methylchrysene is strongly carcinogenic, while its isomers are only moderate carcinogens. Selective determination of closely related isomers is not easily realized with other techniques, such as GC-MS (5.104). In Fig. 4.23 part of the fluorescence and phosphorescence emission spectra of monmethylphenanthrenes and 4,5-methylenephenanthrene in several samples is displayed.

Recently, an interesting example of a quantitative analysis performed with Shpol'skii spectrometry has been reported by Wittenberg et al. (5.106). They utilized their direct calibration method based on the intensity of the scattered mercury excitation line (5.80). The measurements were performed at 10 K by

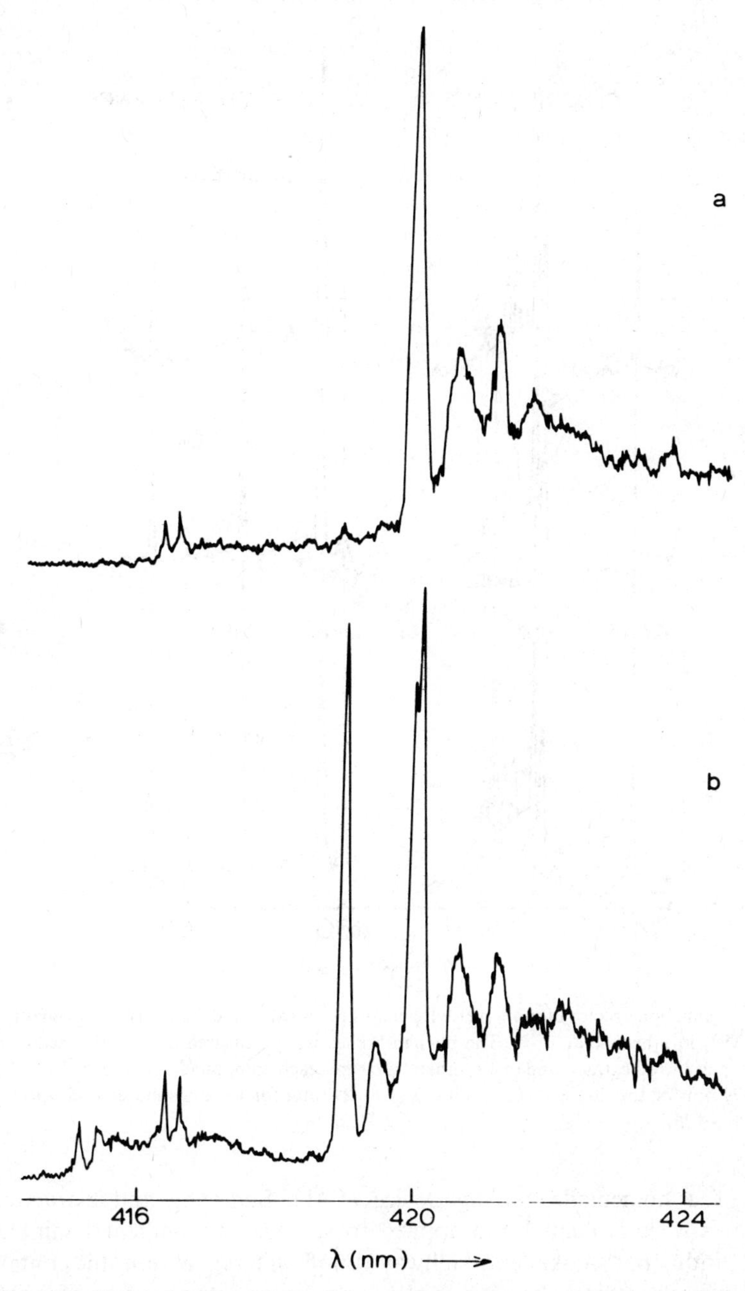

Fig. 4.22. The 0–0 region of the fluorescence spectrum of *o*-durylene at 5 K. (*a*) after photolysis at 20 K without annealing; (*b*) after photolysis at 20 K and annealing (15 min at 80 K).

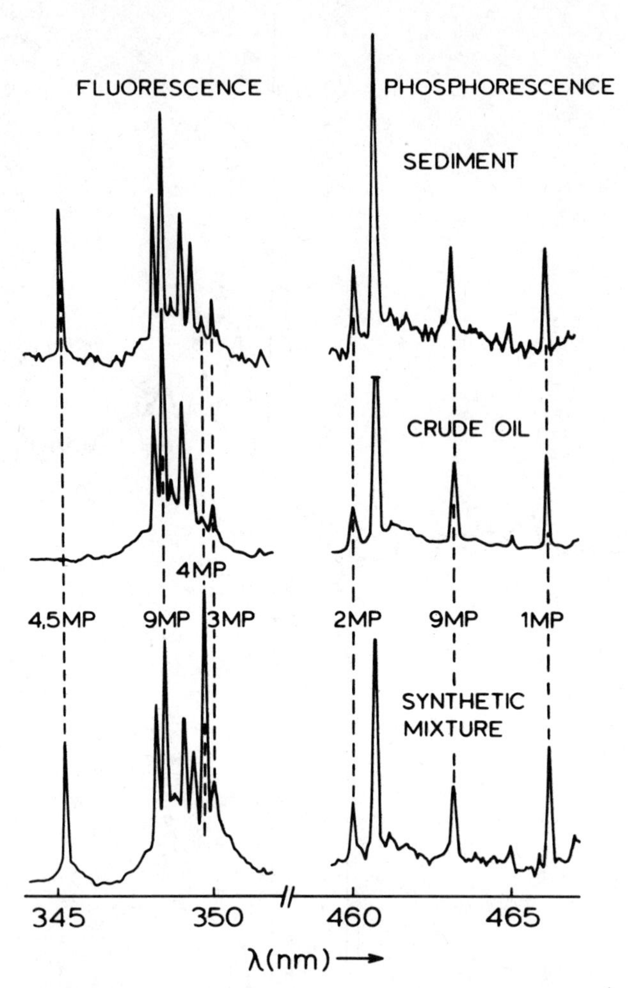

Fig. 4.23. Emission spectra of monomethylphenanthrenes (MP) and 4,5-methylenephenanthrene (4,5 MP) in *n*-hexane at 15 K. The phenanthrenes were measured in a marine sediment extract, in a crude oil extract, and in a synthetic mixture (each compound at $c = 2 \times 10^{-2}$ *M*). λ_{exc} = 299 nm for the fluorescence spectra, λ_{exc} = 297 nm for the phosphorescence spectra. (From ref. 5.105).

means of a homemade cooling device (5.41). Sampling and extraction procedures were evaluated and applied to various environmental samples (i.e., air, industrial smoke, and highway runoff waters). Within this context we also refer to our investigation on the quantitative determination of PAHs in harbor sediment (5.45). In this study a number of PAHs occurring on the priority pollutant list of the Environmental Protection Agency was determined. Deuterated PAHs appeared to be suitable as internal standards; the reproducibility lies between 6 and 20% and the results are in agreement with other methods of analysis (see Table 4.4). No chromatographic separation of the sediment extract was needed before the analysis. In Fig. 4.24 part of the fluorescence spectrum obtained in *n*-octane at 20 K is depicted. Calibration curves obtained for benz[*a*]pyrene are shown in Fig. 4.25. For this compound a linear dynamic range of four decades, from about 5×10^{-6} *M* to the detection limit of 8×10^{-10} *M*, has been found.

Finally, some papers have been reported on the rapid quantitation of the highly carcinogenic benz[*a*]pyrene in water samples (5.107,5.108). The latter publication is especially interesting; a 2-ml aqueous sample is simply mixed with only 0.04 mL of *n*-octane, the resulting emulsion is frozen, and the quasi-linear spectrum of benz[*a*]pyrene can be recorded. The detection limit in water is as low as 3×10^{-9} *M* with a relative standard deviation of 30%.

4.5.5.2. *Laser- and X-Ray-Excited Shpol'skii Spectrometry*

In a series of papers Fassel and co-workers examined the analytical potential of empolying x-ray instead of UV excitation in Shpol'skii spectrometry (5.109–5.111). The aim of the new excitation method is to circumvent crosstalk between exciting and luminescing radiation. The technique was applied successfully to obtain the PAH profiles in the neutral fraction isolated from by-products of coal combustion and conversion and from shale and fuel oils (5.111). A disadvantage of x-ray excitation is that as a result of radiolysis of *n*-alkane solvents, luminescent impurities can be formed resulting in a background emission that worsens the attainable detection limits (5.110).

More important is the development of laser-excited Shpol'skii spectrometry, usually denoted as LESS, by Fassel's group. A comprehensive review on this subject has been published recently (5.38). Laser excitation has also been applied to "pseudo-Shpol'skii spectroscopy," where the Shpol'skii matrix is prepared via matrix isolation procedures with the alkane as matrix gas. This method has been discussed in Section 4.4.

Obviously, with a tunable dye laser selective excitation of different compounds in a mixture or site-selective excitation of a single compound can be realized. The light absorption process is very efficient since the "monochromatic" output can be tuned to a quasi-linear absorption band. Of course, this provides an improvement of selectivity in comparison to conventional Shpol'skii spectrometry so that PAHs can be determined directly in complex mixtures without prior isolation of the PAH fraction. The method was first used to quantitate pyrene and benz[*a*]pyrene directly in coal liquids and shale oil (the sample was only diluted by 5×10^3 with *n*-octane) (5.112).

Table 4.4. Results of the Determination of PAHs in Harbor Sediment by Shpol'skii Fluorometry and Fluorescence-Detected HPLC

Compound	λ_{exc} (nm)	λ_{em} (nm)	Amount Determined with Pyrene-d_{10} as Standard (mg/kg Sediment)	Amount Determined with Perylene-d_{12} as Standard (mg/kg Sediment)	Amount Determined with HPLC (mg/kg Sediment)[a]
Chrysene	272	360.6	1.23 ± 0.13	1.20 ± 0.07	—
Pyrene	339	392.6	2.73 ± 0.24	2.64 ± 0.29	—
Benz[*b*]fluoranthene	304	397.8	1.58 ± 0.19	1.50 ± 0.09	2.1 (4.5)[a]
Benz[*a*]pyrene	389	408.5	1.10 ± 0.17	1.12 ± 0.11	0.93 (6.1)
Benz[*k*]fluoranthene	310	412.8	0.72 ± 0.09	0.72 ± 0.05	0.57 (5.5)
Benz[*ghi*]perylene	304	406.3	0.86 ± 0.10	0.83 ± 0.06	0.53 (8.6)
Indeno[1,2,3-*cd*]pyrene	380	462.7	0.75 ± 0.15	0.70 ± 0.10	1.3 (10.6)

[a] Retention time in minutes given in parentheses.

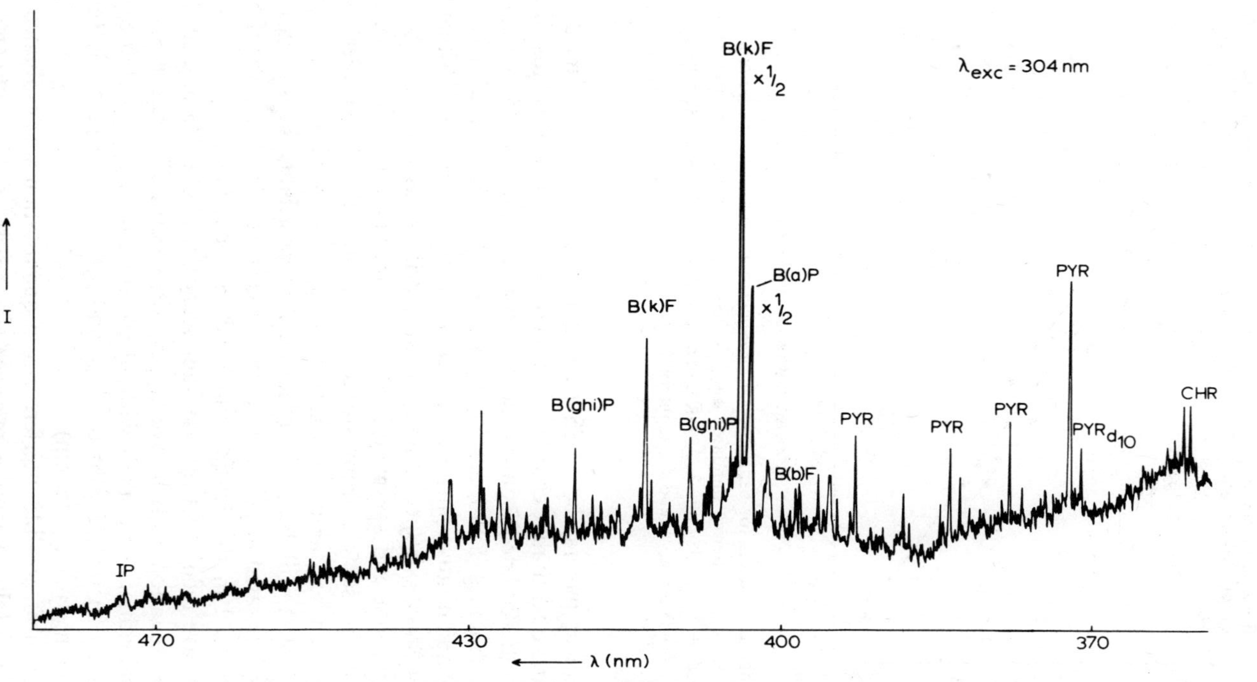

Fig. 4.24. Fluorescence spectrum of a harbor sediment extract in *n*-octane at 20 K. CHR, Chrysene; PYR, pyrene; B(*b*)F, benz[*b*]fluoranthene; B(*a*)P, benz[*a*]pyrene; B[*k*]F benz[*k*]fluoranthene; B(*ghi*)P, benz[*ghi*]perylene; IP, indeno[1,2,3-*cd*]pyrene.

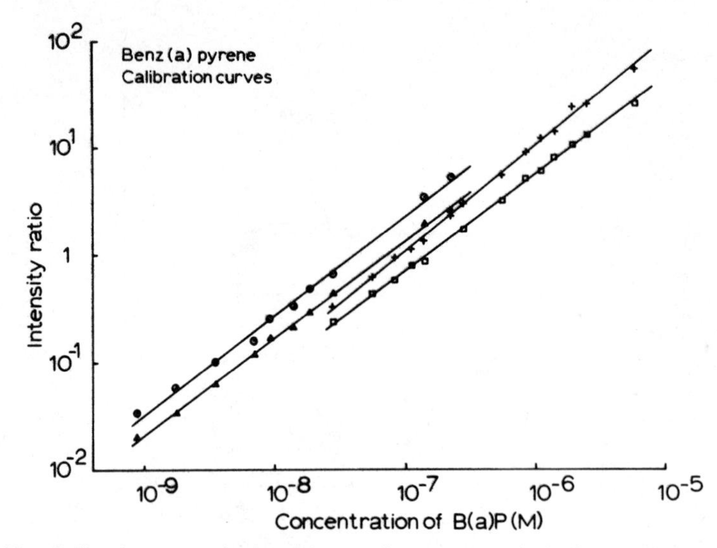

Fig. 4.25. Calibration curves for benz[*a*]pyrene in *n*-octane at 20 K. λ_{exc} = 300 nm, λ_{em} = 403.0 nm; perylene-d_{12} as internal standard. ▲λ_{exc} = 300 nm, λ_{em} = 403.0 nm; pyrene-d_{10} as internal standard. + λ_{exc} = 389 nm, λ_{em} = 408.5 nm; perylene-d_{12} as internal standard. □λ_{exc} = 389 nm, λ_{em} = 408.5 nm; pyrene-d_{10} as internal standard.

In subsequent papers the LESS technique was extended to the direct qualitative and quantitative characterization of very complex PAH mixtures, including multialky lated isomers, samples that cannot be analyzed easily with other techniques (5.113–5.115). It is stressed that the internal reference principle for quantitative purposes is even more important in LESS than in conventional Shpol'skii spectrometry because drift in laser output must be taken into account and sample cell position together with inhomogeneity variations of the sample front surface are more critical. Furthermore, because of the limited sample pretreatment, intermolecular interactions and inner filter effects will be more effective. The possibility of site-specific excitation is illustrated in Fig. 4.26 showing the spectra of a mixture of 6,8-dimethyl- and 11-methylbenz[*a*]anthracene in *n*-octane at 15 K. Both compounds have their most intense fluorescence at about 385 nm, so that nonselective excitation leads to overlapping site spectra. Fluorescence spectra of the individual compounds without site structure can be obtained by selecting excitation wavelengths of 375.0 and 374.7 nm, respectively (note that the difference is only 0.3 nm).

The problem of possible interferences in quantitation was evaluated by determining benz[*a*]pyrene in the presence of excess benz[*k*]fluoranthene; for

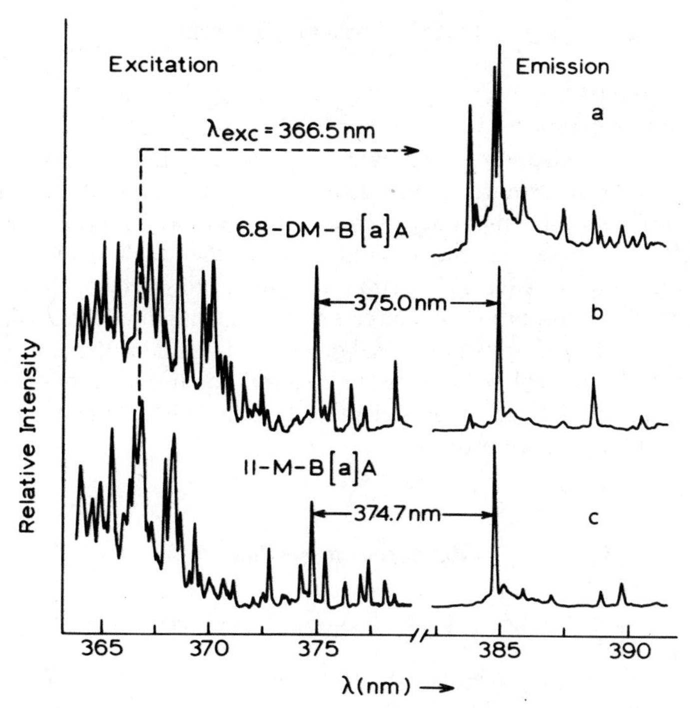

Fig. 4.26. Selective excitation of 6,8-dimethylbenz[*a*]anthracene and 11-methylbenz-[*a*]anthracene in a mixture of these two compounds in *n*-octane at 15 K. In the left-hand side of the figure fluorescence excitation spectra are depicted at the right fluorescence emission spectra. For excitation at 366.5 nm an emission spectrum is obtained that is complicated by site structure and in which both compounds are observed simultaneously (*a*). For excitation at 375.0 and 374.7 nm, respectively, 6,8-DM-B[*a*]A (*b*) and 11-M-B[*a*]A (*c*) are selectively excited. The compounds can be observed separately and, moreover, the multiplet structure of the spectra has disappeared. (Reprinted with permission from Y. Yang, A. P. D'Silva, and V. A. Fassel, *Anal. Chem.*, **53**, 894. Copyright 1981 American Chemical Society).

both compounds the most intense fluorescence lines occur at about 403 nm. By applying benz[*ghi*]perylene as an internal reference and measuring the intensity ratio of benz[*a*]pyrene at 403.1 nm and benz[*ghi*]perylene at 406.5 nm as a function of the benz[*a*]pyrene concentration, an analytical curve with linearity over 3.5 decades was obtained. The procedure used for quantitation was facilitated by empolying perdeuterated PAHs as externally added reference compounds (5.116). Thus the elaborate standard addition procedure with the need of establishing an analytical curve for each analyte and each sample, which is impractical for routine analysis, can be avoided.

4.5.6. Concluding Remarks

In our opinion the developments outlined in Section 4.5.5 imply that Shpol'skii spectrometry is an appropriate and useful analytical method, both in its conventional and sophisticated forms. Also, in the near future its most important application will be the identification and quantitation of a limited number of relevant compounds in complex samples without the need of time-consuming sample treatment. Additionally, the qualitative analysis of mixtures is a relevant application of Shpol'skii spectrometry. On the contrary, for quantitation of a large number of compounds in a mixture the methods developed are still too elaborate due to standardization problems. However, the ideas of Wittenberg et al., calibration by means of the scattered mercury lines (5.80), and of Fassel et al., deuterated PAH analogs as internal references (5.116), are promising.

References to Section 4.5

5.1. E. V. Shpol'skii, A. A. Il'ina, and L. A. Klimova, *Dokl. Akad. Nauk SSSR*, **87**, 935 (1952).
5.2. B. Meyer, *Low Temperature Spectroscopy*, American Elsevier, New York, 1971.
5.3. G. F. Kirkbright and C. G. de Lima, *Analyst*, **99**, 338 (1974).
5.4. G. F. Kirkbright and C. G. de Lima, *Chem. Phys. Lett.*, **37**, 165 (1976).
5.5. H. Sponer, Y. Kanda, and L. A. Blackwell, *Spectrochim. Acta*, **16**, 1135 (1960).
5.6. A. Pellois and J. Ripoche, *Chem. Phys. Lett.*, **3**, 280 (1969).
5.7. A. Moisan-Pellois, R. Collorec, and J. Ripoche, *C. R. Acad. Sci. (Paris) B*, **269**, 1305 (1969).
5.8. R. N. Nurmukhametov, *Russ. Chem Rev.*, **38**, 180 (1969).
5.9. D. M. Grebenshchikov, N. A. Kovrizhnykh,and S. A. Kozlov, *Opt. Spectrosc. (USSR)*, **31**, 392 (1971).
5.10. T. N. Bolotnikova and L. K. Artemova, *Opt. Spectrosc. (USSR)*, **33**, 200 (1972).
5.11. I. I. Abram, R. A. Auerbach, R. R. Birge, B. E. Kohler, and J. M. Stenveson, *J. Chem. Phys.* **63**, 2473 (1975).
5.12. T. N. Bolotnikova, *Opt. Spectrosc. (USSR)*, **7**, 138 (1959).
5.13. E. V. Shpol'skii, *Sov. Phys. Usp*, **5**, 522 (1962).
5.14. E. V. Shpol'skii, *Sov. Phys. Usp*, **6**, 411 (1963).
5.15. C. Pfister, *Chem. Phys.*, **2**, 171 (1973).
5.16. J. W. Hofstraat, I. L. Freriks, M. E. J. de Vreeze, C. Gooijer and N. H. Velthorst, to be published.
5.17. S. Leach, *Pure Appl. Chem.*, **27**, 457 (1971).
5.18. C. Pfister, *J. Chim. Phys.*, **67**, 418 (1970).

5.19. K. Ohno, T. Kajiwara, and H. Inokuchi, *Bull. Chem. Soc. Japan*, **45**, 996 (1972).
5.20. R. I. Personov, I. S. Osad'ko, E. D. Godyaev, and E. I. Al'shits, *Sov. Phys. Solid State*, **13**, 2224 (1972).
5.21. E. V. Shpol'skii and R. I. Personov, *Opt. Spectrosc. (USSR)*, **8**, 172 (1960).
5.22. A. M. Merle, M. Lamotte, S. Risemberg, C. Hauw, J. Gaultier, and J. Ph. Grivet, *Chem. Phys.*, **22**, 207 (1977).
5.23. A. M. Merle, W. M. Pitts, and, M. A. El-Sayed, *Chem. Phys. Lett.*, **54**, 211 (1978).
5.24. M. Lamotte and J. Joussot-Dubien, *J. Chem. Phys.*, **61**, 1892 (1974).
5.25. M. Lamotte, A. M. Merle, J. Joussot-Dubien, and F. Dupuy, *Chem. Phys. Lett.*, **35**, 410 (1975).
5.26. Tuan Vo-Dinh and U. P. Wild, *J. Lumin.*, **6**, 296 (1973).
5.27. G. Jansen, M. Noort, N. van Dijk, and J. H. van der Waals, *Mol. Phys.*, **39**, 865 (1980).
5.28. J. Rima, L. A. Nakhimovsky, M. Lamotte, and J. Joussot-Dubien, *J. Phys. Chem.*, **83**, 4302 (1984).
5.29. J. Rima, M. Lamotte, and A. M. Merle, *Nouv. J. Chim.*, **5**, 605 (1981).
5.30. S. A. Kozlov and D. M. Grebenshchikov, *Opt. Spectrosc. (USSR)*, **44**, 75 (1978).
5.31. L. A. Klimova, G. N. Nersesova, and A. I. Oglobina, *Opt. Spectrosc. (USSR)*, **38**, 538 (1975).
5.32. J. J. Dekkers, G. Ph. Hoornweg, C. MacLean, and N. H. Velthorst, *J. Mol. Spectrosc.*, **68**, 56 (1977),
5.33. E. V. Shpol'skii and L. A. Klimova, *Opt. Spectrosc. (USSR)*, **7**, 499 (1959).
5.34. C. Pfister, *Chem. Phys.*, **2**, 181 (1973).
5.35. W. P. Cofino, G. Ph. Hoornweg, C. Gooijer, C. MacLean, and N. H. Velthorst, *Chem. Phys.*, **72**, 73 (1982).
5.36. W. P. Cofino, G. Ph. Hoornweg, C. Gooijer, C. MacLean, and N. H. Velthorst, *Spectrochim. Acta*, **39A**, 283 (1983).
5.37. E. V. Shpol'skii and T. N. Bolotnikova, *Pure Appl. Chem.*, **37**, 183 (1974).
5.38. A. P. D'Silva and V. A. Fassel, *Anal. Chem.*, **56**, 985A (1984).
5.39. C. S. Woo, A. P. D'Silva, V. A. Fassel, and G. J. Oestreich, *Environ. Sci. Technol.*, **12**, 173 (1978).
5.40. C. S. Woo, A. P. D'Silva, and V. A. Fassel, *Anal. Chem.*, **52**, 159 (1980).
5.41. L. Paturel, J. Jarosz, C. Fachinger, and J. Suptil., *Anal. Chim. Acta*, **147**, 293 (1983).
5.42. B. S. Causey, G. F. Kirkbright, and C. G. de Lima, *Analyst*, **101**, 367 (1976).
5.43. J. J. Dekkers, G. Ph. Hoornweg, C. MacLean, and N. H. Velthorst, *Chem. Phys.*, **5**, 393 (1974).
5.44. J. Rima, M. Lamotte, and J. Joussot-Dubien, *Anal. Chem.*, **54**, 1059 (1982).
5.45. J. W. Hofstraat, H. J. M. Jansen, G. Ph. Hoornweg, C. Gooijer, N. H. Velthorst, and W. P. Cofino, *Int. J. Environ. Anal. Chem.*, **21**, 299 (1985).
5.46. A. Colmsjö, Y. Zebühr, and C. Östman, *Chem. Scripta*, **20**, 123 (1982).
5.47. W. M. Pitts, A. M. Merle, and M. A. El-Sayed, *Chem. Phys.*, **36**, 437 (1979).
5.48. A. Bree and V. V. B. Vilkos, *J. Mol. Spectrosc.*, **48**, 135 (1973).

5.49. R. E. Connors and W. R. Christian, *J. Phys. Chem.*, **86**, 1524 (1982).
5.50. K. M. Danchinov, E. A. Gastilovich, A. M. Radionov, and D. N. Shigorin, *Russ. J. Phys. Chem.*, **57**, 1546 (1983).
5.51. A. L. Colmsjö, Y. U. Zebühr, and C. E. Östman, *Anal. Chem.*, **54**, 1673 (1982).
5.52. O. M. Artamanova, V. P. Klindukhov, A. A. Krashennikov, T. G. Meister, and A. V. Shablya, *Opt. Spectrosc. (USSR)*, **57**, 163 (1984).
5.53. L. A. Klimova, G. N. Nersesova, V. A. Prozorovskaya, and G. S. Ter-Sarkisyan, *Russ. J. Phys. Chem.*, **50**, 1825 (1976).
5.54. W. P. Cofino, S. M. van Dam, D. A. Kamminga, G. Ph. Hoornweg, C. Gooijer, C. MacLean, and N. H. Velthorst, *Mol. Phys.*, **51**, 537 (1984).
5.55. W. P. Cofino, S. M. van Dam, D. A. Kamminga, G.Ph. Hoornweg, C. Gooijer, C. MacLean, and N. H. Velthorst, *Spectrochim. Acta*, **40A**, 219 (1984).
5.56. W. P. Cofino, S. M. van Dam, G. Ph. Hoornweg, C. Gooijer, C. MacLean, and N. H. Velthorst, *Spectrochim. Acta*, **40A**, 251 (1984).
5.57. V. Lejeune, A. Despres, and E. Migirdicyan, *J. Phys. Chem.*, **88**, 2719 (1984).
5.58. B. E. Kohler and T. A. Spiglanin, *J. Chem. Phys.*, **80**, 3091 (1984).
5.59. M. F. Granville, G. R. Holtom, and B. E. Kohler, *J. Chem. Phys.*, **72**, 4671 (1980).
5.60. B. S. Hudson and B. E. Kohler, *J. Chem. Phys.*, **59**, 4984 (1973).
5.61. R. A. Auerbach, R. L. Christensen, M. F. Granville, and B. E. Kohler, *J. Chem. Phys.*, **75**, 4 (1981).
5.62. G. Jansen and M. Noort, *Spectrochim. Acta*, **32A**, 747 (1976).
5.63. M. Noort, G. Jansen, G. W. Canters, and J. H. van der Waals, *Spectrochim. Acta*, **32A**, 1371 (1976).
5.64. A. A. Gorokhovskii, *Opt. Spectrosc. (USSR)*, **40**, 272 (1976).
5.65. T.-H. Huang, W.-H. Chen, K. E. Rieckhoff, and E. M. Voigt, *J. Chem. Phys.*, **80**, 4051 (1984).
5.66. R. J. Platenkamp, H. O. van Osnabrugge, and A. J. W. G. Visser, *Chem. Phys. Lett.*, **72**, 104 (1980).
5.67. K. Palewska and Z. Ruziewica, *Chem. Phys. Lett.*, **64**, 378 (1979).
5.68. W. P. Cofino, M. Engelsma, D. A. Kamminga, G. Ph. Hoornweg, C. Gooijer, C. MacLean, and N. H. Velthorst, *Spectrochim. Acta*, **40A**, 269 (1984).
5.69. E. V. Shpol'skii, *J. Appl. Spectrosc. (USSR)*, **7**, 336 (1967).
5.70. B. Muel and G. Lacroix, *Bull. Soc. Chim. France*, 2139 (1960).
5.71. P. P. Dikun, N. D. Krasnitskaya, and S. G. Chulikin, *Vopr. Onkol.*, **8**, 31 (1962).
5.72. R. I. Personov and T. A. Teplitskaya, *J. Anal. Chem. (USSR)*, **20**, 1176 (1965).
5.73. R. J. Lukasiewicz and J. D. Winefordner, *Talanta*, **19**, 381 (1972).
5.74. R. Farooq and G. F. Kirkbright, *Analyst*, **101**, 566 (1976).
5.75. Y. Yang, A. P. D'Silva, and V. A. Fassel, *Anal. Chem.*, **53**, 894 (1981).
5.76. R. C. Stroupe, P. Tokousbalides, R. B. Dickinson, Jr., E. L. Wehry, and G. Mamantov, *Anal. Chem.*, **49**, 701 (1977).
5.77. W. S. Chen, K.E. Rieckhoff, and E.-M. Voigt, *Can. J. Chem.*, **62**, 2264 (1984).

5.78. P. Tokousbalides, E. L. Wehry, and G. Mamantov, *J. Phys. Chem.*, **81**, 1769 (1977).
5.79. A. L. Colmsjö and C. E. Östman, *Anal. Chem.*, **52**, 2093 (1980).
5.80. M. Wittenberg, J. Jarosz, and L. Paturel, *Anal. Chim. Acta*, **160**, 185 (1984).
5.81. T. Ya. Gaevaya and A. Ya. Khesina, *J. Anal. Chem. (USSR)*, **29**, 1913 (1974).
5.82. T. A. Tepletskaya, T. A. Alekseeva, and M. M. Val'dman, *Atlas of Quasilinear Luminescence Spectra of Aromatic Molecules*, M.G.U., Moscow, 1978.
5.83. A. Colmsjö and C. Östman, *Atlas of Shpol'skii Spectra and Other Low Temperature Spectra of Polycyclic Organic Molecules*, University of Stockholm, Stockholm, 1981.
5.84. E. P. Lai, E. L. Inman, Jr., and J. D. Winefordner, *Talanta*, **29**, 601 (1982).
5.85. E. L. Inman, Jr., A. Jurgensen, and J. D. Winefordner, *Analyst*, **107**, 538 (1982).
5.86. V. I. Vershinin, A. B. Ovechkin, and T. A. Teplitskaya, *J. Anal. Chem. (USSR)*, **37**, 959 (1982).
5.87. G. Sogliero, D. Eastwood, and R. Ehmer, *Appl. Spectrosc.*, **36**, 110 (1982).
5.88. E. L. Inman, Jr., and J. D. Winefordner, *Anal. Chim. Acta*, **141**, 241 (1982).
5.89. M. J. Kerkhoff, L. A. Files, and J. D. Winefordner, *Anal. Chem.*, **57**, 1673 (1985).
5.90. A. Colmsjö and U. Stenberg, *J. Chromatogr.*, **169**, 205 (1979).
5.91. A. Colmsjö and U. Stenberg, In P. W. Jones and P. Labor, Eds., *Polynuclear Aromatic Hydrocarbons*, Ann Arbor Science Publishers, Ann Arbor, Mich., 1979, p. 121.
5.92. A. L. Colmsjö, Y. U. Zebühr, and C. E. Östman, *Anal. Chem.*, **54**, 1673 (1982).
5.93. A. Colmsjö, Y. Zebühr, and C. Östman, *Chem. Scr.*, **24**, 95 (1984).
5.94. A. Colmsjö, Y. Zebühr, and C. Östman, *Chem. Scr.*, **23**, 185 (1984).
5.95. M. Ewald, M. Lamotte, F. Redero, M. J. Tissier, and P. Albrecht, *Adv. Org. Geochem.*, **12**, 275 (1980).
5.96. J. M. Colin, G. Vion, M. Lamotte, and J. Joussot-Dubien, *J. Chromatogr.*, **204**, 135 (1981).
5.97. P. Garrigues, M. Lamotte, M. Ewald, and J. Joussot-Dubien, *C. R. Acad. Sci. (Paris)*, **293**, 567 (1981).
5.98. P. Garrigues, M. Ewald, M. Lamotte, J. Rima, A. Veyres, R. Lapouyade, J. Joussot-Dubien, and G. Bourgeois, *Int. J. Environ. Anal. Chem.*, **11**, 305 (1982).
5.99. J. Rima, M. Lamotte, and J. Joussot-Dubien, *Anal. Chem.*, **54**, 1059 (1982).
5.100. P. Garrigues, R. de Vazelhes, M. Ewald, J. Joussot-Dubien, and G. Guiochon, *Anal. Chem.*, **55**, 138 (1983).
5.101. P. Garrigues and M. Ewald, *Anal. Chem.*, **55**, 2155 (1983).
5.102. M. Ewald, M. Moinet, A. Saliot, and P. Albrecht, *Anal. Chem.*, **55**, 958 (1983).
5.103. P. Garrigues and M. Ewald, *Int. J. Environ. Anal. Chem.*, **21**, 185 (1985).
5.104. P. Garrigues, G. Bourgeois, A. Veyres, J. Rima, M. Lamotte, and M. Ewald, *Anal. Chem.*, **57**, 1068 (1985).
5.105. P. Garrigues, R. de Sury, J. Bellocq, and M. Ewald, *Analysis*, **13**, 81 (1985).

5.106. M. Wittenberg, J. Jarosz, L. Paturel, M. Vial, and M. Martin-Bouyer, *Analysis*, **13**, 249 (1985).
5.107. S. Monarca, B. S. Causey, and G. F. Kirkbright, *Water Res.*, **13**, 503 (1979).
5.108. A. V. Kanyakin, T. S. Sorokina, and N. F. Efimova, *J. Anal. Chem. (USSR)*, **39**, 1357 (1984).
5.109. A. P. D'Silva, G. J. Oestreich, and V. A. Fassel, *Anal. Chem.*, **48**, 917 (1976).
5.110. C. S. Woo, A. P. D'Silva, and V. A. Fassel, and G. J. Oestreich, *Environ. Sci. Technol.*, **12**, 173 (1978).
5.111. C. S. Woo, A. P. D'Silva, and V. A. Fassel, *Anal. Chem.*, **52**, 159 (1980).
5.112. Y. Yang, A. P. D'Silva, V. A. Fassel, and M. Iles, *Anal. Chem.*, **52**, 1351 (1980).
5.113. Y. Yang, A. P. D'Silva, and V. A. Fassel, *Anal. Chem.*, **53**, 894 (1981).
5.114. Y. Yang, A. P. D'Silva, M. Iles, and V. A. Fassel, in J. A. Gelbwachs, Ed., *Laser Spectroscopy for Sensitive Detection*, SPIE, Washington, D.C., 1981, p. 126.
5.115. G. D. Renkes, S. N. Walters, C. S. Woo, M. K. Iles, A. P. D'Silva, and V. A. Fassel, *Anal. Chem.*, **55**, 2229 (1983).
5.116. Y. Yang, A. P. D'Silva, and V. A. Fassel, *Anal. Chem.*, **53**, 2107 (1981).

4.6. FLUORESCENCE AND PHOSPHORESCENCE LINE-NARROWING SPECTROSCOPY

In 1970, Szabo (6.1) for the first time reported the appearance of narrow lines in the fluorescence spectrum of Cr^{3+} ions in ruby upon narrow-band laser excitation at 4.2 K. A few years later Personov et al. (6.2,6.3) observed the same effect for perylene in an ethanol glass at 4.2 K. At present the physical background of optical site selection or fluorescence (phosphorescence) line-narrowing spectroscopy (FLNS or PLNS) is reasonably well understood. It appears that the technique can be applied to diminish inhomogeneous broadening to a large extent in a wide variety of guest–host systems.

First, the principles of FLNS and PLNS will be discussed with attention to the general physical background, the temperature and excitation wavelength dependence of the spectra, and hole-burning effects. Subsequently, some experimental aspects will be briefly treated and a number of applications, especially of FLNS, will be presented; the analytical applicability of PLNS seems limited. To date a number of review articles, concerned mainly with FLNS, have appeared (6.4–6.8).

4.6.1. Principles

4.6.1.1. Physical Background

When a monochromatic light source is used for excitation of molecules in low-temperature matrices, a limited number of molecules—only those molecules whose energy difference between ground and excited state exactly matches the photon energy—can be excited, thus forming an energetically well-defined distribution of excited molecules, denoted as an isochromat. If the isochromat remains sufficiently pure before emission of radiation takes place, a narrow-line luminescence spectrum may be observed.

It is important to stress a fundamental point. The term "(optical) site selection spectroscopy" frequently used to denote LNS suggests incorrectly that only molecules with identical matrix environments are probed. This assumption is not supported by experiment. Griesser and Wild (6.9), for instance, investigating 1,3-dichloroazulene in 3-methylpentane glass, found upon excitation to S_2 a narrow S_2–S_0 fluorescence spectrum but a broad emission from S_2 to S_1. Furthermore, several groups observed upon excitation in the S_1–S_0 transition a broad-banded phosphorescence spectrum but measured at the same time a narrow-line fluorescence spectrum (6.10,6.11). Similar results have been obtained for an inorganic system (6.12). These experimental observations point to the fact that the physical basis of LNS purely lies in *energy* selection: Molecules with an energy *difference* between ground state and excited state which is exactly equal to the energy of the exciting photons will compose the isochromat. Of course, the energy selection criterion is much less stringent than the site selection criterion: All the molecules in a particular site with an energy difference coinciding with the laser photon energy will be excited. However, a number of such sites with different absolute energies exists, which means that the narrow energy distribution created by the selective excitation procedure will consist of molecules in many different matrix sites with a very wide distribution of absolute energies (6.13,6.14). Since the interaction between electronic states and their surroundings is strongly determined by specific properties of those states, it is not surprising that energy levels which are not involved in the energy selection procedure will not show line narrowing.

The width of the bands observed in the spectra depends on various parameters: the homogeneous bandwidth, the inhomogeneous broadening, and the instrumental contributions of light source and monochromator. In LNS only the inhomogeneous broadening can be strongly reduced, so that the homogeneous bandwidth plays an essential role. This explains why LNS is not successful if short-lived vibronic levels are involved [with lifetimes in the low-picosecond range (6.15)] providing a large natural broadening. For transitions between states with longer lifetimes the decisive contribution to the homogeneous linewidth is formed by the interactions between the electronic transitions and the (intra- and intermolecular) vibrations of the matrix molecules. As was pointed out in Section 4.2.3, a strong linear electron–phonon coupling prohibits the observation of a narrow zero-

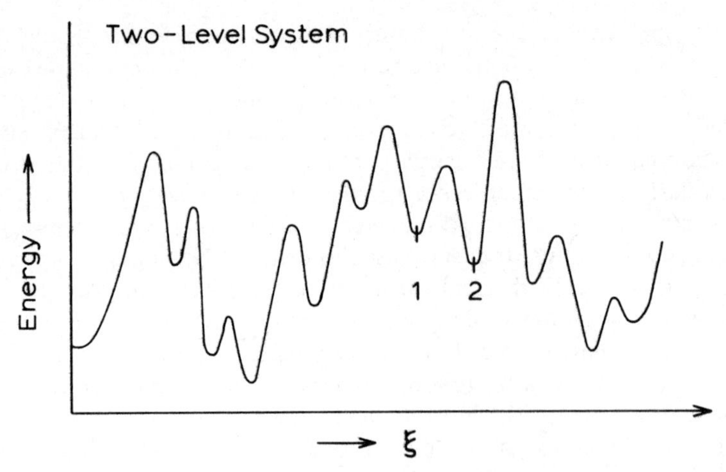

Fig. 4.27. Two-level system model of amorphous solids. According to this model, the electronic structure of such solids can be described by a statistical distribution of double-well potentials, as is illustrated in the figure for one matrix normal coordinate, ξ.

phonon line in the spectrum. By making a judicious choice for the matrix material such a strong interaction between guest and host may be avoided. Other factors that play a role are too high concentration of the guest molecules, which will lead to a broadening of the isochromat via intermolecular energy transfer and local heating of the matrix. The latter effect may be caused by insufficient cooling of the sample, especially when high-intensity laser sources are employed for excitation. Locally, a lot of heat may be released due to radiationless decay processes. This is evident for transitions involving vibrational quanta, but even when the purely electronic 0–0 transition is probed by the exciting source, considerable local heating may occur if the luminescence quantum yield is smaller than 1. Local heating will lead to site interconversion due to diffusion of the matrix molecules, and hence reduction of line narrowing. This reduction may be irreversible if the site interconversion takes place only in the excited state ("nonphotochemical hole burning"; see below).

In general LNS is applied to samples in vitreous matrices, so that it is appropriate to mention some peculiar aspects of the glass structure. The work of Small's group (6.16) on nonphotochemical hole burning has demonstrated that the structure of organic glasses can be regarded as a statistical distribution of asymmetric double-well potentials (see Fig. 4.27). This so-called two-level system (TLS) model was introduced by Anderson et al. (6.17) and Phillips (6.18) to explain the anomalous low-temperature behavior of heat capacity and thermal conductivity of amorphous solids. The interaction of the impurity molecules with the TLS of the amorphous matrix leads to an anomalous temperature dependence of the linewidth of the zero-phonon lines (6.19–6.24). In practice this implies that unless the sample is held at extremely low temperatures [$T < 0.3$ K was observed by Völker and co-workers for dimethyl-*s*-tetrazine, chlorin, and free-base porphin in several organic glasses and polymers (6.24)] the zero-phonon linewidths may be enlarged several orders with respect to crystalline matrices (6.23–6.26). For instance, Thijssen et al. (6.25) in a hole-burning experiment measured a homogeneous linewidth of approximately 6×10^{-3} cm^{-1} for the zero-phonon hole of the 0–0 transition of free-base porphin in *n*-octane at 4.2 K, whereas a value of about 10^{-2} cm^{-1} was found in ethanol. At temperatures of 0.3 K and below, in both crystalline and vitreous matrices, the lifetime limited value of 3×10^{-4} cm^{-1} was almost attained. In spite of the extra broadening caused by the amorphous matrix the linewidths obtained in LNS generally suffice the resolution requirement of 2 Å necessary to discern structural isomers.

The most serious consequence of performing spectroscopy in an amorphous matrix is the risk of site interconversion. This phenomenon, wich gives rise to the nonphotochemical hole-burning effect, can also be explained in terms of the TLS model. The detrimental influence of hole burning on LNS will be discussed in Section 4.6.1.3.

4.6.1.2. Temperature Dependence

At present no thorough investigation of temperature effects on LNS has been reported. The temperature requirements seem to be much more stringent for amorphous than for crystalline matrix materials. Depending on the particular guest–host combination, above 30–50 K no narrow lines are obtained. At 37 K, for instance, the zero-phonon lines in the spectrum of perylene in ethanol have disappeared completely (6.2). The strong dependence of the Debye–Waller factor on temperature in LNS may be understood by realizing that the observed intensity ratio of zero-phonon line and phonon wing is not only due to the centers resonantly excited by the narrow-banded excitation sources. In glassy matrices an important contribution to the phonon wing is due to emission from molecules which were nonresonantly excited in their phonon wings; such molecules rapidly relax to their corresponding phononless state in the excited electronic state and thus give emission to the long-wavelength side of the resonant zero-phonon line (i.e., in the very region where the phonon wing of the resonant centers is found). Thus the Debye–Waller factor observed (α_{obs}) depends on both the Debye–Waller factor of the molecules in the excited state, α_{abs} (which is a measure of the contribution of the nonresonantly excited molecules to the phonon wing) and the Debye–Waller factor of the isochromat in luminescence, α_{lum} (6.8):

$$\alpha_{obs} \approx \alpha_{lum}\alpha_{abs} \approx \alpha^2{}_{lum} \tag{6.1}$$

Hence α_{obs} will in general be much smaller than the actual Debye–Waller factor for the isochromat.

Apart from the intensity of the zero-phonon line its width will also be temperature dependent. As was pointed out before, hole-burning experiments indicate a linewidth for the zero-phonon lines broader in amorphous than in crystalline matrices above 0.3 K. The hole width in amorphous systems is reported to depend on the temperature as T^x, where x is 1.3–1.0 at lower temperatures (6.24,6.25). The thermal broadening of the zero-phonon lines has so far not been investigated in LNS. It seems not to impede the specificity of the method in analytical applications.

4.6.1.3. Excitation Wavelength Dependence

The line-narrowed spectra are strongly dependent on the excitation frequency, ν_{exc}. First, the difference between the photon energy $h\nu_{exc}$ and the molecular 0–0 transition energy plays a role. Secondly, variations of ν_{exc} within an inhomogeneously broadened absorption band affect the spectral features.

As far as the first effect is concerned three excitation regions may be discerned in LNS (6.10,6.27–6.30) excitation (1) in the 0–0 region, (2) to excited vibronic states up to 2000 cm^{-1}, and (3) to higher vibronic states. When excitation is accomplished in the inhomogeneously broadened 0–0 transition region a simple one-sited spectrum emerges. A displacement of the excitation frequency yields a corresponding displacement of the bands in the line-narrowed spectrum, the 0–0 transition coinciding with ν_{exc} (6.10). Since the emission wavelength is exactly equal to the excitation wavelength, this transition cannot be observed in the line-narrowed spectrum unless time-resolved detection is employed.

Excitation to the area of the excited vibronic states (up to approximately 2000 cm^{-1}) results in an increased complexity of the spectrum. In this vibrationally congested region the excitation source generally probes several vibrational subbands, which overlap due to their inhomogeneous broadening. The vibrationally excited states relax to the corresponding vibrational ground state of the first excited electronic state and are subsequently observed in emission. Since different groups of molecules are excited in different vibrationally excited states, different isochromats are formed. As a result, a multiplet structure will be observed in the spectra. The 0–0 transition, however, now is visible in the spectrum. The energy distances of

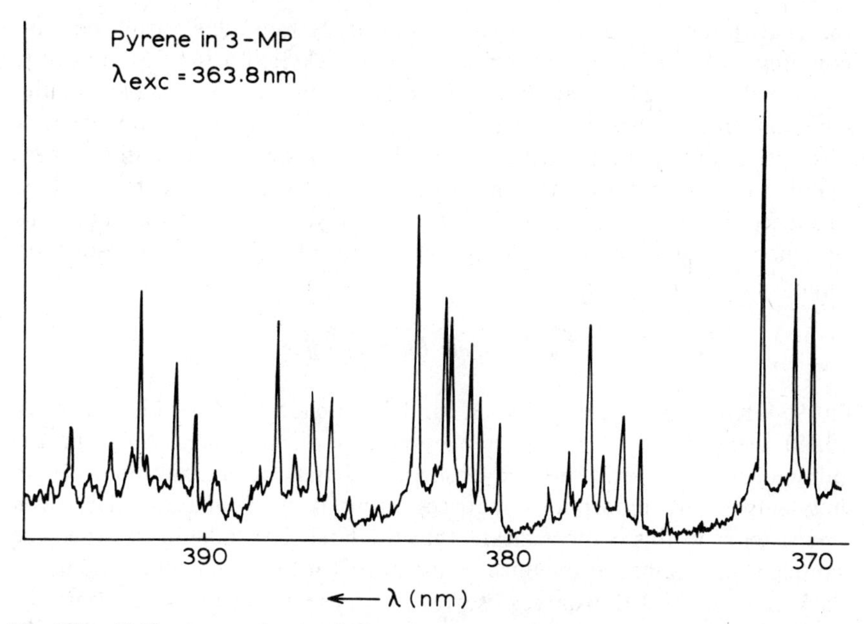

Fig. 4.28. FLN spectrum of pyrene in 3-methylpentane (T = 10 K, λ_{laser} = 363.8 nm). Three sites are discerned in the spectrum with 0–0 transitions at 371.73,370.64, and 370.04 nm, corresponding to vibrational modes in the S_1-state of pyrene with energies of 587, 508, and 464 cm^{-1}, respectively.

the separate lines in the 0–0 multiplet from ν_{exc} correspond to the energy differences of the vibrational quanta in the molecular excited state (6.29,6.30). An example of multiplet structure is shown for pyrene in 3-methylpentane (Fig. 4.28).

Excitation into the region of the higher excited vibronic states (starting at 2000–3000 cm^{-1}) results in a broad-banded spectrum. Here the density of states is so high that many overlapping isochromats will be formed, destroying the line-narrowing effect. Also, local heating of the matrix (6.10) and much faster relaxation of the high-frequency vibrations and combination bands (6.15,6.31) may hamper line narrowing for such excitation. When a short-living higher excited electronic state is involved in the excitation process no narrow zero-phonon lines will be observed in emission (6.32).

As a second effect, variation of ν_{exc} within an inhomogeneously broadened absorption band will not only lead to a corresponding shift of the line-narrowed spectrum, but also to a change of the Debye–Waller factor (6.33–6.35). It appears that if ν_{exc} is increased within the broad absorption band the Debye–Waller factor gradually decreases. Flatscher and Friedrich

(6.33) attributed this decrease to a varying strength of the electron–phonon coupling with site energy. Sapozhnikov and Alekseev (6.34,6.35), however, showed that this effect can easily be explained by taking into account the emission from nonresonantly excited centers within the inhomogeneous absorption distribution function. Namely, the lower the excitation wavelength that is employed, the more nonresonantly excited centers will be produced in the excited state. Optimal Debye–Waller factors are thus obtained for excitation at the long-wavelength side of the broad absorption bands.

4.6.1.4. Hole-Burning Effects

In 1974, Kharlamov et al. (6.36,6.37) and, independently, Gorokhovskii et al. (6.38) were the first to report the burning of persistent holes in the inhomogeneously broadened S_1-absorption bands of organic molecules in low-temperature amorphous matrices upon laser irradiation. The term "persistent" in this context means that the holes last indefinitely when the sample is maintained at the burn temperature (or lower temperatures) in the dark, in contrast with what is observed in transient saturation hole burning.

To date, two types of hole burning are discerned, photochemical (PHB) and nonphotochemical (NPHB). In PHB the holes are formed due to photochemical reactions of the excited molecules. This is not the case in NPHB, where the gaps are formed in the S_1 absorption spectrum of photostable molecules (e.g., perylene, tetracene, and anthracene). Small's group formulated a mechanism for the production of nonphotochemical holes in the framework of the two-level system (TLS) structural model for the amorphous state (6.39–6.42). They supposed that hole formation at a certain burn temperature is due to a subset of the TLSs interacting with the guest molecules in such a way that two conditions are fulfilled. First, the interaction of the TLSs with the excited electronic state must be strong enough to modify the double-well potential so that relaxation between the minima becomes competitive with the normal decay processes of the excited state. Second, the interaction with the electronic ground state must be such that relaxation between the TLSs coupling to this state is slow enough to account for the long hole persistence time. This theory is elucidated in Fig. 4.29.

Little attention has been paid until now to the effect of the hole burning on LNS. Kharlamov et al. (6.37) observed a decrease in the intensity of the zero-phonon lines for perylene in ethanol under prolonged laser irradiation; the decay was found to depend more or less linearly on the laser excitation intensity. Preliminary experiments in our laboratory on tetracene in several matrices suggest that the influence of NPHB on line-narrowed spectra can be minimized by working at slightly increased temperatures (10–20 K), by

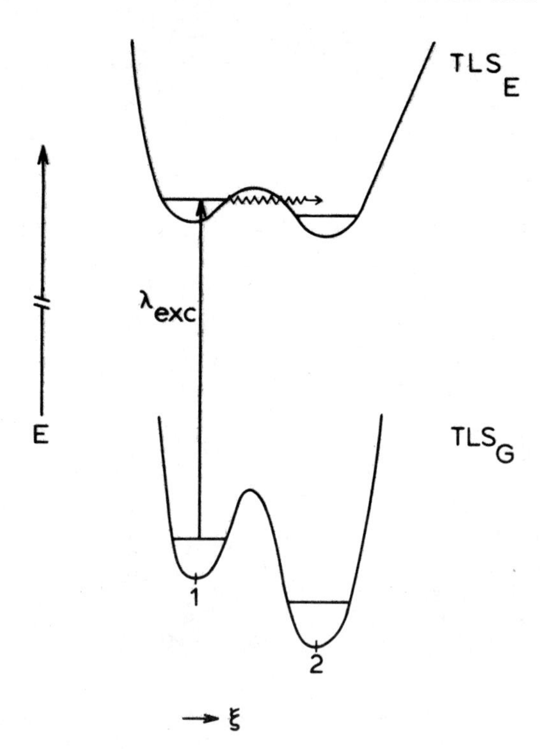

Fig. 4.29. Schematic representation of the mechanism responsible for NPHB according to Small (6.16). The hole-burning effect in this model is attributed to a difference in the interaction of the S_1 and S_0 electronic states of the impurity molecule with the two-level system structure of amorphous solids, sketched in Fig. 4.27. One double-well potential from this figure, corresponding to matrix configuration 1 and 2, is depicted.

making use of excitation into the vibronically excited S_1 states, or by choosing suitable matrix materials (6.43). For tetracene in a polyethylene film, for instance, no NPHB could be observed, whereas NPHB in a decalin glass occurred at a very low rate (6.43). An illustration of hole-burning effects on the FLN spectrum of tetracene is given in Fig. 4.30. No NPHB is observed for photostable molecules embedded in crystalline host matrices. Review articles on hole burning have been published by Friedrich and Haarer (6.44), Rebane et al. (6.45) (both on PHB), and Small (6.16) (on NPHB).

4.6.2. Experimental Aspects

In this section we limit ourselves to experimental aspects typical of LNS. The ideal narrow-line excitation source is of course the laser, which combines high monochromaticity and high power density. Nevertheless, a conven-

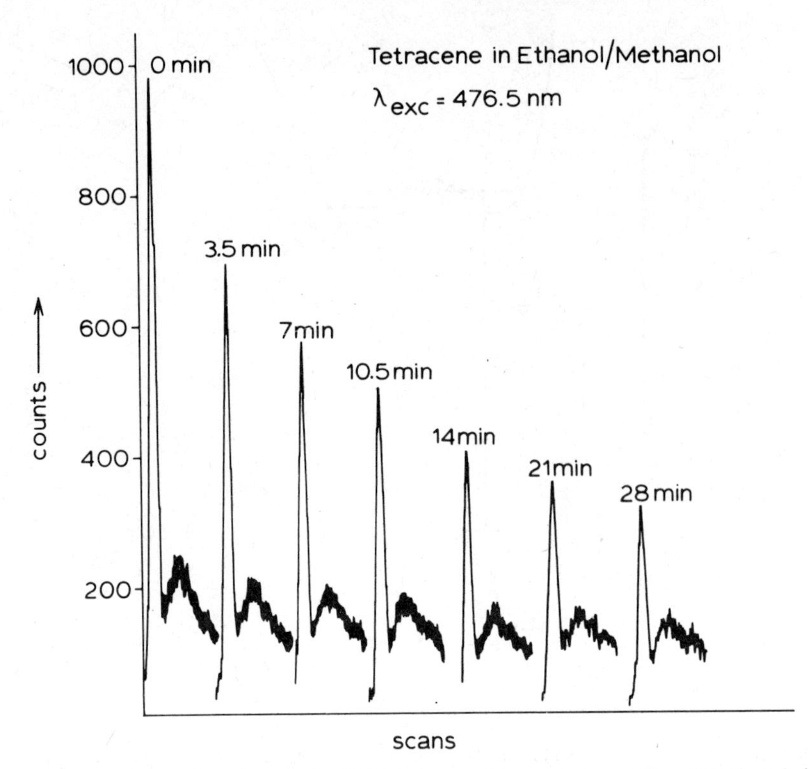

Fig. 4.30. Hole-burning effects in the FLN spectrum of tetracene in an ethanol/methanol glass (4:1, v/v) at 9 K. Excitation was performed at 476.5 nm, in the 0–0 transition of tetracene, with a laser intensity of 2–3 μW/mm². Scans of the 0–314 cm^{-1} vibronic fluorescece band beginning at the times indicated in the figure are depicted.

tional broad-band lamp source in combination with a high-resolution monochromator has also been employed (6.33). The spectral radiant power of the latter system is so low that it seems of no analytical use.

When a laser is applied for excitation, one has to be aware that solvent Raman lines and nonlasing laser plasma lines may appear in the spectrum. The former lines are easily identified by recording a solvent blank or by measuring the spectrum at higher temperatures, since they keep their quasi-linear character. The latter lines can be removed with a prism monochromator. The efficiency of the monochromator for the lasing line will be very high due to its good directionality, so that not much intensity will be lost.

Although any kind of solvent may serve as matrix material, glasses that are clear at low temperatures have the advantage that laser scatter is reduced.

Many organic glasses have been used (3-methylpentane, ethanol, methanol, EPA*, 2-methyltetrahydrofuran, and glycerol/water/ethanol mixtures being the most frequently applied). Care must be taken in the cooling procedure to prevent cracking of the glassy solutions as much as possible.

Probably the best cooling system is a helium bath cryostat since it provides the lowest temperatures and an efficient (contact) cooling. To reduce hole-burning effects, measurement at somewhat higher temperatures (10–20 K) may be useful (6.43). This can be done in a bath cryostat with a variable temperature insert or in a conductance cryostat. When employing the latter type of cryostat, one has to guard against local heating of the matrix.

On the detection side, the main concern is the elimination of laser scatter. To obtain simple line-narrowed spectra, excitation must be in or close to the 0–0 transition area. To gather most information it is necessary to start the measurement as close to the excitation frequency as possible. Optimal detection is achieved with a pulsed laser source and time resolution: not only are laser scatter interferences excluded, but Raman scattering and plasma light effects are removed. Furthermore, if excitation is performed in the 0–0 area, with time resolution this transition may also be observed in emission. When a continuous laser is employed, a double monochromator should be used because of its excellent stray-light rejection capability. To increase the speed of the measurements, photodiode array detectors will be useful.

4.6.3. Applications

A number of general applications of both PLNS and FLNS will be given. Analytical aspects will be discussed in a separate section. Within the context of this chapter it is not possible to achieve completeness.

4.6.3.1. Phosphorescence Line-Narrowing Spectroscopy

To observe line-narrowed phosphorescence spectra, it is necessary to probe the T_1–S_0 transition with the monochromatic excitation source. Since this transition is spin forbidden (the absorption coefficient generally is six to seven orders smaller than for S_1–S_0 transitions), it is very difficult to record PLN spectra. Very few PLN spectra have been reported thus far. In all studies high concentrations (10^{-1}–10^{-2} *M* mostly), very intense laser lines, external or internal heavy atoms, and special phosphoroscopes were used. In the first study of PLNS, spectra of coronene, chrysene, and 1,2-benzpyrene in bromobutane were reported (6.11). Later, narrow-line phosphorescence spectra were obtaned for 5-bromoacenaphthene in bromobutane (6.46,6.47),

* A mixture of ethanol, isopentane, and diethyl ether (2:5:5, v/v/v).

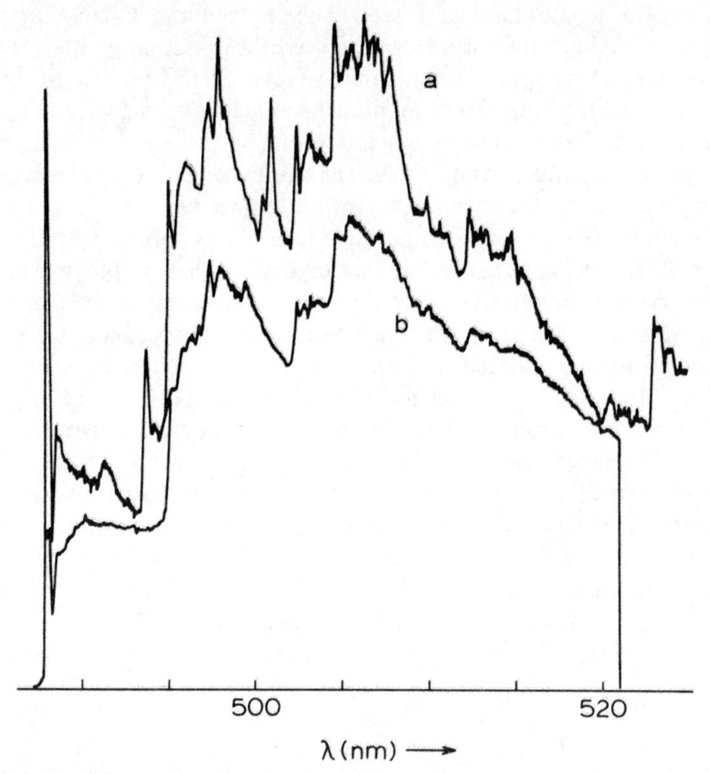

Fig. 4.31. Phosphorescence spectrum of a mixed solution of 1-iodonaphthalene (a, donor) and perfluoronaphthalene (b, acceptor) in ethanol at 4.2 K. The spectra of the two compounds could be obtained separately via temporal resolution. (From ref. 6.48).

1-iodonaphthalene and perfluoronaphthalene in EPA and bromobutane, respectively (6.48), and benzo[*a*]phenazine in bromobutane and 2-methyltetrahydrofuran (6.49).

Of special interest is the possibility to obtain information on the vibronic structure of the T_1 state via PLNS; this can be derived from the site structures of the PLN spectra that are observed when excitation is performed to the vibrationally excited sublevels of the T_1 state. Such information is otherwise very difficult to acquire. Suter and Wild (6.49) showed that it is possible to measure the degree of polarization of the individual phosphorescence bands by combining PLNS and photoselection techniques.

Due to the extremely low absorption coefficients of the T_1–S_0 transition, PLNS probably has little analytical potential. An interesting experiment in this context has been reported by Fünfschilling et al. (6.48), who observed that PLN is conserved during the triplet energy transfer from 1-iodonaphthalene to perfluoronaphthalene (see Fig. 4.31). It may be that sensitized phosphorescence can enhance the scope of PLNS for donors and acceptors with similar T_1–S_0 energy differences.

4.6.3.2. Fluorescence Line-Narrowing Spectroscopy

Much more attention has been devoted to FLNS. Since the first publication on the FLN spectrum of perylene in ethanol in 1972 a large number of compounds in a variety of matrices has been studied. Especially, many PAHs and compounds of biological interest, such as porphyrins and chlorophylls, have been measured. A short survey is presented in Table 4.5. Line-narrowed spectra obtained for samples prepared via matrix isolation procedures have been omitted since they were discussed in Section 4.4. The survey clearly indicates that many very diverse compounds produce narrow-line emission spectra in a wide variety of matrices. Even molecules in lecithin vesicle membranes (6.74) and adducts to DNA (6.66) yield FLN spectra, which illustrates the applicability of this technique to compounds in relatively complex environments. The fact that many polar and even ionic species give FLN spectra in suitable solvents presents a very strong point. For instance, no Shpol'skii spectra of such compounds can be obtained due to solubility and aggregation problems in the apolar *n*-alkane solvents.

Almost all experiments have been performed at liquid-helium temperatures; for some compounds at higher temperatures (5–30 K), FLN spectra have also been reported, indicating that less expensive conductance cooling in a closed-cycle refrigerator is feasible.

Apart from studies of the properties of FLN spectra and the requirements that have to be fulfilled to observe them, most work cited in Table 4.5 is devoted to the study of vibrationally resolved vibronic spectra. For this purpose the FLN technique offers an important advantage over other techniques, as it appears possible to determine the polarization of the individual vibronic transitions in a very easy way. Doping of polymers with the compound of interest and subsequent stretching of the polymer film to orient the molecules renders directly the absolute polarizations of the spectral bands (6.60,6.65) (see also Fig. 4.4). The potential of the FLN method in the biological sciences is illustrated by the work of Heisig et al. on adducts of carcinogenic PAH-metabolites to DNA (6.66) and of Fünfschilling and Walz on chlorophyll *b* in lecithin vesicle membranes (6.74).

4.6.4. Analytical Aspects

The analytical potential of FLNS has been discussed by Wehry and Mamantov (6.86), Brown et al. (6.87), and Bykovskaya et al. (6.80). A short overview of the analytical pros and cons of the technique will be given below.

The main and very important advantage of FLNS above the other low-

Table 4.5. Survey of Systems Studied with Fluorescence Line-Narrowing Spectroscopy

Compound	Solvent	Temperature	References
1. Polynuclear Aromatic Hydrocarbons			
Anthracene	Ethanol	4.2	6.50
	2-MTHF[a]	4.2	6.33
	Glycerol/water (5:4)[b]	4.2	6.51
	Glycerol/water/DMSO[c] (1:1:1)	4.2	6.52
Benz[*a*]pyrene	Glycerol/water/DMSO (1:1:1)	4.2	6.52
Benz[*e*]pyrene	Glycerol/water/DMSO (1:1:1)	4.2	6.52
Benz[*k*]fluoranthene	Glycerol/water/DMSO (1:1:1)	4.2	6.52
3,4,8,9-Dibenzpyrene	Ethanol	4.2	6.53
	Paraffin oil	4.2	6.53
Fluorene	3-Methylpentane	5	6.54
Naphthalene	3-Methylpentane	10	6.54
Perylene	Ethanol	4.2	6.2,6.28
	Polyvinylbutyral	1.5	6.55
	Glycerol/water/DMSO (1:1:1)	4.2	6.52
Phenanthrene	Ethanol	4.2	6.56
	Methylcyclohexane	4.2	6.57
Pyrene	Glycerol/water (5:4)	4.2	6.51
	Glycerol/water/DMSO (1:1:1)	4.2	6.52
	3-Methylpentane	10;20	6.58,6.59
	Silica TLC plate	10;20	6.58,6.59
	Ethanol	10	6.59
Tetracene	2-MTHF	4.2	6.10,6.57
	Polyethylene	5	6.60
	Ethanol/methanol (4:1)	2	6.22
	Diethyl ether/isopropanol (5:2)	2	6.22,6.61
2. Substituted Polynuclear Aromatic Hydrocarbons			
1-Aminoanthracene	Glycerol/water (2 *M* HCl)	4.2	6.62
2-Aminoanthracene	Glycerol/water (2 *M* HCl)	4.2	6.62
1-Aminonaphthalene	Glycerol/water (2 *M* HCl)	4.2	6.63
1-Aminopyrene	Glycerol/water (2 *M* HCl)	4.2	6.62
2,7-Diaminofluorene	Glycerol/water (2 *M* HCl)	4.2	6.63
9,10-Dibromoanthracene	Glycerol/water/DMSO (1:1:1)	4.2	6.52
1-Chloronaphthalene	Ethanol	4.2	6.10
1-Chloropyrene	Silica TLC plate	10	6.59
Dichloropyrene(s)	Silica TLC plate	10	6.59
1-Fluoropyrene	Glycerol/water/ethanol (2:1:1)	10	6.59,6.64
	Silica TLC plate	10	6.59,6,64
1-Methylpyrene	Glycerol/water/DMSO (1:1:1)	4.2	6.52
9-Methylanthracene	Glycerol/water/DMSO (1:1:1)	4.2	6.52
9,10-Dimethylanthracene	Glycerol/water/DMSO (1:1:1)	4.2	6.52
Duryl radical	Polyethylene	5	6.65
P-xylyl radical	Polyethylene	5	6.65
Benz[*a*]pyrene-metabolites (and adducts to DNA)	Glycerol/water/ethanol (45:35:20)	4.2	6.66
Benz[*a*]anthracene, chrysene, 5-methylchrysene, and Benz[*a*]pyrene metabolites	Glycerol/water/ethanol (45:35:20)	4.2	6.67

Compound	Solvent	Temperature	References
3. Compounds of Biological Interest			
Chlorin	Toluene/diethyl ether (1:2)	4.2	6.68
	Polystyrene	4.2	6.68
Isobacteriochlorin	3-Methylpentane	4	6.69
Chlorophyll *a*	2-MTHF	4.2	6.70,6.71
	Diethyl ether	4.2	6.72,6.75
	Methanol/ethanol	5	6.73
Chlorophyll *b*	2-MTH	4.2	6.71,6.74
	Diethyl ether	4.2	6.72,6.75
	Methanol/ethanol	5	6.73
	Ethanol	10	6.74
	n-Butylacetate	10	6.74
	n-Butanol	10	6.74
	Toluene	10	6.74
	Lecithin vesicle membrane	10	6.74
Pheophytin *a*	Methanol/ethanol	5	6.73
Phthalocyanine	Paraffin oil	4.2	6.53
Zn-phthalocyanine	Paraffin oil	4.2	6.56
	n-Butanol	4.2	6.27
Porphin	Toluene/diethyl ether (1:2)	4.2	6.68
	Polystyrene	4.2	6.68
Porphin dication	Trifluoroacetic acid	4.2	6.76
Metalloporphins	Polyvinylbutyral	4.2	6.76
	Toluene/pyridine (5:1)	4.2	6.76
Coproporphyrin III	Ethanol/water (1 *M* HCl)	28.2	6.77
Mesoporphyrin	Ethanol/water (1 *M* HCl)	28.2	6.77
Octaethylporphyrin	Toluene/diethyl ether (1:2)	4.2	6.68
	Polystyrene	4.2	6.68
Protoporphyrin IX	Toluene/diethyl ether (1:2)	4.2	6.68
	Polystyrene	4.2	6.68
	Ethanol	4.2	6.78
	Ethanol/water (1 *M* HCl)	4.2	6.78
4. Other Compounds			
9-Aminoacridine	Ethanol	4.2	6.79
Polymethene dyes	Amyl alcohol	4.2	6.80
Quinizarin	Ethanol/methanol (3:1)	4.2	6.81
Resorufin	Poly(methyl methacrylate)	1.8	6.82
	Ethanol/methanol	4.2	6.83
Dimethyl-*s*-tetrazine	Polyvinylcarbazole	10	6.84
Thioindigo	Paraffin oil	4.2	6.85
	Ethanol	4.2	6.85
	Chloroform	4.2	6.85

[a] 2-Methyltetrahydrofuran.

[b] Ratios refer to volume fractions.

[c] Dimethylsulfoxide.

temperature high-resolution techniques is its versatility in solvent choice. In many polar and apolar and even relatively complex matrices (polymers, lecithin membrane) useful line-narrowed spectra have been observed. Also, water-containing glasses may be used, affording a simple means to analyze aqueous samples. However, as not all solvents yield the same quality of the spectra, some optimization may be necessary. The wide variety of available solvents generally precludes solubility problems and also enables one to obtain a clear sample at low temperatures (provided that it is carefully prepared), leading to a reduction of laser scattering.

Even though many parameters may be varied to optimize experimental conditions, not all fluorescent compounds will provide highly resolved FLN spectra. There are two main prerequisites for the observation of narrow bands. First, the energy selection procedure must result in an isochromat that remains pure until emission takes place. Second, the homogeneous linewidth must be sufficiently narrow. These prerequisites impose a number of conditions on the investigated samples.

In the first place, to ensure conservation of the isochromat, energy transfer between solute molecules and from solute to solvent molecules should be avoided. Solute aggregation effects can be minimized by working at low concentrations. On has to realize here that long-range Förster energy transfer processes can take place over distances of 50 Å or more. Another danger threatening the purity of the isochromat is nonphotochemical hole burning (see Section 4.6.1.4). In this case, changes occur in the direct environment of the excited analytes influencing the site structure of the matrix and thus lead to a decrease in the intensity of the narrow bands in the FLN spectrum. Hole-burning effects can be reduced or removed by, for instance, the choice of a suitable matrix material (6.43).

Second, the homogeneous linewidth in low-temperature matrices is determined by the lifetimes of the states involved in the transition and the strength of the electron–phonon interaction, as was discussed in the introduction. For fluorescence transitions from the S_1 state, the broadening of the lines due to the lifetime of the excited state is generally negligible; the electron–phonon interaction, depending on the particular combination of analyte and solvent, is more important. McColgin et al. (6.83) stated that the Debye–Waller factor varies inversely with the Stokes shift (i.e., the energy difference between the fluorescence and absorption origins) observed for the broad-banded low-temperature spectra. As the Stokes shift is generally large for molecules in which large changes in geometry occur between the electronic ground and excited states, especially well resolved FLN spectra are expected for rigid molecules. This seems to be in line with the results of Flatscher et al., who were unable to obtain a narrow-line fluorescence spectrum for diphenylanthracene (6.50). Nevertheless, it must be emphasized that compounds showing large Stokes shifts in fluid solution do not necessarily also exhibit large shifts in low-temperature solid matrices. Fluoranthene, for instance, has a Stokes shift of several thousands of cm^{-1} in the liquid phase, while the 0–0 transition is the most intense band in the low-temperature Shpol'skii spectrum (6.89).

Finally, it should be realized that the temperature has a strong influence on the Debye–Waller factor, so that an accurate temperature control is needed. Furthermore, special attention has to be paid to possible local heating due to the highly intense laser excitation. In particular, when conductance cooling is employed, such effects may be impressive since the organic matrices are poor heat conductors.

4.6.4.1. Qualitative Analysis

The selectivity of FLNS is high; generally, spectral bandwidths well under 5 cm^{-1} are observed. Hence the measured spectral bandwidths will be determined instrumentally, as some selectivity will be sacrificed to increase the sensitivity. There are, however, some experimental problems. The strong dependence of the appearance of the spectrum on excitation wavelength may complicate identification. Especially when excitation into the higher vibronic S_1 states of the analyte is performed, the spectrum will show a multiplet structure. In complex samples this can be a serious problem. With a tunable laser the appearance of multiplet structure in the spectra usually can be avoided, for instance, by excitation into the S_1–S_0 0–0 transition area. Even there the problem of excitation wavelength dependence remains as the spectrum shifts with λ_{exc}. Furthermore, part of the fluorescence spectrum is obscured due to strong Rayleigh scattering of the laser light. With time-resolved detection the whole emission spectrum can be recorded, but then one has to reckon with strong self-absorption effects of the 0–0 transition. Thus excitation into the lower-energy vibrationally excited S_1 states where the density of states is very low is generally most favorable.

Selectively in FLNS can be increased further by making use of time-resolved gated detection of compounds with different fluorescence lifetimes. In this way pyrene, 1-methylpyrene, and anthracene could be unambiguously detected in a solvent-refined coal sample (6.52).

4.6.4.2. Quantitative Analysis

In FLNS there appears to be no need for standard addition procedures in quantitative analysis (6.51, 6.52). Contrary to the other techniques, in FLNS the samples are not opaque and the outlook of the quasi-linear spectra is not strongly dependent on the preparation of the sample. Calibration curves

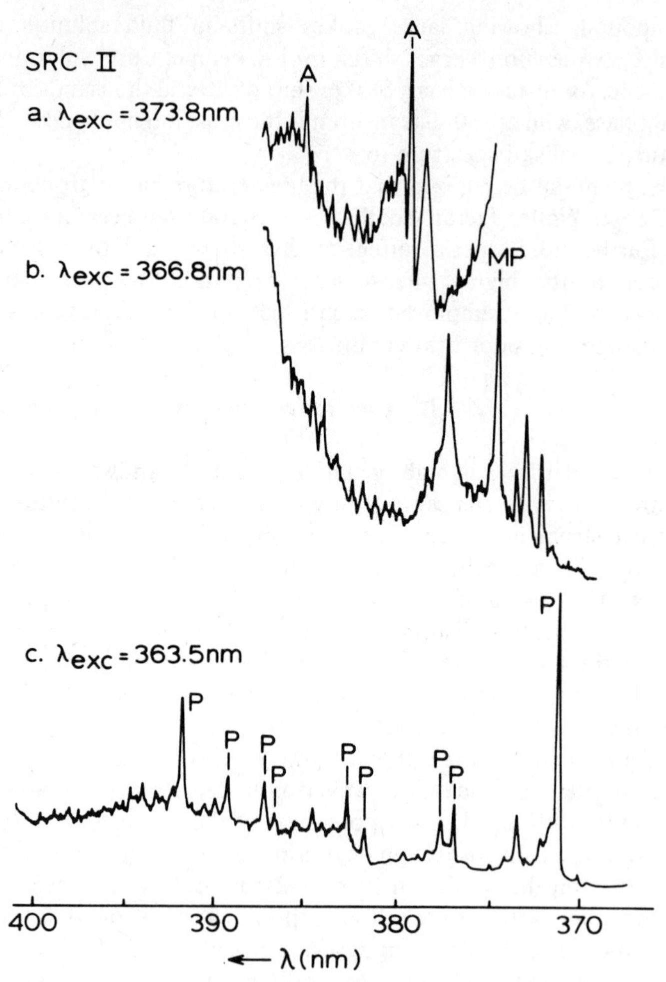

Fig. 4.32. FLN spectra for (*a*) anthracene (A), (*b*) 1-methylpyrene (MP), and (*c*) pyrene (P) in a solvent-refined coal (SRC-II) sample at 4.2 K. Excitation wavelengths were optimized for each PAH. (Reprinted with permission from J. C. Brown, J. A. Dunscanson, Jr, and G. J. Small, *Anal. Chem.*, **52**, 1711. Copyright 1980 American Chemical Society).

which were linear over three orders of magnitude (1 ppb–1 ppm) were measured for pyrene and anthracene in a glycerol–water glass at 4.2 K by simple measurement of the peak height of the 0–0 zero-phonon line above the background. The data were obtained over a 1-week period during which several optical alignments and liquid-helium fills were necessary. However, this did not detract substantially from the reproducibility; between two and five points were taken at each concentration with an average deviation from the mean of 8% (6.51).

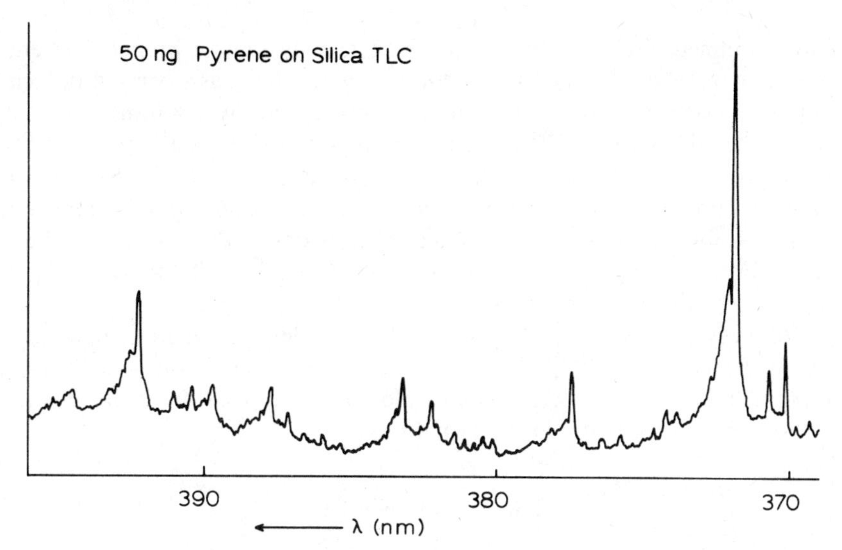

Fig. 4.33. FLN spectrum of pyrene on a silica TLC-plate. 50 ng of this compound was applied in *n*-dodecane (T = 10 K, λ_{exc} = 363.8 nm).

Detection limits in FLNS are expected to be relatively high since mostly glassy matrices are employed which show, even at low temperatures, quite impressive inhomogeneous broadening. So, with narrow-line laser excitation, just a small fraction of the analyte molecules is brought in the excited state. Still, for compounds with near unity fluorescence quantum yield, detection limits in the low ppt range were obtained (6.61,6.88). Bolton and Winefordner in their study on the analytical merits of FLNS reported very much higher detection limits (i.e., four orders of magnitude worse) (6.90). The reason for this discrepancy is not clear, as both groups used similar experimental setups. Hayes and Small critically assessed the paper of Bolton and Winefordner (6.91).

4.6.5. Analytical State of the Art

Few analytical applications of FLNS have appeared in the literature so far, most of them being concerned with academic samples. The potential of FLNS for the analysis of real samples has been illustrated by means of some examples. Pyrene was determined in spiked diluted cigaret tar (6.87); pyrene and anthracene in an extract of soot collected from burning apiezon wax (6.87); benz[a]pyrene, anthracene, 1-methylpyrene, and pyrene in a solvent-refined coal sample (6.52); and benz[a]pyrene and perylene in gasoline (6.88). In all cases the analytes could be unambiguously identified at concentrations in the low-ppb range without any prior cleanup or chromatographic fractionation. In Fig. 4.32 the FLN spectra obtained for several PAHs in a solvent-refined coal sample are shown. The only sample preparation employed consisted of a 1:10,000 dilution with the glass-forming solvent (glycerol–water with some ethanol and 2-methyltetrahydrofuran).

The applicability of FLNS in combination with thin-layer chromatography (TLC) has been studied by Hofstraat et al. (6.58,6.59,6.64). It appeared possible to obtain narrow-line fluorescence spectra of pyrene and halogen-substituted pyrenes on silica TLC plates both with and without prior chromatographic separation (see Fig. 4.33). For quantitation, use of standard addition was required.

Recently, FLNS was employed as an off-line detection method in column liquid chromatography (LC) using TLC plates to store the LC chromatogram (6.92). It is expected that FLNS will be applied on environmental samples more frequently in the near future.

References to Section 4.6

6.1. A. Szabo, *Phys. Rev. Lett.*, **25**, 924 (1970).
6.2. R. I. Personov, E. I. Al'shits, and L. A. Bykovskaya, *Opt. Commun.*, **6**, 169 (1972).
6.3. R. I. Personov, E. I. Al'shits, and L. A. Bykovskaya, *JETP Lett.*, **15**, 431 (1972).
6.4. B. E. Kohler, in C. B. Moore, Ed., *Chemical and Biochemical Applications of Lasers*, Vol. 4, Academic Press, New York, 1979, p. 31.
6.5. R. I. Personov, *J. Lumin.*, **24/25**, 475 (1981).
6.6. M. J. Weber, in W. M. Yen and P. M. Selzer, Eds., *Laser Spectroscopy of Solids*, Springer, New York, 1981, Chap. 6.
6.7. R. I. Personov, *Spectrochim. Acta*, **38B**, 1533 (1983).
6.8. R. I. Personov, in V. M. Agranovich and R. M. Hochstrasser, Eds., *Spectroscopy and Excitation Dynamics of Condensed Molecular Systems*, North-Holland, Amsterdam, 1983, Chap. 10.
6.9. H. J. Griesser and U. P. Wild, *J. Chem. Phys.*, **73**, 4715 (1980).
6.10. K. Cunningham, J. M. Morries, J. Fünfschilling, and D. F. Williams, *Chem. Phys. Lett.*, **32**, 581 (1975).
6.11. E. I. Al'shits, R. I. Personov, and B. M. Kharlamov, *Chem. Phys. Lett.*, **40**, 116 (1976).
6.12. R. Flach, D. S. Hamilton, P. M. Selzer, and W. M. Yen, *Phys. Rev.*, **B15**, 1248 (1977).
6.13. P. M. Selzer, in W. M. Yen and P. M. Selzer, Eds., *Laser Spectroscopy of Solids*, Springer, New York, 1981, Chap. 4.
6.14. H. W. H. Lee, C. A. Walsh, and M. D. Fayer, *J. Chem. Phys.*, **82**, 3948 (1985).
6.15. J. B. Hopkins and P. M. Rentzepis, *Chem. Phys. Lett.*, **117**, 414 (1985).
6.16. G. H. Small, in V. M. Agranovich and R. M. Hochstrasser, Eds., *Spectroscopy and Excitation Dynamics of Condensed Molecular Systems*, North-Holland, Amsterdam, 1983, Chap. 9.
6.17. P. W. Anderson, B. I. Halperin, and C. M. Varma, *Philos. Mag.*, **25**, 1 (1972).

6.18. W. A. Phillips, *J. Low Temp. Phys.*, **7**, 351 (1972).
6.19. P. M. Selzer, D. L. Huber, D. S. Hamilton, W. M. Yen, and M. J. Weber, *Phys. Rev. Lett.*, **36**, 813 (1976).
6.20. S. K. Lyo and R. Orbach, *Phys. Rev.*, **B22**, 4223 (1980).
6.21. H. Morawitz and P. Reineker, *Solid State Comm.*, **42**, 609 (1982).
6.22. J. Friedrich, H. Wolfrum, and D. Haarer, *J. Chem. Phys.*, **77**, 2309 (1982).
6.23. H. P. H. Thijssen, R. E. van den Berg, and S. Völker, *Chem. Phys. Lett.*, **103**, 23 (1983).
6.24. H. P. H. Thijssen, R. van den Berg, and S. Völker, *Chem. Phys. Lett.*, **97**, 295 (1983).
6.25. H. P. H. Thijssen, A. I. M. Dicker, and S. Völker, *Chem. Phys. Lett.*, **92**, 7 (1982).
6.26. S. Völker, *J. Lumin.*, **36**, 251 (1987).
6.27. E. L. Al'shits, R. I. Personov, A. M. Pyndyk, and V. I. Stogov, *Opt. Spectrosc. (USSR)*, **39**, 156 (1975).
6.28. I. I. Abram, R. A. Auerbach, R. R. Birge, B. E. Kohler, and J. M. Stevenson, *J. Chem. Phys.*, **63**, 2473 (1975).
6.29. F. A. Burkhalter and U. P. Wild, *Chem. Phys.*, **66**, 327 (1982).
6.30. V. I. Rakhovski, M. N. Sapozhnikov, and A. L. Shubin, *J. Lumin.*, **28**, 301 (1983).
6.31. B. P. Boczar, E. A. Mangle, S. A. Schwartz, and M. R. Topp, in K. B. Eisenthal, Ed., *Applications of Picosecond Spectroscopy to Chemistry*, D. Reidel, Dordrecht, The Netherlands, 1984.
6.32. K. N. Solovev, I. V. Stanishevskii, A. S. Starukhin, and A. M. Shulga, *Opt. Spectrosc. (USSR)*, **53**, 230 (1982).
6.33. G. Flatscher and J. Friedrich, *Chem. Phys. Lett.*, **50**, 32 (1977).
6.34. M. N. Sapozhnikov and V. I. Alekseev, *Chem. Phys. Lett.*, **87**, 487 (1982).
6.35. M. N. Sapozhnikov and V. I. Alekseev, *Chem. Phys. Lett.*, **107**, 265 (1984).
6.36. B. M. Kharlamov, R. I. Personov and L. A. Bykovskaya, *Opt. Comm.*, **12**, 191 (1974).
6.37. B. M. Kharlamov, R. I. Personov and L. A. Bykovskaya, *Opt. Spectrosc. (USSR)*, **39**, 137 (1975).
6.38. A. A. Gorokhovskii, R. K. Kaarli, and L. A. Rebane, *JETP Lett.*, **20**, 216 (1974).
6.39. J. M. Hayes and G. J. Small, *Chem. Phys.*, **27**, 151 (1978).
6.40. J. M. Hayes and G. J. Small, *Chem. Phys. Lett.*, **54**, 435 (1978).
6.41. J. M. Hayes, R. P. Stout, and G. J. Small, *J. Chem. Phys.*, **73**, 4129 (1980).
6.42. J. M. Hayes, R. P. Stout, and G. J. Small, *J. Chem. Phys.*, **74**, 4266 (1981).
6.43. J. W. Hofstraat, M. Bobeldijk, G. Ph. Hoornweg, C. Gooijer, and N. H. Velthorst, *J. Mol. Struct.*, **141**, 301 (1986).
6.44. J. Friedrich and D. Haarer, *Angew. Chem.*, **96**, 96 (1984).
6.45. L. A. Rebane, A. A. Gorokhovskii and J. V. Kikas, *Appl. Phys.*, **B29**, 235 (1982).
6.46. B. M. Kharlamov, E. I. Al'shits, and R. I. Personov, *Opt. Commun.*, **44**, 149 (1983).
6.47. J. Fünfschilling and I. Zschokke-Gränacher, *Chem. Phys. Lett.*, **110**, 315 (1984).

6.48. J. Fünfschilling, E. Wasmer, and I. Zschokke-Gränacher, *J. Chem. Phys.*, **69**, 2949 (1978).
6.49. G. W. Suter and U. P. Wild, *J. Lumin.*, **24/25**, 497 (1981).
6.50. G. Flatscher, K. Fritz, and J. Friedrich, *Z. Naturforsch.*, **31a**, 1220 (1976).
6.51. J. C. Brown, M. C. Edelson, and G. J. Small, *Anal. Chem.*, **50**, 1394 (1978).
6.52. J. C. Brown, J. A. Duncanson, Jr., and G. J. Small, *Anal. Chem.*, **52**, 1711 (1980).
6.53. R. I. Personov and E. I. Al'shits, *Chem. Phys. Lett.*, **33**, 85 (1975).
6.54. J. W. Hofstraat, G. Ph. Hoornweg, C. Gooijer, and N. H. Velthorst, *Anal. Chim. Acta*, **169**, 125 (1985).
6.55. U. Bogner, *Phys. Rev. Lett.*, **37**, 909 (1976).
6.56. R. I. Personov, E. I. Al'shits, L. A. Bykovskaya, and B. M. Kharlamov, *Sov. Phys. JETP*, **38**, 912 (1974).
6.57. J. H. Eberly, W. C. McColgin, K. Kawaoka, and A. P. Marchetti, *Nature*, **251**, 215 (1974).
6.58. J. W. Hofstraat, M. Engelsma, W. P. Cofino, G. Ph. Hoornweg, C. Gooijer, and N. H. Velthorst, *Anal. Chim. Acta*, **159**, 359 (1984).
6.59. J. W. Hofstraat, H. J. M. Jansen, G. Ph. Hoornweg, C. Gooijer, and N. H. Velthorst, *Anal. Chim. Acta*, **170**, 61 (1985).
6.60. W. P. Cofino, J. W. Hofstraat, G. Ph. Hoornweg, C. Gooijer, C. MacLean, and N. H. Velthorst, *Chem. Phys. Lett.*, **89**, 17 (1982).
6.61. J. M. Hayes, I. Chiang, M. J. McGlade, J. A. Warren, and G. J. Small, in J. A. Gelbwachs, Ed., *Laser Spectroscopy for Sensitive Detection*, SPIE, Washington, D. C., 1981, p. 117.
6.62. I. Chiang, J. M. Hayes, and G. J. Small, *Anal. Chem.*, **54**, 315 (1982).
6.63. M. J. McGlade, J. M. Hayes, G. J. Small, V. Heisig, and A. M. Jeffrey, in W. S. Lyon, Ed., *Analytical Spectroscopy*, Elsevier, Amsterdam, 1984.
6.64. J. W. Hofstraat, H. J. M. Jansen, G. Ph. Hoornweg, C. Gooijer, and N. H. Velthorst, *J. Mol. Struct.*, **142**, 279 (1986).
6.65. W. P. Cofino, J. W. Hofstraat, G. Ph. Hoornweg, C. Gooijer, C. MacLean, and N. H. Velthorst, *Chem. Phys. Lett.*, **98**, 342 (1983).
6.66. V. Heisig, A. M. Jeffrey, M. J. McGlade, and G. J. Small, *Science*, **223**, 289 (1984).
6.67. M. J. Sanders, R. S. Cooper, G. J. Small, V. Heisig, and A. M. Jeffrey, *Anal. Chem.*, **57**, 1148 (1985).
6.68. L. A. Bykovskaya, A. T. Gradyushko, R. I. Personov, Yu. V. Romanovskii, K. H. Solov'ev, A. S. Starukhin, and A. M. Shul'ga, *J. Appl. Spectrosc. (USSR)*, **29**, 1511 (1978).
6.69. F. A. Burkhalter and U. P. Wild, *Chem. Phys.*, **66**, 327 (1982).
6.70. J. Fünfschilling and D. F. Williams, *Appl. Spectrosc.*, **30**, 443 (1976).
6.71. J. Fünfschilling and D. F. Williams, *Photochem. Photobiol.*, **26**, 109 (1977).
6.72. S. S. Dvornikov, V. N. Knyukshto, K. N. Solovev, I. V. Stanishevskii, A. S. Starukhin, and A. E. Turkova, *Opt. Spectrosc. (USSR)*, **57**, 234 (1984).
6.73. R. A. Avarmaa and A. P. Suisalu, *Opt. Spectrosc. (USSR)*, **56**, 32 (1984).
6.74. J. Fünfschilling and D. Walz, *Photochem. Photobiol.*, **38**, 389 (1983).
6.75. R. A. Avarmaa, and K. K. Rebane, *Spectrochim. Acta*, **41A**, 1365 (1905).

6.76. K. N. Solov'ev, I. V. Stanishevskii, A. S. Starukhin, and A. M. Shul'ga, *Bull. Acad. Sci. USSR, Phys. Ser.*, **47**, 144 (1983).
6.77. M. N. Sapozhnikov and V. I. Alekseev, *Chem. Phys. Lett.*, **87**, 487 (1982).
6.78. V. I. Rakhovski, M. N. Sapozhnikov, and A. L. Shubin, *J. Lumin.*, **28**, 301 (1983).
6.79. L. A. Bykovskaya, R. I.Personov, and B. M. Kharlamov, *Chem. Phys. Lett.*, **27**, 80 (1974).
6.80. M. V. Melishchuk, E. A. Tikhonov, and M. T. Shpak, *Opt. Spectrosc. (USSR)*, **56**, 634 (1984).
6.81. F. Graf, H. K. Hong, A. Nazzal, and D. Haarer, *Chem. Phys. Lett.*, **59**, 217 (1978).
6.82. A. P. Marchetti, M. Scozzafava, and R. H. Young, *Chem. Phys. Lett.*, **51**, 424 (1977).
6.83. W. C. McColgin, A. P. Marchetti, and J. H. Eberly, *J. Am. Chem. Soc.*, **100**, 5622 (1978).
6.84. E. Cuellar and G. Castro, *Chem. Phys.*, **54**, 217 (1981).
6.85. E. I. Al'shits, L. A. Bykovskaya, R. I. Personov, Yu. V. Romanovskii, and B. M. Kharlamov, *J. Mol. Struct.*, **60**, 219 (1980).
6.86. E. L. Wehry and G. Mamantov, in E. L. Wehry, Ed., *Molecular Fluorescence Spectroscopy*, Vol. 4, Plenum Press, New York, 1981, p. 193.
6.87. J. C. Brown, J. M. Hayes, J. A. Warren, and G. J. Small, in G. M. Hieftje, J. C. Travis and F. E. Lytle, Eds., *Lasers in Chemical Analysis*, Humana Press, Clifton, N. J. 1981, p. 237.
6.88. L. A. Bykovskaya, R. I. Personov, and Yu. V. Romanovskii, *Anal. Chim. Acta*, **125**, 1 (1981).
6.89. J. W. Hofstraat, G. Ph. Hoornweg, C. Gooijer, and N. H. Velthorst, *Spectrochim. Acta*, **41A**, 801 (1985).
6.90. D. Bolton and J. D. Winefordner, *Talanta*, **30**, 713 (1983).
6.91. J. M. Hayes and G. J. Small, *Talanta*, **31**, 741 (1984).
6.92. J. W. Hofstraat, M. Engelsma, R. J. van de Nesse, C. Gooijer, N. H. Velthorst, and U. A. Th. Brinkman, *Anal. Chim. Acta*, **193** (1987).

4.7. SUPERSONIC JET SPECTROSCOPY

Supersonic jet spectroscopy (SJS) is a method used to obtain highly resolved spectra which is completely different from the three solid-state techniques described earlier. In SJS, experiments are performed in a gas flow. As the reasons for band broadening in the vapor phase and in the low-temperature solid phase differ essentially, attention will first be paid to factors that determine the bandwidth (7.1,7.2).

In the gas phase the molecules are virtually isolated. Inhomogeneous broadening due to nonuniform interactions with solvent or other molecules, the main factor thwarting vibrational resolution in solid-phase spectra, can be neglected. Also, electron–phonon interactions are absent in the gas phase.

The major braodening mechanisms in the gaseous phase are due to rotational and vibrational congestion of the spectra and Doppler broadening; all these types of broadening are strongly temperature dependent. Especially if the vapor pressure of the analytes is low, only by an increase in the sample temperature are sufficient molecules vaporized. Due to this temperature rise, many rotationally and even some vibrationally excited states of the electronic ground state will be populated. In large molecules moments of inertia are very large so that the rotational energy levels lie very close (typical energy separation in the order of 0.1 cm^{-1}). In addition, all the rotational subbands experience Doppler broadening due to the nonuniform velocity distribution of the molecules in the gas (see Section 4.1.1). As the Doppler broadening in general is large compared to the energy separation of the individual rotational bands, every vibronic band shows a rotational contour which may be several cm^{-1} wide and which, of course, increases with temperature.

Generally, temperatures required to get sufficient vapor pressure of the sample molecules are so high that even a number of energetically low-lying vibrational levels are populated. All these levels, which are not populated at the low temperatures used in the solid-state techniques, may be involved in gas-phase spectral transitions and give rise to so-called sequence diffuseness (7.1). This form of congestion is particularly serious for larger molecules. For instance, for naphthalene the vibrational structure of the spectrum is completely obliterated at a temperature of 200°C (7.1).

In the gaseous state the limiting spectral bandwidth is determined only by the lifetime of the states involved in the transition. As the molecules are isolated, one does not have to reckon with the influence of electron–phonon coupling on the vibronic band shapes, which is of utmost importance in the solid phase. Furthermore, in the gas phase the molecules have less opportunity to lose their excitation energy nonradiatively than in the solid state because of absence of solute–solvent interactions. Hence, the emission lifetimes are expected to be longer in the gas phase and the natural linewidths narrower. Nevertheless, especially for transitions involving highly vibrationally excited levels, band broadening is observed. This is due to intramolecular vibrational energy redistribution, a process that involves a scrambling of the vibrational energy of the excited vibronic state over many vibrations to which it may be coupled (7.3). If excitation to relatively high S_1 vibronic states [$E_{vib} > 1800$ cm^{-1} for tetracene in a supersonic jet (7.4)] is performed, appreciable broadening of the emission from the S_1-state is measured, even though photoselective laser excitation is employed.

From the above it is clear that line narrowing in spectra of gaseous samples requires low temperatures. This objective is very efficiently realized in SJS, the mechanism of which will be discussed in Section 4.7.1. Although the supersonic jet was first proposed in 1951 as a molecular beam source by

Kantrowitz and Grey (7.5) and realized experimentally by Kistiakowsky and Slichter (7.6) in the same year, only in 1974 was the first successful spectroscopic application reported. In this year Smalley et al. published a very highly resolved fluorescence excitation spectrum of NO_2 in a supersonic jet (7.7). Recent years, however, have shown a boom in spectroscopic investigations in supersonic jets. Comprehensive reviews of SJS have been given by Levy and co-workers (7.8–7.11).

4.7.1. Principles

When a gas is expanded from a high-pressure region through a narrow orifice (or nozzle) into a vacuum, impressive lowering of the internal energy of the gas molecules may be achieved. A condition that has to be fulfilled in order to realize this cooling is that the width of the orifice (or nozzle) must be significantly smaller than the mean free path of the molecules. Only those molecules that have acquired a large-velocity component in the axial direction will be able to escape, so that a highly directed flow is created ("geometrical cooling"). In addition, when the molecules leave the reservoir, many collisions take place, leading to an averaging of the molecular velocities. Thus a narrowing of the velocity distribution of the gas is achieved, corresponding to an effectively lower translational temperature. The net result is sketched in Fig. 4.34. The peak of the velocity distribution of the gas molecules is shifted to higher velocities with respect to the Maxwell-distribution observed in the reservoir (which is characteristic for the reservoir temperature T_0), but the width of the distribution has become considerably narrower (and is characterized by a much lower effective temperature T_1).

First, we will explain why molecular beams that are constructed as described above are known as supersonic beams or jets. The combined increase of the mass flow velocity u and decrease of the local speed of sound, defined classically as a $= (\gamma kT/m)^{1/2}$, leads to an increase of the Mach number M during the expansion:

$$M \equiv u/a \tag{7.1}$$

In the expression of the speed of sound, γ is the heat capacity ratio C_p/C_v, k the Boltzmann constant, T the absolute temperature, and m the mass of the gas molecules. For an ideal expansion M is equal to 1 at the most constricted point of the orifice and with additional expansion M becomes higher than 1, hence the term "supersonic."

It is possible to estimate which translational cooling can be achieved in the supersonic jet expansion. Under appropriate conditions (7.10) the expansion will be isentropic so that temperature, pressure, density, and Mach number

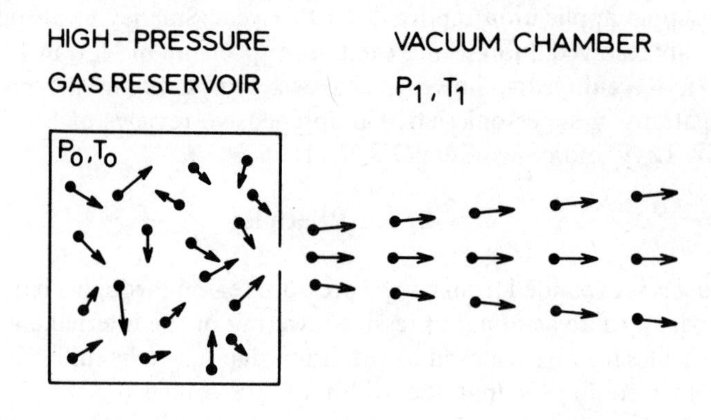

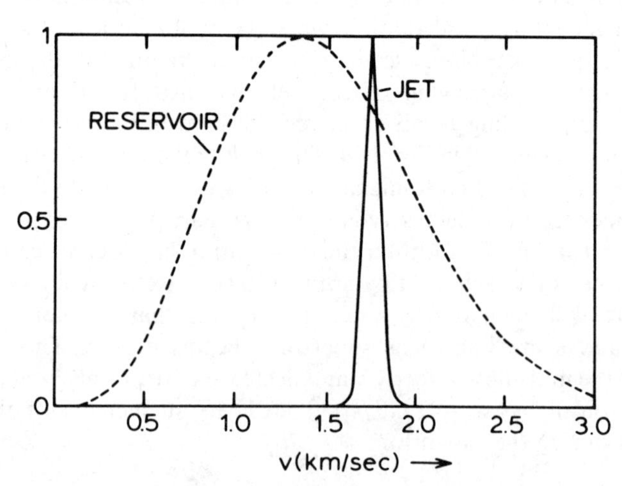

Fig. 4.34. Schematic representation of a supersonic jet experiment. The molecules in the reservoir move at random with a wide distribution of velocities. Via the supersonic expansion a directed mass flow is created with a much narrower velocity distribution. This is illustrated in (b) for an expansion of helium atoms.

can be thermodynamically described. For a perfect gas (7.12),

$$\frac{T_1}{T_0} = \left(\frac{P_1}{P_0}\right)^{(\gamma-1)/\gamma} \equiv \left(\frac{\rho_1}{\rho_0}\right)^{\gamma-1} = \left[1 + \frac{1}{2}(\gamma - 1)M^2\right]^{-1} \quad (7.2)$$

Here T_0, P_0, and ρ_0 are temperature, pressure, and density of the reservoir and T_1, P_1, and ρ_1 the same quantities in the expansion. If the gas is treated as a continuous medium, the value of the Mach number can be expressed in terms of the distance x downstream from the nozzle (7.13).

$$M = A(x/D)^{\gamma - 1} \tag{7.3}$$

where A is a constant depending on γ and D, the diameter of the nozzle. Equation (7.3) holds only for small values of x (i.e., in the region where the density of the molecules is high enough to provide cooling via two-body collision). This is only the case in the first, "continuum," region in the immediate vicinity of the orifice. Farther downstream the collision frequency decreases until in the last stage of the expansion, the free molecular flow region is reached, where the particles move independently. This implies that the Mach number does not increase infinitely. Anderson and Fenn (7.14) have estimated the terminal Mach number, M_T, for expansion of a monoatomic gas:

$$M_T = 133\ (P_0\mathrm{D})^{0.4} \tag{7.4}$$

with P_0 in atm and D in cm. Equation (7.4) does not apply for helium, where quantum effects cause the collisional cross section to increase with decreasing relative energy of the colliding atoms (7.15). Therefore, for helium M_T is considerably higher, so that much lower translational temperatures are expected in helum expansions than in expansions of other monoatomic gases. In conclusion, translational temperatures of monoatomic gases achieve minimal values for reservoirs with high stagnation pressure P_0 and relatively large nozzle diameter D. Unfortunately, these conditions imply high mass throughput, so that high demands are put on the pumping capacity, which has to maintain the vacuum requirements.

A diagrammatic representation of a supersonic expansion is shown in Fig. 4.35. Along the axis, downstream from the nozzle, the onset of the free molecular flow is indicated (x_{FF}). Furthermore, the jet is protected from detrimental interactions with the "hot" background molecules present in the vacuum chamber by a shock wave which is formed around the expansion. Finally, also along the axis of expansion, the shock wave is reached; this point, the Mach disk (x_{MD}) marks the end of the spectroscopically useful part of the jet. Beyond the Mach disk the expanded gas flow is heated via collisional interactions with the background gas in the vacuum chamber (leading to background pressure P_1). On the upstream end the spectroscopically useful area of the jet expansion is limited by the onset of the free-flow region, x_{FF}. Between x_{FF} and x_{MD} the molecules are isolated and have low vibrational and rotational temperatures. The dimensions of this area are

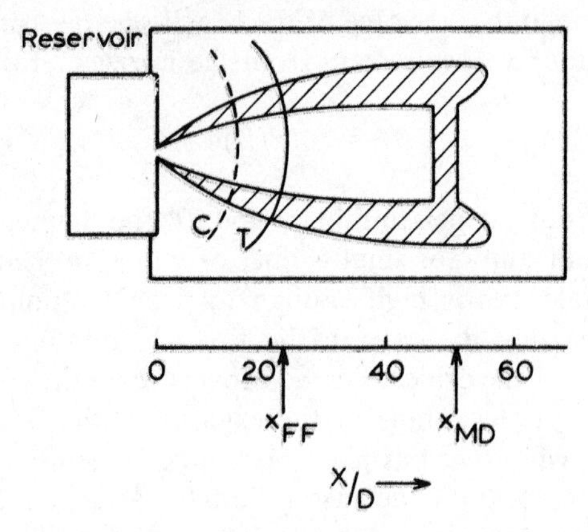

Fig. 4.35. Diagram of a supersonic expansion. In the expansion three regions can be discerned. The first, continuum region C is characterized by relatively high gas densities so that many collisions occur and thermodynamic equilibrium is maintained. In the transition region T much less collisions occur. The expansion then has to be described in terms of effective collision diameters. Beyond the point x_{FF} the free-flow region is reached, where the particles in the expansion move independently. The striped region around the expansion represents the shock wave surrounding the beam. The point x_{MD} indicates the Mach disk, the point along the axis of expansion where the shock wave is reached. Along the horizontal axis the distance x from the nozzle expressed in nozzle diameters D is given.

strongly dependent on the pumping capacity: Low pumping speed leads to a shift of the Mach disk closer to the orifice due to increased background pressure.

Thus far the discussion has been limited to translational cooling. To get highly resolved spectra of gas-phase molecules, reduction of the rotational and vibrational temperatures has to be realized. Rotational and vibrational cooling is achieved mainly in the first, continuum region of the expansion. Here sufficient collisions occur to provide thermal equilibrium continually, so that cooling is adiabatic. In this region there exists thermodynamic equilibrium so that the principle of equipartition of energy among the various degrees of freedom holds. Farther downstream the number of collisions decreases and the cooling of vibrational and rotational degrees of freedom begins to lag behind that of the translational. The cooling of the nontranslational degrees of freedom is now more determined by kinetic than thermodynamic factors. The final rotational and vibrational temperatures that are reached in the jet depend on the rate of equilibration between translations on the one hand, and rotations and vibrations on the other. Since the cross sections for collision-induced rotational transitions are much larger than for vibrational transitions, rotational cooling will be much more effective than vibrational cooling. As the density of the gas-phase molecules decreases, further rotational and vibrational cooling is hindered. In the free molecular flow region the molecules are isolated and frozen at fixed rotational and vibrational temperatures. Typical final temperatures that may be reached in the supersonic jet are 0.05–10 K for translations, 0.2–50 K for rotations, and 10–100 K for vibrations. Although vibrational cooling is not very strong, it is in general sufficient to remove vibrational congestion of the spectra. First, the energies of most vibrational modes are large ($>250\ cm^{-1}$) so that thermal populations are already relatively low, even at ambient temperatures. Second, the lowest-energy vibrational modes tend to have the largest collisional cross sections and are thus most effectively cooled.

The arguments given above are valid for pure gases, but also for gases seeded with low concentrations of guest molecules. By adding small amounts of molecules to a noble gas in the reservoir even relatively large molecules [e.g., pentacene (7.16), ovalene (7.17), and phthalocyanine (7.18)] can be studied in the supersonic jet. The effect of vibrational and rotational cooling on the fluorescence excitation spectrum of phthalocyanine is shown in Fig. 4.36. In theory, all gases should be equally suitable for cooling heavy seeded molecules since the cooling process is a purely kinetic phenomenon. However, cooling is limited by cluster formation between carrier gas atoms and the guest molecules. To minimize cluster formation noble gases are usually chosen as carrier gases. Helium jets especially reach extremely low translational temperatures, but helium expansions have very high velocities. Cooling of seeded molecules that are much heavier, and thus slower, than the atoms of the carrier gas in such high-speed expansions is hampered by the velocity slip effect. Initial collisions between seeded molecules and carrier gas atoms only serve to accelerate (and are not able to cool) the guest molecules until they reach the same speed as the host atoms. Subsequent collisions will lead to cooling. Therefore, decisively more effective vibrational and—to a lesser extent—rotational cooling of heavy seeded molecules is obtained in the heavier noble gases (Ar, Kr, Xe) (7.19).

4.7.2. Experimental Aspects

In general, the practical limit to the performance of the supersonic expansion is formed by the necessity of adequate pumping capacity to handle the gas flow through the nozzle. The necessary pumping capacity is mainly determined by the construction of the orifice used. At first only continuous nozzles were employed in combination with high-speed pumps [30,000–40,000 L/s (7.20)] to obtain low enough background pressures in the vacuum chamber (10^{-3}–10^{-4} torr and less). The objective was to make the mean free path

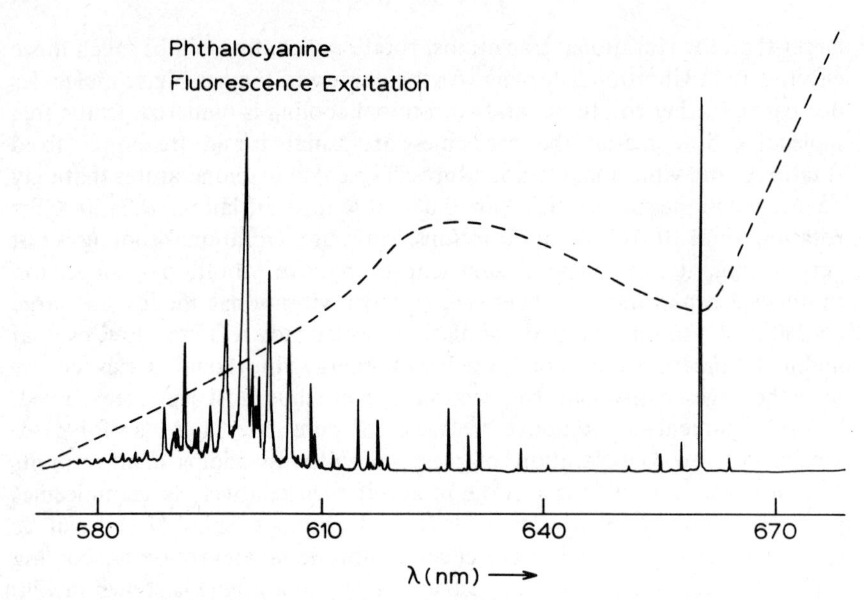

Fig. 4.36. Fluorescence excitation spectrum of free-base phthalocyanine cooled in a supersonic free jet. The dashed curve is the static gas absorption spectrum. (From ref. 7.18).

between collisions of background molecules much larger than the dimensions of the apparatus. However, Campargue realized that at higher background pressures shock waves would be formed around the cold part of the expansion, shielding it from the "hot" background gas in the vacuum chamber (7.21–7.23). Pumping speeds in the range 25–500 L/s are sufficient to reach background pressures of 10^{-2}–1 torr, required for the Campargue-type supersonic beam source (7.23). Typical operating conditions for a Campargue-type beam source are given by Levy (7.10): with a 25-μm-diameter continuous nozzle and a reservoir pressure of 100 atm (76,000 torr) a 0.1-torr background pressure can be attained with a pumping speed between 200–400 L/s. Under these circumstances the cold isentropic core of the expansion extends 1.5 cm downstream of the nozzle. At 0.5 cm from the nozzle the translational temperature is 0.07 K; there, rotational and vibrational temperatures of 0.5 and 20–50 K, respectively, have been observed.

Another way to solve the pumping problem is to make use of a pulsed nozzle (7.24–7.26). In this case the pumping capacity must be sufficient to evacuate the vacuum chamber in between pulses, so that much less pumping power is needed. When a pulsed laser system is employed to probe the jet, the low duty cycle of a pulsed source does not detract from the sensitivity. Several commercial pulsed beam sources are available. An interesting pulsed beam source was developed by Amirav et al. (7.27). By making use of two spinning concentric cylinders they constructed a planar supersonic expansion through a 35 × 0.2 mm slit nozzle, yielding an impressive increase in sensitivity as compared to conventional circular nozzles (with diameters well under 1 mm). With the planar supersonic expansions Amirav et al. were even able to determine absorption spectra of aniline and 9,10-dichloroanthracene (7.27). In conventional supersonic jets, in general, only laser-induced fluorescence emission and excitation spectroscopy can be applied, due to the small numbers of molecules that can be probed.

Sample preparation—even for fairly large and heavy molecules—is easy. For instance, Amirav et al. were able to prepare seeded beams of anthracene, tetracene, pentacene, and ovalene simply by leading the diluent gas through a stainless steel sample chamber which contained the solid sample. The sample chamber was heated to temperatures between 130 and 150°C (for anthracene) to 325–385°C (for ovalene). In this way gaseous mixtures of diluent gas, at a stagnation backing pressure range of 20–8300 torr and the seeded molecule at pressures of 10^{-2}–2×10^{-1} torr, were obtained. The necessity to heat the sample precludes the investigation of thermolabile compounds in the supersonic jet.

When preparing the sample, attention has to be paid to the possible formation of van der Waals molecules. Van der Waals complexes between the seeded molecules and the diluent gas molecules are formed as a consequence of at least three-body collisions. The number of complexes generated therefore varies as $\rho_0 D^2$. The cooling effect is due to two-body collision and varies as $\rho_0 D$, so it is in principle possible to minimize complex formation by modification of the expansion conditions. The presence of van der Waals complexes increases the complexity of the spectra since they give rise to a multiplet structure.

4.7.3. Applications

Supersonic jets form an ideal medium for the study of the spectral properties of isolated molecules. As more and more groups realized the unique advantages of supersonic jets in the last few years a boom in publications in this field was observed. Various subjects have been covered.

1. *Very highly resolved spectral studies of vibronic spectra.* In a number of cases even rotational fine structure of the vibronic bands could be studied. With narrow-line laser excitation single vibronic line fluorescence spectra can be obtained which yield direct information on vibronic coupling. Examples are found in the recent literature for, among others, anthracene (7.28) and β-methylnaphthalene (7.29).
2. *Studies of excited-state dynamics.* The isolated molecules in the super-

sonic jet are very well suited for the study of intramolecular vibrational energy redistribution and other nonradiative processes (7.3,7.30,7.31).

3. *Study of "exotic species."* In this context one may think of molecules which are not stable under normal conditions, such as free radicals and ions. Just as in the low-temperature solid-state techniques, such species can also be prepared and studied in supersonic jets (7.32,7.33). Additionally, species that are difficult or impossible to study with the other techniques can be investigated. Examples of such species are van der Waals complexes of seeded molecules with diluent gas atoms (7.34–7.37) or other weakly bonded compounds (7.38–7.41), dimers and oligomers of atoms and molecules (7.42–7.44), exciplexes (7.45), and clusters of atoms or molecules (7.46–7.48). By careful choice of the jet characteristics, different aggregates can be prepared so that the transition from isolated molecules to molecules in the condensed phase can be examined.

At present many compounds have been investigated. Fundamentally, the range of compounds yielding well-resolved fluorescence spectra is larger than in any of the other high-resolution techniques. Electron–phonon interactions and solvent-induced or solvent-assisted radiationless transitions do not play a role in the supersonic jet. Therefore, higher or at least equal fluorescence quantum yields will be observed, as compared to compounds in condensed media. The only limitation in SJS is the possible thermal decomposition. Another problem in SJS originates from the high speed of the jet, which makes detection of long lived phosphorescence difficult. Abe et al. were able to overcome this problem by making use of sensitized phosphorescence detection (7.49,7.50).

In SJS highly resolved electronic spectra of nonluminescent compounds can be obtained via multiphoton ionization or direct absorption methods. In multiphoton ionization spectroscopy first one photon is used to pump transitions between electronic states of the molecule; subsequently, a second (or more) photons is (are) used to ionize this excited electronic state (7.51,7.52). The absorption between the two bound states of the molecule is then observed indirectly with a charged-particle detector. Direct absorption methods in combination with conventional supersonic beams are only applicable to molecules with large absorption coefficients (7.53,7.54). Sensitivity can be increased by employing planar supersonic expansions through slit nozzles (7.27) or by placing the supersonic jet inside a laser cavity and detecting the effect of absorption on the output power of the laser (7.55).

4.7.4. Analytical Aspects and Applications

Analytical studies on the applicability of SJS have been limited thus far. It is a rather new technique and many factors influencing the jet characteristics are still not fully explored. Reviews on the state of the art have been published recently by Hayes and Small (7.56) and Johnston (7.57).

The prospects of SJS in analytical chemistry, however, seem promising. The technique yields highly resolved fluorescence spectra of a wide range of molecules without interference of solvent molecules and is compatible with dynamic systems, two very strong points in comparison with the low-temperature methods. Furthermore, solubility problems are nonexistent in SJS, although one has to reckon here with possible thermal decomposition effects of nonvolatile materials. As intermolecular interactions are negligible in the beam source, energy transfer effects will not play an important role and, additionally, due to the low concentrations, inner filter effects are absent. On the other hand, it is a relatively expensive technique. Expensive apparatus (vacuum pump, nozzle arrangement, vacuum chamber) is necessary to prepare the jet, a laser should preferably be used for excitation, and finally to detect fluorescence spectra "on the fly" one needs a photodiode array or vidicon detector. Of course, in the solid-state techniques the sample is conserved and can in principle be studied indefinitely.

The first studies on analytical applications of SJS were performed by Small's group (7.58,7.59). They first reported laser-induced fluorescence emission and excitation spectra for naphthalene and the α and β isomers of methylnaphthalene in a supersonic expansion of helium. The high selectivitly of SJS is evident from the excitation spectrum in Fig. 4.37. Linewidths are laser limited at 2 cm^{-1}. Under more optimal conditions, however, reduction of bandwidths down to 0.1 cm^{-1} is possible (7.60) which is at least one order lower than bandwidths attainable with solid-state high-resolution techniques. Even the "large" bandwidth of 2 cm^{-1} allows easy identification of the three compounds in the sample.

In a second paper Hayes and Small reported the coupling of gas chromatography (GC) and SJS (7.59). The experimental setup is shown in Fig. 4.38. They were able to inject reproducible amounts of solute so that quantitative aspects of SJS could be studied. For naphthalene and the methylnaphthalenes detection limits of 14–60 ng were measured; calibration curves were linear over three to four orders. The authors indicate that detection limits of the order of 1 pg must be attainable by obvious modifications in the supersonic jet design. Even though a very simple GC system was used which was not able to separate the two methylnaphthalene isomers chemically, spectral discrimination was easily achieved. That the technique is also applicable to more complex, real samples was demonstrated

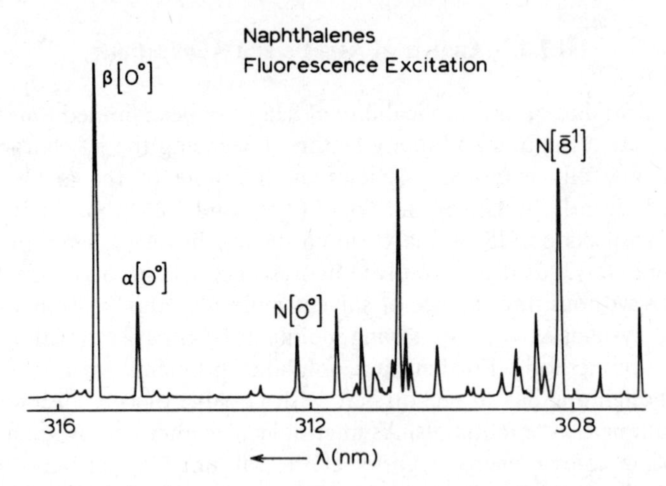

Fig. 4.37. Supersonic jet fluorescence excitation spectrum of naphthalene, α-methylnaphthalene and β-methylnaphthalene. N[0°], α[0°], and β[0°] denote their S_1–S_0 0–0 transitions, respectively. For naphthalene also one vibronic transition is indicated. Monitoring wavelength was 342.5 nm (bandpass 5 nm). (Reprinted with permission from J. A. Warren, J. M. Hayes and G. J. Small, *Anal. Chem.* **54**, 136. Copyright 1982 American Chemical Society.)

by means of a crude oil sample, in which the three compounds could be determined quantitatively.

Other applications have been described by Amirav et al. (7.61). The selectivity of SJS was illustrated with the fluorescence excitation spectra of anthracene and substituted anthracenes and the fluorescence emission spectrum of tetracene in supersonic expansions of argon. Bandwidths of 1.3 cm^{-1} were observed in the spectra. With a conventional circular nozzle the minimum detectable vapor pressure of the guest molecules was estimated to be 10^{-4} torr. Application of a slit nozzle, yielding a planar supersonic expansion, will produce an improvement of three orders in the detection limit according to these authors. Of course, this implies that sufficient vapor pressures of the seeded molecules can be obtained at moderately low temperatures. Another advantage is that lower pressures of argon gas are required for effective intramolecular cooling than in conventional jet configurations (as the cooling is directly proportional to the product of stagnation pressure P_0 and nozzle diameter D). Finally, the authors showed the applicability of SJS for isotopic analysis (e.g., the chlorine isotopes of 9,10-dichloroanthracene).

Imasaka et al. (7.62) report fluorescence detection of perylene and benzo[a]pyrene in supersonic expansions of argon. For perylene a detection limit of 100 ng was found. A relatively simple supersonic jet fluorimeter using a 300-W xenon lamp for excitation and a lock-in amplifier for detection has

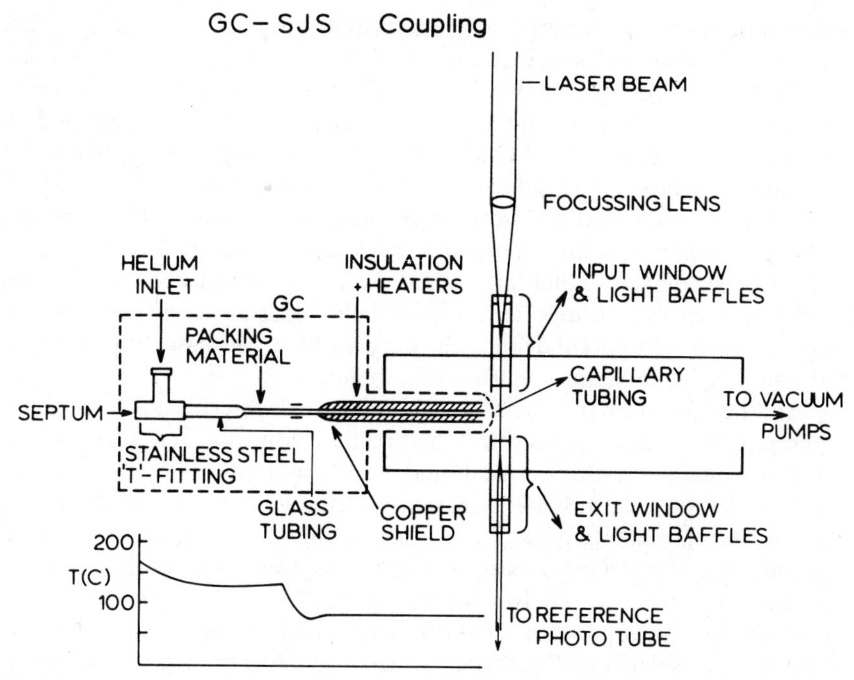

Fig. 4.38. Diagram of the GC-SJS apparatus as used by Hayes and Small. The part enclosed by the dashed line forms a simple GC. The temperature variation along the column is shown in the lower part of the figure. (Reprinted with permission from J. M. Hayes and G. J. Small, *Anal. Chem.*, **54**, 1202. Copyright 1982 American Chemical Society.)

been described by Yamada and co-workers (7.63–7.65). With this setup selective identification of anthracene, 9-methylanthracene, and 9-chloroanthracene in a mixture of anthracenes was accomplished.

For nonfluorescent compounds resonant two- and multiphoton ionization of compounds in supersonic jets has been shown to be a very selective detection method (7.66). This technique, which is also applicable to fluorescent compounds, makes use of the fact that the ionization process is greatly enhanced when the laser is tuned to a real (and not a virtual) intermediate state. As in a supersonic jet also the S_1–S_0 absorption transitions are narrow, very selective ionization can be realized. Detection is performed with a time-of-flight mass spectrometer. The applicability of resonant two- and multiphoton ionization has been illustrated for organic compounds, such as cresol isomers (7.67), disubstituted benzenes (7.68), halogenated hydrocarbons (7.69), and even chlorine and bromine isotopes (7.70).

Future developments in analytical SJS will probably be directed to the combination with chromatography. Gas chromatography (GC)-SJS has already been shown to be feasible by Hayes and Small (7.59). Coupling of SJS to liquid chromatography (LC) may be accomplished. In this context much can be learned from present work on coupling of LC to mass spectrometry, where the transition from liquid phase to vacuum also has to be realized. A number of interfaces for LC-MS coupling have been described in the literature. Examples are (1) direct liquid introduction interface (7.71), which can only be employed for very small amounts of solute; (2) transport interfaces such as moving wires and belts (7.72) on which the solvent can be adequately removed so that large quantities can be introduced into the MS; and (3) a laser vaporization technique, which makes use of molecular beam techniques to transport and ionize the sample (7.73). The latter interface may be especially suitable for spectroscopic applications.

Another approach may be the combination of supercritical fluid chromatography and SJS. A supercritical fluid really is a pressurized gas above its critical temperature. Supercritical fluid chromatography can thus be considered to be an "intermediate" separation method combining aspects of gas and liquid chromatography. The direct combination of supercritical fluid chromatography and mass spectrometry has been shown to work as a highly selective and sensitive analytical method (7.74,7.75). Another interesting prospect for the application of supercritical fluids is the possibility to use them both as solvent and as carrier gas in the cooling process. Thus there is no need to heat the molecules to get the vapor pressure required to seed the carrier gas and the risk of thermal decomposition of the compounds has disappeared. Recently, xenon, for instance, has been shown to be a suitable supercritical solvent for naphthalene (7.76). Fukuoka et al. have published a first study on the applicability of supercritical fluids for sample introduction in SJS (7.77).

References to Section 4.7

7.1. J. P. Byrne and I. G. Ross, *Aust. J. Chem.*, **24**, 1107 (1971).
7.2. M. Stockburger, in J. B. Birks, Ed., *Organic Molecular Photophysics*, Vol. 1, Wiley, London, 1973, p. 57.
7.3. C. S. Parmenter, *Faraday Discuss. Chem. Soc.*, **75**, 7 (1983).
7.4. A. Amirav, U. Even, and J. Jortner, *Chem. Phys. Lett.*, **71**, 12 (1980).
7.5. A. Kantrowitz and J. Grey, *Rev. Sci. Instrum.*, **22**, 328 (1951).
7.6. G. Kistiakowsky and W. Slichter, *Rev. Sci. Instrum.*, **22**, 333 (1951).
7.7. R. E. Smalley, B. L. Ramakrishna, D. H. Levy, and L. Wharton, *J. Chem. Phys.*, **61**, 4363 (1974).
7.8. R. E. Smalley, L. Wharton, and D. H. Levy, *Acc. Chem. Res.*, **10**, 139 (1977).
7.9. D. H. Levy, L. Wharton, and R. E. Smalley, in C. B. Moore, Ed., *Chemical and Biochemical Applications of Lasers*, Vol. 2, Academic Press, New York, 1977, p. 1.

7.10. D. H. Levy, *Ann. Rev. Phys. Chem.*, **31**, 197 (1980).
7.11. D. H. Levy, *Science*, **214**, 263 (1981).
7.12. H. W. Liepmann and A. Rosko, *Elements of Gas Dynamics*, Wiley, New York, 1957.
7.13. H. Ashkenas and F. S. Sherman, in J. H. de Leeuw, Ed., *4th Symposium on Rarefied Gas Dynamics*, Vol. 2, Academic Press, New York, 1966, p. 84.
7.14. J. B. Anderson and J. B. Fenn, *Phys. Fluids*, **8**, 780 (1965).
7.15. J. P. Toennies and K. Winkelmann, *J. Chem. Phys.*, **66**, 3965 (1977).
7.16. A. Amirav, U. Even, and J. Jortner, *Chem. Phys. Lett.*, **72**, 21 (1980).
7.17. A. Amirav, U. Even, and J. Jortner, *Chem. Phys. Lett.*, **69**, 14 (1980).
7.18. P. Fitch, C. A. Haynam, and D. H. Levy, *J. Chem. Phys.*, **73**, 1064 (1980).
7.19. A. Amirav, U. Even, and J. Jortner, *Chem. Phys.*, **51**, 31 (1980).
7.20. J. Deckers and J. B. Fenn, *Rev. Sci. Instrum.*, **34**, 96 (1963).
7.21. R. Campargue, *Rev. Sci. Instrum.*, **35**, 111 (1964).
7.22. R. Campargue, *J. Chem. Phys.*, **52**, 1795 (1970).
7.23. H. C. W. Beijerinck, R. J. F. van Gerwen, E. R. T. Kerstel, J. F. M. Martens, E. J. W. van Vliembergen, M. R. Th. Smits, and G. H. Kaashoek, *Chem. Phys.*, **96**, 153 (1985).
7.24. O. F. Hagena, *Z. Angew. Phys.*, **16**, 183 (1963).
7.25. W. R. Gentry and C. F. Griese, *Rev. Sci. Instrum.*, **49**, 595 (1978).
7.26. R. Byer and M. Duncan, *J. Chem. Phys.*, **74**, 2174 (1981).
7.27. A. Amirav, U. Even, and J. Jortner, *Chem. Phys. Lett.*, **83**, 1 (1981).
7.28. W. R. Lambert, P. M. Felker, and A. H. Zewail, *J. Chem. Phys.*, **81**, 2209 (1984).
7.29. J. A. Warren, J. M. Hayes, and G. J. Small, *J. Chem. Phys.*, **80**, 1786 (1984).
7.30. P. Avouris, W. M. Gelbart, and M. A. El-Sayed, *Chem. Rev.*, **77**, 793 (1977).
7.31. S. A. Rice, in E. C. Lim, Ed., *Excited States*, Vol. II, Academic Press, New York, 1975, p. 111.
7.32. D. E. Powers, J. B. Hopkins, and R. E. Smalley, *J. Phys. Chem.*, **85**, 2711 (1981).
7.33. J. W. Farthing, I. W. Fletcher, and J. C. Whitehead, *J. Phys. Chem.*, **87**, 1663 (1983),
7.34. D. H. Levy, *Adv. Chem. Phys.*, **47**, 323 (1981).
7.35. A. Amirav, U. Even, and J. Jortner, *J. Chem. Phys*, **75**, 2489 (1981).
7.36. D. V. Brumbaugh, J. E. Kenny, and D. H. Levy, *J. Chem. Phys.*, **78**, 3415 (1983).
7.37. T. A. Stephenson and S. A. Rice, *J. Chem. Phys.*, **81**, 1083 (1984).
7.38. S. A. Schwartz and M. R. Topp, *J. Phys. Chem.*, **88**, 5673 (1984).
7.39. M. Schauer and E. R. Bernstein, *J. Chem. Phys.*, **82**, 726 (1985).
7.40. M. Schauer, K. S. Law, and E. R. Bernstein, *J. Chem. Phys.*, **82**, 736 (1985).
7.41. K. S. Law and E. R. Bernstein, *J. Chem. Phys.*, **82**, 2856 (1985).
7.42. C. A. Haynam, D. V. Brumbaugh, and D. H. Levy, *J. Chem. Phys.*, **79**, 1581 (1983).
7.43. L. Young. C. A. Haynam, and D. H. Levy, *J. Chem. Phys.*, **79**, 1592 (1983).
7.44. K. S. Law, M. Schauer, and E. R. Bernstein, *J. Chem. Phys.*, **81**, 4871 (1984).
7.45. O. Anner and Y. Haas, *Chem. Phys. Lett.*, **119**, 199 (1985).

7.46. A. Hermann, E. Schumacher, and L. Wöste, *J. Chem. Phys.*, **68**, 2327 (1978).
7.47. E. L. Quitevis, K. H. Bowen, G. W. Liesegang, and D. R. Herschbach, *J. Phys. Chem.*, **87**, 2076 (1983).
7.48. H. T. Jonkman, U. Even, and J. Kommandeur, *J. Phys. Chem.*, **89**, 4240 (1985).
7.49. A. Abe, S. Kamei, N. Mikami, and M. Ito, *Chem. Phys. Lett.*, **109**, 217 (1984).
7.50. S. Kamei, H. Abe, N. Mikami, and M. Ito, *J. Phys. Chem.*, **89**, 3636 (1985).
7.51. D. Zakheim and P. M. Johnson, *J. Chem. Phys.*, **68**, 3644 (1978).
7.52. P. M. Johnson, *Acc. Chem. Res.*, **13**, 20 (1980).
7.53. V. Vaida and G. M. McClelland, *Chem. Phys. Lett.*, **71**, 436 (1980).
7.54. R. J. Hemley, D. G. Leopold, V. Vaida, and J. L. Roebber, *J. Phys. Chem.*, **85**, 134 (1981).
7.55. W. R. Lambert, P. M. Felker, and A. H. Zewail, *J. Chem. Phys.*, **74**, 4732 (1981).
7.56. J. M. Hayes and G. J. Small, *Anal. Chem.*, **55**, 565A (1983).
7.57. M. V. Johnston, *Trends Anal. Chem.*, **3**, 58 (1984).
7.58. J. A. Warren, J. M. Hayes, and G. J. Small, *Anal. Chem.*, **54**, 138 (1982).
7.59. J. M. Hayes and G. J. Small, *Anal. Chem.*, **54**, 1202 (1982).
7.60. S. M. Beck, D. L. Monts, M. G. Liverman, and R. E. Smalley, *J. Chem. Phys.*, **70**, 1062 (1979).
7.61. A. Amirav, U. Even, and J. Jortner, *Anal. Chem.*, **54**, 1666 (1982).
7.62. T. Imasaka, H. Fukuoka, T. Hayashi, and N. Ishibashi, *Anal. Chim. Acta*, **156**, 111 (1984).
7.63. S. Yamada and J. D. Winefordner, *Anal. Lett.*, **18**, 139 (1985).
7.64. S. Yamada, B. W. Smith, E. Voigtman, and J. D. Winefordner, *Analyst*, **110**, 407 (1985).
7.65. S. Yamada, B. W. Smith, E. Voigtman, and J. D. Winefordner, *Appl. Spectrosc.*, **39**, 513 (1985).
7.66. D. M. Lubman and M. N. Kronick, *Anal. Chem.*, **54**, 660 (1982).
7.67. R. Tembreull and D. M. Lubman, *Anal. Chem.*, **56**, 1962 (1984).
7.68. C. H. Sin, R. Tembreull, and D. M. Lubman, *Anal. Chem.*, **56**, 2776 (1984).
7.69. R. Tembreull, C. H. Sin, P. Li, H. M. Pang, and D. M. Lubman, *Anal. Chem.*, **57**, 1186 (1985).
7.70. D. M. Lubman, R. Tembreull, and C. H. Sin, *Anal. Chem.*, **57**, 1084 (1985).
7.71. P. Krien, G. Devant, and M. Hardy, *J. Chromatogr.*, **251**, 129 (1982).
7.72. N. J. Alcock, C. Eckers, D. E. Games, M. P. L. Games, M. S. Lant, M. A. McDowall, M. Rossiter, R. W. Smith, S. A. Westwood, and H. Y. Wong, *J. Chromatogr.*, **251**, 165 (1982).
7.73. C. R. Blakley, M. J. McAdams, and M. L. Vestal, *J. Chromatogr.*, **158**, 261 (1978).
7.74. R. D. Smith, J. C. Fjeldsted, and M. L. Lee, *J. Chromatogr.*, **247**, 231 (1982).
7.75. R. D. Smith and H. R. Udseth, *Anal. Chem.*, **55**, 2266 (1983).
7.76. V. J. Krukonis, M. A. McHugh, and A. J. Seckner, *J. Phys. Chem.*, **88**, 2687 (1984).
7.77. H. Fukuoka, T. Imasaka, and N. Ishibashi, *Anal. Chem.*, **58**, 375 (1986).

4.8. CONCLUSION

In this chapter we have discussed four major techniques which can be used to obtain highly resolved emission spectra of molecules, with particular emphasis on their applicability in analytical chemistry. At present, of the four techniques, only Shpol'skii spectroscopy can boast an impressive number of analytical applications. Matrix isolation spectroscopy although by now an established technique, has received little attention of analytical chemists and, moreover, the bulk of the MI experiments are, and have been, performed in combination with UV/Vis absorption, IR and Raman, and ESR spectroscopy. Fluorescence (phosphorescence) line narrowing and, particularly, supersonic jet spectroscopy are relatively recent techniques. A few applications have been reported, but still much fundamental work needs to be done until these techniques can be optimally applied. Although much more employed and much longer known, the same holds for Shpol'skii spectroscopy. A number of unclear points regarding the structure and aspects of the preparation of the solid matrix used in the Shpol'skii technique have yet to be elucidated.

From the present state of the art in highly resolved emission spectroscopy, it is clear that all techniques have important advantages for analytical applications. In particular, their extremely high selectivity ensures specific determination of even quite similar compounds. For instance, isomeric alkylated PAHs, compounds that are difficult or impossible to discriminate with other selective methods such as GC-MS, are readily discerned. At the same time, sensitivities can be good since use is made of emission measurements. The techniques have proven to be very useful for qualitative purposes. However, in a number of cases quantitation requires special procedures such as internal standardization or standard addition. Some recent developments, however, indicate that less elaborate quantitation procedures may be realized.

Due to the inherent selectivity of the vibrationally resolved emission techniques, in a number of cases samples can be analyzed without making use of prior cleanup or chromatographic fractionation. For complex samples chromatographic fractionation is generally necessary. As most of the high-resolution techniques require a low-temperature solid sample, the coupling to chromatographic separation methods can only be realized off-line. Only supersonic jet spectroscopy can be coupled on-line to flowing systems; at present several applications of its use as detection method in combination with gas chromatography have been reported. Fluorescence line narrowing can be employed straightforwardly for the determination of compounds separated on thin-layer chromatographic plates. Some examples of semi-on-line coupling of the high-resolution techniques have been reported; in these cases the (gaseous or liquid) chromatographic effluent was immobilized on a solid support so that the necessary cooling could be realized.

It is clear that the application of the techniques described in this chapter requires instrumentation that is still not standard in an average laboratory. Lasers, cryogenic cooling systems, relatively high-resolution monochromators, and the extremely specialized equipment required for supersonic jet spectroscopy are commonly regarded in analytical laboratories as "exotic" apparatus. For the further development and application of the high-resolution emission techniques, the commercial availability of complete setups may be important. In this context it is encouraging to see that especially lasers and cooling systems are becoming available which are very simple to use.

Further developments in high-resolution techniques applied to analytical problems will probably mainly be directed to the improvement of their application for quantitative purposes and to their use in combination with separation techniques. Supersonic jet spectroscopy holds great promise as a detection method in chromatography. The development of optical multichannel detection devices (vidicon cameras, photodiode arrays, charge-coupled devices) with sensitivities comparable to photomultipliers but allowing much faster scan speeds will be very important in this context. Also, scanning of chromatograms immobilized via matrix isolation procedures or on solid supports via adsorption will be greatly facilitated. As fluorescence line narrowing and supersonic jet spectroscopy become more and more common, these techniques will also be applied to real samples. In short, the application of high-resolution emission techniques in analytical spectroscopy is in its beginning stage. Hopefully, we will see a strong development in this promising and exciting field in the near future.

CHAPTER

5

APPLICATIONS OF LANTHANIDE ION LUMINESCENCE FROM INORGANIC SOLIDS

HARRY G. BRITTAIN

Squibb Institute for Medical Research
P.O. Box 191
New Brunswick, NJ 08903

5.1. INTRODUCTION

Modern luminescence spectroscopy dates back to the observations of Brewster in 1833 regarding the emissive properties of the uranyl ion. This information proved to be of great use to Stokes in his formulation of spectroscopic principles and to Becquerel's work on radioactivity (1). Fluorescence was subsequently discovered in aqueous extracts obtained from the wood of *lignum nephriticum* and in many other plant extracts (1). With the discovery that fluorescence could be obtained in a wide variety of organic molecules, most of the subsequent investigations dealt with the luminescence of these compounds. At the beginning of the twentieth century (and accompanying the growth of inorganic chemistry as a discipline), work began again on the luminescence properties of inorganic materials.

The most important advances of this time were made by Nichols and Howes in their detailed investigations of the luminescence properties of uranyl compounds (2). The nature of the phenomena could not be fully understood at the time, since these workers were studying quantum effects in a world that was just receiving the new theory of Bohr. Nevertheless, the sharp-line character of the uranyl luminescence spectra (which stood in marked contrast to the broad-band spectra obtained for organic compounds) was recognized at this time. The uranyl investigations received major attention during the course of the Manhattan Project, where exhaustive spectroscopic studies were carried out in a systematic fashion (3).

That lanthanide compounds could exhibit luminescence was discovered by Crookes during his studies of cathode ray tubes. Urbain found that coating the inside of cathode ray tubes with metal oxides doped with traces of Eu(III) would yield cathodoluminescence featuring exceedingly sharp emission lines in the red portion of the spectrum (4). This situation could be likened to "good news and bad news": The good news was that one could now use cathodoluminescence to identify many new elements, but the bad news was that one obtained so many sharp lines that only through the most careful work could one avoid the discovery of spurious elements (e.g., victorium and incognitum) (5). The results of these early works lay waiting for future workers to develop modern applications in a wide variety of fields.

The number of inorganic compounds that exhibit some type of luminescence phenomena is truly staggering. The sources of the emission have been found to be quite variable, and consequently the theoretical interpretations of the processes vary with the nature of the emissive transitions. For instance, some luminescent processes are intrinsic to the material as a whole, and in other cases the luminescence originates with metal ions present in trace quantities (referred to as "activators"). It is not really possible to totally review the many inorganic materials which have found uses in luminescence, and consequently the author's personal prejudice will limit the majority of the discussions to luminescence phenomena associated with the members of the lanthanide series. Reference to transition metal luminescence will be made whenever necessary, however, so that a semibalanced view of the applications might be achieved.

5.2. LUMINESCENCE SPECTROSCOPY OF LANTHANIDE IONS

The energy levels of lanthanide ions have been understood for quite some time (6–8), and consequently these need only be described briefly here for the particular ions that have found luminescence applications in solid phases. Since covalency is not a factor in lanthanide ion complexation, one invariably finds that the positions of energy levels observed in inorganic solids are scarcely shifted from those known for the free lanthanide ions. These have been illustrated in Fig. 5.1 for the tripositive ions.

Since the lanthanide ions all feature large degrees of spin-orbit coupling, the L and S quantum numbers lose their individual definitions through J–J coupling. The magnitude of crystal field interactions are considerably smaller than those of the interelectronic repulsion terms, and hence one requires high resolution to resolve the crystal field splittings. Lanthanide ion spectra are often deceptively simple, due to the combined effects of small crystal fields and the lack of nephelauxetic shifting of energy levels. The intensities of the absorption and emission transitions are normally quite sensitive to environmental details (6–8).

The emission lifetimes associated with f–f luminescence are normally on the microsecond time scale (10), and would thus be operationally classified as phosphorescence processes. This classification cannot be valid since the spin-orbit coupling precludes a valid definition of a spin state, but a consideration of Fig. 5.1 indicates that the luminescence processes are accompanied by $\Delta S \neq 0$. One concludes that while spin is not a good quantum number, the lanthanide ion energy levels appear to retain some memory of this quantity.

The most efficient luminescence intensities are normally obtained from the ions clustered around Gd(III) in the middle of the series. The large gap between the lowest excited states and the highest member of the ground-state manifold leads to enhanced emission intensitives primarily through a decrease in the probability of nonradiative deactivation. This selection reaches its highest level in aqueous solution, where strong luminescence is normally obtained only for Eu(III) and Tb(III) (9). An exception to this general rule is noted for Nd(III), where strong luminescence out of the $^4F_{3/2}$ level can be observed in the near infrared when this ion is substituted in certain solid phases.

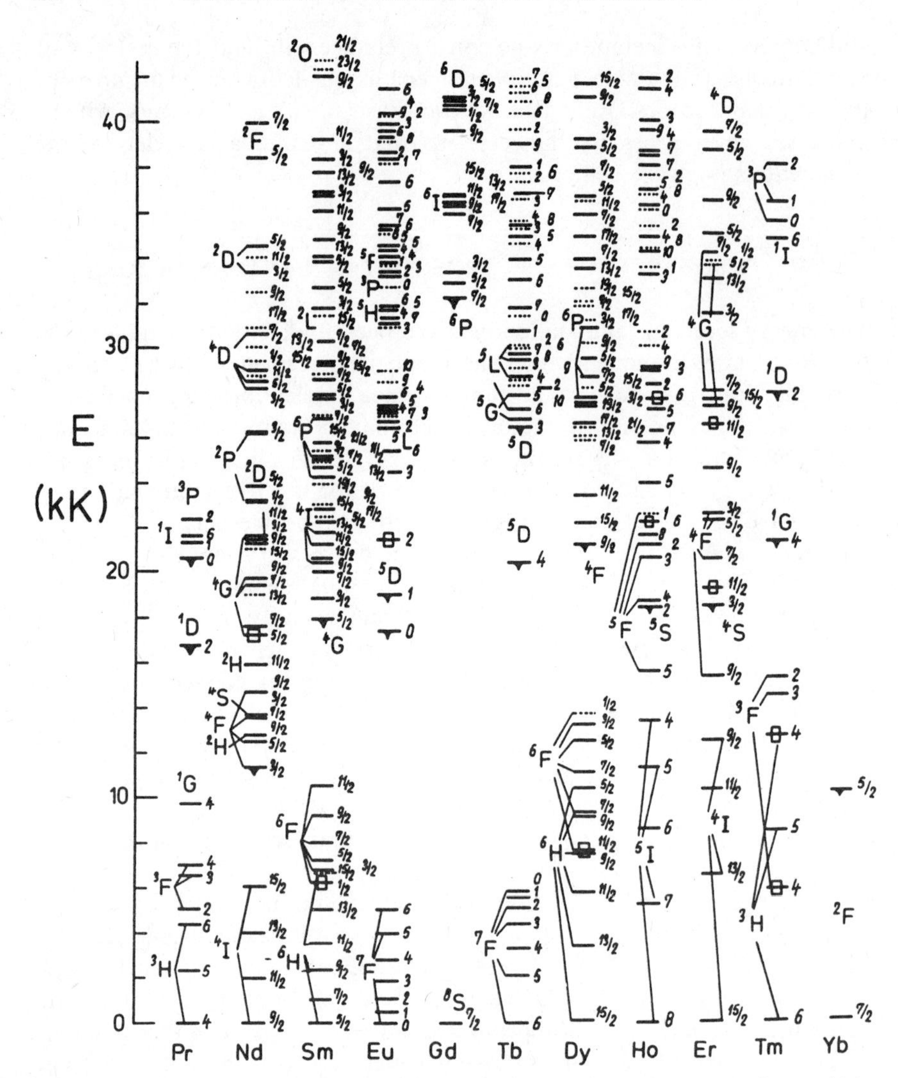

Fig. 5.1. Energy levels of the trivalent lanthanide ions. The excited states from which luminescence normally originates are labeled with black triangles.

The situation can be quite different for lanthanide ions in the dipositive state. Only Eu(II) has found important applications in solid-state luminescence, but these applications have been extremely numerous. Eu(II) is isoelectronic with Gd(III), and hence exhibits a $^8S_{7/2}$ ground state. Unlike Gd(III), however, the lowest excited states are usually derived from the $4f^6 5d^1$ configuration rather than being purely f orbital in character. While

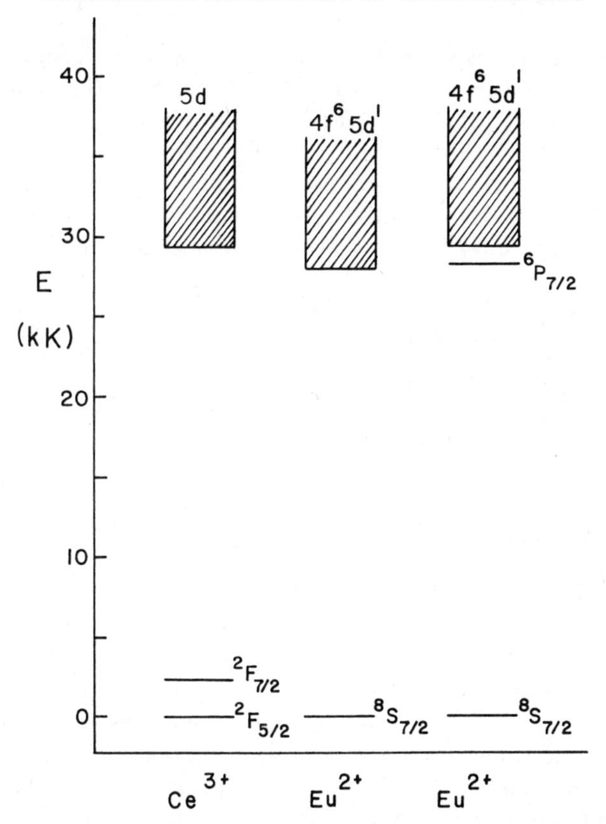

Fig. 5.2. Energy levels of lanthanide activators which exhibit broad-band luminescence. For Eu(II), two situations have been illustrated: (1) where the $4f^4 5d$ manifold of states lies lowest in energy (yielding only broad emisson) and (2) where the $^6P_{7/2}$ level lies lower than the $4f^6 5d$ states (yielding sharp *f–f* emission and broad *d–f* emission).

the *f* orbitals remain nonbonding in their character, the low-lying *d* orbitals are much more sensitive to details of the coordinative environment. The result is that luminescence out of the 4*f*–5*d* levels is broad in character, characterized by much shorter emission lifetimes, and the observed transition energies vary considerably with the nature of the host material. The observations are consistent with the existence of nephelauxetic effects. In certain situations, however, the 4*f*–5*d* excited states are sufficiently high in energy that the 4*f* state lies lowest in energy. In that case, one can observe both sharp line 4*f*–4*f* emission besides the usual broad 4*f*–5*d* emission. (See Fig. 5.2 for an illustration of these possibilites.)

The luminescence observed for Ce(III) compounds tends to resemble the luminescence of Eu(II). Ce(III) features a f^1 configuration, and the spin-orbit coupling splits the ground state into its $^2F_{7/2}$ and $^2F_{5/2}$ components (see Fig. 5.2). The lowest excited states are derived from the $5d$ configuration, and one again finds environmental sensitivities in the luminescence and short emission lifetimes. The lifetimes are normally so short that Ce(III) phosphors are the materials of choice for cathode ray tubes where ultrafast response is required. The luminescence peaks tend to be considerably broader than those associated with pure *f–f* luminescence transitions, and nephelauxetic effects are usually observable.

It should be noted that theoretical analyses of the spectroscopic properties of lanthanide ions are currently performed at high levels of sophistication. A summary of the approaches normally used to analyze lanthanide ion spectral levels is available, and this review also contains an extensive compilation of experimental data for which detailed assignments have been obtained (11). Since it is the purpose of the present review to focus on applications for the luminescence of lanthanide ions in solids, we will not dwell on the theoretical aspects of these phenomena.

5.3. APPLICATIONS

It will become obvious in the following sections that a significant portion of the luminescence work performed on lanthanide ions is oriented toward phosphor applications. These phosphors revolutionized the development of color television, since the major hindrance at the time was the lack of a suitable red phosphor which would function in cathode ray tubes. Picking up on the earliest observations, Palilla demonstrated that Eu(III)-activated phosphors were absolutely ideal for this purpose (12). Similar observations were quickly made in the lamp industry (13). Vastly improved radiography screens were obtained with the replacement of $CaWO_4$ by lanthanide-based phosphors in x-ray-intensifying screens, permitting lower doses of x-rays to be used (14).

Other applications of lanthanide ion luminescence in inorganic solids have not necessarily been so dramatic, but are certainly important in their own right. The display device industry has made considerable use of electroluminescent phosphors, and lanthanide ions have been used as activators in these. Lanthanide phosphors have also been used as up-conversion materials, in attempts to boost the frequencies of light-emitting diodes and diode lasers into the visible region.

These applications summarize the most important applications which have utilized lanthanide ion luminescence in solid oxidic host materials. Each of these will now be discussed in turn.

5.3.1. Cathode-Ray-Tube Phosphor Materials

Cathodoluminescent (CL) phosphors are materials that have been optimized for the conversion of electron beam energy to light energy (15). The energy of the cathode ray beams can vary from a fraction of an electron volt to millions of electron volts, and the bombardment of the phosphor by electron beams creates charge carriers in the material. When these charge carriers encounter the luminescent species, the system relaxes with the emission of a photon. Thus a cathodoluminescent process is quite different from a photoluminescent (PL) process, and one often finds that efficient PL phosphors are actually poor CL phosphors (and vice versa). The most efficient CL phosphors convert only approximately 20% of beam power into radiant energy (16), but lower conversion efficiencies are often tolerated for phosphors possessing other desirable characteristics. The efficiency of a CL phosphor is related to the phonon energies characteristic of the material: Materials possessing high phonon energies (e.g., silicate or vanadate) usually make CL phosphors of low efficiency (17,18).

After efficiency, the second major criterion governing the utility of a CL phosphor is its wavelength of emission. The luminescence lifetime of a phosphor is also of great importance: Television phosphors generally need to decay within milliseconds to avoid smearing of the images (15). Longer persistence times are desired for radar and data displays, while exceedingly short decay times are required for flying-spot scanners. Generally, an increase in the excitation current density yields an increased luminescence output, and hence the linearity of reponse with current density is an important parameter. Finally, CL phosphors must be quite rugged materials: The phosphor materials must be able to withstand the rigors of the tube processing, as well as thousands of hours of electron beam irradiation. Consequently, all phosphors used for CL purposes are based on exceedingly inert inorganic host systems. Almost all usable CL phosphors have employed oxide, oxysulfide, phosphate, vanadate, silicate, tungstate, or borate lattice host systems.

A large number of phosphor systems have been introduced for CL uses, and improvements in the screens quickly followed the luminescence research. As an example, the development of color television systems by RCA will be outlined (15). The original screen used a ZnS : AgCl (blue), Zn_2SiO_4 : Mn (green), and $Zn_3(PO_4)_2$: Mn (red) phosphor blend, but a considerable improvement was obtained with an all-sulfide system composed of ZnS : AgCl (blue), (Zn,Cd)S : AgCl (green), and (Zn,Cd)S : AgCl (red). The major problem with all of these systems was that the red phosphor was actually orange. With the introduction of Eu(III) phosphors for the red (12),

Table 5.1. Commercially Important Cathodoluminescent Phosphors

Phosphor	Color	Application
$YVO_4:Eu$	Red	Color television
$Y_2O_3:Eu$	Red	Color television
$Y_2O_2S:Eu$	Red	Color television
$Gd_2O_2S:Tb$	Green	Avionics
$La_2O_2S:Tb$	Green	Avionics
$Y_2O_2S:Tb$	Green	Avionics
$Y_3Al_5O_{12}:Tb$	Green	Avionics
$Y_3Al_5O_{12}:Ce$	Yellow-green	Flying-spot scanner
$Y_3(Al,Ga)_5O_{12}:Ce$	Yellow	Flying-spot scanner
$Y_2SiO_5:Ce$	Blue	Flying-spot scanner
$Y_2Si_2O_7:Ce$	Blue	Flying-spot scanner

vastly improved color television resulted. For instance, an excellent formulation consisted of ZnS : AgCl (blue), (Zn,Cd)S : Cu : Al (green), and $Y_2O_3:Eu$ (red). The color of the Eu(III) phosphors was essentially perfect for CL applications, and it is highly doubtful that any other activator will ever replace Eu(III) as a red CL phosphor (13).

We will now examine the work that has been carried out in the search for better CL phosphors, and will group the studies according to the chemical composition of the host and activator materials. The lanthanide-activated phosphors which have survived the test of time and which are still of commercial importance are listed in Table 5.1.

5.3.1.1. *Vanadate Host Systems*

CL phosphors based on vanadate materials are of historic importance, since the first red phosphor to replace (Zn,Cd) : Ag was $YVO_4:Eu$ (12). The yttrium vanadate was known to crystallize in the tetragonal zircon ($ZrSiO_4$) structure, and the Eu(III) activator isomorphously replaced Y(III) with no charge compensation problems. The Eu(III)-activated phosphor yielded a spectrum in which almost all of the emitted light was concentrated at 615 nm in the $^5D_0 \rightarrow {}^7F_2$ transition. Brixner carried out detailed structural studies on the Ln_2O_3/V_2O_5 systems, and found that only $LnVO_4$ could be formed (19). All of the $LnVO_4$ compounds were found to be isomorphous, and the luminescence properties of $YVO_4:Eu$ and $GdVO_4:Eu$ were obtained. These observations were confirmed by Palilla, who found that $YVO_4:Eu$, $GdVO_4:Eu$, and $LuVO_4:Eu$ all formed equally efficient CL phosphor systems (20). In this work it was also established that Tm(III), Tb(III), Ho(III), Er(III), Dy(III), and Sm(III) could be used as CL activators, and one could thus obtain phosphors which were cathodoluminescent over the entire visible spectrum.

The interest in vanadate systems as CL phosphors peaked with these early studies, since other host systems were found to be far more promising. $Th_2Li(VO_4)_3$: Eu and $Th_2Na(VO_4)_3$: Eu were found to be good phosphor systems (21), although not superior to the original yttrium vanadate. Effects of impurities and self-quenching were studied in the YVO_4 : Eu system, and it was found that the luminescence efficiencies were very sensitive to the defect structure of the host (22). The effects of other lanthanide ions on the CL efficiency of YVO_4 : Eu was studied; small amounts of Tb(III) and Pr(III) increased the CL yield, while Ce(III) decreased the Eu(III) luminescence (23). $EuNa_2Mg_2(VO_4)_2$: Eu was found to yield CL phosphors six times as bright as YVO_4 : Eu, but the material was found to suffer from thermal quenching at higher temperatures (24).

While the vanadate systems are not currently important as CL phosphors, they have continued to be important as fluorescent lamp phosphors. The developments associated with that sequence of studies will be detailed in a subsequent section.

5.3.1.2. Oxide and Oxysulfide Host Systems

The YVO_4 : Eu phosphor system was replaced as a CL phosphor almost as soon as it was introduced. Y_2O_3 : Eu (25) and Y_2O_2S : Eu (26) were found to be about 50% more efficient in energy conversion, and featured slightly more favorable spectral characteristics. The optimal Eu(III) concentration was also found to be lower in the oxide and oxysulfide hosts, thus lowering the cost of the compounded phosphor. These features have not been improved upon in any other phosphor system, and hence the Eu(III)-activated oxide and oxysulfide systems have remained essentially the only red CL phosphors used commercially to this day. It was found that ABO_2 : Eu phosphors (where A = Li or Na and B = Y or Gd) would function as good CL phosphors (27), but that these decomposed during tube fabrication (28).

In their initial report, Wickersheim and Lefever examined the Eu(III) luminescent behavior in yttrium oxide and many other oxidic host systems (25). Only In_2O_3 : Eu, Sc_2O_3 : Eu, and Gd_2O_3 : Eu were found to yield Eu(III) luminescence under normal excitation conditions, and only Gd_2O_3 : Eu represented a useful phosphor system. The luminescence output of the Y_2O_3 : Eu phosphor was actually found to increase with increasing temperature, with a maximum being reached at 600°C. Essentially the same conclusions were reached in an independent study (29) which studied only the Y, La, and Gd oxidic phosphors activated by Eu(III).

The efficient Y_2O_3 and Gd_2O_3 host systems both exhibit cubic crystal structures, while the inefficient La_2O_3 host system is hexagonal in nature. The effect of host structure was probed by Ropp, who obtained the La_2O_3–Gd_2O_3–Y_2O_3 phase diagram (30). The Eu(III) luminescence properties were obtained when this ion was doped into the pure oxides, and into the mixed oxides (e.g., $LaYO_3$ or $LaGdO_3$). The position of the matrix excitation band was found to depend on the nature of the host material, and that the efficiency of the Y or Gd oxidic hosts was related to the properties of this matrix band.

The crystal structure of Y_2O_3 is such that one would anticipate the existence of more than one replacement site for the Eu(III) activator. These sites would exhibit C_2 and S_6 site symmetries and ought to be present in a 3:1 ratio, respectively. Spectral evidence was presented indicating that emission could be obtained from Eu(III) ions having both symmetries (31), and hence that the Eu(III) activator ions would substitute randomly in the Y_2O_3 structure (32). Recently, time-resolved luminescence spectroscopy was used to obtain the spectra associated with each site (33). Three crystallographically nonequivalent sites were also found for Gd_2O_3:Eu (33), with the spectra of these being resolved through site-selective excitation techniques (34).

While In_2O_3 is isostructural with Y_2O_3, In_2O_3:Eu was found to be a poor phosphor. Through studies of the $(Y,In)_2O_3$: Eu system, it was found that the position of the Eu(III) *f–d* charge transfer band was related to the luminescence efficiency (35). When the charge transfer band was located within the matrix absorption, little Eu(III) emission was noted. This implies that in In_2O_3, the Eu(III) ions cannot function as the deep electron trapping levels which are required for efficience cathodoluminescence. This result is interpretable in light of earlier work where it was shown that the position of the charge transfer band was related to the coordination number of the Eu(III) ion: The band is essentially fixed in octahedral 6-coordination, and variable in 8- and 12-coordinate species (36). In these situations, the band varies as a function of the effective ionic radius of the relevant host ion.

With the report that Y_2O_2S:Eu would make a good red phosphor (26), another major host system received attention. Yttrium oxysulfide crystallizes in the P3m space group, with the Y(III) being 7-coordinate and occupying a site symmetry of C_{3v}. Crystal field effects in the luminescence of Y_2O_2S:Eu were consistent with this point group (37). In YVO_4:Eu and Y_2O_3:Eu essentially all of the Eu(III) emission originates from the $^5D_0$ level, but this was not found to be the case for Y_2O_2S:Eu. Depending on the activator concentration, significant emission could also be obtained from the higher $^5D_1$, $^5D_2$, and $^5D_3$ Eu(III) levels, as has been illustrated in Figure 3. These effects were found to a general property of the Ln_2O_2S host system, as proven through studies of the Ln = Y, La, Gd, and Lu compounds (38). A rise in the barycenters of the ^5D_J states with increasing host cation size was also noted.

The fact that one could obtain emission from multiple Eu(III) levels was found to alter the color of the phosphor, since the luminescence associated with the higher ^5D_J levels was observed at higher energies than the emission from the $^5D_0$ level. Color shifts were obtained as a function of activator concentrations and related to the relative emission yields out of the various ^5D_J lvels (39). It was found that increasing amounts of Eu(III) in Y_2O_2S would decrease the emission yield out of the higher ^5D_J levels. Besides concentration, it was also shown that changes in the mode of excitation could also alter the luminescence yields from the various ^5D_J levels (40). The effect of pressure on the luminescence of La_2O_2S:Eu and Y_2O_2S:Eu was also studied, and it was found that by increasing the system pressure, one could obtain enhanced yields from the higher ^5D_J states (41).

The luminescence yields from the various ^5D_J levels were also found to be strongly affected by temperature. The rate of luminescence decrease (with increasing operating temperature) was found to increase on passing from $J = 0$ to $J = 1$ (42). Quenching from the ^5D_J states was attributed to thermally activated resonance crossovers from the f levels to the charge-transfer states. Through studies of excitation into the charge-transfer states, it was established that the ^5D_j emitting states are directly connected to the charge-transfer states in the oxysulfide phosphors (43). In the oxide phosphors, states higher than ^5D_J were fed after charge-transfer excitation, and these efficiently relax down to the lowest $^5D_0$ level. The strong temperature dependence of these processes greatly affects the color of the phosphor, and these effects have been harnessed in the development of temperature-sensitive fiber optic probes (44).

The nature of these charge-transfer states has been studied in great detail. It was proposed that energy could be lost in the charge-transfer state through thermal dissociation into Eu(II) plus a free hole (45). These holes could be envisioned as energy storage centers, since recombination would yield Eu(III) in the charge-transfer state. Similar effects could be obtained through direct excitation to the $^5D_2$ level, since the Eu(III) ion could reach the charge-transfer state from this level (46). Energy storage effects were used to explain the saturation of cathodoluminescence observed in Y_2O_2S:Eu phosphors under long electron beam pulses (47). Blasse has provided a detailed discussion of the influence of charge transfer and Rydberg states on the luminescence properties of lanthanide ions (48).

The luminescence of activators other than Eu(III) has also been detailed in the oxysulfide host system. The CL efficiency of Er(III) in mixed $(La,Y)_2O_2S$ host systems was followed, and it was observed that increasing Y(III) concentrations led to decreasing CL efficiences (49). This effect was

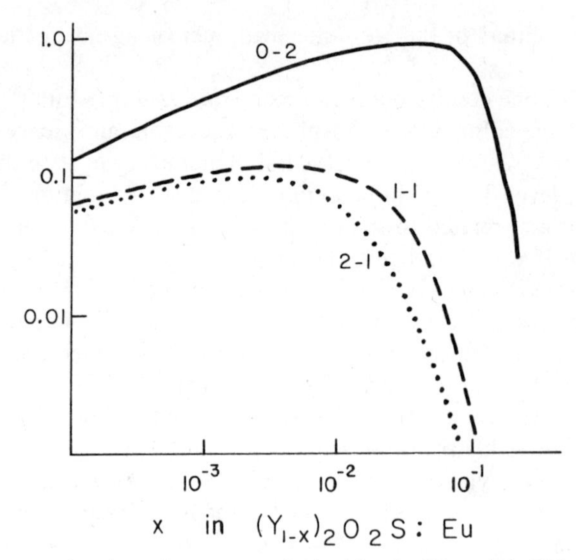

Fig. 5.3. Concentration-dependence curves obtained for the $^5D_0 \rightarrow {}^7F_2$ (626 nm), $^5D_1 \rightarrow {}^7F_1$ (539 nm), and $^5D_2 \rightarrow {}^7F_1$ (476 nm) luminescent band systems of Y_2O_2S : Eu(III). (Adapted from data reported in ref. 50.)

attributed to a lack of energy transfer associated with the Y sites. Y_2O_2S : Pr was shown to exhibit green luminescence, which was mainly due to the $^3P_0 \rightarrow {}^3H_4$ band at 514 nm (50) (Fig. 5.3). It was also found that the CL efficiency of Y_2O_2S : Eu could be enhanced by traces of Tb(III) or Pr(III) through energy transfer mechanisms (51). An exceedingly large stokes shift was observed for Ce(III) in Y_2O_2S or Lu_2O_2S host systems, with the emission being observed around 650–700 nm (52).

The yttrium oxysulfide host system is not the easiest phosphor to synthesize, and consequently much effort has been expended in this area. It is possible to reproducibly obtain good materials through the reaction of H_2S with a blend of lanthanide oxides (53). The phosphor can also be obtained by reducing Eu(III)-doped yttrium sulfite or oxysulfate with carbon monoxide (54). Different particle sizes (in the range 3–50 μm) were obtained in the reaction of lanthanide oxides with a sodium sulfide flux (55).

5.3.1.3. Sulfide-Based Host Systems

The luminescence of material lattices of the ZnS type (group IIB–VIB compounds) has been studied extensively (56). Most of the green and blue phosphors still used in color television are based on activated ZnS or

(Zn,Cd)S materials, and all of the phosphors used in black-and-white television screens are based on IIB–VIB materials. Generally, one excites into material conduction band upon irradiation with electron beams, and the charges quickly become localized at defect or activator sites. Luminescence takes place when these excited states decay back either to the valence band or to some other ground-state level (57).

Most lanthanide ions have been shown to yield efficient luminescence when doped into either ZnS (58) or CdS (59). Consideration of transition energies and energy levels indicated that lanthanide ion defect pairs with some type of acceptor defect were formed: No emission could be observed from a lanthanide ion excited at energies higher than 1.53 eV above the ground state, even though the host band gaps were 2.4–2.5 eV. Detailed studies of charge compensation effects were performed for ZnS : Er materials, and it was noticed that compensation by alkali metals actually interfered with the luminescence process (60). Evidently, the nature of the defects is quite sensitive to the method of charge compensation. In a recent study of ZnS : Tm phosphors, an analysis of energy levels and crystal field parameters permitted the deduction of a T_d site symmetry for the lanthanide ion (61).

Lehmann carried out detailed studies of lanthanide activators in CaS and found that efficient luminescence could be obtained for nearly all members of the lanthanide series (62). By contrast, CaO was found to be a considerably less versatile phosphor host system, although several lanthanide ions were found to luminescence in CaO (63). Site-selective spectroscopy was used to examine the defect chemistry of CaO : Eu, and it was found that the Eu(III) ion could occupy cubic and orthorhombic sites (64).

Efficient green luminescence was observed for CaS : Ce(III), and red luminescence was observed for CaS : Eu(II) (65). The very fast decay times associated with these phosphors made them useful for flying-spot scanning applications. (Ca,Sr)S : Ce phosphors were prepared by a flux method, and crystallites resistant to atmospheric attack were produced by this method (66). The efficiency of these phosphors was found to be comparable to those obtained after H_2S reactions. In another study it was found that under certain condition MgS and CaS were miscible, and thus one could obtain (Ca,Mg)S : Eu or (Ca,Mg)S : Ce phosphors which worked equally well (67).

Cathodoluminescence in thiogallate phosphors has been studied with a number of lanthanide ion activators. Host materials having the formula MGa_2S_4 (where M = Ca, Sr, Ba, and Pb) were propared with Ce(III) and Eu(III) as activators (68). Strong luminescence under ultraviolet and cathod ray excitation was observed. The Ce(III) phosphors were examined in more detail for flying-spot scanner applications, and it was found that $SrGa_2S_4$: (Ce,Na) was a superior phosphor featuring an exceedingly short decay time (69). Efficient green Er(III) luminescence was also obtained in the thiogallate system, and the effects of Er charge transfer bands on the luminescence efficiencies were examined (70).

5.3.1.4. Aluminate Host Systems

The host characteristics of aluminate host systems have received enormous attention with respect to laser applications. Yttrium aluminum garnet ($Y_3Al_5O_{12}$, abbreviated YAG) has been found to make an excellent host for Nd(III), and highly efficient neodymium YAG laser systems have become common in laser laboratories. Investigations concerned with these applications are exceedingly numerous, and a discussion of these is beyond the intended scope of this review. At the present time, only applications relating to phosphors will be detailed. A large fraction of the aluminate work has concerned fast phosphors based on Ce(III) activation, and a discussion of these materials follows in Section 3.1.E.

Er(III) was found to make an acceptable CL activator in YAG, and accompanying photoluminescence measurements were used to deduce crystal field splittings in the excited ${}^2P_{3/2}$ level (71). Concentration dependencies in the CL of Ce-, Eu-, Gd-, or Tb-activated YAG phosphors have also been studied (72). The observed trends were interpretable using general theories developed earlier for oxysulfide phosphors, and involve competitive recombination at bulk lattice defect centers. It had been reported that lanthanide ions could substitute at octahedral and dodecahedral sites in YAG hosts (73,74), and consequently the site selection in YAG:Eu phosphors was examined (75). Analysis of the data revealed that Eu(III) did occupy two different sites, experiencing crystal fields of orthorhombic (*d*-sites) and axial (*a* sites) symmetry. Examples of the differing spectra may be found in Fig. 5.4.

The Eu(III) work nicely confirmed an earlier study involving YAG:Tb, where the orthorhombic sites were studied in detail (76). In the YAG host, Tb(III) was found to emit from both the ${}^5D_3$ and ${}^5D_4$ excited states, as it does in the oxysulfide host system. An improved synthesis of the YAG:Tb phosphor involved ion implantation techniques, and the resulting phosphors were found to yield variable colors upon CL excitation (77). This phenomenon undoubtably reflects varying population of the Tb(III) ${}^5D_J$ luminescent excited states. Addition of V_2O_3 to YAG:Tb phosphors produced materials exhibiting greatly increased lifetimes, indicating supression of lattice defect sites (78).

Di Bartolo and co-workers have carried out extremely detailed investigations of energy transfer and fluorescence characteristics of lanthanide-activated YAG materials. The infrared luminescence obtained from

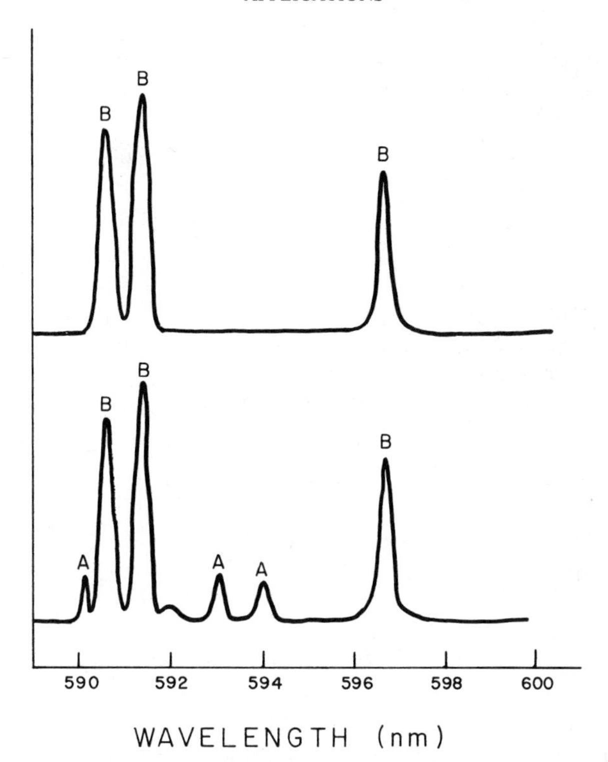

Fig. 5.4. $^5D_0 \rightarrow {}^7F_1$ luminescence spectra obtained for the two Eu(III) sites in $Y_3Al_5O_{12}$: Eu. The bands due to site B were obtained through excitation of tis $^7F_0 \rightarrow {}^5D_0$ absorption at 17.214 kK, while the spectra of sites A + B were obtained upon excitation at 17.232 kK. (Adapted from data reported in ref. 75.)

YAG : Er and YAG : Ho was studied in detail and spectroscopic assignments provided for the observed transitions (79). Small amounts of Ho(III) in YAG : Er, Ho materials was found to totally deactivate the Er(III) luminescence, indicating exceedingly efficient energy transfer which did not exhibit appreciable temperaure dependence. Sensitization of Er(III) luminescence by Yb(III) in ytterbium aluminum garnet was found to take place, and the resonant Yb → Er energy transfer processes were studied (80). Subsequently, the Er → Ho and Er → Tm energy transfer processes were studied in the erbium aluminum garnet host system, and rate equations were obtained which could explain the observed dynamics (81).

Substitution of lanthanide ions at both dodecahedral and octahedral sites has also been confirmed for yttrium gallium garnets (82). Emission out of both $^5D_3$ and $^5D_4$ levels was observed in yttrium gallium garnet activated with Tb(III), and the observed spectra were consistent with the existence of a low site symmetry (83). The luminescence of Tb(III) in each site symmetry has been obtained, and it was reported that strontium gallium garnet doped with Tb(III) exhibited CL decay times nearly three times those observed for YAG:Tb (84). Since YAG:Tb (designated as phosphor P53) is used as a phosphor in avionics displays, the longer decay time of Tb-activated strontium gallium garnet would make this phosphor superior for those applications.

Yttrium orthoaluminate is the only known solid-state material (other than YAG) which has potential as a laser host system (85). While the structure of YAG is cubic and isotropic, $YAlO_3$ is orthorhombic and hence anisotropic. The $LaAlO_3$:Tb phosphor was found to exhibit strong cathodoluminescence, with variable colors being obtainable at different Tb(III) doping levels (86). The Tb(III) luminescence can be sensitized by Ce(III) in the $GdAlO_3$:(Ce,Tb) phosphor; this process taking place by the usual dipole–dipole coulombic interaction mechanism (87). Crystal field splittings and luminescence properties of Nd-, Er-, and Dy-activated $YAlO_3$ have been reported, and quite high luminescence polarizations were obtained in this anisotropic host system (88).

5.3.1.5. Ce(III)-Activated CL Phosphors

While color television requires phosphors that exhibit reasonably short decay times, certain other applications require phosphors that decay exceedingly rapidly (89). The first application relates to the luminescent screen in the cathode ray tube of a flying-spot scanner. These devices are used in the generation of television signals for the transmission of films and slides. The second application relates to the phosphor that produces the sychronizing signal for the electron stream in a beam indexing tube. Ce(III)-activated phosphors exhibit the short lifetimes necessary for these applications and are sufficiently stable as to be used in CL applications. In addition, the afterglow associated with these materials can essentially be eliminated (89). The major Ce(III) phosphors used for these purposes were listed earlier as part of Table 5.1.

The first detailed studies concerned Ce(III)-activated YAG phosphors (90). Large crystal field splittings and Stokes shifts were noted in YAG:Ce and SrY_2O_4:Ce, and it was noted that the luminescence lifetimes were on the nanosecond time scale. Normally, Ce(III) phosphors emit in the ultraviolet or blue, but in the YAG host emission was obtained in the green. Stadler and co-workers studied the luminescence of Ce(III) in yttrium aluminum and yttrium gallium garnet systems and found that the wavelengths of the luminescence were greatly affected by the host system used (91). In all of these materials the lifetime of the Ce(III) luminescence was exceedingly

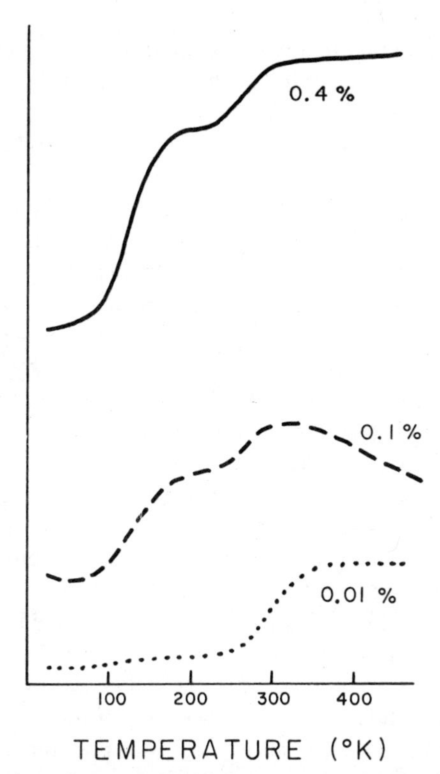

Fig. 5.5. Temperature dependence of the luminescence associated with varying activator concentrations in $Y_3Al_5O_{12}$:Ce. The intensity increases noted above 50 and 220°K were attributed to thermalization of the competitive recombination centers associated with the two Ce(III) sites in the YAG host. (Adapted from data reported in ref. 95.)

short, and no significant afterglow was detected (92). This behavior stood in sharp contrast to the results obtained for CaS: Ce, where an afterglow of several milliseconds could not be eliminated (65). Through an extensive series of substitutions in the basic garnet structure, it proved possible to obtain fast Ce phosphors emitting in the red as well as in the green spectral regions (93).

Quite interesting results were obtained from studies of the temperature dependence of YAG: Ce emission properties (94). Luminescence was observed from three characteristic lattice defects as well as from the ^2D excited states of Ce(III). Upon raising the temperature of the system, higher luminescence yields were obtained out of the pure Ce(III) states, primarily by the thermal emptying of the traps. These effects have been illustrated in Fig. 5.5. However, this increase in emission efficiency was offset by the introduction of afterglow (94). After study of the temperature dependence of energy transfer processes involving other lanthanide ions, the defects were identified as hole trapping centers (95). In another study it was learned that the Ce(III) ion experiences a tetragonal crystal field in YAG: Ce, and these crystal field effects could be used to explain the temperature dependence of the phosphor efficiencies (96).

The $YAlO_3$: Ce phosphor was found to emit at 370 nm with a decay time of 16 ns (97), but apparently was difficult to keep from coverting into YAG: Ce over extended periods of time (98). These problems were traced to an inability to obtain the phosphor completely free of other phases, but through improved synthetic methods $YAlO_3$: Ce was prepared in a single phase (99). This material did not show any tendency to decompose, and exhibited an efficiency of 7% with a decay time of 30 ns.

Since a flying-spot scanner must be responsive to the entire visible spectrum, one cannot make such a device solely with green-emitting YAG: Ce. Ce(III)-activated silicates emit in the blue and are normally compounded in with the YAG: Ce to make the cathode ray tube. The first useful system proposed was Y_2SiO_5 : Ce, with an activator level of 1–2% (100). This material was found to emit at 415 nm, exhibiting a decay time of only 30 ns with an efficiency of 6% (101). Equally efficient are the Ce(III) phosphors based on $Ca_2MgSi_2O_7$ or $Y_2Si_2O_7$ (93). A series of Ce(III)-activated phosphors were prepared in the BaO–SrO–SiO_2 system, with all of these materials exhibiting luminescence at 390 nm (102). Characterization was reported for $MSiO_3$: Ce and M_2SiO_4 : Ce phosphors (where M = Ba or Sr). Phosphors having slightly more efficiency were obtained through the incorporation of halogens into the silicate structure (producing $Y_3Si_2O_8Cl$: Ce), but this phosphor system was found to deteriorate too rapidly under CL conditions to be of use (103).

Ce(III) activation of lanthanide phosphates has been found to yield phosphors of slightly lower efficiency, but with shorter lifetimes than the silicate-based systems (104). The luminescence wavelengths of the Ce phosphors was found to depend on the identity of the lanthanide ion used to prepare the host. $LaPO_4$: Ce emits at 316 and 337 nm, while YPO_4 : Ce emits at 330 and 365 nm; these differences can be ascribed to differences in the crystal structures of the hosts. It was subsequently shown that slight excesses of included phosphate could shift the YPO_4 material (tetragonal xenotime structure) into the $LaPO_4$ structure (monoclinic monazite structure), altering the wavelength of the Ce(III) emission (105).

5.3.1.6. *Other CL Phosphor Systems*

Several other host systems have been examined as possible CL materials, although the attention given to these has been significantly smaller. For instance, $Sr_3(PO_4)_2:Tb$ was found to be a bluish-white emitting CL phosphor with a long decay time (106). This material might have found use in avionics displays had not Tb(III)-activated oxysulfide phosphors been found to be more efficient.

Cathodoluminescence in borate systems has received some attention. Green-emitting $InBO_3:Tb$ was found to be equally efficient as $Zn_2SiO_4:Mn$, and considerably more efficient than other Tb(III)-activated borates (107). The CL efficiences of Tb(III)- and Eu(III)-activated orthoborates series (MBO_3, where M = Sc, Y, La, Gd, and Lu) was examined in detail. While all host systems yielded acceptable CL efficiencies, the In(III)-based hosts were particularly effective for Tb(III) and Eu(III) activation (108). The CL properties of $YAl_3B_4O_{12}:Tb$ were investigated, but it was concluded that this phosphor was of only moderate efficiency and could not therefore replace the more commonly used green phosphors (109).

A series of Tb(III)-activated silicate phosphors was found to function excellently well at high current densities (110). These were based on the Ln_2O_3–SiO_2 phase diagram, and studies on $Y_2SiO_5:Tb$, $Y_2Si_2O_7:Tb$, and $Y_4(SiO_4)_3:Tb$ were reported in the greatest detail. The CL emission wavelengths of Ce(III)- and Eu(III)-activated feldspars ($MAl_2Si_2O_8$, where M = Ca, Sr, or Ba) were found to depend strongly on the identity of the alkaline earth ion used to prepare the phosphor host (111). $In_2Si_2O_7:Tb$ was found to yield a phosphor as efficient as any Tb(III) silicate, and it was also observed that neither Ce, Pr, Eu, Dy, or Tm could act as an activator in this host system (112).

Lanthanide-activated tungstates have been surveyed for use as CL phosphors (113), but relatively few tungstates have actually found use as CL phosphors. $Y_2W_3O_{12}:Eu$ has been found to be a highly efficient red phosphor and was equivalent in efficiency to the commonly used $Y_2O_2S:Eu$ phosphor (114). Refinements in the preparation of this phosphor have been reported, as well as more definitive estimates of its efficiency (115).

The use of $SnO_2:Eu$ phosphors in low-energy electron beam excitation applications has been reported (116). Under these excitation conditions, all emission originates from the $^5D_0$ level, giving the phosphor only red emissive properties. The solubility limit of Eu(III) in SnO_2 was 0.05–0.06 atomic percent, and the material exhibits an excitation maximum at 300 nm attributable to matrix excitation (117).

5.3.2. Fluorescent Lamp Phosphor Materials

The excitation processes that take place in fluorescent lamps are essentially those of photoluminescence (PL) and are very different in nature from cathodoluminescence. Ultraviolet light is generated in a gas discharge tube (usually with the addition of small amounts of mercury), and hence a good phosphor is one that features much of its absorbance around 250 nm. Should the activator species possess sufficient absorptivity at this wavelength, it can be excited directly. The other excitation mode involves absorption of the UV by the host material (through some sort of matrix band), and subsequent nonradiative transfer of this absorbed energy to the activator. Detailed descriptions of these various processes have been reviewed (118).

A very detailed overview of the science and technology of fluorescent lamp phosphors has been provided by Butler (119). Between 20 and 30 phosphors are of commercial importance, with halophosphate materials accounting for about 85% of the total weight produced. Barium titanium phosphate and three tin-activated phosphates account for another 10% of the total, and the remaining 5% are materials that are of specialized use (119). Important activators are Pb(III) (in strontium hexaborate or barium disilicate), Sn(II) (in strontium pyrophosphate or calcium and strontium orthophosphates), and Mn(II) (in magnesium gallate or zinc orthosilicate). Other phosphors involve host activation (calcium, calcium-lead, or magnesium tungstates), and double activation (e.g., calcium halophosphate activated by Mn + Pb or Mn + Sb). Earlier phosphors have used essentially the same activators as just mentioned, but have differed in the nature of the host material (119).

The most important lanthanide activator is certainly Eu(II), although specialized applications require the use of Eu(III), Dy(III), and Ce(III). Eu(II) is particularly useful as an activator since it generally produces strong photoluminescence regardless of its host material. The most useful phosphor systems have been based on phosphate, silicate, and aluminate host systems, although PL behavior has been investigated in essentially every matrix system for which one can substitute lanthanide ions. In many of these studies, other lanthanide ions have been introduced as activators so that detailed information regarding the PL processes could be obtained. For instance, the PL properties of Eu(III) (120)- and Tb(III) (121)-activated phosphors have been compared for a variety of host systems. High PL efficiencies and stabilities of lanthanide-activated phosphors appears to require the use of oxygen-dominated host lattices (122,123).

In our survey of materials, we will attempt to review investigations covering a wide range of materials and will not be restricted to hosts for which successful commercial use has been attained.

5.3.2.1. Borate Host Systems

Activated borates were among some of the first phosphor systems for which lanthanide activation was investigated. It has been shown that Tb(III)-activated borates exhibit efficiencies of more than 40% upon excitation in the

250–270 nm region (124). The most extensive work was performed for orthoborates ($M_3B_2O_6$, M = Ca, Sr, Ba), but pyroborates ($M_2B_2O_5$), metaborates (MB_2O_4), and tetraborates (MB_4O_7) were all found to function as good host systems. Tb(III) was also found to exhibit strong PL when substituted into alkaline earth borate systems, exhibiting luminescence lifetimes of several milliseconds (125). The luminescence of Tb(III) in a new borate system, $X_2ZBO_3)_2$ (where X = Ba, Sr, Ca and Z = Ca, Mg), has been reported (126). The Tb(III) luminescence efficiencies were reported as high as 75% (126), but were found to be strongly affected by the method of charge compensation.

Eu(III) activation of borates has also yielded efficient phosphors. Strong red luminescence was obtained when doping gadolinium and lanthanum borates with Eu(III), and evidence was obtained indicating that nonradiative transfer of energy could take place from Gd(III) to Eu(III) (127). At elevated temperatures the luminescence intensity of the borate phosphors actually exceeded that of Gd_2O_3 : Eu. Further studies of the temperature dependence of Eu(III) emission in $ScBO_3$ and $LaBO_3$ hosts were performed, and the observed loss in intensity with temperature was correlated with the position of matrix absorption bands (128). Renewed interest in Eu(III)-activated borate phosphors has come about with investigations on phosphor systems suitable for gas discharge displays. These display devices require phosphors excitable in the region 160–170 nm, and hence require slightly different phosphors than those used in conventional lamps. (Y,Gd)BO_3 : Tb has been found to function as a reasonably good material, but (Y,Gd)BO_3 : Eu has been found to be a superior phosphor for such purposes (129,130).

Energy transfer phenomena in lanthanide-activated borates has been studied to a significant extent. Sensitization of Tb(III) luminescence by Ce(III) has been observed in $ScBO_3$, although under ultraviolet excitation the transfer was not sufficiently efficient as to completely quench the Ce(III) luminescence (131). It has also been shown that the presence of Gd(III) can enhance the sensitivity of matrix-to-activator energy transfer processes, acting primarily as an intermediate (132). In $LaMgB_5O_{10}$ borates, the lanthanide polyhedra form isolated chains, and thus self-quenching by means of one-dimensional energy transfer was followed in $LaMgB_5O_{10}$: Eu and $LaMgB_5O_{10}$: Tb (133). Further studies were carried out in $GdMgB_5O_{10}$ activated by Mn(II), Bi(III), Ce(III), and Tb(III) (134). Excitation energy could migrate along the Gd sublattice, but dilution of Gd with La could block this transfer. Very efficient phosphors were obtained using Ce as a sensitizer, the Gd sublattice as an intermediate, and Tb as the activator (134). Three main Eu(III) sites were detected in $GdAl_3B_4O_{12}$: Eu, and energy transfer among these could be followed using time-resolved site-selection techniques (135).

Eu(III) activation of borate hosts has also been reported. While most alkaline earth borates doped with Eu(II) yielded only weak emission, BaB_8O_{13} : Eu was found to yield efficient blue emission after 254-nm excitation (136). Efficient emission was also obtained from SrB_4O_7 : Eu, and this observation was explained by means of the unique crystallographic properties of the host (137). Synthesis of this phosphor under high-pressure conditions yielded SrB_2O_4 : Eu phosphors having different structures and drastically increased quantum efficiencies (138). The wavelength of the Eu(II) emission was also found to shift from 367 nm (low-pressure phases) to 395 nm (high-pressure phases), indicating rearrangement of the borate polyhedra.

5.3.2.2. *Aluminate Host Systems*

Phosphors based on group IIIA oxide systems (aluminates and gallates) have found important commercial applications as lamp phosphors, and Eu(II) activation of these has been particularly useful. Activation by Eu(II) or Ce(III) yields blue phosphors, activation by co-doped Ce(III) and Tb(III) yields green phosphors, and activation by Eu(III) or Mn(II) yields red phosphors (139). The most important aluminate host materials are $M^{2+}Al_{12}O_{19}$ (found in the magnetoplumbite structure) and $M^{+}Al_{11}O_{17}$ (crystallizing in the β-alumina structure).

The first report concerning Eu(II) activation focused on cubic spinel MAl_2O_4 : Eu and $MAl_{12}O_{19}$: Eu phosphors, where M = Ca, Sr, or Ba (140). All of the $MAl_{12}O_{19}$: Eu phosphors were blue in their emission, but more variation was obtained in the MAl_2O_4 : Eu system. While the emission of $CaAl_2O_4$: Eu was also blue, the luminescence from $BaAl_2O_4$: Eu and $SrAl_2O_4$: Eu was green. Interestingly, the luminescence from $MgAl_2O_4$: Eu was found to lie intermediate in color relative to the Ca- or Sr,Ba-activated host systems (141). These effects have all been illustrated in Fig. 5.6. It was later found that the optical and electrical properties of $SrAl_2O_4$: Eu varied with the Sr/Eu ratio, and evidence was presented indicating that holes could be trapped at lattice defects (142).

Other host systems were investigated at this time, but evidently these did not prove to be as useful as the magnetoplumtibe or β-alumina types of host systems. Eu(II) activation of $EuAlO_3$ was found to yield a green phosphor, and the Eu(III) ions of the host were only weakly emissive under PL excitation (143). It was concluded that Eu(III)–Eu(III) concentration quenching deactivated the host, and that there was little Eu(II)–Eu(III) energy transfer. Eu(II) activation of the ternary $EuO–Al_2O_3–SiO_2$ system was systematically investigated, and several useful phosphors were obtained in aluminosilicate host systems (144,145).

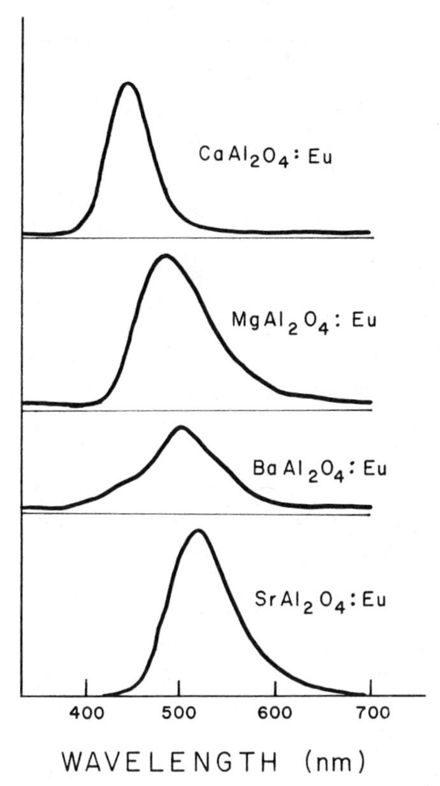

Fig. 5.6. Luminescence spectra obtained from Eu(II)–activated alkaline earth aluminates. (Adapted from data reported in ref. 141.)

$LaMgAl_{11}O_{19}$:Ce was found to make a highly efficient blue/UV phosphor, which exhibited little concentration quenching (146). Activation by Tb did not yield very efficient phosphors, but coactivation with Ce and Tb yielded efficient green phosphors. This effect indicated energy transfer from Ce(III) to Tb(III), and such transfer was investigated in great detail (147). In a general review, the utility of hexagonal aluminate and gallate host lattices was considered, and it was concluded that activation by Eu(II), Ce(III), or Ce(III) + Tb(III) could yield many efficient phosphor systems (148). In $SrAl_{12}O_{12}$: Eu(II), it was found that the $4f^7\,{}^6P_{7/2}$ excited state was slightly lower in energy than the levels derived from the $4f^6 5d^1$ configuration, and one could obtain Eu(II) line emission (rather than the usual broad bands) under certain conditions (149).

Luminescence decay studies were used to follow energy transfer from Ce(III) to Tb(III) and Eu(III) in the $LaMgAl_{11}O_{19}$ host system (150). While energy transfer from Ce to Tb yielded efficient luminescence, energy transfer from Ce to Eu yielded no activator emission even though the Ce emission was quenched. The Eu(II) activation in hexagonal aluminates was finally systematized: In the magnetoplumbite structure Eu(II) occurs at blue wavelengths, in the La-aluminates Eu(II) emission occurs at green wavelengths, and in the β-alumina structure the Eu(II) luminescence occurs at wavelengths between blue and green (151). The same general trends were found to hold for Ce(III) activation of hexagonal aluminates containing large divalent or trivalent cations (152).

The effect of defects induced by substitution of small amounts of Mg for Al in $SrAl_{12}O_{19}$: Eu was examined, and it was found that all Eu(II) emission could be quenched by the oxidic defects (153). Nonstoichiometry in the host materials was found to strongly influence the luminescence spectrum and quantum efficiency of Eu(II) activated aluminate phosphors. Changes in local crystal structure (from β-alumina type to magnetoplumbite type) could occur when preparing either Al_2O_3-rich or Al_2O_3-poor phosphors, and these changes were found to greatly infuence the Eu(II) luminescence properties (154).

Eu(III) activation of $CaLaGaO_4$ hosts was found to yield red phosphors, whose efficiency depended greatly upon the morphology of the host (155). Eu(III) luminescence in the α-phase (olivine structure) was found to be much more efficient than luminescence from the β-phase (K_2NiF_4 structure). The PL from yttrium gallium garnet hosts activated by Tb(III) has been reported, and increases in emission efficiencies were obtained by the addition of small amounts of V_2O_3 (156).

5.3.2.3. *Silicate Host Systems*

Eu(II) activation of Sr, Mg, Ba, and Ca silicate host systems has yielded lamp phosphors of high efficiency and commercial importance. The luminescence of $MSiO_4$: Eu, M_3SiO_5 : Eu, $M_2MgSi_2O_7$: Eu, and $M_3MgSi_2O_8$: Eu (where M = Ca, Sr, or Ba) phosphors was studied in detail (157). The efficiency of these phosphors was generally quite high, and the emission maxima were found to vary between 435 and 600 nm. At the same time, an extremely detailed study of similar systems was undertaken by Barry. The luminescence yields of $M_3MgSi_2O_8$: Eu (M = Ca, Sr, or Ba) were found to be quite high, with the emission maxima being found at increasingly shorter wavelengths as the radius of the alkaline earth cation decreased (158). Eu(II) activation in binary alkaline earth orthosilicate systems was also examined, and it was found that the $(Ba,Sr)_2Si_4$: Eu system yielded the most efficient phosphors (159). Energy transfer from Eu(II) to Mn(II) was found to occur in

$BaMg_2Si_2O_7$: Eu,Mn, producing an exceedingly efficient red phosphor (160).

It has also been reported that Mg_2SiO_4 can be activated by Tb(III) [using Li(I) as a flux and charge compensator], yielding efficient green phosphors (161). In a similar vein, $Lu_2Si_2O_7$ can be activated by Eu(III), yielding efficient red phosphors (162). Eu(II) activation in $MBe_2Si_2O_7$ (M = Sr or Ba) yielded phosphors which exhibited both broad $4f$–$5d$ band emission and sharp $4f$–$4f$ line emission (163). Eu(II) luminescence within the monoclinic phase of ferromagnetic Eu_2SiO_4 has been reported, exhibiting intense yellow emission characterized by short decay times (164).

The luminescence properties of several lanthanide-activated ternary silicate systems has also been studied. Efficient PL phosphors have been obtained through the Eu(II) activation of $BaZrSi_3O_9$, with an emission maximum being observed at 475 nm (165). The family of feldspars, $MAl_2Si_2O_8$ (m = Ca, Sr, or Ba), can be activated by Eu(II), with the resulting emission maxima ranging from 370 to 425 nm (166). Reasonably efficient red phosphors have been obtained through Eu(III) activation of the silicate oxyapatites $M_2R_8Si_6O_{26}$ (M = Ca or Mg, and R = Gd, La, or Y) (167). Pr(III) has been shown to yield strong ultraviolet $5d$–$4f$ emission and weak $4f$–$4f$ emission in $BaR_4Z_5O_{17}$ (R = Y or Gd and Z = Si or Ge) (168). In this system, the $5d$–$4f$ emission was ascribed to Pr(III) on the R sites, and the $4f$–$4f$ emission to Pr(III) on the Ba sites.

Lanthanide activation of alkaline earth halosilicates has also been reported. Activation of $Ca_3SiO_4Cl_2$ by Eu(II) yielded a material exhibiting an emission maximum at 510 nm and a quantum efficiency of about 25% (169). Hetereogeneous halide-silica phosphors containing about 90% SiO_2 and 10% $CaCl_2$ can be activated by Eu(II), or Eu(II) plus Mn(II) (170). In these materials, the emission took place from small segregations of luminescent halides dispersed in larger nonluminescent SiO_2 particles. $RSiO_3X$ (R = La or Y and X = F, Cl, or Br) can be activated by trivalent lanthanide ions or Eu(II) (171). $Ba_2SiO_6Cl_6$: Eu has been shown to yield a phosphor characterized by a small Stokes shift and narrow bandwidth and by a high quenching temperature (172).

Several studies focusing exclusively on germanate-based host systems have appeared. $Ba_2MgGe_2O_7$ has been activated by Eu(III), Tb(III), and Nd(III) in both crystal and glass matrices, and crystal field analyses reported for the observed sharp lines (173). Eu(III) activation of Ca_2GeO_4, $CaGeO_3$, and Ca_3GeO_3 has been demonstrated, and efficiencies approaching that of Y_2O_3 : Eu have been achieved (174). Detailed analysis of the Eu(III) luminescence within $Bi_4Ge_3O_{12}$ enabled a determination of the site symmetry of the Eu(III) ion as being C_{3v} (175). In a similar study, the site symmetry of Eu(III) in $Bi_2Ge_3O_9$ was found to be C_3 (176). In neither of these latter two studies was the phosphor efficiency estimated, and the primary focus of the work was in the determination of crystal field parameters.

5.3.2.4. Phosphate Host Systems

As mentioned earlier, halophosphate phosphors (based on variations in the apatite composition) are certainly the most important lamp phosphors. The activation of these by lanthanide ions is generally restricted to specialized lamp applications, although a wide variety of studies involving many phosphate systems have been reported.

Eu(II) activation in alkaline earth pyrophosphates, $M_2P_2O_7$ (M = Ca or Sr), has been found to yield efficient phosphors emitting in the deep-blue or ultraviolet spectral regions (177). In addition, energy transfer from Eu(II) to Mn(II) is possible in these systems, yielding phosphors emissive in the red. Orthophosphate systems can also be activated by Eu(II), and the emission maxima exhibit shifts with the structure of the host system. The luminescence maximum of α-$Ca_3(PO_4)_2$: Eu was found at 480 nm, while the corresponding maximum for the β-structure was located at 415 nm (178). A variety of phases can be obtained for $(Mg,Sr)_3(PO_4)_2$, and Eu(II) activation of these has yielded phosphors exhibiting emission maxima between 395 and 440 nm (179). The phase diagrams between all possible alkaline earth oxides and P_2O_5 have been investigated, and Eu(II) activation has been achieved in a wide variety of orthophosphate and pyrophosphate compositions (180–182).

The binary lanthanide phosphates, RPO_4, where R = La, Y, Gd, or Lu, form good host systems for phosphor work. Structurally, these are related to monazite, which is basically $(Ce,La)PO_4$. Essentially any of the trivalent lanthanide ions can be used to prepare phosphors emitting in any desired spectral region, with Gd(III), Eu(III), and Tb(III) yielding the most efficient photoluminescence (183). Vacuum UV excitation spectra of the RPO_4 materials have been reported, and a narrow band at the absorption edge (145–160 nm) has been assigned to an intramolecular transition of the phosphate anions (184). These conclusions indicate that matrix absorptions (except when usng the 180-nm Hg line) are not highly effective in producing activator luminescence. In $GdPO_4$, one can observe energy transfer among activators, with studies on Dy(III) sensitization by Bi(III) and Sb(III) having been reported (185).

A variety of different phases can be prepared by the partial or full replacement of the trivalent ion in MPO_4 materials by a M′(I)/M″(II) pair. In this manner, an efficient red phosphor having the composition $LiBaLaPO_4$: Eu(III) was reported (186). The Eu(II) activation of $M'M''PO_4$ (where M′ = Li, Na, or K and M″ = Ca, Sr, or Ba) has been demonstrated,

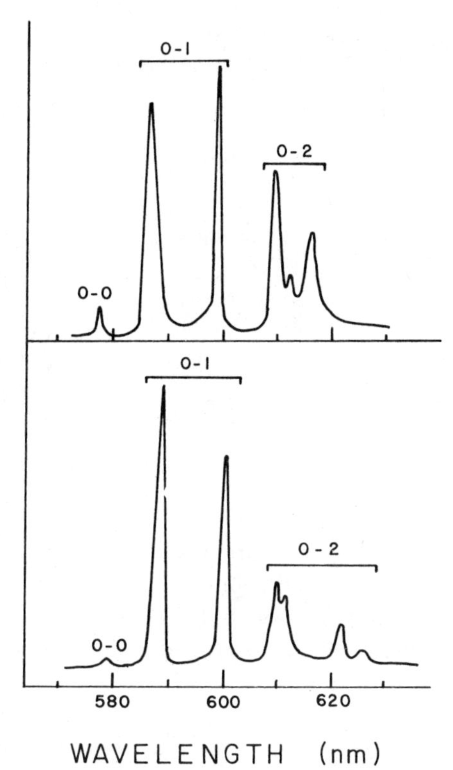

Fig. 5.7. Eu(III) luminescence spectra obtained from $NaSrEu(PO_4)_2$ (upper trace) and $KCaEu(PO_4)_2$ (lower trace), illustrating the effect of site symmetry on the observed luminescence. (Adapted from data reported in ref. 190.)

and emission maxima ranging from 430 to 505 nm were reported (187). Ce(III) activation of various phases in the Li_2O–SrO–P_2O_5 system yielded efficient blue phosphors, and these may be used in turn to produce red phosphors through energy transfer from Ce(III) to Mn(II) (188). Energy transfer from Ce(III) to Tb(III) has also been demonstrated in similar systems (189). Studies of crystal field splittings observed in Nd(III), Eu(III), and Gd(III) luminescence spectra when these were substituted in $(Na,Sr)_3(PO_4)_2$ indicated the presence of particularly large interactions (190). Two examples of Eu(III) luminescence as affected by different host structures are shown in Fig. 5.7.

Other phosphate host systems can be formed which are based on anionic lanthanide phosphates. Compounds of the general formula $M_3R(PO_4)_3$ (where M = Sr or Ba and R = La, Nd, Gd, Y, or Lu) can be activated by Ce(III) to yield ultraviolet phosphors (191). Tb(III) activation of $LiYP_4O_{12}$

yields efficient green phosphors, while Eu(III) activation was found to yield good red phosphors (192). In this study it was noted that energy transfer from Ce(III) to Tb(III) could take place in both $LiCeP_4O_{12}$:Tb and CeP_3O_9:Tb. Efficient phosphors having the compositions $Na_3Ce(PO_4)_2$:Tb and $Na_3La(PO_4)_2$:Gd have been used as green and UV phosphors, and energy transfer from Gd(III) to Tb(III) has been followed in $Na_3Gd(PO_4)_2$:Tb (193). The luminescence of Ce(III)-, Eu(III)-, and Tb(III)-activated $LiLaP_4O_{12}$ has been reported, and interestingly only the Ce(III) phosphor was found to suffer from concentration quenching (194). The depreciation of $Na_3Ce(PO_4)_2$:Tb in low-pressure discharge lamps has been studied in detail, and it was established that the loss in phosphor efficiency was due to ion-electron recombination processes (195).

It is surprising to note that although halophosphate phosphors are exceedingly important to the lamp industry, lanthanide activation of these has received relatively little attention. $Sr_3(PO_4)_2Cl$ and $Ba_3(PO_4)_2Cl$ can be activated by Eu(II), and a wide variety of spectral properties were obtained depending on the material formulations (196). In a search for better yellow-emitting phosphors, Ce(III)- and Mn(II)-activated fluoroapatites were prepared and studied (197). Although good efficiencies were obtained, the quality of the phosphor materials was found to degrade as a result of the lamp-making process.

5.3.2.5. *Molybdate and Tungstate Host Systems*

Although lanthanide-activated molybdate and tungstate phosphors have not found extensive commercial use, studies on these materials have yielded information important to the characterization of lanthanide ion luminescence. Much of the earliest work was performed by Van Uitert and co-workers at Bell Laboratories. The intensity of Eu(III) or Tb(III) luminescence was shown to vary with the electrostatic binding energy in tungstates, and increasing the polarization of the lattice oxygens appeared to decrease the total binding energies of the ions (198). That energy transfer could take place between lanthanide ions was first realized in tungstate host systems having the scheelite structure (199), and the mechanism of this transfer was investigated in great detail for a variety of ions and lattice structures (200–201). The results of these investigations indicated that dipole–dipole and dipole–quadrupole interactions predominate among activators doped in the scheelite structure (202).

In Z_2WO_6 or Z_2MoO_6 compounds (Z = any trivalent ion), any lanthanide ion can replace the Z(III) species without needing charge compensation. Intrinsic host emission may be observed, and energy transfer from tungatate ions to lanthanide activators has been observed (203). Interestingly, in $Y_2(W,Mo)O_6$: Eu the molybdate group acts as a fluorescence killer under short-wavelength UV excitation, but as a sensitizer under long-wave UV excitation. The R_2MoO_6 : Eu(III) (where R = La, Gd, or Y) system has been examined in great detail, with a C_2 site symmetry being deduced for the lanthanide ion in each case (204). The energy levels obtained for La_2MoO_6 : Eu appear to be similar to those observed in oxyhalides, and in the other compounds Eu(III) occupancy of multiple sites was observed. The luminescence properties of $(Bi,La)_2WO_6$ as activated by Pr, Sm, Eu, Tb, Dy, or Er have been reported, and in several instances energy transfer from the tungstate group to the activator ion was observed (205). The trivalent cation in most hosts can also be replaced by a M(I)/M(II) couple, and consequently the luminescence properties of Sr_2NaWO_6 : R and Sr_2NaUO_6 : R (R = La, Gd, or Eu) have been obtained (206).

A wide variety of polymeric molybdates and tungstates exist, and activation of these materials by lanthanide ions usually yields strongly emissive compounds. The low-temperature luminescence in Eu(III)-doped tungstates of the perovskite stacking-variant type ($Ba_2La_2Mg(WO_6)_2$ and $Ba_6Y_2(WO_6)_3$) has been obtained (207). These are layered compounds, and the luminescence spectra and energy transfer parameters obtained from 12- and 18-layered materials were found to be strikingly different. Further studies of the 18-layered compounds were carried out for the $Ba_6RW_3O_{18}$ system (R = Y, Er, or Yb) when employing Ho(III) as an activator ion (208). Considerable enhancement of Ho(III) infrared emission was obtained through Er–Ho or Yb–Ho energy transfer. The position of the Eu(III) charge transfer was found to have a profound effect on the emissive properties of $Sr_8SrGd_2W_4O_{24}$: Eu, since the most efficient excitation process occurred primarily through this band and not through energy transfer from the tungstate group (209).

Additional studies have been performed on lanthanide-activated tungstates having the Scheelite structure. The luminescence associated with Tb(III)-, Dy(III)-, and Eu(III)-activated $MMoO_4$ or MWO_4 (M = Ca, Sr, Pb, or Cd) phosphors has been detailed with respect to the infuence of charge compensation on the emission properties (210). The Eu(III) luminescence of doped $M_2(WO_4)_3$ (M = Al, Sc, or Lu) materials proved to be sensitive to the hydration state of these hygroscopic compounds, and a large number of defect sites greatly complicated the analysis (211). The luminescence properties of lanthanide-activated and unactivated $Gd_2(MoO_4)_3$, as well as those of $Eu_2(MoO_4)_3$ and $Tb_2(MoO_4)_3$, have been reported (212). Energy transfer from the molybdate groups to the lanthanide activators was observed, and in $Eu_2(MoO_4)_3$ Eu–Eu energy migration was not observed [although such migration could take place in $Tb_2(MoO_4)_3$]. Site-selective spectroscopy was used to study the various Eu(III) sites in $M_3MoO_4)_3$ (M = La, Bi, or Gd) and $Gd_2(WO_4)_3$ (213).

The luminescence properties of lanthanide-activated halotungstates have been investigated in great detail. The RWO_4Cl series of compounds (R can range from La to Sm) all crystallize in orthorhombic symmetry, with the W atom being in the form a trigonal bipyramid (214). The luminescence of Sm(III), Tb(III), and Eu(III) activators has been studied in the $LaWO_4Cl$ host material (215), and tungstate sensitization of the activator emission was found to take place (216). In $GdWO_4Cl$, the W atom is tetrahedrally coordinated, and the luminescence of Eu(III) within this material was found to reflect this structural difference (217). In $La_3WO_6Cl_3$, the W atom is found coordinated as a trigonal prism, and activation by Sm(III), Eu(III), Tb(III), Dy(III), and Tm(III) is possible (218). Energy transfer from the tungstate group to the lanthanide activator ions was found to be inefficient in this system, largely due to the localized nature of the tungstate excitation (219).

5.3.2.6. Group VB Oxidic Host Systems

The members of group VB (V, Nb, or Ta) form a variety of oxidic structures, and the luminescence properties of lanthanide-activated materials have received a great deal of attention. The most important host systems have contained the vanadate ion, and a significant amount of work was discussed earlier when considering the cathodoluminescent phosphor materials (Section 5.3.1.1). However, the inertness of the vanadate systems has led to additional applications as lamp phosphors, and the studies concerning the latter aspect will be examined now.

Activation of YVO_4 by Bi(III) yields efficient blue phosphor materials, and co-doping the Bi with Eu(III) results in substantial Bi–Eu energy transfer (220,221). A general survey of the photoluminescence associated with RVO_4 materials (where R = Y, Gd, or La) was performed, and it was learned that Sm(III), Eu(III), and Dy(III) activation yielded the most efficient phosphors (222). The efficiency of these materials was found to be due to energy transfer from the host to the activator ion, and the YVO_4 : Dy phosphor was noted to be particularly efficient. These YVO_4 : Dy materials have become very important as yellow phosphors, and consequently their efficiency has been evaluated for a broad range of host materials and operating conditions (223).

The quenching of Dy(III) emission by Eu(III) tin the YVO_4 host was not found to be important until the Eu(III) concentration exceeded that of the Dy(III), but Tb(III) quenching of the Dy(III) emission was found to be

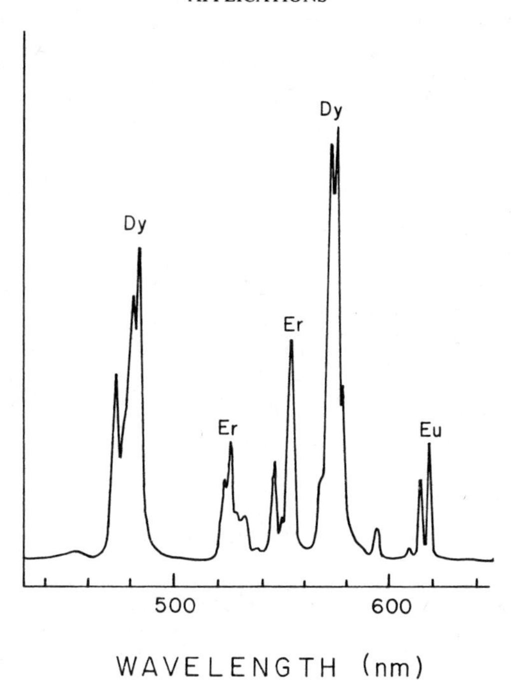

Fig. 5.8. Luminescence spectrum of YVO_4: Dy,Er,Eu. Although the color of a blended phosphor was attained using this single-host system, the gaps in the emission spectrum did not yield good color rendition. (Adapted from data reported in ref. 225.)

extremely efficient (224). Since the Dy–Tb energy transfer is much closer to resonance than that of Dy–Eu, these observations are quite reasonable. Multiple lanthanide ion activation of YVO_4 materials has been used to develop phosphors which yield essentially white photoluminescence (225). In these compounds, blue, green, and red ion luminescence all take place simultaneously, as has been illustrated in Figure 5.8. A mixed YVO_4 : Eu,Tb has been found to function as an efficient high-pressure mercury lamp phosphor (226). An energy transfer pathway consisting of the vanadate–Tb–Eu sequence was used to explain the high efficiency of the phosphor. The luminescence yield of YVO_4 : Eu phosphors can be enhanced by using hetereogeneous materials, in which the emissive particles are insulated by vitreous silica (227).

The analogous $RNbO_4$ and $RTaO_4$ compounds crystallize in the Fergusonite structure (for R = Nd to Er), and the luminescence of lanthanide activators differs in this host system from that in the vanadate systems. In $RTaO_4$ (R = La, Gd, Y, or Lu) it was found that a NbO_4 center could be an efficient emitter, and that $GdTaO_4$ suffered complete concentration quenching (228). Eu(III) and Tb(III) activation was found to yield efficient phosphors, and it was found that Eu(III) could accept energy from the TaO_4 centers through its charge-transfer band. The energy transfer phenomena associated with $GdTaO_4$: Tb has been studied, and it was learned that the population of Tb(III)-excited ^5D_J states depended on the excitation energy and the energy transfer mechanism (229). A different crystal structure can be obtained for $RTaO_4$ (where R = Tm to Lu), and lanthanide activation of materials having this M′ structure was found to be superior to that of those having the Fergusonite structure (230).

The niobate and tantalate polymers also form good phosphor hosts which accept lanthanide ion activators. The luminescence of Eu(III)-activated compounds having a general formula of RM_2O_6 (R = Y, La, or Gd and M = Ti and Nb, Sb, or Ta) has been reported and the resulting spectra explained in terms of the known crystal structures (231). Perturbations in the luminescence of Sm(III) and Eu(III) $KTaO_3$ (as induced by the addition of Nb) were used to determine the nature of the site substitution associated with the activators (232). Essentially all of the lanthanide ions can be used to activate (Ca,Cd)Nb_2O_6 (233), and cryogenic high-resolution spectral studies were subsequently used to study the defect structures in the Eu(III)-activated materials (234). $NaCa_2Mg_2V_3O_{12}$ has been shown to have the garnet structure, and replacement of a Ca/Mg pair by a Eu/Li pair yields strong red luminescence (235). $Li_{14}ZnTa_4O_{18}$ can also be activated by Eu(III), and the presence of multiple emitting sites was deduced from the broadness of the observed bands (236). A large number of polycrystalline tantalate–based oxides were considered as host materials for Nd(III) and Eu(III) emission, and aspects of synthesizing efficient phosphor materials have been discussed (237).

5.3.2.7. *Sulfate-Based Host Systems*

The luminescence properties of lanthanide-activated oxygen–sulfur compounds have already been mentioned in conjunction with the cathodoluminescent phosphors. The emissive properties of Eu(III)-activated oxysulfide phosphors are extremely favorable, and the brightness of this phosphor system is essentially unmatched by any other red phosphor. As part of the initial work on this system, the luminescence properties of a variety of other Eu(III)-activated sulfur–oxygen compounds were examined, and it was found that only the oxysulfate materials yielded phosphors of sufficient brightness (238). These workers also concluded that neither hydrated nor anhydrous sulfates could yield sufficient Eu(III) luminescence to be considered further.

The luminescence properties of lanthanide-activated oxysulfate phosphors have been studied in detail. The introduction of Eu(III) into $R_2O_2SO_4$ host materials (where R = La, Gd, Y, or Lu) yields phosphors of relatively low CL efficiency, but which emit quite well under short-wave UV excitation (239). Activation by Eu(III) or Tb(III) gives efficiently emitting materials, but Ce(III) activation produces phosphors of low efficiency. Site symmetries and crystal field parameters were deduced from extremely careful studies on $R_2O_2SO_4$:Eu (where R = La, Gd, or Y) (240). In the thermal decomposition of hydrated lanthanide sulfates, one forms anhydrous sulfates, oxysulfates, and finally oxides. These steps were followed through studies of the Eu(III) luminescence obtained at each stage in the thermal decomposition of $Eu_2(SO_4)_3 \cdot 8H_2O$, and the presence of a new oxysulfate [$Eu_2O(SO_4)_2$] was deduced from the spectra (241). Representative examples of Eu(III) spectra obtained from the various compounds are located in Fig. 5.9.

Although the anhydrous sulfates do not make good phosphor hosts, alkaline earth sulfates show more promise. The $4f$–$5d$ excited state of Eu(II) was raised sufficiently high in energy when this ion was substituted in $BaMg(SO_4)_2$ so that the fine structure associated with the $4f$–$4f$ emission could be observed (242). The decay time of this emission was also found to be exceedingly short relative to the lifetimes normally observed for Eu(II) in sulfate host systems. The luminescence properties of $NaR(SO_4)_2 \cdot H_2O$ (where R = Ce, Sm, Eu, Dy, Gd, or Tb) have been detailed, but only the Ce, Gd, and Tb materials were found to show efficient emission (243). This situation was explained in terms of an energy gap law, since the host lattice was characterized by high-energy vibrations.

Tellurate materials have been found to form good host systems for phosphors. R_2TeO_6 (where R = La, Gd, Y, or Lu) can be activated by Eu(III), and $LaTeO_6$ could be activated by Dy(III), Ho(III), and Er(III) (244). The fine structure within the Eu(III) luminescence was used to evaluate the structural differences among the various host systems. Alkaline earth tellurate perovskites of the general formula $M'M''TeO_6$ (where both M′ or M″ can be Ba, Sr, Ca, or Cd) can be activated by Eu(III), Tb(III), Dy(III), or Er(III) along with Na(I) as a charge compensator (245). The Eu(III) luminescence was found to be particular sensitive to changes from the ordered to disordered perovskite structure, and only the Ca_2MgTeO_6 host was found to be effective in promoting activator luminescence.

5.3.2.8. Group IVB Oxidic Host Systems

Several investigations have considered the oxidic systems based on members of group IVB, Ti, Zr, Hf. One could also consider Ce and Th host materials as being part of this group, although most workers would consider Ce as an activator species and not a host cation. Eu(III) activation of $NaRTiO_4$

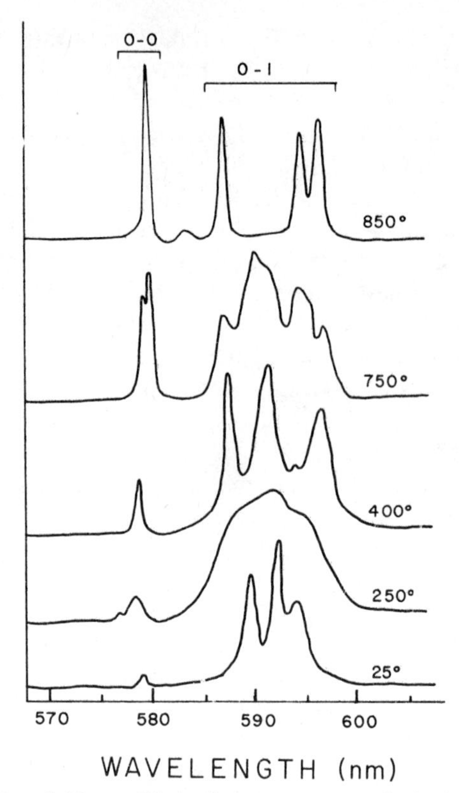

Fig. 5.9. $^5D_0 \rightarrow {}^7F_0$ and $^5D_0 \rightarrow {}^7F_1$ luminescence spectra obtained during calcination of hydrated europium sulfate. The 25°C spectrum corresponds to the $Eu_2(SO_4)_3 \cdot 8H_2O$ starting material, while the 250°C spectrum is indicative of partial dehydration. The 400°C spectrum is that of anhydrous $Eu_2(SO_4)_3$. At 750°C, the observed spectrum exhibits features due to both $Eu_2O(SO_4)_2$ and $Eu_2O_2SO_4$, while at 850°C only the emission of $Eu_2O_2SO_4$ can be detected. (Adapted from data reported in ref. 241.)

materials (where R = La, Gd, Y, or Lu) has been reported, and the relative intensities of the Eu(III) emission were found to depend on the identity of the R ion (246). Although reasonable efficiencies were obtained, the anisotropic nature of the crystal structure (and the orientation of the UV-absorbing TiO_6 groups) was not conducive toward strong Eu(III) luminescence. The energy migration in $NaGdTiO_4$: Eu was further examined, and transfer from intrinsic Eu(III) ions to extrinsic Eu(III) traps was observed (247). The characteristics of the energy migration processes were explained using a two-dimensional diffusion model, and the temperature dependence was explained assuming phonon-assisted energy transfer.

The Eu(III) luminescence spectra of lanthanide titanate, titanate–

stannate, stannate, hafnate, and zirconate pyrochlores, $A_2B_2O_7$, indicated that a linear relationship existed between the splitting of the $^5D_0 \rightarrow {}^7F_1$ transition and the size of the cation present (248). Small A(III) cations and larger B(IV) cations both produced a lattice with a smaller degree of deviation from octahedral symmetry. The Eu(III) luminescence in monoclinic ZrO_2 (baddeleyite) has been reported, and only the lowest doping level (0.1%) was sufficient to prevent the occupation of defect sites by the Eu(III) ions (249). The Eu(III) and Tb(III) activation of $R_2Hf_2O_7$ materials (R = all lanthanides, plus Sc, Y, and La) has been reported (250). While the compounds for which R = La to Tb are ordered, those for Dy to Lu crystallize in a disordered calcium fluoride structure.

The slight radioactivity associated with thorium has essentially prevented its use in phosphor systems, but this situation is unfortunate since Th(IV) normally yields host materials capable of lanthanide ion sensitization. The Ce(III) luminescence in YPO_4 : Ce can be greatly improved by the addition of small amounts of Th (251). Substitution of lanthanide ions into ThO_2 enables the direct determination of sub-ppm quantities of these metals by means of x-ray-excited optical fluorescence (252). The luminescence of ThO_2 : Eu can also be produced through suitable catalytic reactions taking place on the oxide surface, and such catala luminescence can be used to follow the course of chemical oxidations (253). The various Eu(III) sites existing in ThO_2 have been characterized by site-selection spectroscopy: One cubic site and two charge compensated sites were identified (254). Hydrolysis of aqueous Th(IV) ions results in the precipitation of hydrated thorium oxide, and this material exhibits photoluminescence at cryogenic temperatures (255). This luminescence is due to the presence of defect sites, and may be quenched by doped Eu(III).

5.3.3. Lanthanide-Activated Phosphors Used in X-Ray-Intensifying Screens

The passage of x-rays through an object produces a radiologic image which must be converted into an optical image by some process. In most applications this is performed through the use of a luminescent intensifying screen. The modulated x-rays are absorbed by the screen and produce visible photons, and this emitted light is used to develop a plate containing photographic film. The assembly of a typical cassette is shown in Fig. 5.10, and one may consider the phosphor screen to be acting as an amplifier and imager for the x-ray photons.

Initially, the blue luminescence of x-ray-excited $CaWO_4$ was used as the screen phosphor. In an effort to reduce patient exposure, workers have continually sought to lower the intensity of x-ray irradiation. This can be achieved by improving the absorption of x-ray photons and by improving the

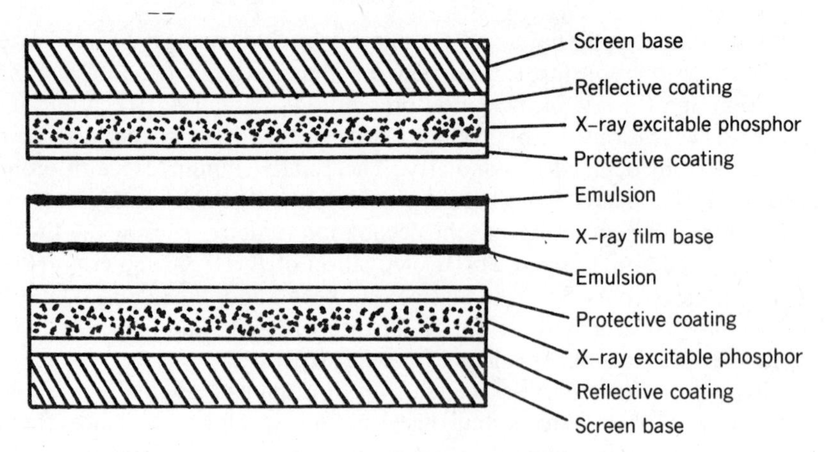

Fig. 5.10. Schematic diagram of a typical double-screen-film X-ray intensifying screen.

quantum yield of luminescence within the phosphor screens. As a result of these investigations, the traditional $CaWO_4$ materials have been replaced (for the most part) by lanthanide-activated oxysulfide, oxyhalide, or halide host materials. The gree-to-blue emission of the R_2O_2S: Tb(III) phosphors (R = La, Gd, or Y) can be extremely efficient under x-ray excitation, and the Gd phosphor features particularly high absorptivities for x-ray photons.

Since the luminescence properties of lanthanide-activated oxysulfide phosphors were discussed in detail during the consideration of cathodoluminescent phosphors (see Section 5.3.1.2), further review of these particular compounds will not be duplicated here.

5.3.3.1. Oxyhalide Host Materials

The most important oxyhalide phosphors which have found use in x-ray-intensifying screens are LaOBr : Tb and LaOBr : Tm (256). The luminescence associated with Eu(III)-activated lanthanide oxyhalides, ROX (R = Y, La, or Gd and X = Cl or Br) was investigated in an attempt to learn more about this host system (257). All of the ROX materials except for the fluorides were known to be isomorphous (having the PbFCl structure) and hence form an isostructural series suitable for systematic work. It was found that the spectral energy distribution of the Eu(III) emission depended strongly on the halide atom, and to a lesser extent on the choice of lanthanide ion. These variations could be accounted for in an qualitative fashion by assuming a strongly varying linear component in the crystal field (257). A similar investigation was carried out by the same workers examining the properties of ROX : Tb phosphors, and it was learned that the dependence of spectral energy distribution upon the Tb(III) surroundings was much weaker than had been noted in the case of Eu(III) (258). This difference was related to a lack of suitable charge transfer bands in the UV region for the Tb(III)-activated materials.

In a series of reports, Rabatin has studied the excitation and luminescence properties of a series of lanthanide-activated oxybromide phosphors. LaOBr : Tb phosphors are efficient under UV (259), cathode ray, and x-ray excitation (260). LaOBr : Tm phosphors are efficient under cathode ray and x-ray excitation, and exhibit broad-band emissions around 300 and 370 nm (261). LaOBr activated by Ce(III) (262) or Dy(III) (263) can also be relatively efficient under cathode ray and x-ray excitations. Due to the lanthanide contraction, the LuOBr materials are less stable than the LaOBr materials, but it has been found that Tb(III), Tm(III), Ce(III), Sm(III), or Dy(III) activation of LuOBr yields phosphors whose behavior is very similar to the corresponding LaOBr phosphors (264).

The concentration quenching in LaOBr : Tb phosphors has been studied, with dipole–dipole and dipole–quadrupole interactions being implicated in the energy-transfer phenomena (265). The blue luminescence of LaOBr : Ce may be quenched by small amounts of codoped Tb(III), indicating that Ce–Tb energy transfer is an efficient process in the oxybromide host (266). In a similar study, the quenching efficiency of a series of lanthanide ions on Ce(III) emission in LaOBr : Ce was examined, and it was noted that Sm(III), Eu(III), Tm(III), and Yb(III) all quenched the Ce(III) emission without the appearance of activator luminescence (267). Crystal field parameters were evaluated for ROBr : Eu (R = Y, La, Gd) and LaOI : Eu, and it was noted that the magnitude of the crystal field appears to increase with the ionic radius of the R host cation (268). The crystal field associated with LaOI : Eu is essentially the same as for LaOBr : Eu.

Although the oxychloride host system has not been used in the x-ray-intensifying screens, a variety of spectroscopic studies have been carried out which have contributed to an understanding of the dynamics of the oxyhalide systems in general. LaOCl, LuOCl, and ScOCl all exhibit different crystal structures, and activation of each sytem has been achieved using Sm(III), Eu(III), Tb(III), Dy(III), and Tm(III) (269). The activator ion emission spectra all accurately reflect the structural changes associated with the host systems. Crystal field parameters have been obtained for the YOCl : Eu(III) (270) and ROCl : Tb(III) (where R = La, Gd, or Y) (271) systems, all indicative of the existence of rather strong crystal field strengths. An anti-Stokes luminescence has been observed upon dye laser excitation of LaOCl : Eu; direct pumping of the $^5D_0$ level led to the observation of luminescence out of the higher $^5D_{1,2,3}$ excited states (272).

5.3.3.2. Alkaline Earth Halide Host Materials

Lanthanide ion-doped alkaline earth fluorides have been of extreme historical interest due to their unusual defect equilibria. As the concentration of the dopant is lowered, the fraction of lanthanide ions experiencing local fluoride interstitial charge compensation increases relative to those with distant compensation (273). This behavior is exactly opposite to that expected from the conventional models for defect behavior. CdF_2 is a related and interesting compound, since suitable treatment procedures can yield either insulating, near-insulating, or semiconductor materials. Activation of CdF_2 by lanthanide ions yields luminescence spectra characterized by cooperative phenomena (274,275) and interionic energy transfer (276). The nature of the defect structures has been examined through the use of site-selective spectroscopy in CdF_2 : Eu (277).

The optical studies of lanthanide ion–activated CaF_2 materials have normally focused on obtaining details of the defect structure. Unless extreme care is exercised, the doping of Eu(III) into CaF_2 can yield a variety of europium–oxygen complexes (278). The photophysical properties associated with Pr(III) (279–281)-, Sm(III) (282)-, Eu(III) (273)-, and Er(III) (284,285)-activated CaF_2 have been used to study the nature of the defect sites. In addition, it has been shown that Eu(II) will sensitize the luminescence of Tb(III) in CaF_2 (286), and that Yb(III) can receive energy from Mn(II) (287).

It has now become established that BaFCl : Eu(II) is the most promising phosphor for working with standard blue-sensitive x-ray films (288). Detailed structural and fluorescence data have been reported for EuFCl, an material isostructural with SrFCl (289). At room temperature, an emission maximum at 416 nm was observed, and no $4f$–$4f$ luminescence was noted. This situation may be contrasted to that of $BaCl_2$ doped with Eu(II), where both broad $5d$–$4f$ and sharp $4f$–$4f$ emission bands were observed (290). A detailed investigation of the luminescence properties of MCl_2 (M = Ca, Ba, or Sr) and MFCl (M = Ba or Sr) doped with Eu(II) has been carried out in an attempt to determine whether broadly tunable lasers could be developed using the broad $5d$–$4f$ emission (291). It was determined that while the laser threshold could easily be achieved, strong excited-state absorption prohibited laser oscillations.

Recently, a completely new radiographic technique has been reported (292). In Fuji-computed radiography, the x-ray patterns are converted into an image by means of photostimulable phosphors. The materials used are BaFX : Eu(II) (where X = Cl, Br, or I) phosphors, where the photostimulation is caused by the liberation of electrons trapped at F centers created by the x-ray exposure. These F centers correspond to trapped electrons in F^-

and Cl^- ion vacancies. The patterns are converted into digital signals by means of He–Ne laser scanning of the image plate. The photostimulatable emission peaks were located between 385 and 405 nm, while the photoluminescence peaks were found between 430 and 630 nm. The response of the photostimulated luminescence intensity to the exposing x-ray energy showed good linearity in more than a 10^5 range of x-ray dosage (292).

5.3.4. Electroluminescent Lanthanide Phosphors

The desirability of developing flat panel display devices has spurred the development of phosphors stimulable by the passage of electric current. These electroluminescent (EL) devices use either thin or thick films, and the design of the two devices varies somewhat with the film thickness (see Fig. 5.11 for a comparison). The thick-film devices have shorter lifetimes and lower intensities of emission but are easier to fabricate than are the thin-film devices. Both types of devices can operate with either dc or ad stimulation, but the ac devices have been found to exhibit greater efficiency. The most commonly used EL phosphor is yellow-emitting ZnS : Mn. Depending on the mode of preparation, ZnS : Cu phosphors can be produced which exhibit EL in either green or blue colors, although the efficiencies of these are much less than those of the ZnS : Mn phosphors.

The use of lanthanide activators in EL phosphors has been relatively limited. The introduction of Eu(III) into CaS : Cu EL phosphors led to the observation of red luminescence, and at moderate doping levels the blue emission of CaS : Cu was completely suppressed (293). The difference between EL and host-excited photoluminescence was examined in thin films of ZnS : Tb, and substantial differences in concentration quenching associated with the two types of emission processes were observed (294). EL has been measured for the $5d$–$4f$ emission of Ce(III) in ZnS : CeF_3 thin films, where it was noted that the EL emission maxima shifted from green to bluish white as the concentration of Ce(III) increased (295). The presence of strong Ce–Ce interactions was used to account for the low EL efficiencies.

Zinc chalcongenides may be doped with lanthanide ions to yield EL phosphors. The EL excitation of ZnSe doped with lanthanide trifluorides (RF_3, where R = Nd, Sm, Tb, Dy, Ho, Er, Tm, or Yb) has been studied at both room and liquid-nitrogen temperatures (296). Electroluminescence was observable with all activators, with the exception of Eu. The EL behavior of Er(III)- and Tm(III)-doped ZnSe light-emitting diodes has also been studied, and important differences in carrier generation mechanisms and phonon-assisted excitation processes were noted (297).

EL behavior has also been studied in oxidic host materials. Y_2O_3, La_2O_3, and Y_2O_2S can each be activated by Eu(III), Tb(III), or Tm(III), with the

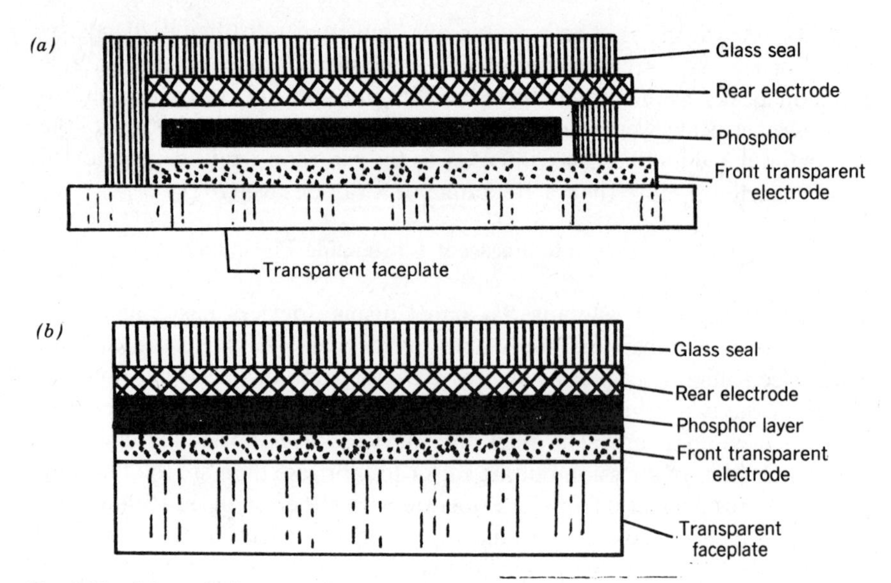

Fig. 5.11. Schematic diagrams of (a) a thin-film electroluminescent device (drawing) and (b) a thick-film electroluminescent device.

most efficient EL spectra being obtained under dc conditions in the La_2O_3 host material (298). Oxysulfide host materials (Y_2O_2S and La_2O_2S) have been examined, and with the use of Tm(III), Tb(III), and Eu(III) as dopants devices have been produced with emit in the three basic colors (blue, green, and red, respectively) (299). The doping of Dy, Yb, Nd, Pr, Gd, La, Sm, and Er into ZnO has been studied, with both photoluminescence and EL spectra being obtained (300). The characteristic EL of undoped ZnO was substantially modified (and generally quenched somewhat), with a donor–acceptor model being used to explain the experimental results.

5.3.5. Materials Suitable for Infrared-to-Visible Up-conversion

Phosphors capable of converting infrared radiation into visible light are of great importance since they can be combined with solid-state diode lasers to generate useful display devices. These devices could be considered as alternatives to light-emitting diodes which provide visible light directly. This up-conversion is termed an anti-Stokes luminescence process, since the usual Stokes luminescence involves the absorption of high-energy photons and the emission of photons having lower energies.

The production of anti-Stokes luminescence is characterized by two-photon absorption phenomena. The simplest energy-level diagram illustrat-

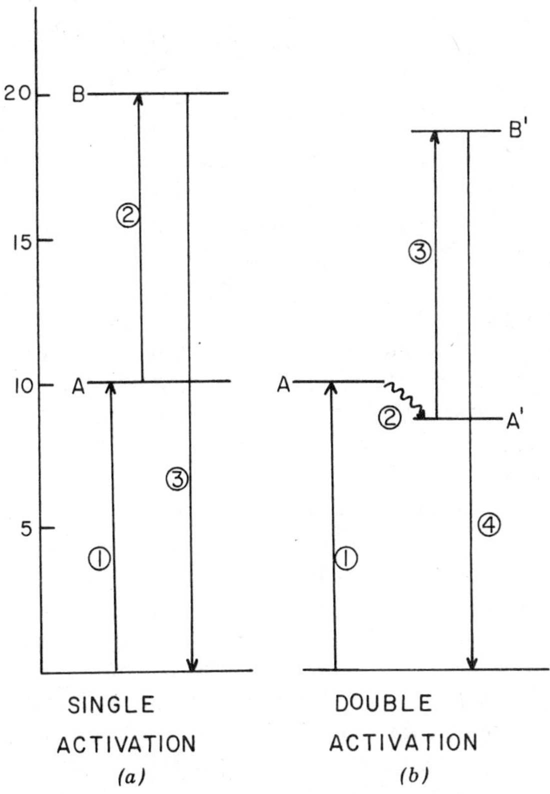

Fig. 5.12. General energy-level schemes for up-conversion luminescence processes. In the single-activator sequence, two infrared photons are sequentially absorbed by the lanthanide ion (steps 1 and 2), and the luminescence then takes place from level B (step 3). In the double-activator sequence, the first activator is excited in step 1. Energy transfer to the second activator takes place immediately (step 2), whereupon another absorption of an infrared photon takes place (step 3). Luminescence finally occurs from level B′ of the second activator in step 4.

ing the conversion of infrared to visible light is provided as Fig. 5.12a. The molecular species absorbs two infrared quanta (steps 1 and 2), populating state B as long as the lifetime of state A is sufficiently long. An efficient up-conversion process requires that the infrared absorptivity of the required transition be as high as possible. The ion returns to the ground state, G, through the emission of a single photon of visible light (step 3). Energy transfer phenomena can also aid the up-conversion process, as has been illustrated in Fig. 5.12b. One ion is raised to its lowest excited state, A (step 1), and energy transfer from donor to acceptor takes place immediately (step 2). The excited state, A′, of the acceptor ion is now raised to its second excited state, B′ (step 3), and emission of visible light takes place when the ion returns from B′ to its ground state (step 4).

Essentially all useful lanthanide-activated up-conversion phosphors have employed the energy-transfer pathway (301). The relatively strong Yb(III) $^2F_{7/2} \rightarrow {}^2F_{5/2}$ absorption at 970 nm is used during the first absorption step, and energy transfer from this level to a suitable activator subsequently occurs. As an example, we will consider Er(III) as the acceptor ion. The $^4I_{11/2}$ state of Er(III) is essentially resonant with the $^2F_{5/2}$ state of Yb(III), and the energy transfer process is very efficient. Absorption of a second IR quantum raises the Er(III) ion to the $^4F_{7/2}$ state. From this level a nonradiative decay to the $^4F_{9/2}$ state takes place, followed by a luminescent $^4F_{9/2} \rightarrow {}^4I_{15/2}$ transition at 660 nm. Alternatively, the $^4F_{7/2}$ state can relax nonradiatively to the $^4S_{3/2}$ state. This level is also capable of undergoing a luminescent transition to the ground state, with light being emitted at 550 nm. Besides Er(III), Tm(III), Pr(III), and Ho(III) have been used successfully as acceptor ions for up-conversion processes. The pioneering work is usually accredited to Auzel, who reported the sensitization of Er(III) and Tm(III) luminescence by Yb(III) and observed the nonlinear dependencies of emission intensitives upon excitation power which are characteristic of two-photon processes (302). In a general survey of fluoride and oxidic host materials, it was found that the decay times of the green and red emissions obtained after IR excitation of Yb,Er-doped compounds were determined by the depopulation of the Yb levels and not by that of the Er levels (303). These workers were also able to estimate an upper limit of about 10% for the up-conversion process.

5.3.5.1. Fluoride Host Materials

The most efficient lanthanide-activated up-conversion phosphors have employed anhydrous fluoride host materials (301). The luminescence of lanthanide ions in these compounds has been studied for quite some time as part of pure spectroscopic studies, with RF_3 (R = La, Y, Gd, or Lu), $AMgF_3$ (A = K, Rb, or Cs), $LiYF_4$, ARF_5 (where A = Ba, or Sr and R = Y or La), KR_3F_{10} (R = La, Y, or Lu), or $A_2R_2F_{10}$ (A = Ca, Sr, or Ba and R = Lu, Y, or La) being used as important host materials. Theories of interionic energy transfer have often been tested in anhydrous fluoride systems (304–309). Detailed probing of up-conversion phenomena were carried out for RF_3 : Yb,R′ materials (where R = La, Y, Gd, or Lu and R′ = Er, Ho, or Tm), for which it was firmly established that the excitation process involved a Yb(III) absorption followed by energy transfer to the R′ ions (310). The effects of multiphonon sidebands on the energy transfers among lanthanide ions was subsequently investigated through excitation spectral studies (311).

The largest number of reported works have concerned Yb,Er-doped fluorides. The excitation mechanisms associated with $NaYF_4$: Yb,Er materials have been examined through 365- and 970-nm excitation spectra, following the intensities of green and red emissions (312). Substantial variation in light output were obtained upon systematic variation of the concentration of lanthanide ion dopants. The $NaRF_4$: Yb, Er system (R = Y, Gd, or La) was investigated in detail, and it was found that this phosphor system was four to five times as bright as the LaF_3 : Yb,Er phosphor (313). A large variation in intensity and color of emission was found for a series of Yb,Er-doped fluorides, trends that could be related to the surroundings of the lanthanide ions in the different crystal lattices (314). The most efficient host materials were found to be RF_3, $BaYF_5$, and α-$NaYF_4$. The efficiency of the best phosphor [α-$NaYF_4$: Yb(0.2),Er(0.03)] was found to be about 6% (315).

Since the method of preparation apparently dictated the efficiency of the up-conversion process, significant efforts have been placed on this aspect. The original preparation involved using the highest-purity fluorides as starting materials, and with the addition of BeF_2 as a flux, encapsulant, and oxygen getter one could obtain well-crystallized grains (316). Fluorides precipitated by aqueous HF were dehydrated in a stream of HF at 900 °C (also converting oxyfluorides to fluorides), and phosphors prepared in this manner functioned as efficient phosphors (317). The fluorides prepared by aqueous HF precipitation could be dried directly if extensive washing with NH_4F was employed after precipitation, and one could thermally dehydrate these without formation of the oxyfluorides (318). That the use of ammonium fluoride could prevent formation of oxyfluorides was investigated through high-resolution luminescence studies of EuF_3 (319). In this work it was learned that two types of lattice water (removable at low and high temperatures) existed, and that the function of the NH_4F was to replace the high-temperature water. Only this water is involved in the pyrohydrolysis, which results in the formation of oxyfluorides.

The blue emission from Yb(III)-sensitized Tm(III) involves three photon-assisted energy transfers from Yb(III), followed by a fast multiphonon relaxation (302,320). Thus it was thought that the intensity of Tm(III) emission would exhibit a strong dependence on the nature of the host material. Examination of a series of fluoride and oxide host systems revealed this to indeed be the case, and the strongest emission was detected in α-$NaYF_4$ (321). The three-photon Yb(III) sensitization of Ho(III) can yield green emission, but the efficiency of this process is rather low (322,323). The dynamics of YOF : Yb,Ho and YF_3 : Yb,Ho phosphors have been studied, and essentially pure green emission at 545 nm can be obtained under certain situations (324). This may be considered as an improvement over the yellow-green output associated with the Yb,Er phosphors for applications where color rendition is crucial.

CdF_2 doped with $(Yb,Er)F_3$ has been shown to function efficiently as an up-conversion phosphor material, capable of being pumped by the 930-nm output of a Si:GaAs diode (325). Either the green or red Er(III) luminescence was obtainable, depending on the Yb(III) concentration. The successive absorption of two IR photons to produce a single emitted visible photon (scheme A in Fig. 5.12) was realized in CdF_2:Er and $(Sr,Cd)F_2$:Er phosphors (326).

5.3.5.2. *Chloride Host Materials*

Studies of up-conversion phenomena in anhydrous chloride host materials have been rather few, presumably due to the relatively poor efficiencies of the visible luminescence. In materials characterized by high-energy lattice vibrations, the activator luminescence is too weak to be observed even when the interionic energy transfer is efficient. Due to this situation, all successful work has focused on halides isomorphous with elpasolite (K_2NaAlF_6), a host material featuring an exactly octahedral lanthanide ion site. Extremely detailed analyses of the luminescence obtained from Cs_2NaYCl_6 activated by Pr(III) (327), Eu(III) (328–330), Tb(III) (331,332), Ho(III) (333), and Er(III) (334) have been reported.

Up-conversion phenomena in Cs_2NaYCl_6:Yb,Er were interpreted in terms of a stepwise energy transfer mechanism (335). While highly efficient materials were not obtained, the Er(III) luminescence was found to increase with the Yb(III) pumping energy to the limit of the tungsten lamp used as the source. Both direct two-photon excitation and energy-transfer processes led to the visible emission (336). It was interesting to note that only the green ${}^4S_{3/2} \rightarrow {}^4I_{15/2}$ Er(III) emission was obtainable, and that the red ${}^4F_{9/2} \rightarrow {}^4I_{15/2}$ emission was not observable at any combination of activator concentrations.

5.3.5.3. *Oxidic Host Materials*

Due to the lower up-conversion yields, oxidic host materials have not been used as up-conversion phosphors to the extent that the fluoride-based phosphors have been used. Nevertheless, the phenomena are observable, and many studies of the conversion of infrared photons to visible photons have been carried out in oxidic hosts. One important difference that exists when comparing fluoride and oxidic materials is that the red Er(III) emission usually predominates in the oxidic host materials.

As mentioned earlier, during the course of work on fluoride-based Yb,Er phosphors, it was found that the preparative mode and composition of the materials could change the red/green emission yields. As a result, the properties of Y_2O_3:Yb,Er and Gd_2O_3:Yb,Er (prepared by extensive calcination of the respective trifluorides) were studied in detail (337). Unlike the fluoride materials where the green Er(III) luminescence normally dominates, in the oxidic host systems the red Er(III) emission was observed to be the most intense. A general survey of Yb (20% doping level) and Er (3% doping level) was carried out in the oxides ARO_2 (A = Li or Na and R = Gd, Y, or Lu), R_2O_3 (R = La, Gd, Y, or Lu), and RBO_4 (R = La or Y and B = P, V, As, or Nb) (338). The up-conversion efficiencies were uniformly lower than those of Yb,Er-doped fluoride materials, and the color of the Er(III) depended strongly on the nature of the host. In a subsequent work, the same group greatly extended the range of oxidic host materials studied to learn more about the influence of the host material on the color of the Yb(III)-sensitized Er(III) emission (339). Trends relating the green-to-red ratio with cationic radius and valence were established, and the effect of Er(III) site symmetry was also discussed.

Exceedingly efficient up-conversion phosphors were obtained using lanthanide ion–doped vitroceramics as the host materials (340). Using a composition of metal oxide = 27.18%, PbF_2 = 67.57%, Yb_2O_3 = 4.85%, and Er_2O_3 = 0.39%, phosphor materials could be synthesized whose efficiencies approached that of YF_3 : Yb,Er. The existence of microcrystalline zones was established, and it appeared that the lanthanide ion dopants concentrated in these regions.

During studies of Y_2O_2S:Yb,Er, an Er(III) excitation maximum at 810 nm was detected when observing the green Er(III) luminescence (341). This band could easily be affected by the inclusion of small impurities, upon where one observed the more conventional Yb,Er excitation maximum at 970 nm. The connection between these two transitions implied the existence of several possibilities for multiphoton lanthanide ion excitation. Detailed studies of the red emission of Er(III) in Y_2O_2S:Yb,Er supported the multiphoton mechanism proposed by Sommerdijk (312), but indicated that the model needed to be modified somewhat to accommodate new information. Evidence was presented indicating that the relaxation of the first photon absorption was dominatly radiative rather than being due to a multiphonon process (342).

IR up-conversion has also been achieved in oxychloride host materials. It was found that while YOCl : Yb,Er crystals yielded the red Er(III) luminescence, crystals of Y_3OCl_7 : Yb,Er emitted primarily at the green Er(III) wavelengths (343). However, the color of Y_3OCl_7 : Yb,Er changed from green to red as the excitation power was increased. Through a different preparative method, Y_3OCl_7 : Yb,Er and YOCl : Yb,Er were obtained without the direct use of Cl_2 or HCl gas (344). It was found that variation of the

Y/Yb ratio in $Y_3OCl_7:Yb,Er$ could drastically alter the green-to-red ratio, with increasing concentrations of Yb favoring the red emission.

5.3.6. Concluding Remarks

The use of lanthanide ion activators in solid matrices has been employed extensively in phosphor applications. Certainly, there would be no satisfactory color television without the use of Eu(III) phosphors. Lanthanide ion luminescence is essential to many specialized phosphor applications; one can only assume that their use will become more extensive as developments continue to be made. The use of the term "lanthanide" rather than "rare earth" has been stressed throughout this chapter since the ample quantities of these materials negates the concept of rarity. Thus it may be anticipated that more applications for lanthanide ions in phosphors will be forthcoming in the near future.

During the course of this review, the materials were generally divided according to the chemistry and application for the phosphor. In the United States, one often hears that phosphors are a "mature" science and that the development of new phosphors need not be carried out any longer. This situation is, of course, only approximately true, since researchers are constantly working to advance the state of the art. However, many new phosphors have been developed in Japan, and this work is proceeding at a furious pace. Unfortunately, in most cases the only mention of these in the open literature is a citation dealing with a patent application. We have compiled a list of the new lanthanide ion–activated phosphors patented by Japanese groups since 1982, and this list is provided as Table 5.2. Presumably, these workers have only begun to realize the scope of their possible applications, and further work on the discovery and characterization of new lanthanide-activated phosphors will continue unabated in the future.

Table 5.2. New Phosphors Patented by Japanese Workers

Material	Japan Kokai Tokkyo Koho JP	Chemical Abstracts Citation
$Y_2SiO_5:Ce,Tb$	81,116,778	96:43701d
$(Y,Gd)OF–MgF_2:Tb$	81,116,779	96:43702e
$Gd_2O_3–P_2O_5–B_2O_3:Eu$	81,115,382	96:43703f
$(Zn,Mg)_3(PO_4)_2$: lanthanides	81,136,875	96:60731f
$Ba_5SiO_4(F,Cl):Eu$	81,115,381	96:77396y
$(Zn,Mg,Ca,Sr,Ba)O–Al_2O_3:Eu$	81,152,882	96:94785s
$SnO_2:Eu$	81,135,590	96:113313e
$(Y,La,Gd,Lu)_2O_3–SiO_2:Ca,Tb$	81,155,283	96:113318k

Table 5.2. (*Continued*)

Material	Japan Kokai Tokkyo Koho JP	Chemical Abstracts Citation
$Gd(B,Ga)O_3 : Eu$	81,155,282	96 : 113319m
$Gd(B,Al)O_3 : Eu$	81,155,281	96 : 113320e
$(Mg,Ba)O–Al_2O_3 : Mn,Eu$	81,152,883	96 : 132982p
$CeAl_3(BO_3)_4 : Tb$	81,163,185	96 : 132984r
$(Sr,Ba,Ca)O–P_2O_5–B_2O_3 : Eu$	81,157,482	96 : 152525v
$YScSiO_5 : Tb$	81,167,783	96 : 152527x
$(Na,La)Ga_2S_4 : Ce$	81,166,284	96 : 152528y
$2Y_2O_3–GeO_2 : Eu$	82, 02,387	96 : 152529z
$Y_2O_3–GeO_2 : Eu$	82, 02,388	96 : 152530t
$(Y,Gd,La,Lu)_2O_2S : Tb,Dy$	82, 13,651	96 : 190460c
$Y_2O_3–7HfO_2 : Eu$	82, 10,677	96 : 190464g
$Al_2O_3–2Y_2O_3 : Eu$	82, 10,678	96 : 190465h
$3Y_2O_3–WO_3 : Eu$	82, 23,676	96 : 190467k
$(Ce,La)PO_4 : Tb$	82, 23,674	96 : 208178d
$(Y,Gd,La)_2O_3–(Zn,Mg,Ca,Sr,Ba)O–SiO_2 : Tb_2O_3–(In,Bi,Ce,Tl)O_3$	82, 30,782	97 : 14661f
$9Y_2O_3–4WO_3 : Eu$	82, 44,695	97 : 31074j
$3BaO–2Y_2O_3 : Eu$	82, 51,784	97 : 82514k
$(In,Y,Sc,La,Gd)BO_3 : Eu$	82, 51,783	97 : 82515m
$4(Sr,Ba)O–(Y,Gd)_2O_3–(Ta,Nb)_2O_5 : Eu$	82, 80,477	97 : 118036x
$(Gd,Y,Ce)_2O_3–3B_2O_3 : Zn,Tb$	82, 83,581	97 : 118048c
$(Y,La,Gd)_2O_3–SiO_2–P_2O_5–Al_2O_3 : Ce,Tb$	82, 90,086	97 : 153760a
$(Sr,Ca,Ba)O–P_2O_5–(Sr,Ca,Ba)(Cl,F,Br)_2–B_2O_3 : Eu$	82,102,984	97 : 191087u
$BaO–(Y,Gd)_2O_3 : Eu$	82,115,483	97 : 191099z
$3[(Sr,Ca,Ba)_3(PO_4)_2]–(Ca,Sr),Cl_2 : Eu$	82,125,285	97 : 227227f
$(Y,Gd,La,Sc)_2O_2S : Tb,Pr$	82,139,172	97 : 227223e
$[(Y,La)_2O_3–(Y,La)F_3]–GeO_2 : Eu$	82,137,381	97 : 227235g
$(Y,Gd,La)_2O_2S : Eu$	82,133,180	98 : 9903c
$(Y,Gd,La)_2O_2S : Tb$	82,133,181	98 : 9904d
$(Mg,Be,Zn,Cd,Ca,Sr,Ba)O–B_2O_3 : Ce,Tb$	82,128,755	98 : 9909j
$(Y,Gd,La,Lu)_2O_2S : Tb,Dy$	82,141,482	98 : 25336a
$Y_2O_2S : Tb,Gd,Dy$	82,143,389	98 : 25340x
$(Y,Gd,La,Lu)_2O_2S–ZnO : Tb,Dy$	82,147,581	98 : 25341y
$(Y,Gd,La)_2O_3–SiO_2 : Tb$	82,131,278	98 : 43983p

Table 5.2. (*Continued*)

Material	Japan Kokai Tokkyo Koho JP	Chemical Abstracts Citation
$(Y,La,Gd,Lu,Sc)_2O_2S-SnO_2-In_2O_3:Tb$	82,151,684	98:63147x
$(Y,La,Gd,Lu,Sc)_2O_2S-SnO_2-In_2O_3:Eu$	82,151,683	98:63148y
$(Y,La,Gd,Lu,Sc)_2O_2S-SnO_2-In_2O_3:Er$	82,151,680	98:63150t
$(Y,La,Gd,Lu,Sc)_2O_2S-SnO_2-In_2O_3:Tm$	82,151,679	98:63151u
$(Y,Gd)_2O_2S:Eu,Er$	82,158,283	98:81304a
$(Y,Gd)(Nb,Ta)O_4:Eu$	82,168,982	98:135041h
$(Y,Gd)(Nb,Ta)O_4-SnO_2-In_2O_3:Eu$	82,180,688	98:152657f
$(Y,La,Gd)_2O_3-(Li,Na,K,Rb,Cs)O-P_2O_5-Al_2O_3:Ce,Tb$	82,179,278	98:152661c
$3Ga_2O_3-2In_2O_3-2Y_2O_3:Eu$	82,180,687	98:170216r
$(Y,Gd)(Nb,Ta)O_4:Tb$	82,182,381	98:170218t
$(Y,Gd)(Nb,Ta)O_3S:Tb$	82,182,382	98:170219u
$(Y,La,Gd,Lu)PO_4-(Hf,Zr)O_2:Ce,Tb$	82,187,383	98:170220n
$Y_2O_2S-(Zn,Cd)(Al,Ga)_2O_4:Cr,Eu$	82,192,485	98:188810q
$(Y,Gd,La)VO_4-SbO_2:Eu$	82,198,781	98:225045s
$(Y,Gd)_2O_3-GeO_2:Eu$	82,195,785	98:225048v
$Y_2W_3O_{12}:Eu$	82,212,286	98:225051r
$(Ba,Sr,Ca)(F,Cl,Br)_2-K(Cl,Br):Eu$	83, 57,108	98:225054u
$Y_2O_3-Ta_2O_5:Eu$	82,209,984	99:30572z
$Y_2O_3-Nb_2O_5:Eu$	82,209,983	99:30573a
$(Y,Al)_2O_3:Eu$	82, 52,382	99:61583h
(Pr,Nd,Sm,Eu,Tb,Dy,Ho,Er,Tm,Yb,Bi)O(Cl,Br):Ce	83, 69,281	99:79841a
$(Ca,Mg,Ba)WO_4:Eu$	83, 21,477	99:113596m
ZnS:Ce,Er,Cu,Au	83, 27,777	99:131139u
$(Ba,Mg)Al_2O_4-Y_2O_3:Eu$	83, 76,486	100:15139j
$(Y,La,Gd,Lu,Sc)_2O_2S:Eu,Bi$	83, 87,186	100:28014q
$(Al,La)PO_4:Ce,Tb$	83,187,483	100:42808h
$CaCl_2:Ce,Li,Sb$	83,210,990	100:77133m
$(Sr,Ca,Ba)O-P_2O_5-(Sr,Ca,Ba)(Cl,F,Br)_2-B_2O_3-(Sc,Y,La)_2O_3:Eu$	83,204,089	100:77134n
$(Yu,Gd,La)_2O_2S-(Ca,Mg,Sr,Ba)O:Eu$	83,164,681	100:94327v
Ca(S,Se):Eu,Ce	83,125,781	100:111987e
$Y_2O_3-(Mg,Ca,Sr,Ba,Zn)O:Eu$	83,127,777	100:111988f
$(Sr,Ca,Ba)O-P_2O_3-(Sr,Ca,Ba)(Cl,F,Br)_2-B_2O_3:Eu$	83,157,889	100:111992c

Table 5.2. (*Continued*)

Material	Japan Kokai Tokkyo Koho JP	Chemical Abstracts Citation
$Ln_2O_2SO_4$ (Ln = any lanthanide)	83,167,426	100:119993d
$(Ba,Ca,Mg)(F,Cl,Br)(PO_4)_3:Eu$	83,162,689	100:131005m
$ZnAl_2O_4:Mn,Tb,Ce$	83,154,786	100:148291z
$BaF(Cl,Br,I)–(Li,Na,Rb,Cs,Mg, Ca,Sr,Ba,Zn,Mn)(Si,Ti,Zr)F_6:Eu$	83,217,582	100:148339w
$(Sr,Ca,Ba,Mg,Zn,Cd)O–P_2O_5–(Sr,Ca,Ba)(Cl,F,Br)_2–B_2O_3:Eu$	83,138,773	100:165219z
[Ca,K, (Y,Gd)]S: Eu,Ce	84, 47,291	101:30914x
$(Sr,Ca,Mg,Ba)O–P_2O_5–(Sr,Ca,Mg,Ba)(Cl,F,Br)_2–B_2O_3$	84, 53,581	101:63403h
$Y_2O_3–GeO_2:Eu$	84, 86,685	101:63410h
$Y_3Al_5O_{12}:Tb,Dy$	84, 24,786	101:101003n
$(In,Y)BO_3:Eu$	84, 56,481	101:120235f
$(Ba,Sr,Ca)O–(Al,B)_2O_3:Eu$	84,128,211	101:140869u
$(Ba,Mg,Ca)_{10}(PO_4)_6F:Eu$	84, 24,785	101:161013a
$2(SrO)–MgO–(Al,B,Ga,Y,La)_2O_3:Eu$	84,102,979	101:180893d
$Y_2O_2S:Tb$	83, 43,090	101:201162g
(Ba;Be,Mg,Ca,Sr,Zn,Cd)(F;Cl, Br,I)O: Eu	84, 93,782	101:219465t
$(Y,La,Gd)_2O_2S$: Eu, (Ca,Mg,Sr,Ba)	84,102,981	101:219466u
(Ba;Be,Mg,Ca,Sr,Zn,Cd)(F;Cl, Br,I): Eu,Ce	84,102,980	101:219467v
$(La,Gd)_2O_3–P_2O_5–AlPO_4:Eu$	84,108,082	101:219469x
$(La,Gd)_2O_3–P_2O_5–SiO_2:Eu$	84,113,084	101:219470r
$(La,Gd,Lu)_2O_3–MgO–P_2O_5:Ce,Tb$	84,129,287	101:219473u
$Y_2O_2S:Tb,Ge$	84,138,292	101:219475w
(Ca,Sr)(S,Se): lanthanides	84,166,584	101:237923v
$(Sr,Ca,Ba,Mg,Zn,Cd)O–(P_2O_5)–(Sr,Ca,Ba)(Cl,F,Br)_2–B_2O_3–(Al,Ga,In,Tl,Sc,Y,La)_2O_3:Eu$	84, 47,288	102:36426g
BaFBr: Eu	84,149,978	102:36431e
$(Y,La,Gd,Lu)_2O_3–BaO–SiO_2:Tb,Dy,Pr$	84,193,983	102:53711p
(Ba,Be,Mg,Ca,Sr,Zn,Cd)F(Cl,Br,I): Eu	84,230,088	102:157713g
YF_3–(Ba,Mg,Ca,Sr,Zn,Cd)F(Cl, Br,I): Eu	84,226,090	102:158264y
LuF_3–(Ba,Mg,Ca,Sr,Zn,Cd)F(Cl, Br,I): Eu	84,221,379	102:194897r
$SrO–MgO–Al_2O_3:Eu$	85, 93,783	102:229262j

Source: From *Chem. Abstr.*, **96** (Jan. 1982) through **102** (June 1985).

REFERENCES

1. P. Pringsheim, *Fluorescence and Phosphorescence*, Interscience, New York, 1949.
2. E. Nichols and H. Howes, *Fluorescence of the Uranyl Salts*, Carnegie Institute, Washington, D.C., 1919.
3. E. Rabinowitch and R. L. Belford, *Spectroscopy and Photochemistry of Uranyl Compounds*, Pergamon Press, Oxford, 1964.
4. G. Urbain, *Ann. Chim. Phys. (Paris)*, **18**, 222, 289 (1909).
5. G. Urbain, *Chem. Rev.*, **1**, 143 (1925).
6. B. G. Wybourne, *Spectroscopic Properties of Rare Earths*, Wiley-Interscience, New York, 1965.
7. H. M. Crosswhite and H. W. Moos, *Optical Properties of Ions in Crystals*, Wiley-Interscience, New York, 1967.
8. G. H. Dieke, *Spectra and Energy Levels of Rare Earth Ions in Crystals*, Wiley-Interscience, New York, 1968.
9. F. S. Richardson, *Chem. Rev.*, **82**, 541 (1982).
10. G. E. Peterson, *Trans. Met. Chem.*, **3**, 202 (1966).
11. C. A. Morrison and R. P. Leavitt, in K. A. Gschneider and L. Eyring, Eds., *Handbook on the Physics and Chemistry of Rare Earths*, Vol. 5, North-Holland, Amsterdam, 1982, Chap. 46.
12. A. K. Levine and F. C. Palilla, *Appl. Phys. Lett.*, **5**, 118 (1964).
13. W. A. Thornton, in *Industrial Applications of Rare Earth Elements*, American Chemical Society, Washington, D.C., 1981, Chap. 11.
14. J. C. Rabatin, in *Industrial Applications of Rare Earth Elements*, American Chemical Society, Washington, D.C., 1981, Chap. 12.
15. P. N. Yocom, *RCA Eng.*, **29**, 78 (1984).
16. A. Bril and H. Klasens, *Philips Res. Rep.*, **7**, 401 (1955).
17. G. Blasse and A. Bril., *J. Electrochem. Soc.*, **115**, 1067 (1968).
18. G. Blasse, *J. Solid State Chem.*, **27**, 3 (1979).
19. L. H. Brixner and E. Abramson, *J. Electrochem. Soc.*, **112**, 70 (1965).
20. F. C. Palilla, A. K. Levine, and M. Rinkevics, *J. Electrochem. Soc.*, **112**, 776 (1965).
21. H. L. Burrus and A. G. Paulusz, *J. Electrochem. Soc.*, **115**, 976 (1968).
22. T. Kano and Y. Otomo, *J. Electrochem. Soc.*, **116**, 64 (1969).
23. E. J. Mehalchick, F. F. Mikus, and J. E. Mathers, *J. Electrochem. Soc.*, **116**, 1017 (1969).
24. H. Yamamoto, S. Seki, J.-P. Jesser, and T. Ishiba, *J. Electrochem. Soc.*, **127**, 694 (1980).
25. K. A. Wickersheim and R. A. Lefever, *J. Electrochem. Soc.*, **111**, 47 (1964).
26. A. E. Hardy, *IEEE Trans. Electron Devices*, **ED15**, 868 (1968).
27. L. H. Brixner, *J. Electrochem. Soc.*, **114**, 252 (1967).
28. P. M. Jaffe and J. D. Konitzer, *J. Electrochem. Soc.*, **116**, 633 (1969).
29. A. Bril, W. L. Wanmaker, and C. D. J. C. deLaat, *J. Electrochem. Soc.*, **112**, 111 (1965).

30. R. C. Ropp, *J. Electrochem. Soc.*, **112**, 181 (1965).
31. H. Forest and G. Ban, *J. Electrochem. Soc.*, **116**, 474 (1969).
32. H. Forest and G. Ban, *J. Electrochem. Soc.*, **118**, 1999 (1971).
33. A. T. Rhys Williams and M. J. Fuller, *Comput. Enhanced Spectrosc.*, **1**, 145 (1983).
34. J. G. Daly, J. A. Schmidt, and J. B. Gruber, *Phys. Rev. B*, **27**, 5250 (1983).
35. H. Yamamoto and K. Urabe, *J. Electrochem. Soc.*, **129**, 2069 (1982).
36. H. E. Hoefdraad, *J. Solid State Chem.*, **15**, 175 (1975).
37. O. J. Sovers and T. Yoshioka, *J. Chem. Phys.*, **49**, 4945 (1968).
38. O. J. Sovers and T. Toshioka, *J. Chem. Phys.*, **51**, 5330 (1969).
39. H. Forest, *J. Electrochem. Soc.*, **120**, 695 (1973).
40. L. Ozawa and H. N. Hersh, *J. Electrochem. Soc.*, **122**, 1222 (1975).
41. G. Webster and H. G. Drickamer, *J. Chem. Phys.*, **72**, 3470 (1980).
42. W. H. Fonger and C. W. Struck, *J. Chem. Phys.*, **52**, 6364 (1970).
43. C. W. Struck and W. H. Fonger, *J. Lumin.*, **1/2**, 456 (1970).
44. K. A. Wickersheim and R. B. Alves, *Ind. Res. Dev.*, 232 (1979).
45. W. H. Fonger and C. W. Struck, *J. Electrochem. Soc.*, **118**, 273 (1971).
46. H. Forest, A. Cocco, and H. Hersh, *J. Lumin.*, **3**, 25 (1970).
47. D. J. Robbins, *J. Electrochem. Soc.*, **123**, 1219 (1976).
48. G. Blasse, *Struct. Bonding*, **26**, 43 (1976).
49. R. E. Schrader and P. N. Yocom, *J. Lumin.*, **1/2**, 814 (1970).
50. L. Ozawa and P. M. Jaffe, *J. Electrochem. Soc.*, **117**, 1297 (1970).
51. H. Yamamoto and T. Kano, *J. Electrochem. Soc.*, **126**, 305 (1979).
52. S. Yokono, T. Abe, and T. Hoshina, *J. Lumin.*, **24/25**, 309 (1981).
53. D. W. Ormond and E. Banks, *J. Electrochem. Soc.*, **122**, 152 (1975).
54. M. Koskenlinna, M. Leskela, and L. Niinisto, *J. Electrochem. Soc.*, **123**, 75 (1976).
55. L. Ozawa, *J. Electrochem. Soc.*, **124**, 413 (1977).
56. S. Shionoya, in *P. Goldberg, Ed., Luminescence of Inorganic Solids*, Academic Press, New York, 1966, Chap. 4.
57. S. Larach and A. E. Hardy, *Proc. IEEE*, **61**, 915 (1973).
58. W. W. Anderson, S. Razi, and D. J. Walsh, *J. Chem. Phys.*, **43**, 1153 (1965).
59. W. W. Anderson, *J. Chem. Phys.*, **44**, 3283 (1966).
60. S. Larach, R. E. Schrader, and P. N. Yocom, *J. Electrochem. Soc.*, **116**, 471 (1969).
61. Y. Charreire and P. Porcher, *J. Electrochem. Soc.*, **130**, 175 (1983).
62. W. Lehmann, *J. Lumin.*, **5**, 87 (1975).
63. W. Lehmann, *J. Lumin.*, **6**, 455 (1973).
64. L. C. Porter and J. C. Wright, *J. Chem. Phys.*, **77**, 2322 (1982).
65. W. Lehmann and F. M. Ryan, *J. Electrochem. Soc.*, **118**, 477 (1971).
66. F. Okamoto and K. Kato, *J. Electrochem. Soc.*, **130**, 432 (1983).
67. H. Kasano, K. Megumi, and H. Yamamoto, *J. Electrochem. Soc.*, **131**, 1953 (1984).
68. T. E. Peters and J. A. Baglio, *J. Electrochem. Soc.*, **119**, 230 (1972).
69. T. E. Peters, *J. Electrochem. Soc.*, **119**, 1720 (1972).

70. A. Garcia, C. Fouassier, and J. P. Dougier, *J. Electrochem. Soc.*, **129**, 2063 (1982).
71. J. E. Ralph, *J. Phys. Chem. Solids*, **31**, 507 (1970).
72. D. J. Robbins, B. Cockayne, B. Lent, and J. L. Glasper, *J. Electrochem. Soc.*, **126**, 1556 (1979).
73. M. L. Keith and R. Roy, *Am. Mineral.*, **39**, 1 (1954).
74. R. S. Roth, *J. Res. Nat. Bur. Stand.*, **58**, 75 (1957).
75. M. Asano and J. A. Koningstein, *Chem. Phys.*, **42**, 369 (1979).
76. B. D. Joshi and A. G. Page., *J. Lumin.*, **6**, 441 (1973).
77. D. J. Robbins, B. Cockayne, A. G. Cullis, and J. L. Glasper, *J. Electrochem. Soc.*, **129**, 816 (1982).
78. T. Takamori, S. Fine, and D. B. Dove, *J. Electrochem. Soc.*, **129**, 2638 (1982).
79. J. T. Karpick and B. DiBartolo, *J. Lumin.*, **4**, 309 (1971).
80. D. Pacheco and B. DiBartolo, *J. Lumin.*, **14**, 19 (1976).
81. D. Pacheco and B. DiBartolo, *J. Lumin.*, **16**, 1 (1978).
82. L. Suchow, M. Kokta, and V. J. Flynn, *J. Solid State Chem.*, **2**, 137 (1970); *ibid.*, **5**, 329 (1972).
83. B. D. Joshi and A. G. Page, *J. Lumin.*, **15** 29 (1977).
84. P. Avouris, I. F. Chang, P. H. Duvigneaud, E. A. Giess, and T. N. Morgan, *J. Lumin.*, **26**, 213 (1982).
85. W. Koechner, *Solid State Laser Engineering*, Vol. 1, Springer-Verlag, New York, 1976, p. 36.
86. J. Loriers, R. Heindl, F. Clerc, and J. C. Bourcet, *J. Electrochem. Soc.*, **123**, 1882 (1976).
87. J. Fava, G. LeFlem, J. C. Bourcet, and F. Gaume-Mahn, *Mater. Res. Bull.*, **11**, 1 (1976).
88. K. K. Deb, *J. Phys. Chem. Solids*, **43**, 819 (1982).
89. A. Bril, G. Blasse, A. H. Gomes de Mesquita, and J. A. dePoorter, *Philips Tech. Rev.*, **32**, 125 (1971).
90. G. Blasse and A. Bril, *Appl. Phys. Lett.*, **11**, 53 (1967); J. Chem. Phys., **47**, 5139 (1967).
91. T. Y. Tien, E. F. Gibbons, R. G. DeLosh, P. J. Zacmanidis, D. E. Smith, and H. L. Stadler, *J. Electrochem. Soc.*, **120**, 278 (1973).
92. E. F. Gibbons, T. Y. Tien, R. G. DeLosh, P. J. Zacmanidis, and H. L. Stadler, *J. Electrochem. Soc.*, **120**, 835 (1973).
93. J. M. Robertson, M. W. VanTol, W. H. Smits, and J. P. H. Heynen, *Philips J. Res.*, **36**, 15 (1981).
94. D. J. Robbins, B. Cockayne, J. L. Glasper, and B. Lent, *J. Electrochem. Soc.*, **126**, 1213 (1979).
95. D. J. Robbins, B. Cockayne, J. L. Glasper, and B. Lent, *J. Electrochem. Soc.*, **126**, 1221 (1979).
96. D. J. Robbins, *J. Electrochem. Soc.*, **126**, 1550 (1979).
97. M. J. Weber, *J. Appl. Phys.*, **44**, 3205 (1973).
98. J. S. Abell, I. R. Harris, and B. Cocyayne, *J. Mater. Sci.*, **7**, 1088 (1972); **9**, 527 (1974).

99. T. Takeda, T. Miyata, F. Muramatsu, and T. Tomiki, *J. Electrochem. Soc.*, **127**, 438 (1980).
100. A. H. Gomes de Mesquita and A. Bril, *Mater. Res. Bull.*, **4**, 643 (1969).
101. A. H. Gomes de Mesquita and A. Bril, *J. Electrochem. Soc.*, **116**, 871 (1969).
102. P. V. Kelsey and J. J. Brown, *J. Electrochem. Soc.*, **123**, 1384 (1969).
103. H. Yamada, T. Kano, and M. Tanabe, *Mater. Res. Bull.*, **13**, 101 (1978).
104. R. C. Ropp, *J. Electrochem. Soc.*, **115**, 531 (1968).
105. R. C. Ropp, *J. Lumin.*, **3**, 152 (1970).
106. A. Bril, W. L. Wanmaker, and J. W. ter Vrugh, *J. Electrochem. Soc.*, **115**, 776 (1968).
107. F. J. Avella, *J. Electrochem. Soc.*, **113**, 1225 (1966).
108. F. J. Avella, O. J. Sovers, and C. S. Wiggins, *J. Electrochem. Soc.*, **114**, 613 (1967).
109. T. Takahashi and O. Yamada, *J. Electrochem. Soc.*, **124**, 955 (1977).
110. T. E. Peters, *J. Electrochem. Soc.*, **116**, 985 (1969).
111. K. R. Laud, E. F. Gibbons, T. Y. Tien, and H. L. Stadler, *J. Electrochem. Soc.*, **118**, 918 (1971).
112. Y. Tsujimoto, Y. Fukoda, S. Sugai, and M. Fukai, *J. Lumin.*, **9**, 475 (1975).
113. J. Hans and J. Borchardt, *J. Chem. Phys.*, **38**, 1251 (1963); *ibid.*, **42**, 3743 (1965).
114. T. Kano, K. Kinameri, and S. Seki, *J. Electrochem. Soc.*, **129**, 2296 (1982).
115. T. Kano, S. Seki, and S. Z. Zhung, *J. Lumin.*, **29**, 163 (1984).
116. T. Matsuoka, Y. Kashara, M. Tsuchiya, T. Nitta, and S. Hayakawa, *J. Electrochem. Soc.*, **125**, 102 (1978).
117. T. Matsuoka, T. Tohda, and T. Nitta, *J. Electrochem. Soc.*, **130**, 417 (1983).
118. G. Blasse and A. Bril, *Philips Tech. Rev.*, **31**, 304, 314, 324 (1970).
119. K. H. Butler, *Fluorescent Lamp Phosphors*, Pennsylvania State University Press, University Park, P., 1980.
120. W. L. Wanmaker, A. Bril, J. W. ter Vrught, and J. Broos, *Philips Res. Rep.*, **21**, 270 (1966).
121. G. Blasse and A. Bril, *Philips Res. Rep.*, **22**, 481 (1967).
122. H. J. Borchardt, *J. Chem. Phys.*, **42**, 3743 (1965).
123. G. Blasse, *J. Chem. Phys.*, **45**, 2356 (1966).
124. W. L. Wanmaker and A. Brill, *Philips Res. Rep.*, **19**, 479 (1964).
125. W. L. Wanmaker, A. Bril, and J. W. ter Vrugh, *J. Electrochem. Soc.*, **112**, 1147 (1965).
126. J. M. P. J. Verstegen, *J. Electrochem. Soc.*, **121**, 1631 (1974).
127. A. Bril and W. L. Wanmaker, *J. Electrochem. Soc.*, **111**, 1363 (1964).
128. E. P. Riedel, *J. Lumin.*, **1/2**, 176 (1970).
129. A. W. Veenis and A. Bril, *Philips J. Res.*, **33**, 124 (1978).
130. J. Koike, T. Kojima, R. Toyonaga, A. Kagami, T. Hase, and S. Inaho, *J. Electrochem. Soc.*, **126**, 1008 (1979).
131. G. Blasse and A. Bril, *J. Lumin.*, **3**, 18 (1970).
132. J. Th. de Hair, *J. Lumin.*, **18/19**, 797 (1979).
133. C. Fouassier, B. Saubat, and P. Hagenmuller, *J. Lumin.*, **23**, 405 (1981).
134. M. Leskela, M. Saakes, and G. Blasse, *Mater. Res. Bull.*, **19**, 151 (1984).

135. F. Kellendonk and G. Blasse, *J. Chem. Phys.*, **75**, 561 (1981).
136. G. Blasse, A. Bril, and J. de Vries, *J. Electrochem. Soc.*, **115**, 977 (1968).
137. K. Machida, G. Adachi, and J. Shiokawa, *J. Lumin.*, **21**, 101 (1979).
138. K. Machida, G. Adachi, J. Shiokawa, M. Shimada, and M. Koizumi, *J. Lumin.*, **21**, 233 (1980).
139. J. L. Sommerdijk and A. L. N. Stevels, *Philips Tech. Rev.*, **37**, 221 (1977).
140. G. Blasse and A. Bril, *Philips Res. Rep.*, **23**, 201 (1968).
141. F. C. Palilla, A. K. Levine, and M. R. Tomkus, *J. Electrochem. Soc.*, **115**, 642 (1968).
142. V. Abbruscato, *J. Electrochem. Soc.*, **118**, 930 (1971).
143. P. M. Jaffe, *J. Electrochem. Soc.*, **117**, 918 (1970).
144. A. Wachtel, *J. Electrochem. Soc.*, **116**, 61 (1969).
145. P. M. Jaffe, *J. Electrochem. Soc.*, **116**, 629 (1969).
146. J. M. P. J. Verstegen, J. L. Sommerdijk, and J. G. Verriet, *J. Lumin.*, **6**, 425 (1973).
147. J. L. Sommerdijk and J. M. P. J. Verstegen, *J. Lumin.*, **9**, 415 (1974).
148. J. M. P. J. Verstegen, *J. Electrochem. Soc.*, **121**, 1623 (1974).
149. J. M. P. J. Verstegen, J. L. Sommerdijk, and A. Bril, *J. Lumin.*, **9**, 420 (1974).
150. J. L. Sommerdijk, J. S. W. Der Does De Bye, and P. H. J. M. Verberne, *J. Lumin.*, **14**, 91 (1976).
151. A. L. N. Stevels and A. D. M. Schrama-de Pauw, *J. Electrochem. Soc.*, **123**, 6916 (1976).
152. A. L. N. Stevels, *J. Electrochem. Soc.*, **125**, 588 (1978).
153. A. L. N. Stevels and A. D. M. Schrama-de Pauw, *J. Lumin.*, **14**, 147 (1976).
154. A. L. N. Stevels and A. D. M. Schrama-de Pauw, *J. Lumin.*, **14**, 153 (1976); *ibid.*, **17**, 121 (1978).
155. H. Ronde, D. M. Krol, and G. Blasse, *J. Electrochem. Soc.*, **124**, 1276 (1977).
156. T. Takamori and D. B. Dove, *J. Lumin.*, **28**, 485 (1983).
157. G. Blasse, W. L. Wanmaker, J. W. ter Vrught, and A. Bril, *Philips Res. Rep.*, **23**, 189 (1968).
158. T. L. Barry, *J. Electrochem. Soc.*, **115**, 733 (1968).
159. T. L. Barry, *J. Electrochem. Soc.*, **115**, 1181 (1968).
160. T. L. Barry, *J. Electrochem. Soc.*, **117**, 381 (1970).
161. W. A. McAllister, *J. Electrochem. Soc.* **113**, 226 (1966).
162. F. Bretheau-Raynal, N. Tercier, B. Blanzat, and M. Drifford, *Mater. Res. Bull.*, **15**, 639 (1980).
163. J. M. P. J. Verstegen and J. L. Sommerdijk, *J. Lumin.*, **9**, 297 (1974).
164. E. Kaldis, P. Streit, and P. Wachter, *J. Phys. Chem. Solids*, **32**, 159 (1971).
165. G. Blasse and A. Bril, *J. Solid State Chem.*, **2**, 105 (1970).
166. T. J. Isaacs, *J. Electrochem. Soc.*, **118**, 1009 (1971).
167. T. J. Isaacs, *J. Electrochem. Soc.*, **120**, 654 (1973).
168. J. Th. de Hair, *J. Solid State Chem.*, **33**, 33 (1980).
169. W. L. Wanmaker and J. G. Verriet, *Philips Res. Rep.*, **28**, 80 (1973).
170. W. Lehmann, *J. Electrochem. Soc.*, **122**, 748 (1975).
171. W. Lehmann and T. J. Isaacs, *J. Electrochem. Soc.*, **125**, 445 (1978).

172. A. Garcia, B. Latorurrette, and C. Fouassier, *J. Electrochem. Soc.*, **126**, 1734 (1979).
173. M. Munashighe, *J. Electrochem. Soc.*, **119**, 902 (1972).
174. T. J. Isaacs and A. A. Price, *J. Electrochem. Soc.*, **120**, 997 (1973).
175. F. Raynal, B. Blanzat, J. P. Denis, J. Louriers, and C. Dannel, *Mater. Res. Bull.*, **11**, 731 (1976).
176. M. Sekita, *J. Lumin.*, **22**, 335 (1981).
177. W. L. Wanmaker and J. W. ter Vrugt, *Philips Res. Rep.*, **22**, 355 (1967).
178. W. L. Wanmaker and J. W. ter Vrugh, *Philips Res. Rep.*, **23**, 362 (1968).
179. M. V. Hoffman, *J. Electrochem. Soc.*, **115**, 590 (1968).
180. C. C. Lagos, *J. Electrochem. Soc.*, **115**, 1271 (1968); *ibid.*, **117**, 1189 (1970).
181. J. R. Looney and J. J. Brown, *J. Electrochem. Soc.*, **118**, 470 (1971).
182. R. A. McCauley, F. A. Hummel, and M. V. Hoffman, *J. Electrochem. soc.*, **118**, 755 (1971).
183. R. C. Ropp, *J. Electrochem. Soc.*, **118**, 841 (1971).
184. E. Nakazawa and F. Shiga, *J. Lumin.*, **15**, 255 (1977).
185. J. Th. de Hair and W. L. Konijnendijk, *J. Electrochem. Soc.*, **127**, 161 (1980).
186. W. A. McAllister, *J. Electrochem. Soc.*, **115**, 535 (1968).
187. M. S. Waite, *J. Electrochem. Soc.*, **121**, 1122 (1974).
188. A. Wachtel, *J. Electrochem. Soc.*, **128**, 208 (1981).
189. P. Bochu, C. Parent, A. Daoudi, G. Le Flem, and P. Hagenmuller, *Mater. Res. Bull.*, **16**, 883 (1981).
190. C. Parent, P. Bochu, A. Daoudi, and G. Le Flem, *J. Solid State Chem.*, **43**, 190 (1982).
191. G. Blasse, *J. Solid State Chem.*, **2**, 27 (1970).
192. Y. Tsujimoto, Y. Fukuda, and R. Fukai, *J. Electrochem. Soc.*, **124**, 553 (1977).
193. J. Fava, A. Perrin, J. C. Bourcet, R. Salmon, C. Parent, and G. Le Flem, *J. Lumin.*, **18/19**, 389 (1979).
194. G. Blasse and G. J. Dirksen, *Phys. Status. Solidi B*, **110**, 487 (1982).
195. W. Lehmann, F. M. Ryan, A. S. Manocha, and W. A. McAllister, *J. Electrochem. Soc.*, **130**, 171 (1983).
196. F. C. Palilla and B. E. O'Reilly, *J. Electrochem. Soc.*, **130**, 1076 (1968).
197. R. G. Pappalardo, J. Walsh, and R. B. Hunt, *J. Electrochem. Soc.*, **130**, 2087 (1983).
198. L. G. Van Uitert, *J. Electrochem. Soc.*, **110**, 46 (1963).
199. L. G. Van Uitert, *J. Electrochem. Soc.*, **114**, 1048 (1967).
200. L. G. Van Uitert, E. F. Dearborn, and J. J. Rubin, *J. Chem. Phys.*, **46**, 420 (1967); *ibid.*, **47**, 547, 1595, 3653 (1967).
201. L. G. Van Uitert, E. F. Dearborn, and W. H. Grodhiewicz, *J. Chem. Phys.*, **49**, 4400 (1968).
202. L. G. Van Uitert, *J. Lumin.*, **4**, 1 (1971).
203. G. Blasse and A. Bril, *J. Chem. Phys.*, **45**, 2350 (1966).
204. J. Huang, L. Loriers, and P. Porcher, *J. Solid State Chem.*, **43**, 87 (1982).
205. G. Blasse and G. J. Dirksen, *J. Solid State Chem.*, **42**, 163 (1982).
206. G. Blasse and D. M. Krol, *J. Lumin.*, **22**, 289 (1981).

207. G. Blasse and S. Kemmler-Sack, *Ber. Bunsenges. Phys. Chem.*, **87**, 352 (1983).
208. D. Wolf and S. Kemmler-Sack, *Phys. Status. Solidi*, **86A**, 685 (1984).
209. G. Blasse and S. Kemmler-Sack, *Ber. Bunsenges. Phys. Chem.*, **87**, 698 (1983).
210. W. Viehmann, *J. Chem. Phys.*, **47**, 875 (1967).
211. G. Blasse and M. Ouwerkerk, *J. Electrochem. Soc.*, **127**, 429 (1980).
212. M. Ouwerkerk, F. Kellendonk, and G. Blasse, *J. Chem. Soc. Faraday Trans. II*, **78**, 603 (1982).
213. J. Huang, J. Loriers, and P. Porcher, *J. Solid State Chem.*, **48**, 333 (1983).
214. L. H. Brixner, H.-Y. Chen, and C. M. Foris, *Mater. Res. Bull.*, **12**, 1545 (1982).
215. L. H. Brixner, H.-Y. Chen, and C. M. Foris, *J. Solid State Chem.*, **45**, 80 (1982).
216. G. Blasse, G. Bokkers, G. J. Dirksen, and L. H. Brixner, *J. Solid State Chem.*, **46**, 215 (1983).
217. G. Blasse and L. H. Brixner, *J. Solid State Chem.*, **47**, 368 (1983).
218. L. H. Brixner, H.-Y. Chen, and C. M. Foris, *J. Solid State Chem.*, **44**, 99 (1982).
219. G. Blasse, G. J. Dirksen, and L. H. Brixner, *J. Solid Chem.*, **46**, 294 (1983).
220. S. Z. Toma, F. F. Mikus, and J. E. Mathers, *J. Electrochem. Soc.*, **114**, 953 (1967).
221. R. K. Datta, *J. Electrochem. Soc.*, **114**, 1057 (1967).
222. R. C. Ropp, *J. Electrochem. Soc.*, **115**, 940 (1968).
223. J. L. Sommerdijk and A. Bril, *J. Electrochem. Soc.*, **122**, 952 (1975); *Philips Res. Rep.* **32**, 149 (1977).
224. S. Faria and D. T. Palumbo, *J. Electrochem. Soc.*, **116**, 157 (1969).
225. S. Faria and D. T. Palumbo, *J. Electrochem. Soc.*, **117**, 124 (1970).
226. S. Faria and D. T. Palumbo, *J. Electrochem. Soc.*, **121**, 305 (1970).
227. A. Wachtel, *J. Electrochem. Soc.*, **123**, 246 (1976).
228. G. Blasse and A. Bril, *J. Lumin.*, **3**, 109 (1970).
229. M. J. J. Lammers and G. Blasse, *Mater. Res. Bull.*, **19**, 759 (1984).
230. L. H. Brixner and H.-Y. Chen, *J. Electrochem. Soc.*, **130**, 2435 (1983).
231. G. Blasse and A. Bril, *Philips Res. Rep.*, **22**, 46 (1967).
232. F. M. Lay, *J. Phys. Chem. Solids*, **29**, 2043 (1968).
233. W. A. McAllister, *J. Electrochem. Soc.*, **131**, 1207 (1984).
234. H. G. Brittain and W. A. McAllister, *Spectrochim. Acta*, **41A**, 1041 (1985).
235. G. Blasse and A. Bril, *J. Electrochem. Soc.*, **114**, 250 (1967).
236. A. Wachtel and T. J. Isaacs, *J. Electrochem. Soc.*, **124**, 247 (1977).
237. F. A. Rozhdestvenskii and M. G. Zuev, *J. Lumin.*, **28**, 465 (1983).
238. J. W. Haynes and J. J. Brown, *J. Electrochem. Soc.*, **115**, 1060 (1968).
239. G. Blasse and A. Bril, *Philips Res. Rep.*, **23**, 461 (1968).
240. P. Porcher, D. R. Svoronos, M. Leskela, and J. Holsa, *J. Solid State Chem.*, **46**, 101 (1983).
241. H. G. Brittain, *J. Less Common Met.*, **93**, 97 (1983).
242. F. M. Ryan, W. Lehmann, D. W. Feldman, and J. Murphy, *J. Electrochem. Soc.*, **121**, 1475 (1974).
243. H. Zhiran, G. J. Dirksen, and G. Blasse, *J. Solid State Chem.*, **52**, 130 (1984).
244. S. Natansohn, *J. Electrochem. Soc.*, **116**, 1250 (1969).
245. R. R. Neurgaonkar, L. E. Cross, and W. B. White, *J. Electrochem. Soc.*, **125**, 1130 (1978).

246. G. Blasse and A. Bril, *J. Chem. Phys.*, **48**, 3652 (1968).
247. P. A. M. Berdowski and G. Blasse, *J. Lumin.*, **29**, 243 (1984).
248. R. A. McCauley and F. A. Hummel, *J. Lumin.*, **6**, 105 (1973).
249. J. Dexpert-Ghys, M. Faucher, and P. Caro, *C. R. Acad. Sci. (Paris)*, **298**, 621 (1984).
250. L. H. Brixner, *Mater. Res. Bull.*, **19**, 143 (1984).
251. K. Awazu and K. Muto, *J. Electrochem. Soc.*, **116**, 282 (1969).
252. T. R. Saranathan, V. A. Fassel, and E. L. Dekalb, *Anal. Chem.*, **42**, 325 (1970).
253. M. Breysse and L. Faure, *J. Lumin.*, **26**, 107 (1981).
254. L. C. Porter and J. C. Wright, *J. Lumin.*, **27**, 237 (1982).
255. H. G. Brittain, L. Tsao, and D. L. Perry, *J. Lumin.*, **28**, 257 (1983).
256. J. G. Rabatin, in K. A. Gschneider, Jr., Ed., *Industrial Applications of Rare Earth Elements, Am. Chem. Soc. Symp.* **164**, 203 (1967).
257. G. Blasse and A. Bril, *J. Chem. Phys.*, **46**, 2579 (1967).
258. G. Blasse and A. Bril, *Philips Res. Rep.*, **22**, 481 (1967).
259. J. G. Rabatin, *Abstract 78*, p. 189, The Electrochemical Society Extended Abstracts, New York, NY, May 4–9, 1969.
260. J. G. Rabatin, *Abstract 102*, p. 250, The Electrochemical Society Extended Abstracts, San Francisco, May 12–17, 1974.
261. J. G. Rabatin, *Abstract 198*, p. 467, The Electrochemical Society Extended Abstracts, Toronto, Ontario, Canada, May 11–16, 1975.
262. J. G. Rabatin, *Abstract 218*, p. 563, The Electrochemical Society Extended Abstracts, Boston, May 6–11, 1979.
263. J. G. Rabatin, *Abstract 306*, p. 788, The Electrochemical Society Extended Abstracts, Los Angeles, Oct. 14–19, 1979.
264. J. G. Rabatin, *J. Electrochem. Soc.*, **129**, 1552 (1982).
265. J. Holsa, M. Leskela, and L. Niinisto, *Mater. Res. Bull.*, **14**, 1403 (1979).
266. J. Holsa, M. Leskela, and L. Niinisto, *J. Solid State Chem.*, **37**, 267 (1981).
267. L. Y. Mo, F. Guillen, C. Fouassier, and P. Hagenmuller, *J. Electrochem. Soc.*, **132**, 717 (1985).
268. J. Holsa and P. Porcher, *J. Chem. Phys.*, **76**, 2790 (1982).
269. L. H. Brixner, J. F. Ackerman, and C. M. Foris, *J. Lumin.*, **26**, 1 (1981).
270. O. L. Malta, *Chem. Phys. Lett.*, **88**, 353 (1982).
271. J. Holsa and P. Porcher, *J. Chem. Phys.*, **76**, 2798 (1982).
272. J. Holsa, *Chem. Phys. Lett.*, **112**, 246 (1984).
273. E. Secemski and W. Low, *J. Chem. Phys.*, **64**, 4240 (1976).
274. S. Benci and G. Schianchi, *J. Lumin.*, **11**, 349 (1976).
275. M. Bancie-Grillow, *J. Lumin.*, **12/13**, 681 (1976).
276. J. P. Jouart, *J. Lumin.*, **21**, 153 (1980).
277. S.-I. Mho and J. C. Wright, *J. Chem. Phys.*, **77**, 1183 (1982).
278. R. L. Amster and C. S. Wiggins, *J. Electrochem. Soc.*, **116**, 68 (1969).
279. V. P. Bhola, *J. Lumin.*, **10**, 185 (1975); *ibid.*, **14**, 115 (1976).
280. P. Evesque, J. Kliava, and J. Duran, *J. Lumin.*, **18/19**, 646 (1979).
281. J. Chrysochoos, P. W. M. Jacobs, M. J. Stillman, and A. V. Chadwick, *J. Lumin.*, **28**, 177 (1983).
282. P. F. Weller, J. D. Axe, and G. D. Pettit, *J. Electrochem. Soc.*, **112**, 74 (1965).

283. R. J. Hammers, J. R. Wietfeldt, and J. C. Wright, *J. Chem. Phys.*, **77**, 683 (1982).
284. V. P. Bhola, *J. Lumin.*, **9**, 121 (1974).
285. M. D. Kurz and J. C. Wright, *J. Lumin.*, **15**, 169 (1977).
286. R. L. Amster, *J. Electrochem. Soc.*, **117**, 791 (1970).
287. J. A. DeLuca and F. S. Ham, *J. Electrochem. Soc.*, **124**, 1592 (1977).
288. A. L. N. Stevels, *J. Lumin.*, **12/13**, 97 (1976).
289. L. H. Brixner and J. D. Bierlein, *Mater. Res. Bull.*, **9**, 99 (1974).
290. L. H. Brixner and A. Ferretti, *J. Solid State Chem.*, **18**, 111 (1976).
291. T. Kobayasi, S. Mroczkowski, J. F. Owen, and L. H. Brixner, *J. Lumin.*, **21**, 247 (1980).
292. K. Takahashi, J. Miyahara, and Y. Shibahara, *J. Electrochem. Soc.*, **132**, 1492 (1985).
293. A. Wachtel, *J. Electrochem. Soc.*, **107**, 199 (1960).
294. J. I. Pankove, M. A. Lampert, J. J. Hanak, and J. E. Berkeyheiser, *J. Lumin.*, **15**, 349 (1977).
295. S. Tanaka, H. Kobayashi, M. Shiiki, T. Kunov, V. Shanker, and H. Sasakura, *J. Lumin.*, **31/32**, 945 (1984).
296. I. Szcaurek and H. Lozykowski, *J. Lumin.*, **14**, 389 (1976).
297. F. J. Bryant, W. E. Hagston, and A. Krier, *J. Lumin.*, **31/32**, 948 (1984).
298. S. Tanaka, Y. Maruyama, H. Kobayashi, and H. Sasakura, *J. Electrochem. Soc.*, **123**, 1917 (1976).
299. J. Benoit, P. Benalloul, and B. Blanzat, *J. Lumin.*, **23**, 175 (1981).
300. S. Bhushan, A. N. Pandey, and B. R. Kaza, *J. Lumin.*, **20**, 29 (1979).
301. J. L. Sommerdijk and A. Bril, *Philips Tech. Rev.*, **34**, 24 (1974).
302. F. Auzel, *C. R. Acad. Sci. (Paris)*, **262B**, 1016 (1966); *ibid.*, *263B*, 819 (1966).
303. J. L. Sommerdijk, A. Bril, J. A. de Poorter, and R. E. Breemer, *Philips Res. Rep.*, **28**, 475 (1973); *ibid.*, **29**, 13 (1974).
304. M. R. Brown, J. S. S. Whiting, and W. A. Shand, *J. Chem. Phys.*, **43**, 1 (1965).
305. H. E. Rast, H. H. Caspers, and S. A. Miller, *J. Chem. Phys.*, **47**, 3874 (1967).
306. P. M. Selzer, D. S. Hamiltton, R. Flach, and W. M. Yen, *J. Lumin.*, **12/13**, 737 (1976).
307. G. Blasse, *Phys. Status Solidi*, **A73**, 205 (1982).
308. B. R. Reddy and P. Venkateswarlu, *J. Chem. Phys.*, **77**, 2862 (1982).
309. J. Hegarty, D. L. Huber, and W. M. Yen, *Phys. Rev. B*, **25**, 5638 (1982).
310. R. A. Hewes, *J. Lumin.*, **1/2**, 778 (1970).
311. F. Auzel, *J. Lumin.*, **12/13**, 715 (1976).
312. J. L. Sommerdijk, *J. Lumin.*, **4**, 441 (1971).
313. T. Kano, H. Yamamoto, and Y. Otomo, *J. Electrochem. Soc.*, **119**, 1561 (1972).
314. J. L. Sommerdijk, *J. Lumin.*, **6**, 61 (1973).
315. A. Bril, J. L. Sommerdijk, and A. W. de Jager, *J. Electrochem. Soc.*, **122**, 660 (1975).
316. L. G. Van Uitert, L. Pictroski, and W. H. Grodkiewicz, *Mater. Res. Bull.*, **4**, 777 (1969).
317. S. G. Parker and R. E. Johnson, *J. Electrochem. Soc.*, **119**, 610 (1972).
318. N. M. P. Low and A. L. Major, *Mater. Res. Bull.*, **7**, 203 (1972).
319. J. V. Posluszny and H. G. Brittain, *Thermochim. Acta*, **83**, 271 (1985).

320. R. A. Hewes and J. R. Sarver, *Phys. Rev.*, **182**, 427 (1969).
321. J. L. Sommerdijk, *J. Lumin.*, **8**, 126 (1973).
322. R. K. Watts, *J. Chem. Phys.*, **53**, 3552 (1970).
323. R. K. Watts and H. J. Richter, *Phys. Rev. B* **4**, 1584 (1972).
324. J. Wojciechowski, I. Pawelska, R. Grodecki, and L. Szymanski, *J. Electrochem. Soc.*, **122**, 312 (1975).
325. M. Greenblatt and E. Banks, *J. Electrochem. Soc.*, **124**, 409 (1977).
326. J. P. Jouart, C. Bissieux, and G. Mary, *J. Lumin.*, **29**, 261 (1984).
327. J. P. Morley, T. R. Faulkner, F. S. Richardson, and R. W. Schwartz, *J. Chem. Phys.*, **77**, 1734 (1982).
328. O. A. Serra and L. C. Thompson, *Inorg. Chem.*, **15**, 504 (1976).
329. C. D. Flint and F. L. Stewart-Darling, *Mol. Phys.*, **44**, 61 (1981).
330. J. P. Morley, T. R. Faulkner, and F. S. Richardson, *J. Chem. Phys.*, **77**, 1710 (1982).
331. R. W. Schwartz, H. G. Brittain, J. P. Riehl, W. Yeakel, and F. S. Richardson, *Mol. Phys.*, **34**, 361 (1977).
332. T. R. Faulkner and F. S. Richardson, *Mol. Phys.*, **36**, 193 (1978).
333. J. P. Morley, T. R. Faulkner, F. S. Richardson, and R. W. Schwartz, *J. Chem. Phys.*, **75**, 539 (1981).
334. W. Rybe-Romanowski, Z. Mazurak, and B. Jezowska-Trzebiatowska, *J. Lumin.*, **27**, 177 (1982).
335. Z. Mazurak, E. Lukowiak, B. Jezows ka-Trzebiatowska, and W. Ryba-Romanowski, *J. Lumin.*, **29**, 47 (1984).
336. Z. Mazurak, E. Lukowiak, and B. Jezowska-Trzebiatowska, *J. Lumin.*, **33**, 159 (1985).
337. N. M. P. Low and A. L. Major, *J. Lumin.*, **4**, 357 (1971).
338. J. L. Sommerdijk, W. L. Wanmaker, and J. G. Verriet, *J. Lumin.*, **4**, 404 (1971).
339. J. L. Sommerdijk, W. L. Wanmaker, and J. G. Verriet, *J. Lumin.*, **5**, 297 (1972).
340. F. Auzel, D. Pecile, and D. Morin, *J. Electrochem. Soc.*, **122**, 101 (1975).
341. G. F. J. Garlick and C. L. Richards, *J. Lumin.*, **9**, 424 (1974).
342. G. F. J. Garlick and C. L. Richards, *J. Lumin.*, **9**, 432 (1974).
343. L. G. Van Uitert, H. J. Levinstein, and W. H. Levinstein, *Mater. Res. Bull.*, **4**, 381 (1969).
344. T. Matsubara, *Mater. Res. Bull.*, **7**, 963 (1972).

CHAPTER

6

PROTON TRANSFER KINETICS OF ELECTRONICALLY EXCITED ACIDS AND BASES

RICHARD N. KELLY AND STEPHEN G. SCHULMAN

College of Pharmacy
University of Florida
Gainesville, FL 32610

6.1. INTRODUCTION

Potentially fluorescent molecules have lifetimes in the lowest excited singlet state (S_1), typically of the order 10^{-11}–10^{-7} s. Although these lifetimes are extremely short, a number of chemical phenomena are sufficiently rapid to compete with the photophysical processes which return the excited molecule to the ground electronic state (S_0) or convert it to the lowest triplet state (T_1). Some of these reactions, such as isomerization, are unimolecular but some, such as proton transfer and aggregation, are bimolecular and if solvation and desolvation are considered as part of the reaction, may even be considered to be polymolecular. The temporal competition of these chemical reactions with fluorescence of the excited species often results in multiple luminescences from solutions in which only a single species is directly excited.

In those reactions that are polymolecular, the fluorescence spectra will vary with the concentrations of the various reactants. One can determine the relative quantum yields of fluorescence by steady-state fluorescence methods or the fluorescence decay profile by time-resolved fluorescence spectroscopy of the excited reactants and products as a function of the chemical variables of the system. The kinetics of the reaction in the S_1 state can then often be evaluated.

Notwithstanding that the kinetics of ultrafast reactions studied by fluorescence spectroscopy invariably apply to processes occurring in the S_1 state, the conclusions that can be drawn from these studies can be inferred, in a general sense, to apply to reactions of the same kind in any given electronic state, including S_0. Consequently, fluorescence spectroscopy represents a powerful (and relatively economical) tool for the study of some of the most fundamental of chemical reactions. In this chapter then we shall consider some of the more important fast reactions and the ways in which they have been studied by fluorescence spectroscopy.

6.2. PROTON TRANSFER IN THE LOWEST EXCITED SINGLET STATE

Proton transfer in the lowest excited singlet state is a phenomenon exhibited by certain aromatic acids and bases which have their acidic or basic groups attached to the aromatic ring. Because the lowest excited singlet state has an electronic distribution which is generally very different from the ground electronic state of the same molecule, the excited state may be much more or less acidic or basic than the ground state. This means that in, say, a phenolic molecule, where the excited state is more acidic than the ground state, emission may be observed from the phenolate anion even at pH so low that the undissociated phenol is the sole absorbing (ground-state) species.

It is now more than three decades since Förster (1) recognized the "anomalous" pH dependence of fluorescence first observed by Weber (2), as the result of proton transfer occurring during the lifetimes of the lowest excited singlet states of organic acids and bases. Only a few years later Weller (3), Förster's former student, put the phenomenon on a quantitative basis, relating the pH dependences of the fluorescence afficiencies of conjugate acids and bases to the competitive kinetics of proton transfer in and photophysical deactivation of the lowest excited single state. A few years later, Eigen and his co-workers (4–8) developed the field-dissociation technique extending the range of application of ultrafast reaction kinetics and led the way to a substantial understanding of the rates and mechanisms of reactions that occur with velocities at or near the theoretical upper limits of reactions rates in aqueous solution.

Weller's experiments yielded rate constants whose correspondence with results obtained from Eigen's work indicated that the same chemistry that governed the acid–base behavior of the ground-state species was applicable to the excited species, as well.

By using the field dissociation effect, Eigen would disrupt an acid–base system at equilibrium with an intense electric field change of short duration and subsequently, follow the return of the system to equilibrium. This approach, which is typical of relaxation methods, had the advantage that a wide range of reaction times could be followed, but the disadvantage was also inherent—that only the bimolecular protonation reaction of conjugate bases or proton abstraction by hydroxide ion from conjugate acids could be followed kinetically, leaving the reciprocal pseudo-first-order reaction rate constant to be calculated from the equilibrium constant and the measurable second-order rate constant. Weller's fluorescence method did not have the time range of applicability of Eigen's approach, being restricted to kinetics occurring in the time scale of fluorescence events (liberally, 10^{-11}–10^{-7} s); however, Weller's approach had the advantage of being able to extract forward and reverse reaction rate constants directly from a fluorimetric pH titration and independently of the vagaries of methods used to determine equilibrium contants. Although other methods have come into use for the study of proton-exchange kinetics, each has its optimal range of applicability, and fluorimetry remains one of the simplest methods for the study of proton-exchange kinetics in the nanosecond time region.

In the years since Weller initially developed the kinetic approach to excited-state proton transfer there have been a number of applications of his approach, most of which were reviewed from time to time in the literature (9–14). With the advent of pulsed lasers in recent years, instrumentation available for the quantitation of proton-transfer kinetics has become better and a complement to Weller's steady-state approach, in the form of time-correlated single-photon counting, has extended the treatment to some systems which were difficult to handle by the straightforward steady-state approach (15–18). Even more recently phase-resolved fluorimetry has been applied to excited-state proton transfer kinetics (19). The techniques of time- and phase-resolved fluorimetry have been discussed in Chapter 2 and will not be elaborated further here; however, the algebra used to evaluate the kinetics of excited-state proton transfer by the steady-state, time-resolved, and phase-resolved methods is substantially different, and since each approach has its own merits, the formalism of each will be considered.

6.2.1. Properties of Acids and Bases in the Lowest Excited Singlet State

Upon excitation to the S_1 state the electronic distribution of an aromatic acid or base changes. Acidic and basic functional groups may, depending on their nature, become either enriched or deficient in electronic charge when the molecule is excited. This may lead to substantial differences between the acidities of the same functional groups in S_0 and S_1 as reflected by the respective equilibrium constants pK_a and pK_a^*. Usually, only those functional groups bonded directly to a aromatic ring will experience changes in charge distribution sufficient to cause detectable differences between pK_a and pK_a^*.

Functional groups that possess low-lying vacant π orbitals can, upon electronic excitation, accept electronic charge from the aromatic system. Groups of this kind include carbonyl, carboxyl, carboyxlate, and amide groups. The increase in electronic charge density at one of these groups subsequent to excitation makes it more difficult to remove a proton (or easier to add a proton). As a result, pK_a^* > pK_a, and these groups are, therefore, less acidic (or more basic) in S_1 than they are in S_0.

Electron-donating groups have lone electron pairs which upon excitation may be transferred to the lowest-lying vacant π orbitals of the aromatic system. Examples are hydroxy, sulfhydryl, and amino groups. Pyrrolic nitrogen atoms (which are really secondary arylamino groups) also fall into this category. Upon excitation, the electronic charge density at these groups decreases, weakening bonds, and thus a proton may be more readily lost (or less readily gained) from the group in S_1 than in S_0. Consequently, pK_a^* < pK_a, and these groups are thus more acidic (or less basic) in S_1 than they are in S_0.

Pyridinic nitrogen atoms (such as those found in pyridine and quinoline) do not possess low-lying vacant orbitals but are more electronegative than are the carbon atoms to which they are bonded. Upon excitation, therefore, some of the electronic charge in the molecule is localized on the nitrogen atom. This results in the pyridinic nitrogen atom being more basic in the excited state than it was in the ground electronic state.

Since the magnitude of K_a^* is determined, in part, by the strength of the bond between the acidic hydrogen atom and the rest of the molecule, it is not surprising that different classes of aromatic compounds (e.g., aromatic carboxylic acids, phenols, pyridines, etc.) are characterized by pK_a^* values which are similar within a given class. Substitutents (e.g., nitro groups, alkyl groups, etc.) on the aromatic system of a given aromatic acid may change the pK_a^* of that acid in accordance with the electron-withdrawing or electron-donating nature of the substitutent and its location (orientation) relative to the acidic hydrogen or site of protonation. Substitutent effects are often qualitatively or semiquantitatively predictable and may, therefore, be taken into account when comparing pK_a^* values of aromatic acids and bases. The effects of substitutents on ground-state acid–base chemistry are well known, and discussions of them may be found in a variety of organic chemistry texts. In some cases, however, the influence of substitutent orientation on pK_a^* may appear to be anomalous. An example of this is found in the cyanophenols

(20). The pK_a and pK_a^* values of *ortho*-, *meta*-, and *para*-cyanophenol have been determined (20), and it was found that pK_a^* (ortho) < pK_a^* (meta) < pK_a^* (para). The order of these pK_a^* values is different from the order of ground state pK_a values (ortho < para < meta). This can be rationalized in terms of the directions of movement of electronic charge in the various isomers upon excitation, relative to the orientations of the substitutent groups. It also demonstrates that it is risky to make predictions of excited-state chemistry based on valence-bond (resonance) models.

6.2.2. Relationship between Electronic Spectra and State of Protonation

In molecules containing an electron-withdrawing group, the transfer of electronic charge to that group results from excitation from the S_0 to the S_1 state. This causes the S_1 state to be stabilized to a greater degree than is the S_0 state due to the acquisition of positive charge accompanying protonation and decreases the energy difference between S_0 and S_1 in the protonated molecule. The spectral result is the shifting of the longest-wavelength absorption band and the fluorescence spectrum (both of which arise from electronic transition between ground and lowest excited singlet state) to longer wavelengths upon protonation and to shorter wavelengths upon dissociation. This kind of behavior is exhibited by carboxylic acids and N-heterocyclics, such as acridine.

In molecules containing electron donor groups, the loss of electronic charge from the functional group which accompanies excitation from the S_0 to S_1 state is inhibited by protonation and facilitated by dissociation. This causes the energy difference between S_0 and S_1 to be greater in the conjugate species having the higher state of protonation. As a result, the longest wavelength absorption and fluorescence bands of these molecules (e.g., arylamines and hydroxyaromatics) shift to shorter wavelengths upon protonation and to longer wavelengths upon dissociation.

The shifting of the long-wavelength absorption and fluorescence bands of molecules possessing electron acceptor groups to longer wavelengths upon protonation and to shorter wavelengths upon dissociation is related to the increase in basicity and decrease in acidity upon excitation from the ground state to the lowest excited singlet state. Similarly, in molecules having electron donor groups, the shifting of the long-wavelength absorption and fluorescence bands to shorter wavelengths upon protonation and to longer wavelengths upon dissociation is indicative of a decrease in basicity and an increase in acidity upon going from the ground state to the lowest excited singlet state. Förster (1) was the first to recognize this and to make a quantitative statement of the relationship between the shifting of the long-wavelength absorption and fluorescence spectra accompanying protonation and dissociation and the change in acidity or basicity upon going from S_0 to S_1.

6.2.3. The Förster Cycle

It is possible to estimate, from the shifting of the long-wavelength absorption and fluorescence spectra accompanying protonation or dissociation, the pK_a^* corresponding to an excited-state proton exchange, provided that the pK_a of the corresponding ground-state process is known. This approach was developed by Förster (1) and is based on the thermodynamic equivalence of alternate paths from the ground state of the acid to the thermally equilibrated, lowest electronically excited singlet state of the conjugate base (Fig. 6.1).

From Fig. 6.1 it can be seen that there are two energetically equivalent ways of converting the ground-state acid A to the excited base B*. The first pathway entails the absorption of electronic energy E_A by A, to form A*, followed by dissociation of A* to B* with attendant enthalpy of dissociation ΔH_d^*. The total energy involved in the latter process is $E_A + \Delta H_d^*$. The second route from A to B* entails dissociation of A to B in the ground state, with the enthalpy of dissociation ΔH_d, following by excitation of B to B* as a result of absorption of energy E_B. The second pathway requires a total energy of $E_B + \Delta H_d$. Since both pathways are energetically equivalent,

$$E_A + \Delta H_d^* = E_B + \Delta H_d \tag{6.1}$$

In Eq. (6.1) E_A and E_B are, respectively, equivalent to $Nhc\bar{\nu}_A$ and $Nhc\bar{\nu}_B$, where N is Avogadro's number, h is Planck's constant, c is the velocity of light, and $\bar{\nu}_A$ and $\bar{\nu}_B$ are the frequencies of radiation (in wave numbers) absorbed or emitted in the transitions between A and A* and B and B*, respectively. If it is assumed that the entropies of protonation in ground and excited states are identical, then Eq. (6.1) becomes

$$\Delta G - \Delta G^* = Nhc(\bar{\nu}_A - \bar{\nu}_B) \tag{6.2}$$

For very dilute solutions $\Delta G - \Delta G^* = \Delta G^0 - \Delta G^{0*}$ and it immediately follows that

$$pK_a - pK_a^* = \frac{Nhc}{2.303RT}(\bar{\nu}_A - \bar{\nu}_B) \tag{6.3}$$

where R is the universal gas constant and T is the absolute temperature. Consequently, if the transition energies of acid and conjugate base are known

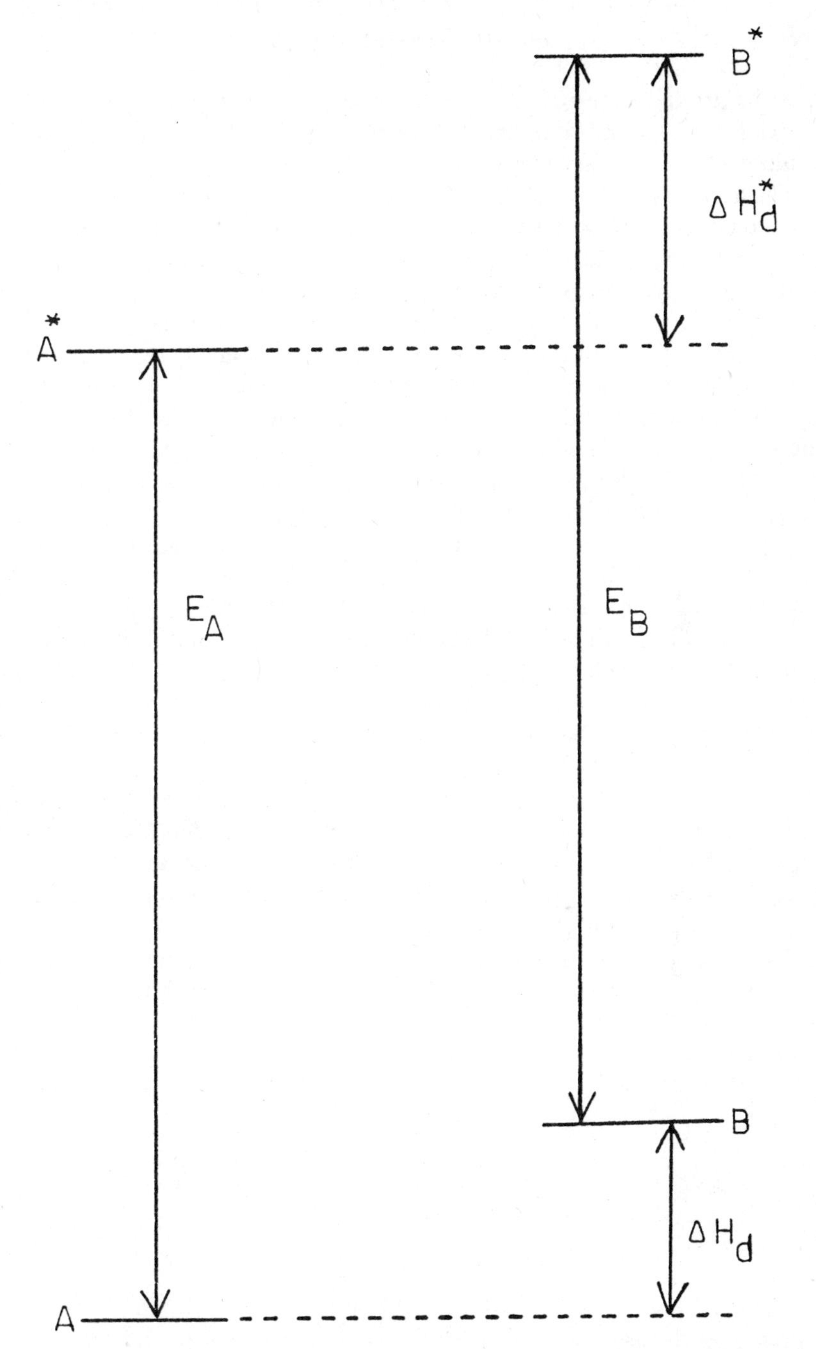

Fig. 6.1. Förster cycle (ideal case). A, A*, B, and B*, represent the ground and excited states of an acid and its conjugate base. E_A and E_B are the spectroscopic energies of the transitions A ↔ A* and B ↔ B*, respectively, and ΔH_d and ΔH_d^* are the enthalpies of protonation in the ground and electronically excited states, respectively.

and the ground-state pK_a can be evaluated, the excited-state pK_a^* for the corresponding equilibrium can be calculated (regardless of whether or not equilibrium in S_1 is attained).

The accuracy with which pK_a^* can be calculated from the Förster cycle is dependent on the way in which $\bar{\nu}_A$ and $\bar{\nu}_B$ are chosen. These frequencies must represent adiabatic transitions if the Förster cycle is truly to represent the difference in free energy between proton transfer in S_0 and S_1. Strictly speaking, $\bar{\nu}_A$ and $\bar{\nu}_B$ should correspond to the hypothetical 0–0 vibronic bands of absorption or fluorescence of conjugate acid and base, which should coincide for each species. The term "hypothetical" is included here because the 0–0 band is a truly adiabatic transition only in molecules that have identical geometries and solvent cages in the S_0 and S_1 states. Many acidic and basic molecules have functional groups that alter their positions with respect to the rest of the molecule and virtually all are solvated somewhat differently in S_1 than in S_0. The thermal relaxation of the molecular geometry and the solvent cage following excitation from S_0 to S_1 and again following emission cause the 0–0 bands of fluorescence and absorption to be substantially displaced from one another and to correspond to nonadiabatic transitions. This is illustrated in Fig. 6.2. If the 0–0 bands of the fluorescence of acid and conjugate base were used to calculate pK_a^* from Eq. (6.3), the result would actually be an over- or underestimation of the true value. If the 0–0 bands of the long-wavelength absorption spectra were used, the result would be an under- or overestimation of pK_a^* (opposite to the error using fluorescence spectra alone). Moreover, in hydroxylic solvents (e.g., water) the vast majority of acids and bases have diffuse absorption and fluorescence spectra, so that the problem of locating even the strictly vibronic 0–0 bands is experimentally difficult or impossible.

If, however, it is assumed that the vibrational compositions of the S_0 and S_1 states are identical, so that absorption and fluorescence from the same species have the same band shape, it is possible to reformulate, in a more elaborate form, the Förster cycle using Fig. 6.2, to obtain from the absorption spectra

$$E_A^a - \Delta H_{A^*}^t + \Delta H_d^* = \Delta H_d + E_B^a - \Delta H_{B^*}^t \tag{6.4}$$

or

$$pK_a - pK_a^* = \frac{Nhc(\bar{\nu}_A^a - \bar{\nu}_B^a) + \Delta H_{B^*}^t - \Delta H_{A^*}^t}{2.303RT} \tag{6.5}$$

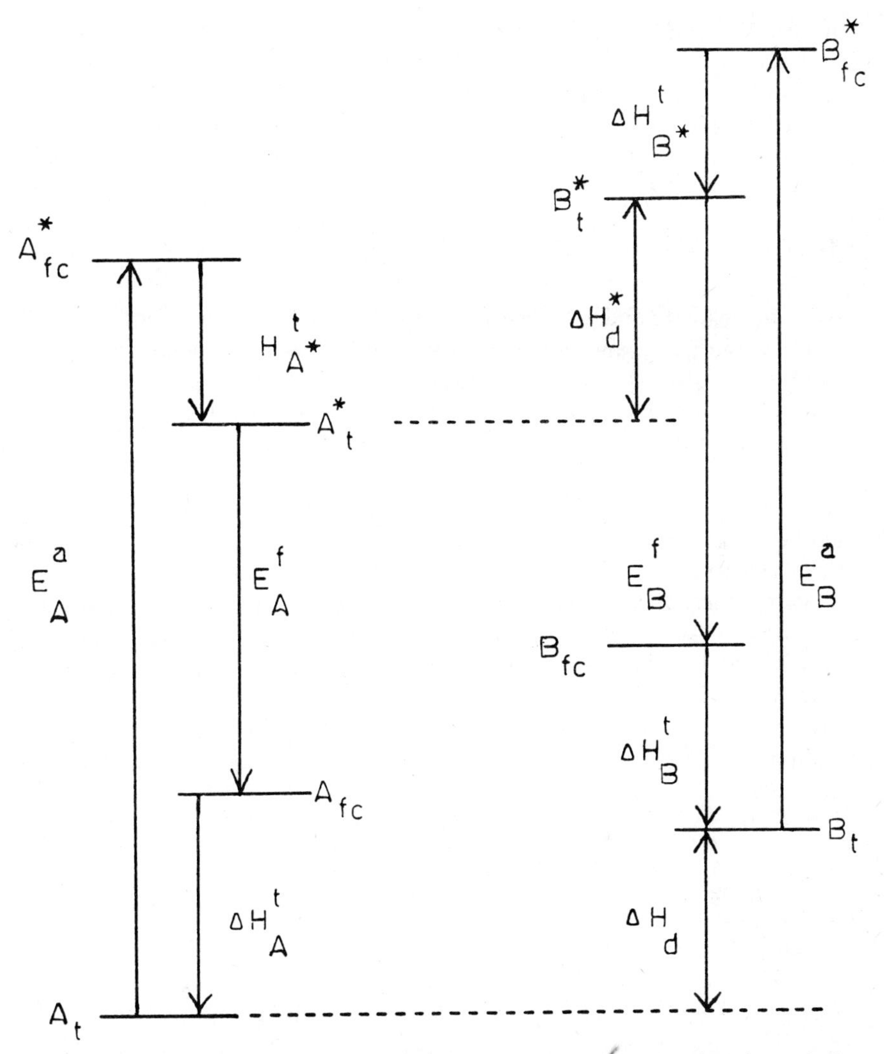

Fig. 6.2. Förster cycle for the case of conformational and/or solvent relaxation after excitation and emission. A_t, B_t, A_t^*, and B_t^* represent the thermally equilibrated ground and excited states of an acid and its conjugate base. A_{fc}, B_{fc}, A_{fc}^*, and B_{fc}^* represent the Franck–Condon ground and excited states of an acid and its conjugate base. E_A^a and E_B^a are the spectroscopic energies of the transitions $A_t \rightarrow A_{fc}^*$ and $B_t \rightarrow B_{fc}^*$, respectively, and E_A^f and E_B^f are the spectroscopic energies of the transitions $A_t^* \rightarrow A_{fc}$ and $B_t^* \rightarrow B_{fc}$, respectively. ΔH_d and ΔH_d^* are the enthalpies of protonation in the ground and electronically excited states, respectively, and ΔH_A^t, ΔH_B^t, ΔH_{A*}^t, and ΔH_{B*}^t are the enthalpies of thermal relaxation of the ground state and excited state acid and base species, respectively.

and from the fluorescence spectra,

$$E_{\mathrm{A}}^{f} - \Delta H_{\mathrm{B}}^{t} + \Delta H_{d}^{*} = \Delta H_{d} + E_{\mathrm{B}}^{f} - \Delta H_{\mathrm{A}}^{t} \tag{6.6}$$

or

$$\mathrm{p}K_{a} - \mathrm{p}K_{a}^{*} = \frac{Nhc(\bar{\nu}_{\mathrm{A}}^{f} - \bar{\nu}_{\mathrm{B}}^{f}) + \Delta H_{\mathrm{A}}^{t} - \Delta H_{\mathrm{B}}^{t}}{2.303RT} \tag{6.7}$$

where E_{A}^{a}, E_{B}^{a}, $\bar{\nu}_{\mathrm{A}}^{a}$, and $\bar{\nu}_{\mathrm{B}}^{a}$ correspond to the absorption band maxima of A and B, E_{A}^{f}, E_{B}^{f}, $\bar{\nu}_{A}^{f}$, and $\bar{\nu}_{\mathrm{B}}^{f}$ correspond to the fluorescence band maxima of A and B, $\Delta H_{\mathrm{A}^*}^{t}$ and $\Delta H_{\mathrm{B}^*}^{t}$ are the enthalpies of thermal relaxation of A* and B* in S_1, and $\Delta H_{\mathrm{A}}^{t}$ and $\Delta H_{\mathrm{B}}^{t}$ are the enthalpies of thermal relaxation of A and B in S_0. In Eq. (6.5)

$$\frac{\Delta H_{\mathrm{B}^*}^{t} - \Delta H_{\mathrm{A}^*}^{t}}{2.303RT}$$

which is generally unknown, represents the major part of the error in estimating $\mathrm{p}K_a - \mathrm{p}K_a^*$ from absorption maxima alone and $(\Delta H_{\mathrm{A}}^{t} - \Delta H_{\mathrm{B}}^{t})/2.303RT$ is the corresponding error in estimating $\mathrm{p}K_a - \mathrm{p}K_a^*$ from fluorescence band maxima alone.

In practice the Förster cycle is usually applied by averaging $\bar{\nu}_{\mathrm{A}}^{a}$ and $\bar{\nu}_{\mathrm{A}}^{f}$ to obtain a "best" value for the hypothetical 0–0 transition frequency of A:

$$\bar{\nu}_{\mathrm{A}}^{0-0} = \frac{\bar{\nu}_{\mathrm{A}}^{a} + \bar{\nu}_{\mathrm{A}}^{f}}{2} \tag{6.8}$$

and for B:

$$\bar{\nu}_{\mathrm{B}}^{0-0} = \frac{\bar{\nu}_{\mathrm{B}}^{a} + \bar{\nu}_{\mathrm{B}}^{f}}{2} \tag{6.9}$$

Equations (6.8) and (6.9) are based on the supposition that the nonadiabatic portions of $\bar{\nu}_{\mathrm{A}}^{a}$ and $\bar{\nu}_{\mathrm{A}}^{f}$ will subtract from each other and then be cut in half (as will those of $\bar{\nu}_{\mathrm{B}}^{a}$ and $\bar{\nu}_{\mathrm{B}}^{f}$). The use of $\bar{\nu}_{\mathrm{A}}^{0-0}$ and $\bar{\nu}_{\mathrm{B}}^{0-0}$ for $\bar{\nu}_{\mathrm{A}}$ and $\bar{\nu}_{\mathrm{B}}$ in Eq. (6.3) is equivalent to averaging Eqs. (6.5) and (6.7) to yield

$$\mathrm{p}K_a - \mathrm{p}K_a^* = \frac{Nhc[(\bar{\nu}_{\mathrm{A}}^{a} + \bar{\nu}_{A}^{f}) - (\bar{\nu}_{\mathrm{B}}^{a} + \bar{\nu}_{\mathrm{B}}^{f})] + (\Delta H_{\mathrm{B}^*}^{t} - \Delta H_{\mathrm{B}}^{t}) - (\Delta H_{\mathrm{A}^*}^{t} - \Delta H_{\mathrm{A}}^{t})]}{4.606RT} \tag{6.10}$$

Values of pK_a^* calculated in this way usually give the best agreement (to within 0.3 log unit) with values calculated from kinetic measurements (see below). However, occasionally the vibrational compositions and/or solvation energies of A, A*, B, and B* are so different that the averaging procedure introduces more error than it eliminates. In this case the use of absorption or fluorescence maxima alone may give the best results provided that an intelligent assessment of which kind of spectra to use can be made.

It should be noted that it is not necessary for equilibrium to be attained in the S_1 state in order to calculate pK_a^* from the Förster cycle. The importance of the Föster cycle to the subject of this chapter is that it permits the independent estimation of a chemical property of the S_1 state that can also be determined by kinetic means. Since there are few independent approaches to the study of the lowest excited singlet state, the Förster cycle is a powerful tool for validating kinetic measurements.

6.2.4. Determination of Proton-Transfer Rate Constants

6.2.4.1. The Steady-State Approach

The steady-state approach to the determination of proton-transfer rate constants of acids or bases in the S_1 state is based on the assumption of the attainment of a steady state involving the various photophysical and photochemical processes deactivating S_1. In the absence of buffer species in aqueous solutions, two overall mechanisms of proton transfer are possible. In the first of these (Scheme I) water is the proton acceptor and the solvated proton, the proton donor:

$$\begin{array}{ccc} A^* + rH_2O & \underset{k_b}{\overset{k_a}{\rightleftharpoons}} & H^+ + B^* \\ \downarrow 1/\tau_0 & & \downarrow 1/\tau_0' \\ A & \rightleftharpoons & B \end{array}$$

Scheme I

In the second (Scheme II) water is the proton donor and the solvated hydroxide ion the proton acceptor.

$$\begin{array}{ccc} B^* + rH_2O & \underset{k_b}{\overset{k_a}{\rightleftharpoons}} & A^* + OH^- \\ \downarrow 1/\tau_0' & & \downarrow 1/\tau_0 \\ B & \rightleftharpoons & A \end{array}$$

Scheme II

where r is the difference between the number of water molecules solvating the products and that solvating the reactant and τ_0 and τ_0' are the respective lifetimes of A* and B* in S_1 in the absence of proton transfer (i.e., at very low and very high pH) which are also the reciprocals of the rate constants for photophysical deactivation of S_1.

From Scheme I the relative quantum yields of fluorescence of excited acids (A*) and bases (B*) may be obtained as a function of the hydrogen ion concentration $[H^+]$ and the photophysical and photochemical properties of the excited singlet states by simultaneously solving the following two equations for the disappearance of A* and B* from the excited state under steady-state conditions:

$$-\int_{\alpha_A}^{0} d[A^*] = \int_0^{\infty}\left(\frac{1}{\tau_0} + k_a\right)[A^*]\,dt - \int_0^{\infty} k_b[H^+][B^*]\,dt \qquad (6.11)$$

and

$$-\int_{\alpha_B}^{0} d[B^*] = \int_0^{\infty}\left(\frac{1}{\tau_0'} + k_b[H^+]\right)[B^*]\,dt - \int_0^{\infty} k_a[A^*]\,dt \qquad (6.12)$$

In Eqs. (6.11) and (6.12) [A*] and [B*] are the respective probabilities of finding an A* or B* molecule in the excited state at time t after excitation. k_a and k_b are the rate constants (or probabilities) of dissociation of A* and protonation of B*, respectively; and $[H^+]$ is the molar concentration of hydrogen ion. The lower limits of integration of d[A*] and d[B*], α_A and α_B, are the fractions of directly excited A and B (the fractions of total light quanta at the nominal wavelength of excitation absorbed by A and B), respectively, and are defined by

$$\alpha_A = \frac{\varepsilon_A[A]}{\varepsilon_A[A] + \varepsilon_B[B]} \quad \text{and} \quad \alpha_B = \frac{\varepsilon_B[B]}{\varepsilon_A[A] + \varepsilon_B[B]}$$

where ε_A and ε_B are the molar absorptivities of A and B at the wavelength of excitation and [A] and [B] are the equilibrium ground-state concentrations of A and B at the pH at which the experiment is carried out. We can also write

$$\alpha_A = \frac{\varepsilon_A[H^+]}{\varepsilon_A[H^+] + \varepsilon_B K_a} \quad \text{and} \quad \alpha_B = \frac{\varepsilon_B K_a}{\varepsilon_A[H^+] + \varepsilon_B K_a}$$

where K_a is the equilibrium constant for ground-state dissociation of A. The relative quantum yields of fluorescence are defined by the integrals (3)

$$\int_0^\infty [A^*]\,dt = \tau = \frac{\varphi}{\varphi_0}\tau_0 \qquad \text{and} \qquad \int_0^\infty [B^*]\,dt = \tau' = \frac{\varphi'}{\varphi_0'}\tau_0'$$

where τ and τ' are the actual lifetimes of A* and B* (i.e., at pH where proton transfer in S_1 occurs) and are equivalent to relative fluorescence intensity measurements when excitation is effected at an isosbestic point (i.e., when $\varepsilon_A = \varepsilon_B$). Integration and rearrangement of Eq. (6.11) and (6.12) yields

$$\frac{\varphi}{\varphi_0} = \frac{\alpha_A + k_b\tau_0'[H^+]}{1 + k_a\tau_0 + k_b\tau_0'[H^+]} \tag{6.13}$$

and

$$\frac{\varphi'}{\varphi_0'} = \frac{\alpha_B + k_a\tau_0}{1 + k_a\tau_0 + k_b\tau_0[H^+]} \tag{6.14}$$

Figure 6.3 shows typical titration curves for a hydroxyaromatic in which α_A Eq. (6.13) is equal to unity. Since $\alpha_A + \alpha_B = 1$, Eqs. (6.13) and (6.14) can be combined to yield

$$\frac{\varphi/\varphi_0}{\varphi'/\varphi_0' - \alpha_B} = \frac{1}{k_a\tau_0} + \frac{k_b\tau_0'[H^+]}{k_a\tau_0}\,\frac{\varphi'/\varphi_0'}{\varphi'/\varphi_0' - \alpha_B} \tag{6.15}$$

which can be used to determine k_a and k_b when A is more acidic in S_1 than in S_0 (or if B* is less basic in S_1 than in S_0). A plot of

$$\frac{\varphi/\varphi_0}{\varphi'/\varphi_0' - \alpha_B} \qquad \text{versus} \qquad [H^+]\frac{\varphi'/\varphi_0'}{\varphi'/\varphi_0' - \alpha_B}$$

should have a slope of $k_b\tau_0'/k_a\tau_0$ and a vertical axis intercept of $1/k_a\tau_0$, so that if the lifetimes of the excited state acid and conjugate base can be measured or estimated, the rate constants of proton exchange, k_a and k_b, can be calculated. Of course, if protonation of the excited conjugate base occurs at $[H^+] > K_a$, $\alpha_B = 0$ and Eq. (6.15) reduces to Weller's equation (3):

$$\frac{\varphi/\varphi_0}{\varphi'/\varphi_0'} = \frac{1}{k_a\tau_0} + \frac{k_b\tau_0'}{k_a\tau_0}[H^+] \tag{6.16}$$

In Eq. (6.16) k_b is often typical of diffusion-controlled reactions ($k_b \leq 5 \times 10^{10}\ M^{-1}\ s^{-1}$) and generally $\tau_0 \leq 1 \times 10^{-7}$ s, so that $k_b\tau_0 < 10^3$.

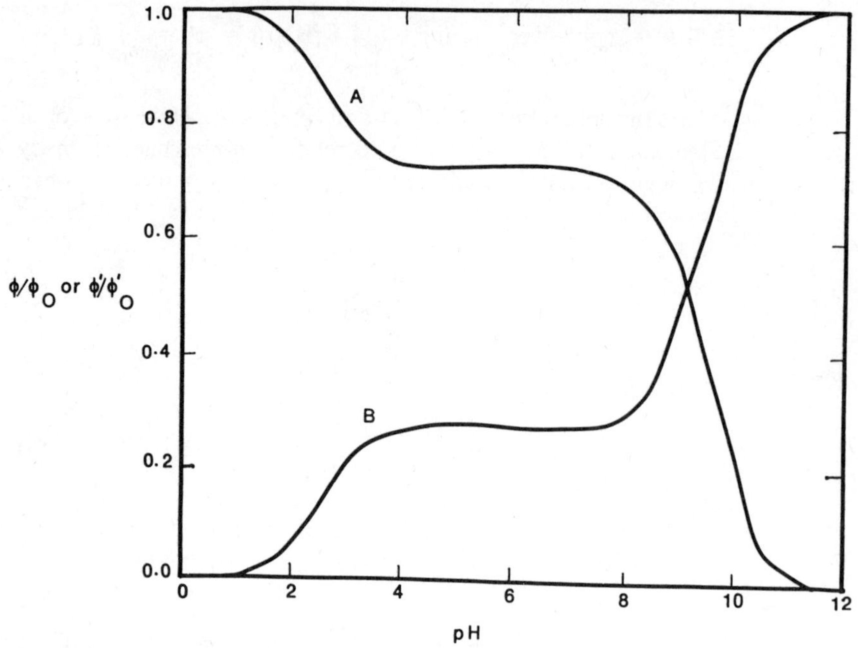

Fig. 6.3. Variation with pH of the relative fluorescence intensities of the excited-state conjugate acid (curve A) and base species (curve B) (φ/φ_0 and φ'/φ'_0, respectively) of β-naphthol.

Consequently, at pH > 5, $k_b\tau'_0[H^+]/(k_a\tau_0) \to 0$ and φ/φ_0, φ'/φ'_0, and therefore $\varphi'/\varphi'_0/\varphi/\varphi_0$ become independent of pH; that is,

$$\frac{\varphi'/\varphi'_0}{\varphi/\varphi_0} = \frac{(\varphi'/\varphi'_0)_{\text{const}}}{(\varphi/\varphi_0)_{\text{const}}} = k_a\tau_0 \tag{6.17}$$

which permits a quick determination of k_a. For excited bases with $\tau'_0 < 10^{-7}$ s and/or with $k_b < 5 \times 10^{-10}\ M^{-1}\ s^{-1}$, $[H^+]$ can be greater than $10^{-5}\ M$ and still the acid and base may demonstrate $\varphi'/\varphi'_0/\varphi/\varphi_0$ independent of pH. In other words, the smaller $k_b\tau'_0$, the lower is the pH at which the reprotonation of B*, as reflected by the dependence of φ/φ_0 and φ'/φ'_0 on $[H^+]$, will be observed.

It is important to note that in Eq. (6.15) and (6.16), k_a and k_b are somewhat dependent on the composition of the reaction medium (i.e., they generally vary with $[H^+]$). The true constants of the reaction $k_a(0)$ and $k_b(0)$ are the rate constants in the pure reference solvent, in this case, water, at infinite dilution of solutes (i.e., when $[H^+] = a_{H^+}$, where a_{H^+} is the activity of the hydrogen ion). In relatively dilute acidic solutions of pH 1–4 the correction of Eq. (6.16) for medium effects entails inclusion of the Brønsted kinetic activity factor (21–23) based on the Debye–Hückel type of ionic screening treatment. Of course, this approach entails no correction at all where the conjugate base is neutral and the conjugate acid singly charged. Weller (24,25) has additionally developed an elegant treatment to correct for the partial dependence of the observed relative quantum yields of fluorescence, at low pH, on the occurrence of transient protonation of the excited conjugate base occurring prior to the onset of the steady state. In cases where transient reaction occurs, plots of $\varphi/\varphi_0/\varphi'/\varphi'_0$ versus $[H^+]$ corresponding to Eq. (6.16) tend to become nonlinear at low pH unless the correction is performed. This correction, however, is really necessary only for conjugate bases whose reprotonation rates are diffusion controlled, as a substantial activation barrier to reprotonation would diminish the probability of reaction between an excited conjugate base and a proton located within its diffusion volume. The vast majority of bases that become protonated in fairly concentrated acids probably do so by activation-controlled processes.

In more concentrated acid solutions, variations on Eq. (6.15) which include the activity of water (which is less than unity in concentrated acid or electrolyte solutions) and represent the activity of the proton by the appropriate acidity function (26–28) have been used with some success.

When A is less acidic in S_1 than in S_0 (or B is more basic in S_1 than in S_0) it is preferable to weight the fluorimetric titration data by rearranging Eq. (6.15) to

$$\frac{\varphi'/\varphi'_0[H^+]}{\varphi/\varphi_0 - \alpha_A} = \frac{1}{k_b\tau'_0} + \frac{k_a\tau_0}{k_b\tau'_0}\frac{\varphi/\varphi_0}{\varphi/\varphi_0 - \alpha_A} \tag{6.18}$$

When $\alpha_A = 0$ (at high pH) Eq. (6.18) becomes

$$\frac{\varphi'/\varphi'_0[H^+]}{\varphi/\varphi_0} = \frac{1 + k_a\tau_0}{k_b\tau'_0} \tag{6.19}$$

from which $(1 + k_a\tau_0)/k_b\tau'_0$ may be determined but k_a and k_b cannot be explicitly evaluated. In some cases, where $k_a\tau_0 \ll 1$, as when A* is nonfluorescent (so that τ_0 is very small) k_b can be determined directly from a single fluorimetric titration (29) with the aid of Eq. (6.19). However, if $k_a\tau_0$ is appreciable but τ_0 can be varied by the addition of a quencher of the fluorescence of A*, a series of fluorimetric pH titrations, each in the presence of a different concentration of the quencher, can be used to generate a family of titration curves, each having a different value of $(1 + k_a\tau_0)/k_b\tau'_0$. If τ'_0 and τ_0 are determined at each quencher concentration, $(\varphi'/\varphi'_0/\varphi/\varphi_0)\tau'_0[H^+]$ can be plotted against τ_0 to give a straight line of slope k_a/k_b and vertical axis intercept $1/k_b$ (30,31). It should be noted that unlike the situation described by Eqs. (6.16) and (6.17), there is no pH region, corresponding to Eq. (6.19) in which pH-independent proton transfer occurs. For the circumstances described by Eq. (6.19) when $[H^+]$ is very small no proton transfer in S_1 occurs. When $[H^+]$ is appreciable, φ'/φ'_0 and φ/φ_0 are continuously dependent on $[H^+]$.

In the event that $k_a\tau_0 \gg 1$, Eqs. (6.16) and (6.19) both become

$$\frac{\varphi'/\varphi_0[H^+]}{\varphi/\varphi_0} = \frac{k_a\tau_0}{k_b\tau'_0} = K_a^* \frac{\tau_0}{\tau'_0} \tag{6.20}$$

where $K_a^* = k_a/k_b$ is the equilibrium constant for proton exchange in S_1. $k_a\tau_0 \gg 1$ defines the condition of equilibrium in S_1 (27,32) and occurs when the rate of dissociation of A* is much greater than the rate of photophysical deactivation (fluorescence, internal conversion, and singlet–triplet intersystem crossing) of S_1. Equilibrium in S_1 is favored when A* is a very strong acid (high k_a) and when S_1 is very long-lived (high τ_0). Equilibrium in S_1 is a rare occurrence for acids whose conjugate bases are protonated in the pH region 1–4 but is fairly common among acids whose conjugate bases are protonated in concentrated acid media (27,33).

Since Eq. (6.20) is valid regardless of whether A* is stronger or weaker in S_1 than in S_0, the shapes of the fluorimetric titration curves for both situations will be identical (like that in Fig. 6.4) when equilibrium occurs. In this case, k_a and k_b cannot be determined from a single fluorimetric titration. However, by addition of a quencher, τ_0 may be diminished and the condition of equilibrium removed, so that k_a and k_b can than be determined. In logarithmic form Eq. (6.20) becomes

$$\mathrm{pH} = \log\frac{\varphi'/\varphi'_0}{\varphi'/\varphi_0} + \mathrm{p}K_a + \log\frac{\tau'_0}{\tau_0} \tag{6.21}$$

which differs from the Henderson–Hasselbach equation in that the vertical-axis intercept contains $\log \tau'_0/\tau_0$ as well as $\mathrm{p}K_a^*$.

Excited bases and acids that undergo proton transfer in alkaline aqueous solutions in S_1 react according to scheme II, with water as a proton donor and the hydroxide ion as a proton acceptor, respectively. In these solutions the steady-state equations that must be solved in order to obtain the rate constants k_a and k_b for proton abstraction from A* by OH^- and protonation of B* by water, respectively, in S_1 are

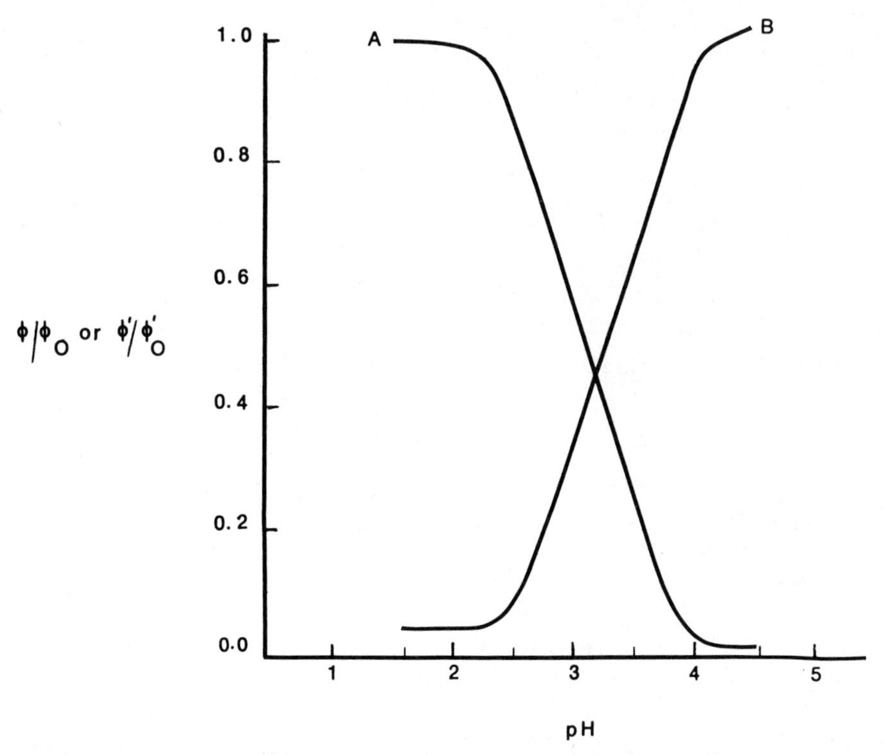

Fig. 6.4. Variation with pH of the relative fluorescence intensities of the excited-state conjugate acid (curve A) and base (curve B) species (φ/φ_0 and φ'/φ'_0, respectively) for a fluorescent acid or base which attains equilibrium in the excited state.

$$\int_{\alpha_B}^{0} d[B^*] = \int_{0}^{\infty}\left(\frac{1}{\tau_0} + k_b\right)[B^*]\,dt - \int_{0}^{\infty} k_a[OH^-][A^*]\,dt \qquad (6.22)$$

and

$$\int_{\alpha_A}^{0} d[A^*] = \int_{0}^{\infty}\left(\frac{1}{\tau'_0} + k_a[OH^-]\right)[A^*]\,dt - \int_{0}^{\infty} k_b[B^*]\,dt \qquad (6.23)$$

Integration of Eqs. (6.22) and (6.23) and subsequent rearrangement leads to

$$\frac{\varphi'}{\varphi'_0} = \frac{\alpha_B + k_a\tau_0[OH^-]}{1 + k_b\tau'_0 + k_a\tau_0[OH^-]} \qquad (6.24)$$

and

$$\frac{\varphi}{\varphi_0} = \frac{\alpha_A + k_b\tau_0'}{1 + k_b\tau_0' + k_a\tau_0[OH^-]} \tag{6.25}$$

which can be combined to give

$$\frac{\varphi'/\varphi_0'}{\varphi/\varphi_0 - \alpha_A} = \frac{1}{k_b\tau_0} + \frac{k_a\tau_0[OH^-]}{k_b\tau_0'} \frac{\varphi/\varphi_0}{\varphi/\varphi_0 - \alpha_A} \tag{6.26}$$

From Eq. (6.26) $k_a\tau_0$ and $k_b\tau_0'$ can be determined by plotting

$$\frac{\varphi_0'/\varphi_0'}{\varphi/\varphi_0 - \alpha_A} \qquad \text{versus} \qquad [OH^-]\frac{\varphi/\varphi_0}{\varphi/\varphi_0 - \alpha_A}$$

When $[H^+] \ll K_a$, $\alpha_A = 0$ and Eq. (6.26) becomes

$$\frac{\varphi'/\varphi_0'}{\varphi/\varphi_0} = \frac{1}{k_b\tau_0'} + \frac{k_a\tau_0[OH^-]}{k_b\tau_0'} \tag{6.27}$$

a relationship also first derived by Weller (34).

Several N-heterocyclic bases such as the quinolines and acridines which are more basic in S_1 than in S_0 demonstrate the fluorimetric pH titration behavior described by Eqs. (6.24) to (6.27), and one example is shown in Fig. 6.5. The pH-dependent φ/φ_0 and φ'/φ_0' seen at low pH in Fig. 6.5 correspond to the pH region in which A is converted to B in the ground state (i.e., when $\alpha_A \sim \alpha_B$). The flat portions of the titration curves have $\varphi/\varphi_0 = (\varphi/\varphi_0)_{const}$ and $\varphi'/\varphi_0' = (\varphi'/\varphi_0')_{const}$, where $(\varphi/\varphi_0)_{const}$ and $(\varphi'/\varphi_0')_{const}$ are independent of pH and correspond to the situation where $\alpha_A = 0$ and $k_a\tau_0[OH^-] \ll 1$. This will occur invariably when pH < 9 as $[OH^-]$ is then very small. In this case Eq. (6.20) can be simplified and rearranged to

$$\frac{(\varphi/\varphi_0)_{const}}{(\varphi'/\varphi_0')_{const}} = k_b\tau_0' \tag{6.28}$$

which permits a rapid evaluation of k_b. The pH-dependent portion of Fig. 6.5, at high pH, results from proton abstraction from A* by OH^- as $k_a\tau_0[OH^-]$ becomes significant at high pH and corresponds to Eq. (6.27), from which both k_a and k_b may be determined. As in the case of Eq. (6.16), corrections to Eq. (6.27) will often be necessary to account for ionic strength and/or other medium effects in solutions of high alkalinity.

Certain acids which are too weak to ionize in aqueous media in the ground

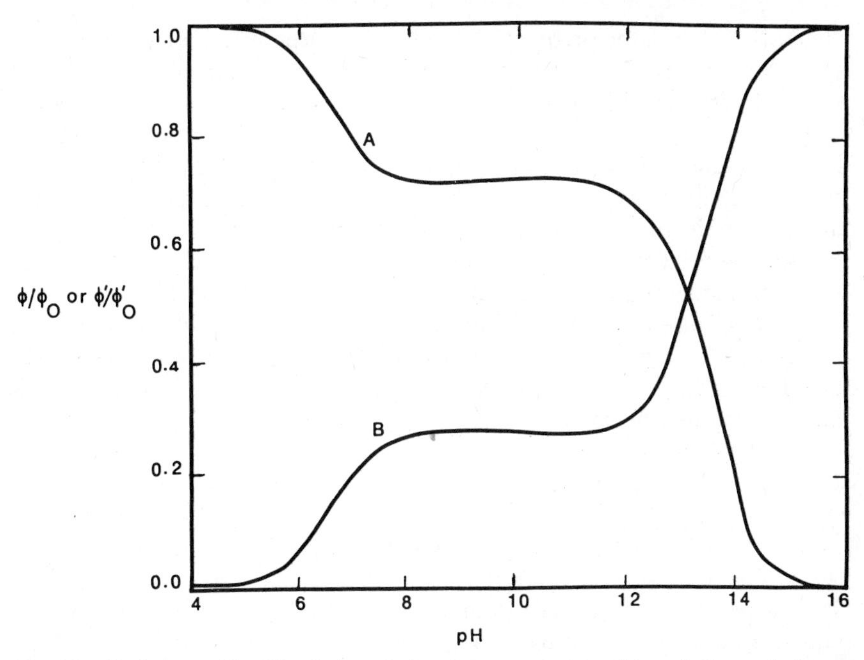

Fig. 6.5. Variation with pH of the relative fluorescence intensities of the excited-state conjugate acid (curve A) and base species (curve B) (φ/φ_0 and φ/φ_0', respectively) of 8-methoxyquinaldine.

electronic state may become sufficiently acidic in S_1 to dissociate in aqueous alkali solutions. Some well-known examples are the neutral arylamines (35) and pyrrolic N-heterocycles such as carbazole (36) whose fluorescence spectra shift to longer wavelengths with increasing pH in the region near pH 12. In the same pH region no change is observed in the absorption spectra of these compounds, as they are too weakly acidic to dissociate in S_0. Their fluorimetric titration behavior is best described by rearranging Eq. (6.26) to

$$\frac{\varphi/\varphi_0[\mathrm{OH}^-]}{\varphi'/\varphi_0' - \alpha_\mathrm{B}} = \frac{1}{k_a\tau_0} + \frac{k_b\tau_0'}{k_a\tau_0}\frac{\varphi'/\varphi_0'}{\varphi'/\varphi_0' - \alpha_\mathrm{B}} \tag{6.29}$$

Since dissociation does not occur in the ground state $\alpha_\mathrm{B} = 0$ and

$$\frac{\varphi/\varphi_0[\mathrm{OH}^-]}{\varphi'/\varphi_0'} = \frac{1 + k_b\tau_0'}{k_a\tau_0} \tag{6.30}$$

which gives $(1 + k_b\tau_0')/k_a\tau_0$ from the fluorimetric titration curve but to evaluate k_b and k_a unambiguously, several fluorimetric titrations must be conducted each in the presence of a different concentration of a quencher of the fluorescence of B*. This will cause τ_0' to vary and will displace the titration curve on the pH scale. A plot of $(\varphi/\varphi_0/\varphi'/\varphi_0')\tau_0[OH^-]$ against τ_0' will have a slope of k_b/k_a and a vertical-axis intercept of $1/k_a$. As k_a and k_b are defined in scheme II, k_b/k_a is the hydrolysis constant K_b^* in S_1 and is related to the dissociation constant K_a^* of A* by the usual $K_a^* K_b^* = K_W$, where K_W is the autoprotolysis constant of water. Some typical rate constants for proton transfer between excited acids and bases and water, the hydronium ion, and the hydroxide ion are given in ref. 9.

If buffer ions D and DH^+ are present in the solutions in which A and/or B are excited, it is necessary to consider that the former may act as proton acceptors from A* and proton donors to B* (25,34,37–39). These reactions are depicted in scheme III.

$$A^* + D \underset{k_d}{\overset{k_c}{\rightleftharpoons}} DH^+ + B^*$$

Scheme III

If Scheme III is temporally competitive with scheme I, the steady-state equations may be reformulated and integrated to yield

$$\varphi/\varphi_0 = \frac{1 + k_b\tau_0'c + k_d\tau_0'[DH^+]}{1 + k_a\tau_0 + k_c\tau_0[D] + k_b\tau_0'c + k_d\tau_0'[DH^+]} \tag{6.31}$$

$$\varphi'/\varphi_0' = \frac{k_a\tau_0 + k_c\tau_0[D]}{1 + k_a\tau_0 + k_c\tau_0[D] + k_b\tau_0'c + k_d\tau_0'[DH^+]} \tag{6.32}$$

when $c = [H^+]$ and A* is exclusively excited so that $\alpha_A = 1$ and $\alpha_B = 0$, and

$$\varphi'/\varphi_0' = \frac{1 + k_a\tau_0 c + k_c\tau_0[D]}{1 + k_b\tau_0' + k_d\tau_0'[DH^+] + k_a\tau_0 c + k_c\tau_0[D]} \tag{6.33}$$

$$\varphi/\varphi_0 = \frac{k_b\tau_0' + k_d\tau_0'[DH^+]}{1 + k_b\tau_0' + k_d\tau_0'[DH^+] + k_a\tau_0 c + k_c\tau_0[D]} \tag{6.34}$$

when $c = [OH^-]$ and B* is exclusively excited so that $\alpha_B = 1$ and $\alpha_A = 0$. Usually, one carries out buffer experiments at pH where $c \to 0$ and pH $\ll$ pK_{DH^+} or pH $\gg$ pK_{DH^+}. Then when pH $\ll$ pK_{DH^+}: $[D] \to 0$ and

$$\frac{\varphi}{\varphi_0} = \frac{1 + k_d\tau_0'[DH^+]}{1 + k_a\tau_0 + k_d\tau_0'[DH^+]} \tag{6.35}$$

$$\frac{\varphi}{\varphi'_0} = \frac{k_a\tau_0}{1 + k_a\tau_0 + k_d\tau'_0[\mathrm{DH}^+]} \tag{6.36}$$

or

$$\frac{\varphi/\varphi_0}{\varphi'/\varphi'_0} = \frac{1 + k_d\tau'_0[\mathrm{DH}^+]}{k_a\tau_0} \tag{6.37}$$

which permits evaluation of k_d when H_2O is the proton acceptor and DH^+ the proton donor

$$\frac{\varphi'}{\varphi'_0} = \frac{1}{1 + k_b\tau'_0 + k_d\tau'_0[\mathrm{DH}^+]} \tag{6.38}$$

$$\frac{\varphi}{\varphi_0} = \frac{k_b\tau'_0 + k_d\tau'_0[\mathrm{DH}^+]}{1 + k_b\tau'_0 + k_d\tau'_0[\mathrm{DH}^+]} \tag{6.38a}$$

or

$$\frac{\varphi/\varphi_0}{\varphi'/\varphi'_0} = k_b\tau'_0 + k_d\tau'_0[\mathrm{DH}^+] \tag{6.39}$$

which permits evaluation of k_d when H_2O and DH^+ are both proton donors. When $\mathrm{pH} \gg \mathrm{p}K_{\mathrm{DH}^+}$ $[\mathrm{DH}^+] \rightarrow 0$ and

$$\frac{\varphi}{\varphi_0} = \frac{1}{1 + k_a\tau_0 + k_c\tau_0[\mathrm{D}]} \tag{6.40}$$

$$\frac{\varphi'}{\varphi'_0} = \frac{k_a\tau_0 + k_c\tau_0[\mathrm{D}]}{1 + k_a\tau_0 + k_c\tau_0[\mathrm{D}]} \tag{6.41}$$

or

$$\frac{\varphi'/\varphi'_0}{\varphi/\varphi_0} = k_a\tau_0 + k_c\tau_0[\mathrm{D}] \tag{6.42}$$

which permits evaluation of k_c when H_2O and D are both proton acceptors.

$$\frac{\varphi'}{\varphi'_0} = \frac{1 + k_c\tau_0[\mathrm{D}]}{1 + k_b\tau'_0 + k_c\tau_0[\mathrm{D}]} \tag{6.43}$$

$$\frac{\varphi}{\varphi_0} = \frac{k_b\tau_0'}{1 + k_b\tau_0' + k_c\tau_0[\mathrm{D}]} \tag{6.44}$$

or

$$\frac{\varphi'/\varphi_0'}{\varphi/\varphi_0} = \frac{1}{k_b\tau_0} + \frac{k_c\tau_0[\mathrm{D}]}{k_b\tau_0'} \tag{6.45}$$

which permits evaluation of k_c when H_2O is the proton donor and D the proton acceptor. Of course, in those solutions containing H^+, OH^-, or buffer concentrations in excess of 10^{-4} *M*, Brønsted kinetic activity (21–23) factor corrections must be made. The second-order rate constants k_c and k_d usually correspond in magnitude to diffusion-controlled reactions (40,41) but tend to be somewhat smaller than k_b (for reaction of B* with H^+) or k_a (for reaction of A* with OH^-). This is a consequence of the fact that the buffer species migrate by frictional diffusion, whereas H^+ and OH^- ions move through aqueous solutions by the much faster structural diffusion processes (4–8). Some typical rate constants for reactions of excited acids and bases with various buffer ions are given in ref. 9.

6.2.4.2. *Time-Resolved Fluorimetry*

Excited-state proton-transfer rate constants are evaluated in time-resolved fluorimetry by direct measurement of the decay of fluorescence emission after pulsed excitation followed by numerical analysis. The time-resolved decay is reconstructed by either the pulse sampling (stroboscopic) method or the photon counting method. A description of these methods and the requisite instrumentation is given in Chapter 2.

The first step in the numerical analysis of fluorescence decay is the postulation of a kinetic model and the derivation of rate equations based on the proposed model. Excimer and exciplex formation, excited-state isomerization and tautomerism, and both intra- and intermolecular excited-state proton-transfer reactions are examples of coupled two-state reactions conforming to Scheme IV:

$$\begin{array}{ccc} M^* + \sum N_i & \underset{k_R}{\overset{k_M}{\rightleftharpoons}} & R^* + \sum S_i \\ \downarrow 1/\tau_0 & & \downarrow 1/\tau_0 \\ M & & R \end{array}$$

Scheme IV

where M* and R* are the excited-state species M and R at time t, τ_0 and τ_0' are the respective lifetimes of M* and R* in S_1 in the absence of excited-state reaction, N_i and S_i are all chemical species which react with M* and R*, respectively, which in the cases of excited-state isomerization and tautomerism are equal to zero, and k_M and k_R represent the sums of the probabilities of reaction of species M* and R* with reactants N_i and S_i, respectively, that is, $k_M = \sum_{i=1}^{n} k_{m_i}[N_i]$, $k_R = \sum_{i=1}^{n} k_{r_i}[S_i]$. An infinitesimally narrow excitation pulse width is assumed so that excitation is instantaneous.

It is possible, furthermore, to classify all intermolecular excited-state proton-transfer reactions (IESPTs) according to whether they can be placed into one of two general classes.

To the first class belong all IESPTs which react according to scheme I, in which the excited conjugate acid, characterized by an excited-state pK_a^* smaller than the ground-state pK_a, reacts with a or any of a combination of competitive proton acceptors to produce excited conjugate base and the proton donor species. To the second class belong all IESPTs which react according to scheme II, in which the excited conjugate base characterized by an excited-state pK_a^* greater than the ground-state pK_a reacts with a or any combination of proton donors to produce excited conjugate acid and the proton acceptor species.

The differential equations describing excited-state reaction according to Scheme I are

$$-\frac{d[A^*]}{dt} = \left(\frac{1}{\tau_0} + k_a\right)[A^*] - k_b[B^*][H^+] \tag{6.46}$$

$$-\frac{d[B^*]}{dt} = \left(\frac{1}{\tau_0'} + k_b[H^+]\right)[B^*] - k_a[A^*] \tag{6.47}$$

If the boundary conditions are chosen such that $[A^*] = [A_0^*]$ and $[B^*] = 0$ at $t = 0$, then as originally demonstrated by Birks et al. (42) for the case of excimer formation and subsequently, by Laws and Brand (43) for the case of excited-state proton transfer, solutions to Eqs. (6.46) and (6.47) are obtained by performing the following operations.

First we solve for [A*].

Step 1. Equations (6.46) and (6.47) are reexpressed as

$$-\frac{d[A^*]}{dt} = \gamma_A[A^*] - k_b[B^*][H^+] \tag{6.48}$$

$$-\frac{d[B^*]}{dt} = \gamma_B[B^*] - k_a[A^*] \tag{6.49}$$

where $\gamma_A = 1/\tau_0 + k_a$ and $\gamma_B = 1/\tau_0' + k_b\,[H^+]$.

Step 2. Equation (6.48) is solved for [B*] and the resulting expression is substituted in Eq. (6.49) and rearranged to give the following quadratic equation:

$$\{D^2 + (\gamma_A + \gamma_B)D + (\gamma_A\gamma_B - k_ak_b[H^+])\}[A^*] = 0 \tag{6.50}$$

where D^2 is the operator d^2/dt^2, D is the operator d/dt, and the roots of the quadratic are

$$D = \tfrac{1}{2}[(\gamma_A + \gamma_B) \pm \{(\gamma_B - \gamma_A)^2 - 4k_ak_b[H^+]\}^{1/2}] \tag{6.51}$$

Step 3. γ_1 and γ_2 are defined to be

$$\gamma_1 = T_1^{-1} = \tfrac{1}{2}[(\gamma_A + \gamma_B) + \{(\gamma_B - \gamma_A)^2 + 4k_ak_b[H^+]\}^{1/2}] \tag{6.52}$$

and

$$\gamma_2 = T_2^{-1} = \tfrac{1}{2}[(\gamma_A + \gamma_B) - \{(\gamma_B - \gamma_A)^2 + 4k_ak_b[H^+]\}^{1/2}] \tag{6.53}$$

Equation (6.50) is then reexpressed as

$$[(D + \gamma_1)(D + \gamma_2)][A^*] = 0 \tag{6.54}$$

which yields, after integration,

$$[A^*] = C_1e^{-\gamma_1 t} + C_2e^{-\gamma_2 t} \tag{6.55}$$

As at $t = 0$, $[A^*] = [A_0^*]$, Eq. (6.55) at $t = 0$ is

$$[A_0^*] = C_1 + C_2 \tag{6.56}$$

Step 4. When Eq. (6.55) is substituted into Eq. (6.48) the result is

$$(C_1e^{-\gamma_1 t} + C_2e^{-\gamma_2 t})\gamma_A - k_b[B^*][H^+] = \gamma_1C_1e^{-\gamma_1 t} + \gamma_2C_2e^{-\gamma_2 t} \tag{6.57}$$

which after imposition of the previously defined boundary conditions and the equality expressed in Eq. (6.56) leads to Eq. (6.58):

$$[A_0^*]\gamma_A = \gamma_1 C_1 + \gamma_2 C_2 \tag{6.58}$$

Step 5. Equations (6.58) and (6.56) are now solved simultaneously giving the following expressions for C_1 and C_2:

$$C_1 = [A_0^*]\frac{\gamma_A - \gamma_2}{\gamma_1 - \gamma_2} \tag{6.59}$$

$$C_2 = [A_0^*]\frac{\gamma_1 - \gamma_A}{\gamma_1 - \gamma_2} \tag{6.60}$$

which when substituted into Eq. (6.55) yields the solution for the time-dependent concentration of [A*].

$$[A^*] = \frac{[A_0^*]}{\gamma_1 - \gamma_2}[(\gamma_A - \gamma_2)e^{-\gamma_1 t} + (\gamma_1 - \gamma_A)e^{-\gamma_2 t}] \tag{6.61}$$

The solution for the time-dependent concentration of B* is similarly obtained and found to be

$$[B^*] = \frac{-k_a[A_0^*]}{\gamma_1 - \gamma_2}(e^{-\gamma_1 t} - e^{-\gamma_2 t}) \tag{6.62}$$

The time-dependent intensities of fluorescence emission of A* and B* are related to [A*] and [B*] by

$$I_A(\bar{v}, t) = C_A(\bar{v})[A^*]k_{fA} \tag{6.63}$$

$$I_B(\bar{v}, t) = C_B(\bar{v})[B^*]k_{fB} \tag{6.64}$$

where k_{fA} and k_{fB} are the probabilities of radiative deactivation of A* and B*, respectively, $C_A(\bar{v})$ and $C_B(\bar{v})$ are the emission spectra of species A* and B* normalized to unit area

$$C_A(\bar{v}) = \frac{\theta_A(\bar{v})}{\int_0^\infty \theta_A(\bar{v})\,dv} \tag{6.65}$$

$$C_B(\bar{v}) = \frac{\theta_B(\bar{v})}{\int_0^\infty \theta_B(\bar{v})\,dv} \tag{6.65}$$

and $\theta_A(\bar{v})$ and $\theta_B(\bar{v})$ are the number of quanta emitted at frequency $\bar{v}$ by the A* and B* species, respectively.

The solutions of the differential equations describing the excited-state proton transfer of those species which react according to reaction scheme I when A is exclusively excited and of those species which react according to reaction scheme II when B is exclusively excited, differ only in the definition of their respective decay times and preexponential factors. Both have as solutions

$$I_A(\bar{\nu}, t) = \alpha_1(\bar{\nu})e^{-\gamma_1 t} + \alpha_2(\bar{\nu})e^{-\gamma_2 t} \tag{6.67}$$

$$I_B(\bar{\nu}, t) = \beta_1(\bar{\nu})e^{-\gamma_1 t} + \beta_2(\bar{\nu})e^{-\gamma_2 t} \tag{6.68}$$

However, for those species that react according to scheme I,

$$\gamma_1, \gamma_2 = T_1^{-1}, T_2^{-2} = \tfrac{1}{2}[(\gamma_A + \gamma_B] \pm \{(\gamma_B - \gamma_A)^2 + 4k_a k_b[H^+]\}^{1/2}] \tag{6.69}$$

$$\alpha_1(\bar{\nu}) = C_A(\bar{\nu})[A_0^*]\frac{\gamma_A - \gamma_2}{\gamma_1 - \gamma_2}k_{fA} \tag{6.70}$$

$$\alpha_2(\bar{\nu}) = C_A(\bar{\nu})[A_0^*]\frac{\gamma_1 - \gamma_A}{\gamma_1 - \gamma_2}k_{fA} \tag{6.71}$$

$$-\beta_1(\bar{\nu}) = \beta_2(\bar{\nu}) = \frac{C_B(\bar{\nu})k_A[A_0^*]k_{fB}}{\gamma_1 - \gamma_2} \tag{6.72}$$

whereas for those species that react according to reaction scheme II,

$$\gamma_1, \gamma_2 = T_1^{-1}, T_2^{-2} = \tfrac{1}{2}[(\gamma_A + \gamma_B] \pm \{(\gamma_A - \gamma_B)^2 + 4k_a k_b[OH^-]\}^{-1/2}] \tag{6.73}$$

$$-\alpha_1(\bar{\nu}) = \alpha_2(\bar{\nu}) = C_A(\bar{\nu})\frac{k_b[B_0^*]k_{fA}}{\gamma_1 - \gamma_2} \tag{6.74}$$

$$\beta_1(\bar{\nu}) = C_B(\bar{\nu})(B_0^*]\frac{\gamma_B - \gamma_2}{\gamma_1 - \gamma_2}k_{fB} \tag{6.75}$$

$$\beta_2(\bar{\nu}) = C_B(\bar{\nu})[B_0^*]\frac{\gamma_1 - \gamma_B}{\gamma_1 - \gamma_2}k_{fB} \tag{6.76}$$

These results are readily understood when Eqs. (6.46) and (6.47) are compared to the differential rate equations which describe the reaction according to scheme II.

$$-\frac{d[B^*]}{dt} = \left(\frac{1}{\tau_0'} + k_b\right)[B^*] - k_a[A^*][OH^-] \tag{6.77}$$

$$-\frac{d[A^*]}{dt} = \left(\frac{1}{\tau_0} + k_a[OH^-]\right)[A^*] - k_b[B^*] \tag{6.77a}$$

If wavelengths can be selected where spectral overlap does not occur such that emission from A* and/or B* can be monitored exclusively, the rate equations in either integrated or derivative forms can be used directly to determine the kinetic parameters of interest, thus allowing the investigator to proceed to the second stage of numerical analysis, the actual data reduction.

Several conceptually different methods exist which can be applied to the problem of extracting values for the rate parameters from the data obtained from the fluorescence decay measurements (44). Three of these methods will now be discussed.

Difference of Lifetimes of Single Exponentials. In this method, advantage is taken of the simplications that result in the differential rate equations (6.46), (6.47), (6.76), and (6.77), when solution conditions are chosen such as to render the observed fluorescence emission monoexponential.

For those species reacting according to scheme I when the pH is adjusted such as to render forward reaction so highly improbable as to be considered nonexistent, Eq. (6.46) simplifies to

$$-\frac{d[A^*]}{dt} = \frac{1}{\tau_0}[A^*] \tag{6.78}$$

At the opposite extreme, when the hydronium concentration is made so small as to render back reaction negligible, then the proton-transfer reaction is irreversible and is adequately described by the equation

$$-\frac{d[A^*]}{dt} = \left(\frac{1}{\tau_0} + k_a\right)[A^*] \tag{6.79}$$

Both Eqs. (6.78) and (6.79) are first-order differential equations having monoexponential solutions with lifetimes of τ_0 and $(1/\tau_0 + k_a)^{-1}$, respectively. If τ_0 is independent of the concentrations of A and H^+, the difference between the reciprocals of these two lifetimes yields the forward rate constant, k_a.

Similarly, for those species reacting according to reaction scheme II, if the solution is made sufficiently basic such that only the conjugate base exists in solution, then Eq. (6.76) reduces to

$$-\frac{d[B^*]}{dt} = \frac{1}{\tau'_0}[B^*] \tag{6.80}$$

whereas if the pH is adjusted such that the reaction becomes irreversible (i.e., when the hydroxide ion concentration becomes negligibly small), Eq. (6.76) takes the form

$$-\frac{d[B^*]}{dt} = \left(\frac{1}{\tau'_0} + k_b\right)[B^*] \tag{6.81}$$

The pseudo-first-order rate constant k_b is again obtained upon subtraction of the reciprocals of the lifetimes characterizing the exponential decay of B* under the solution conditions for which Eqs. (6.80) and (6.81) apply.

Estimations of the rate constants for the back reaction of excited-state proton-transfer reactions can be made using the forward rate constants and the excited-state pK_a^*s values estimated by means of the Förster cycle (see Section 6.2.3) as $k_b = k_a/K^*$.

Instantaneous Derivative Intensity Graphs. A second method of calculating rate constants is based on a rearrangement of the original rate equations. Division of Eq. (6.46) by [A*] and Eq. (6.47) by [B*] yields

$$-\frac{d[A^*]}{dt[A^*]} = \left(\frac{1}{\tau_0} + k_a\right) - k_b\frac{[B^*][H^+]}{[A^*]} \tag{6.82}$$

$$-\frac{d[B^*]}{dt[B^*]} = \left(\frac{1}{\tau'_0} + k_b[H^+]\right) - k_a\frac{[A^*]}{[B^*]} \tag{6.83}$$

Since fluorescence lifetime techniques measure the time-dependent changes in intensity rather than concentration, it is necessary to introduce the proper proportionality factors relating intensity to concentration and therefore to the rates

$$\frac{-dI_A(\bar{\nu}, t)}{dtI_A(\bar{\nu}, t)} = \left(\frac{1}{\tau_0} + k_a\right) - k_b[H^+]\frac{I_B(\bar{\nu}, t)}{I_A(\bar{\nu}, t)}\frac{C_B}{C_A} \tag{6.84}$$

$$\frac{-dI_B(\bar{\nu}, t)}{dtI_B(\bar{\nu}, t)} = \left(\frac{1}{\tau'_0} + k_b[H^+]\right) - k_a\frac{I_A(\bar{\nu}, t)}{I_B(\bar{\nu}, t)}\frac{C_A}{C_B} \tag{6.85}$$

The excited-state rate constants are obtained from plots of the instantaneous intensity derivatives as functions of the corrected intensity ratios. The resulting plots are lines of slope $k_b[H^+]$ and intercept $(1/\tau_0 + k_a)$ in the case of the application of Eq. (6.84) and slope k_a and intercept $(1/\tau_0' + k_b[H^+])$ in the case of the application of Eq. (6.85).

When multiple proton donors and/or acceptors are available for reaction with excited state species A* and B* then the forward and reverse rate constants are composite terms equal to

$$k_a' = \sum_{i=1}^{n} k_{a_i}[\text{acceptor}_i] \tag{6.86a}$$

$$k_b' = \sum_{i=1}^{n} k_{b_i}[\text{donor}_i] \tag{6.86b}$$

The individual rate constants k_{a_i} and k_{b_i} corresponding to the probability of excited-state reaction of A* and B*, respectively, with the ith reactant can be determined by studying the decay of fluorescence emission as a function of the concentration of the ith reactant. For each concentration studied apparent values of k_a, $(k_a + 1/\tau_0)$, k_b, and $(k_b + 1/\tau_0')$ are generated from the derivative plots, Eq. (6.84) being used in this case. A plot of the k_b values so obtained as a function of the concentration of the ith reactant is a line with slope k_{b_i} and intercept $(\sum_{i=1}^{n-1} k_b[\text{donor}_i])$. A plot of the corresponding intercept values $(k_a + 1/\tau_0)$ as a function of the concentration of the ith species is a line with slope k_{a_i} and intercept $(1/\tau_0 + \sum_{i=1}^{n-1} k_a[\text{acceptor}_i])$.

Consider the case of excited-state reaction according to reaction scheme III, in which the buffer species D competes with water for the role of proton acceptor. The derivative intensity equations for this reaction corresponding to Eqs. (6.48) and (6.85) are

$$\frac{dI_A(\nu, t)}{dtI_A(\nu, t)} = \left(k_a + k_c[D] + \frac{1}{\tau_0}\right) - (k_b[H^+] + k_d[DH^+])\frac{I_B C_B}{I_A C_A} \tag{6.87a}$$

$$\frac{-dI_B(\nu, t)}{dtI_B(\nu, t)} = \left(k_b[H^+] + k_d[DH] + \frac{1}{\tau_0}\right) - (k_a + k_c[D])\frac{I_A C_A}{I_B C_B} \tag{6.87b}$$

When the data from the fluorescence decay measurement at a fixed concentration of D, DH^+, and H^+ are substituted into Eq. (6.87*a*), the slope and intercept received from the derivative plot are $-(k_b[H^+] + k_d[DH^+])$ and $(k_a + k_c[D] + 1/\tau_0)$, respectively. By making a series of measurements in which the concentrations of D, DH^+, and H^+ are varied independently and then plotting the slopes and intercepts so obtained against these concentrations, k_a, k_c, k_b, and k_d can be determined as described previously.

The ratio of C_A and C_B required to calculate the correct intensity ratio is determined from the amplitude factors of the equations describing the time course of fluorescence decay for A* and B* when the solution conditions are chosen such as to make A* and B* the only emitting species. The integrated rate equations for these two solutions can be written as

$$[A^*] = [A_0^*]e^{-t/\tau_0} = I_A C_A e^{-t/\tau_0} \tag{6.88}$$

$$[B^*] = [B_0^*]e^{-t/\tau'_0} = I_B C_B e^{-t/\tau'_0} \tag{6.89}$$

The ratio of the number of molecules initially placed in excited states [A*] and [B*] is equal to the ratio of the light intensities absorbed by these two solutions.

$$\frac{[A_0^*]}{[B_0^*]} = \frac{\alpha_A}{\alpha_B} \tag{6.90}$$

The relative light intensity absorbed by the two solutions can be calculated from the spectral output of the excitation source after passage through an interference filter and from the absorbance spectra of the solutions over the range of the transmission of the filter. This can be expressed as

$$\text{relative light intensity absorbed} = \int_{\lambda_1}^{\lambda_2} I(\lambda)\varepsilon(\lambda)\,d\lambda \tag{6.91}$$

where $\varepsilon(\lambda)$ is the sample absorbance and $I(\lambda)$ is the relative photon flux incident on the cuvette after passage through the interference filter with significant transmission only in the spectral region λ_1 to λ_2.

Relative values for I_{A_0} and I_{B_0} can be obtained for these same solutions *normalized to the same light intensity* from an analysis using the method of moments. A calculation of the ratio C_A/C_B can then be obtained from

$$\frac{\alpha_A}{\alpha_B} = \frac{I_{A_0} C_A}{I_{B_0} C_B} \tag{6.92}$$

The instantaneous derivative of intensity $I_A(\bar{\nu}, t)$ at time t can be calculated as demonstrated by Brand by fitting a number of consecutive points to a polynomial expression $f(t)$ and differentiating this expression to provide $df(t)/dt$ for the central point. Succeeding derivatives are obtained by successively shifting the interval of calculation along the decay curve.

Exponential Analysis of Decay Curves. In the exponential analysis of decay curves the decay curves are fit to a sum of exponential terms. The numerical methods used for this purpose include the method of moments, iterative reconvolution, Fourier transform, Laplace transform, and the multiple series method. From the analysis values for the preexponential factors and the time constants are obtained. These values are used to calculate the excited state rate constants. For example, for those species reacting according to reaction scheme I, $\gamma_1 + \gamma_2$ is equal to the sum $1/\tau_0 + 1/\tau_0' + k_a + k_b[H^+]$. Therefore, when $\gamma_1 + \gamma_2$ is plotted as a function of the hydrogen ion concentration k_b is obtained as the slope of the resulting line. k_a is obtained from Eq. (6.72) when the substitution $[A_0^*] = \alpha_1(\bar{\nu}) + \alpha_2(\bar{\nu})$ is made.

Alternatively, if τ_0 can be independently determined, the solutions for k_a, k_b, and τ_0' can be expressed in terms of γ_A, γ_B, γ_1, γ_2, $\alpha_1(\bar{\nu})$, and $\alpha_2(\bar{\nu})$ as

$$k_a = \alpha_1(\bar{\nu})\gamma_A + \alpha_2(\bar{\nu})\gamma_B - \frac{1}{\tau_0} \tag{6.93}$$

$$k_b = \frac{\gamma_1\gamma_2 - [\alpha_1(\bar{\nu})\gamma_1 + \alpha_2(\bar{\nu})\gamma_2](\alpha_1(\bar{\nu})\gamma_2 + \alpha_2(\bar{\nu})\gamma_1[H^+]}{(1/\tau_0) - \alpha_1(\bar{\nu})\gamma_1 - \alpha_2(\bar{\nu})\gamma_2} \tag{6.94}$$

$$\tau_0' = \frac{(1/\tau_0) - \alpha_1(\bar{\nu})\gamma_1 - \alpha_2(\bar{\nu})\gamma_2}{(1 - \alpha_1(\bar{\nu})\gamma_1 + \alpha_2(\bar{\nu})\gamma_2)(1/\tau_0) - \gamma_1\gamma_2} \tag{6.95}$$

These equations simplify in the case of $\tau_0 = \tau_0'$.

$$k_a = \alpha_1(\bar{\nu})(\gamma_1 - \gamma_2) \tag{6.96}$$

$$k_b = \alpha_2(\bar{\nu})(\gamma_1 - \gamma_2) \tag{6.97}$$

$$\frac{1}{\tau_0} = \frac{1}{\tau_0'} = \gamma_2 \tag{6.98}$$

Complications. In the preceding discussion of the methods for determining the excited-state rate parameters from the observed fluorescence decay several assumptions have been made. It has been assumed that emission wavelengths exist at which spectral overlap is absent such that the decay of emission from A* and B* can be exclusively monitored. It has been assumed that only one conjugate species exists in the ground state and that this species is exclusively excited. It has been assumed that dynamic quenching does not occur and that the excitation function $E(t)$ is approximated by a Dirac delta function. When these conditions are not met, the solutions to the differential rate equations (6.46) and (6.47) must be modified and/or corrections must first be applied.

Spectral Overlap. When spectral overlap of the emission of A* and B* occurs, the time-dependent intensity of the observed fluorescence is described by the summation of the separate solutions for $I_A(\bar{\nu}, t)$ and $I_B(\bar{\nu}, t)$.

$$I'(\bar{\nu}, t) = [\alpha_1(\bar{\nu}) + \beta(\bar{\nu})]e^{-\lambda_1 t} + [\Lambda_2(\bar{\nu}) + \beta(\bar{\nu})]e^{-\lambda_2 t} \qquad (6.99)$$

The observed decay curves $I_A(\bar{\nu}, t)$ and $I_B(\bar{\nu}, t)$ can, however, be readily corrected for spectral overlap if the assumption is made that the spectral distribution of A* and B* is pH independent. The required correction procedure is identical to the correction applied in the case of spectral overlap in steady-state measurements as outlined in the review of excited-state acid–base properties by Ireland and Wyatt (12).

The first step in this procedure is reexpressing $I'_A(\bar{\nu}, t)$ and $I'_B(\bar{\nu}, t)$, the observed fluorescence intensities, as

$$I'_A(\bar{\nu}, t) = I_A(\bar{\nu}, t) + k_2 I_B(\bar{\nu}, t) \qquad (6.100)$$

$$I'_B(\bar{\nu}, t) = I_B(\bar{\nu}, t) + k_1 I_A(\bar{\nu}, t) \qquad (6.101)$$

where

$$k_1 = \frac{I_A(\bar{\nu}_B)}{I_A(\bar{\nu}_A)} \quad \text{when A is isolated} \qquad (6.102)$$

$$k_2 = \frac{I_B(\bar{\nu}_A)}{I_B(\bar{\nu}_B)} \quad \text{when B is isolated} \qquad (6.103)$$

That is, k_1 and k_2 are the overlap ratios of A* and B* evaluated with solution conditions chosen such that only A or B exist. k_1 is the ratio of the fluorescence intensity of A* measured at the analytical wavelength for B* to the intensity of fluorescence of A* measured at the analytical wavelength of A*. k_2 is the ratio of fluorescence intensity of B* measured at the analytical wavelength of A* to the intensity of fluorescence of B* measured at the analytical wavelength of B*.

By rearranging Eqs. (6.100) and (6.101), the true time dependence of the fluorescence intensity of A and B can be expressed in terms of $I'_A(\bar{\nu}, t)$, $I'_B(\bar{\nu}, t)$, k_1, and k_2,

$$I_A(\bar{\nu}, t) = \frac{I'_A(\bar{\nu}, t) - k_2 I'_B(\bar{\nu}, t)}{1 - k_1 k_2} \tag{6.104}$$

$$I_B(\bar{\nu}, t) = \frac{I'_B(\bar{\nu}, t) - k_1 I'_A(\bar{\nu}, t)}{1 - k_1 k_2} \tag{6.105}$$

Simultaneous Excitation of A and B. When A and B both exist in significant amounts in the ground state, their absorption spectra completely overlap and no wavelength exists where $\varepsilon_A \gg \varepsilon_B$, then the boundary condition $[A] = [A_0]$, $[B] = 0$ at $t = 0$ is not satisfied and the solutions for the time dependence of I_A and I_B are no longer valid.

Although the general solutions expressed in Eqs. (6.67) and (6.68) still apply, the amplitude expressions for those species reacting according to Scheme I are defined under these conditions as

$$\alpha_1(\bar{\nu}) = \frac{C_A(\bar{\nu})(1/\tau_0)}{\gamma_1 - \gamma_2}(\alpha_A(\gamma_B - \gamma_1) + \alpha_B k_b[H^+]) \tag{6.106}$$

$$\alpha_2(\bar{\nu}) = \frac{C_A(\bar{\nu})(1/\tau_0)}{\gamma_1 - \gamma_2}(\alpha_A(\gamma_B - \gamma_2) + \alpha_B k_b[H^+]) \tag{6.107}$$

$$\beta_1(\bar{\nu}) = \frac{C_B(\bar{\nu})(1/\tau_0)}{\gamma_1 - \gamma_2}(\alpha_A k_a + \alpha_B(\gamma_2 - \gamma_B)) \tag{6.108}$$

$$\beta_2(\bar{\nu}) = \frac{C_B(\bar{\nu})(1/\tau_0)}{\gamma_1 - \gamma_2}(\alpha_A k_a + \alpha_B(\gamma_1 - \gamma_B)) \tag{6.109}$$

where all terms retain their previous definitions.

Nonzero-Excitation Pulse Width. In the measurement of the excited-state rate parameters for those excited-state events that occur on a time scale comparable to the duration of the excitation pulse width or the instrumental response time, the assumption can no longer be made that $E(t)$, the excitation function, is approximated by a delta function. The observed fluorescence response $I'(t)$ under these conditions is the convolution of the overall time function of the excitation source and detector $E(t')$ and the time response function of the fluorescence as would be observed if excitation had been achieved with a delta pulse $I(\bar{\nu}, t - t')$; specifically,

$$I'(\bar{\nu}, t) = \int_0^t E(t')I(\bar{\nu}, t - t')\,dt \tag{6.110}$$

In order, therefore, to obtain $I(\bar{\nu}, t - t')$, $I'(\bar{\nu}, t)$ must be deconvoluted. The principal methods of deconvolution are

1. Iterative reconvolution
2. The method of moments
3. Fourier transform
4. Laplace transform
5. The multiple series method

For a discussion of these techniques and their relative merits the reader is referred to the paper by Ware et al. (45) and the references contained therein.

Dynamic Quenching. The measurement of decay curves as a function of hydrogen ion concentration is required in many methods for determining excited-state rate parameters (see above). Proton quenching of fluorescence, however, is a common photochemical event, and when this or any other form of dynamic quenching occurs, this means of deactivation must be accounted for in the rate equations (46). The reaction scheme corresponding to the general case of dynamic quenching of an excited-state deprotonation reaction is

$$\begin{array}{ccccc} & A^* & \underset{k_b}{\overset{k_a}{\rightleftharpoons}} & B^* + H^+ & \\ (1/\tau_0 + k_{q_{ai}}[Q_i]) & \downarrow & & \downarrow & (1/\tau_0' + k_{q_{bi}}[Q_i]) \\ & A & & B & \end{array}$$

Scheme V

where $1/\tau_0$ and $1/\tau_0'$ are the inverses of the lifetimes of A* and B* in the absence of quenching and excited-state reaction, k_{qa} *and* k_{qb} are the rate constants for photochemical deactivation by quencher Q_i of A* and B*, respectively, and all remaining terms retain their previous definitions.

The rate equations describing reaction according to Scheme V with $Q_i = [H^+]$ are

$$-\frac{d[A^*]}{dt} = \left(\frac{1}{\tau_0} + k_a + k_{qa}[H^+]\right)[A^*] - k_b[B^*][H^+] \qquad (6.111)$$

$$-\frac{d[B^*]}{dt} = \left(\frac{1}{\tau_0'} + k_b[H^+] + k_{qb}[H^+]\right)[B^*] - k_a[A^*] \qquad (6.112)$$

If proton quenching is significant, neither the method of the difference of lifetimes of single exponentials nor the results from instantaneous derivative intensity graphs can be used to determine k_a and k_b unless k_{qa} and k_{qb} are first evaluated, as can be accomplished via Stern–Volmer plots. This limitation is the result of the fact that the application of these methods requires the determination of parameters τ_0 and C_A, respectively, which are evaluated at low pH. For instance, the application of the method of the difference of reciprocal decay times would yield $k_a + k_{qa}[H^+]$ rather than k_a when proton quenching is extant.

Difficulties are also encountered in the application of the method of exponential analysis. Integration of Eqs. (6.112) and (6.113) leads to the general solutions expressed in Eqs. (6.67) and (6.68); however, the parameters γ_A and γ_B are now defined as

$$\gamma_A = \frac{1}{\tau_0} + k_a + k_{qa}[H^+] \tag{6.113}$$

$$\gamma_B = \frac{1}{\tau_0'} + k_b + k_{qb}[H^+] \tag{6.114}$$

At neutral pH where $[H^+]$, $k_{qa}[H^+]$, $k_{qb}[H^+]$, and $k_b[H^+]$ tend toward zero, the decay of fluorescence emission is described by

$$I(\bar{\nu}, t)_A = I_A(\bar{\nu}, t = 0)e^{-\gamma_1 t} \tag{6.115}$$

$$I(\bar{\nu}, t)_B = \frac{I_A(\bar{\nu}, t = 0)k_a^f}{\gamma_1 - \gamma_2}(e^{-\gamma_2 t} - e^{-\gamma_1 t}) \tag{6.116}$$

where $\gamma_1 = 1/\tau_0 + k_a$ and $\gamma_2 = 1/\tau_0'$. If a value of $1/\tau_0$ could be obtained, k_a could be determined. However, as already stated, τ_0 is usually evaluated at low-solution pH where proton quenching is a competitive means of photophysical deactivation, thus requiring application of the Stern–Volmer treatment.

At low solution pH, $\gamma_1\gamma_2 = \gamma_a\gamma_b - k_a k_b[H^+]$, with γ_A and γ_B defined by Eqs. (6.113) and (6.114), respectively. Substitution and rearrangement give

$$\frac{\gamma_1\gamma_2 - \tau_0^{-1}\tau_0'^{-1}}{[H^+]} = \tau_0^{-1}(k_b + k_{qb}) + \tau_0'^{-1}k_{qa} - k_a k_b + k_{qb}(k_a + k_{qb})[H^+] \tag{6.117}$$

The left-hand side of Eq. (6.117) may be plotted against H^+ to give a linear plot of slope $k_{qa}(k_b + k_{qb})$ and intercept $\tau_0^{-1}(k_B + k_{qb}) + \tau_0'^{-1}k_{qa} - k_a k_b$. In addition, at low pH,

$$\gamma_1 + \gamma_2 = \gamma_A + \gamma_B = \frac{1}{\tau_0} + \frac{1}{\tau_0'} + k_a + (k_b + k_{qb} + k_{qa})[H^+] \quad (6.118)$$

The left-hand side of Eq. (6.119) when plotted against $[H^+]$ gives a linear plot of slope $(k_b + k_{qb} + k_{qa})$ and intercept $(1/\tau_0' + 1/\tau_0 + k_A)$.

At low pH the fluorescence decay is monoexponential, being described by

$$[A] = [A^*]e^{-\gamma_1 t} \qquad \text{where } \gamma_1 = \frac{1}{\tau_0} + k_{qa}[H^+] \quad (6.119)$$

When γ_1 is plotted as a function of hydrogen ion concentration the resultant line has a slope equal to k_{qa} and intercept equal to $1/\tau_0$. This is, of course, the Stern–Volmer plot.

If the values obtained for k_a from measurements at very low and neutral pH and for k_{qa} from the Stern–Volmer plot are subsituted into Eqs. (6.117) and (6.118) a value for $(k_a + k_{qb})$ can be obtained; however, only if k_{qb} is negligible can the value for k_b be obtained. Fortunately, this is often the case.

6.2.4.3. *Phase Shift Demodulation Fluorimetry*

Phase shift demodulation fluorimetry is the oldest technique available for the measurement of fluorescence lifetimes. The first phase fluorimeter was constructed by Gaviola (47) in 1926, the detailed mathematical description of its operating principles having been provided by Dushinsky (48) in 1933. Until the advent of single-photon counting fluorimetry, which has largely supplanted phase fluorimetry because of the ease with which the observed decay can be interpreted, phase fluorimetry was the preferred method of lifetime measurement.

The phase shift demodulation method is based on the fact that when a fluorescent sample is excited with light whose intensity is modulated in a sinusoidal manner, the emission is a forced response having the same circular frequency ($\omega = 2\pi\nu$) as the excitation but being phase shifted and demodulated with respect to the excitation because of the time lag between absorption and emission. The quantities measured in a phase fluorimetric measurement are the phase angle (θ) and the relative demodulation factor M, where $M = m_{\text{luminescence}}/m_{\text{excitation}}$ and m is defined as $(F_{max}\text{-}F_{min})/(F_{max} + F_{min})$ (see Fig. 6.6).

For simple monoexponential emission θ and m are related to the radiative lifetime of a single-component sample through.

$$\tau_p = \frac{\tan\theta}{\omega} \quad (6.120)$$

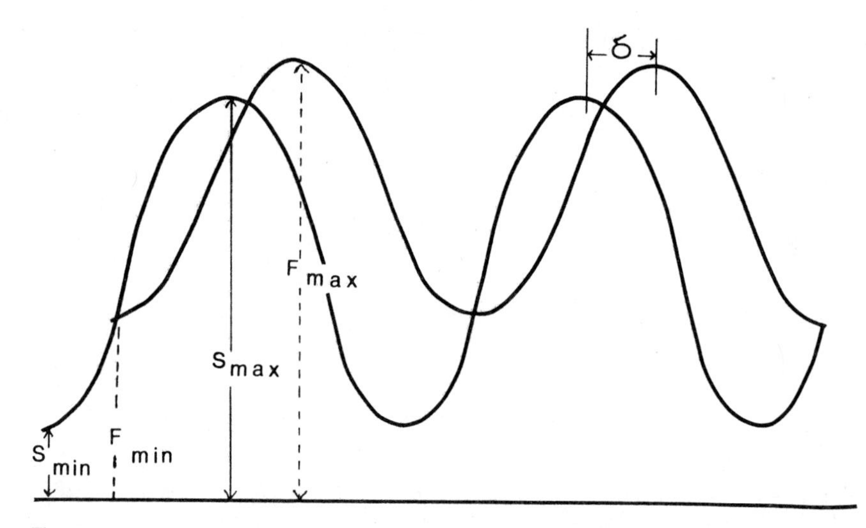

Fig. 6.6. Relative demodulation (M) and phase lag (δ) existing between the fluorescence of a scattering solution and that of a fluorophore when both are exposed to a sinusoidally modulated excitation source. $M = m_{\text{luminescence}}/m_{\text{excitation}}$, where $m_{\text{luminescence}} = (F_{\max} - F_{\min})/(F_{\max} + F_{\min})$. S is the photocurrent due to scattering of the excitation and F is the photocurrent due to the luminescence of the fluorescent sample. (From ref. 42.)

$$\tau_m = \frac{(1/m^2 - 1)^{1/2}}{\omega} \tag{6.121}$$

where $\tau_p = \tau_m = \tau_{\text{radiative}}$.

There are several routes one can take to derive these equations. The following derivation is taken from the paper of Birks et al. (42). Suppose that the response of a fluorescent system to a light pulse represented by a delta function at time $t = 0$ is $i(t)$ and is linear (i.e., proportional to the light intensity). The fluorescence response to a general exciting light function $p(t)$ is given by the superposition integral:

$$f(t) = \int_{\infty}^{t} p(t')i(t - t')\,dt' \tag{6.122}$$

In the phase modulation fluorimeter $p(t)$ is a periodic function that may be expressed as a Fourier series in terms of the fundamental angular frequency ω. Application of Eq. (6.122) to such a series yields a similar Fourier series in which the phases and amplitudes of the various frequency components depend only on those of the same frequency components in $p(t)$ and $i(t)$. For the present purpose $p(t)$ can be regarded as being given by

$$p(t) = c_0(1 + m_p e^{-i\omega t}) \tag{6.123}$$

where only the components actually observed are included, m_p is the degree or amplitude of modulation of the exciting light, and the phase is defined as zero since all measurements are referenced to the exciting light.

For simple monoexponential decay described by the fluorescence response function $i(t) = c_N e^{-\gamma_n t}$, application of Eq. (6.123) yields

$$f(t) = c_N c_0 e^{-\gamma_n t}\left[\int_{-\infty}^{t} e^{\gamma_n t'}\,dt'\, m_p \int_{-\infty}^{t} e^{(\gamma_n t + i\omega)t'}\,dt'\right] \tag{6.124}$$

$$= \frac{c_n c_0}{\gamma_n}\left(1 + m_p \frac{\gamma_n}{\gamma_n + i\omega} e^{i\omega t}\right) \tag{6.125}$$

The phase and amplitude of the modulation relative to the corresponding quantities of $p(t)$ are thus given by

$$M_n = m_n e^{-i\theta_n} = \gamma_n/(\gamma_n + i\omega) \tag{6.126}$$

$$= m_n \cos\theta_n - m_n i \sin\theta_n \tag{6.127}$$

$$m_n \cos\theta_n = \gamma_n^2/(\gamma_n^2 + \omega^2) \tag{6.128}$$

$$m_n i \sin\theta_n = i\gamma_n\omega/(\gamma_n^2 + \omega^2) \tag{6.129}$$

$$m_n = \gamma_n/(\gamma_n^2 + \omega^2)^{1/2} = \cos\theta_n \tag{6.130}$$

from which it follows that

$$\tau_m = \frac{(1/m^2 - 1)^{1/2}}{\omega}, \qquad \tau_p = \frac{\tan\theta_n}{\omega} \tag{6.131}$$

It is to be emphasized that these simple relationships hold only for the case of monoexponential decay. If the equations above are used to calculate the radiative lifetime without verification of the form of the fluorescence decay law, then the significance of the resulting values is questionable. The observation of $\tau_m \neq \tau_p$ indicates that a more complex decay law is operative.

From Eqs. (6.126) to (6.131) it is obvious that if phase fluorimetry is to be used to determine a radiative lifetime, then the frequency of modulation employed must be comparable to the magnitude of the fluorescence lifetime. Classical electrodynamics shows that for emission in the visible and ultraviolet regions of the spectrum that k_f is of the order of 10^8 s^{-1} (or of order

10^{-8} s); therefore, the frequencies to be employed in the excitation must be of the order of $10^8/2\pi$ or in the region of 1–50 MHz. All commercial phase fluorimeters employ Debye–Sears modulators allowing the selection of one from three frequencies, typically 6, 18, or 30 MHz, these values corresponding to the harmonic and the second and fourth overtones of the modulator's quartz crystal.

From the first derivatives of Eqs. (6.120) and (6.121),

$$d\tau_m/dm_n = \frac{-1}{(1/m^2 - 1)^{1/2} m^3 \omega} \tag{6.132}$$

$$d\tau_p/d\theta_n = \sec\theta_n/\omega \tag{6.133}$$

it is seen that errors made in the measurement of a small demodulation value, for instance in the measurement of short radiative lifetimes, lead to large errors in the calculated modulation lifetime. Conversely, the largest errors in τ_p occur when errors are made in the measurement of large phase angles such as those encountered in the measurement of long radiative lifetimes.

Birks et al. (42) in their study of the excited-state kinetics of pyrene eximer formation and dissociation (see Scheme VI) were the first investigators to apply phase fluorimetry to the analysis of fluorescence time functions more complex than that of simple exponential decay.

$$M^* + M \underset{k_1}{\overset{k_1}{\rightleftharpoons}} D^*$$

Scheme VI

D^* = excited-state dimer
M^* = excited-state monomer
M = ground-state monomer

In this reaction, which is representative of reversible excited-state reactions subject to the boundary condition of zero product concentration at time = 0, the dimer, the product of this reaction, is characterized by a fluorescence response function the form of which is identical to the solution of the differential equation describing the time dependence of the emission of the product species of a reversible excited-state reaction after pulsed excitation [see Eqs. (6.67) to (6.72)].

$$i_D(t) = c_D(e^{-\gamma_1 t} - e^{-\gamma_2 t}) \tag{6.134}$$

$$\gamma_1 = \tfrac{1}{2}[(\gamma_M + \gamma_D) + \{(\gamma_D - \gamma_M)^2 + 4k_1k_2[\mathrm{m}]\}^{1/2}] \tag{6.135}$$

$$\gamma_2 = \tfrac{1}{2}[(\gamma_M + \gamma_D) - \{(\gamma_D - \gamma_M)^2 + 4k_1k_2[m]\}^{1/2}] \qquad (6.136)$$

$$\gamma_M = \frac{1}{\tau_m} + k_1[m] \qquad (6.137)$$

$$\gamma_D = \frac{1}{\tau_D} + k_2 \qquad (6.138)$$

where [m] = ground-state monomer concentration.

Application of Eq. (6.122) yields

$$f_D(t) = c_D c_0 \frac{\gamma_2 - \gamma_1}{\gamma_2\gamma_1}\left(1 + mp\frac{\gamma_1\gamma_2}{(\gamma_1 + i\omega)(\gamma_2 + i\omega)}e^{i\omega t}\right) \qquad (6.139)$$

Following the same steps as presented in Eqs. (6.126) to (6.131), the following expressions are reached for the phase and amplitude of the modulation relative to the corresponding quantities of $p(t)$.

$$M = m_D e^{-\theta_D} \qquad (6.140)$$

$$\theta_D = \theta_1 + \theta_2 \qquad (6.141)$$

$$m_D = m_1 m_2 \qquad (6.142)$$

where the quantities on the right-hand side are given by the expressons in Eqs. (6.126) to (6.131) with $n = 1$ and 2. Substitution and rearrangement give

$$\tan\theta_D = \tan(\theta_1 + \theta_2) = \frac{\alpha\omega}{\beta - \omega^2} \qquad (6.143)$$

$$m_D = \frac{\beta}{\sqrt{(\beta - \omega^2)^2 + \alpha^2\omega^2}} \qquad (6.144)$$

$$\alpha = \tau_M^{-1} + \tau_D^{-1} + k_1 + k_2 \qquad (6.145)$$

$$\beta = \tau_M^{-1}\tau_D^{-1} + \tau_D^{-1}k_1 + \tau_M^{-1}k_2 \qquad (6.146)$$

The fluorescence response function for the excited pyrene monomer population is

$$i_m(t) = c_M(e^{-\gamma_1 t} + Ae^{-\gamma_2 t}) \qquad (6.147)$$

where

$$A = (\gamma_m - \gamma_1)/(\gamma_2 - \gamma_m) \tag{6.148}$$

which is similar in form to the solution of the differential equation describing the time dependence of the emission of the reactant species of an excited-state reaction subject to the boundary condition of zero product concentration at time $t = 0$ after pulsed excitation.

Application of Eq. (6.123) yields

$$f_M(t) = c_M c_0 \left[\frac{1}{\gamma_1} + \frac{A}{\gamma_2} + m_p \left(\frac{1}{\gamma_1 + i\omega} + \frac{A}{\gamma_2 + i\omega} \right) e^{i\omega t} \right] \tag{6.149}$$

The phase and amplitude of the modulation relative to the corresponding quantities of $p(t)$ are given by

$$M_M = m_M e^{-i\theta_M} = \frac{m_1/\gamma_1 + Am_2/\gamma_2}{1/\gamma_1 + A/\gamma_2} \tag{6.150}$$

where m_1 and m_2 are given by the expressions in Eqs. (6.126) to (6.131) with $n = 1$ and so that after substitution and rearrangement the following expressions result:

$$m_M = \frac{\beta}{\gamma_D} \frac{\sqrt{\gamma_D^2 + \omega^2}}{(\beta - \omega^2)^2 + \alpha^2 \omega^2} \tag{6.151}$$

and

$$\tan \theta_M = \frac{\omega(\alpha\gamma_D - \beta + \omega^2)}{\alpha\omega^2 + (\beta - \omega^2)\gamma_D} \tag{6.152}$$

As θ and m in Eqs. (6.140) to (6.152) depend on all four kinetic parameters $(k_1, k_2, \tau_M, \tau_D)$, the solutions they yield for k_1 and k_2 are complex.

Weber (49) realized that the evaluation of k_2 could be simplified by taking advantage of the fact that the apparent phase angle of the product of an excited state reaction is, as indicated in Eqs. (6.140) to (6.142), actually the sum of the measured phase angles of the excited-state reactant species and the intrinsic phase angle of the product species (i.e., the phase angle that would be measured if the product species could be and was exclusively excited).

This being the case, Eqs. (6.140) to (6.146) can be reexpressed as

$$\theta_{D0} = \theta_D - \theta_M \tag{6.153}$$

$$\tan \theta_{D0} = \tan(\theta_D - \theta_M) = \frac{\tan \theta_D + \tan \theta_M}{1 - \tan \theta_D \tan \theta_M} \tag{6.154}$$

$$\tan \Delta\theta = \frac{\omega}{\tau_D^{-1} + k_2} \tag{6.155}$$

It is seen that the complexity of the individual phase angles is not present in the phase difference. If θ_D is known or can be determined (see below) and if wavelengths in the emission spectra of the excited-state reactant and product which are uncomplicated by spectral overlap are available so that the apparent phase angles of the excited-state species can be determined accurately, this often being the case at the blue and red extrema of the spectra, k_2 can also be accurately determined.

Advantage can similarly be taken in the determination of k_2 of the multiplicative property of the demodulation factor of the product state. The demodulation of the product state is the product of the demodulation of the reactant state and the instrinsic demodulation of the product state.

$$m_D = m_M m_{D_0} \tag{6.156}$$

If m_D and m_M can be determined, then m_{D_0} can be calculated and related to the reverse rate constant through

$$m_{D_0} = \frac{m_D}{m_M} = \frac{\tau_D + k_2}{\sqrt{(\tau_D + k_2)^2 + \omega^2}} \tag{6.157}$$

Therefore, k_2 can be determined if τ_D is known.

By proper selection of the solvent conditions many reversible reactions can be forced to take on the characteristics of an irreversible reaction. When this is possible the evaluation of k_1 is facilitated. For instance, the excited-state ionization of 2-naphthol-6-sulfonate is reversible. However, within the middle pH region (pH 4–9), due to the low concentration of hydrogen ions at these pH values, the reaction is rendered irreversible and is adequately described by reaction Scheme VII.

$$A^* \xrightarrow{k_a} B^* + H^+$$

Scheme VII

Under these conditions the fluorescence response function for the acid is given by $i(t) = C_A e^{-\gamma t}$, where γ in this case is equal to the sum, $\gamma = \tau_0^{-1} + k_a$. The expressions relating the measured phase and modulations values of the conjugate acid to k_a are

$$\tan \theta_A = \omega(\tau_0^{-1} + k_a)^{-1} \tag{6.158}$$

$$m_A = (\tau_0^{-1} + k_a)/[(\tau_0^{-1} + k_a)^2 + \omega^2]^{1/2} \tag{6.159}$$

If a wavelength is chosen in the emission spectra where emission is exclusively from the conjugate acid so that m_A and or θ_A can be determined and if τ_0 is known or can be independently determined, then k_a can be calculated from either expression in Eqs. (6.158) and (6.159).

When phase fluorimetric measurements are made at wavelengths within the emission spectra of the two-state excited-state process where spectral overlap is significant, the phase angles and demodulation factors are functions of the individual phase angles and demodulation factors of the contributing species [see Eqs. (6.160–6.164)] and therefore cannot be used to determine the rate parameters.

$$\tan \theta(\bar{\nu}) = N(\bar{\nu})/D(\bar{\nu}) \tag{6.160}$$

$$m(\bar{\nu}) = [N(\bar{\nu})^2 + D(\bar{\nu})^2]^{1/2} \tag{6.161}$$

$$N(\bar{\nu}) = m(\bar{\nu}) \sin \theta(\bar{\nu}) = f_A(\bar{\nu}) m_A \sin \theta_A + f_B(\bar{\nu}) m_B \sin \theta_B \tag{6.162}$$

$$D(\bar{\nu}) = m(\bar{\nu}) \cos \theta(\bar{\nu}) = f_A(\bar{\nu}) m_A \cos \theta_A + f_B(\bar{\nu}) m_B \cos \theta_B \tag{6.163}$$

where

$$f_A = 1 - f_B = \frac{I_A(\bar{\nu}, t = 0)}{I_A(\bar{\nu}, t = 0) + I_B(\bar{\nu}, t = 0)} \tag{6.164}$$

For spectra that overlap completely, the only means by which the rate parameters can be evaluated is by a technique introduced by Veselova and co-workers (50,51) known as phase suppression or phase-sensitive spectroscopy. In this technique the phase of the voltage input function (θ_{in}) of a cross-correlation fluorimeter is chosen so as to be in quadrature (i.e., 90° out of phase), with the phase of one of the two contributing species. This operation, when the phase angle is properly chosen, results in complete suppression of the output voltage of the species with which the input voltage of the detector is in quadrature, so that the output voltage is proportional to the unsuppressed component. This result can be demonstrated mathematically as follows.

In the cross-correlation technique as modified by Spencer and Weber (52) the photomultiplier gain is varied by modulating the voltage on one or more of the dynodes, the phase of this voltage being varied linearly with time so that $\theta_{mod} = 2\pi\,\Delta\nu\tau = \Delta\omega\tau$. The response of the PMT thus varies according to

$$R(t) = a + b\sin(\omega t + \theta_{mod}) \tag{6.165}$$

where a and b are constants. The resulting photocurrent $[C(t)]$ can be determined by the integrated intensity over one complete cycle.

$$C(t) = \int_{t}^{t+2\pi/\omega} i(t')R(t')\,dt' \tag{6.166}$$

For a monoexponential input function, Eq. (6.166) gives

$$C(t) = aA + \tfrac{1}{2}bB\cos\theta_{in}\cos(\theta_{mod} - \theta_{in}) \tag{6.167}$$

where A and B are constants originating from the input function. When the input function is of the form of Eqs. (6.126) to (6.131) (i.e., double exponential), then Eq. (6.166) gives

$$C(t) = C_1(t) + C_2(t) \tag{6.168}$$

It follows from Eqs. (6.167) and (6.168) that when $\theta_{mod} = 90^{0} + \theta_{in1}$ then

$$C(t) = a_2A_2 + \tfrac{1}{2}B_2\cos\theta_{in2}\sin(\theta_{in1} - \theta_{in2}) \tag{6.169}$$

and when $\theta_{mod} = 90^{0} + \theta_{in2}$,

$$C(t) = a_1A_1 + \tfrac{1}{2}B_1\cos\theta_{in1}\sin(\theta_{in2} - \theta_{in1}) \tag{6.170}$$

From these equations it is seen that when θ_{mod} is in quadrature with θ_{in}, the resulting signal current is proportional to the phase and demodulation of the emission of component 2, and vice versa, when the emission of component 2 is suppressed, the signal current is proportional to the phase and demodulation of component 1.

If the phase angle of one component is known, the phase angle and demodulation factor of the second is easily calculated. The values obtained can then be related to the corresponding rate parameters, as discussed previously. In practice it is often the case that neither phase angle is known and it is necessary by trial and error to vary the detector modulating voltage and to record the phase fluorimetric spectra at each setting. This process is continued until a setting is found where the phase spectra is equivalent or proportional to the steady-state spectra of either one of the components, equivalence of the spectra indicating that a setting of θ mod $= 90 + \theta_i$ has been found.

It has been assumed in the previous discussion that the composition of the fluorescent sample and the form of the decay law are known a priori. This is not always the case, and when such situations occur it is necessary to be able to distinguish between the possible emissive events that could give rise to the measured phase angles and demodulation factors. Fortunately, criteria exist that can be applied toward this purpose.

The oldest and most useful of these criteria, introduced by Galanin (1955) (50), makes use of the ratio $m(\bar{\nu})/\cos\theta(\bar{\nu})$. In the case of emission from a sample exhibiting monoexponential decay, $m(\bar{\nu})/\cos\theta(\bar{\nu})$ is equal to unity, this value being wavelength independent. For a heterogeneous sample the components of which exhibit solely monoexponential decay the value of $m(\bar{\nu})/\cos\theta(\bar{\nu})$ is less than unity. Consider a two-component system from this class. In this case

$$m\bar{\nu})/\cos\theta(\bar{\nu}) = (m_1 + m_2)/\cos(\theta_1 + \theta_2) \tag{6.171}$$

$$= \frac{\cos\theta_1 + \cos\theta_2}{\cos\theta_1\cos\theta_2 - \sin\theta_1\sin\theta_2} < 1 \tag{6.172}$$

For an irreversible excited-state reaction the value of $m(\bar{\nu})/\cos\theta(\bar{\nu})$ will be wavelength dependent. When a wavelength for phase analysis is chosen within the emission spectra of the excited-state reactant which is free from spectral overlap, then $m(\bar{\nu})/\cos\theta(\bar{\nu})$ will equal unity, as it will be recalled that the decay of the reactant species is monoexponential. If, on the other hand, a wavelength of phase analysis is chosen within the emission spectra of the excited-state reactant, then $m(\bar{\nu})/\cos\theta(\bar{\nu})$ will exceed unity, again barring significant spectral overlap. Mathematically, this is easily demonstrated. Substitution of the values of m and θ from Eqs. (6.141) to (6.143) yields

$$\frac{m(\bar{\nu})}{\cos\theta(\bar{\nu})} = \frac{m_1 m_2}{\cos(\theta_1 + \theta_2)} = \frac{\cos\theta_1 + \cos\theta_2}{\cos\theta_1\cos\theta_2 - \sin\theta_1\sin\theta_2} \tag{6.173}$$

$$= \frac{1}{1 - \tan\theta_1\tan\theta_2} \tag{6.174}$$

$$= \frac{1}{1 - \omega^2 \tau_1 \tau_2} > 1 \tag{6.175}$$

The observance of $m(\bar{\nu})/\cos\theta(\bar{\nu})$ greater than unity is sufficient but not necessary proof that the emission is from the product of an excited-state reaction. Spectral overlap can obscur this observance by imparting heterogeneous character [see Eqs. (6.162) to (6.166). The typical spectrum of a sample that undergoes an irreversible excited-state reaction with minimal spectral overlap at the extrema is then characterized by a value of $m(\bar{\nu})/\cos\theta(\bar{\nu}) = 1$ at the blue edge, $m(\bar{\nu})/\cos\theta(\bar{\nu}) < 1$ within the midregion of the spectra where spectral overlap is significant, and $m(\bar{\nu})/\cos\theta(\bar{\nu}) > 1$ at the red edge.

In the case of a reversible excited-state reaction as the emission from the reactant is no longer monoexponential, the value of $m(\bar{\nu})/\cos\theta(\bar{\nu})$ obtained from the reactant emission is no longer equal to unity. In addition, the value for $m(\bar{\nu})/\cos\theta(\bar{\nu})$ for the product emission will be reduced in proportion to the magnitude of the reverse rate constant and so for large values of k_R, $m(\bar{\nu})/\cos\theta(\bar{\nu})$ may not exceed unity.

The frequency dependence of the measure phase angles and demodulation factors may also be used as an indicator of the nature of the emission process. If the sample consists of fluorophores exhibiting monoexponential decay, regardless of heterogeneity, as the value of τ_p and τ_m for each component is given by $\tau_p = \tan\theta/\omega$, $\tau_m = [(1/m^2) - 1]^{1/2}/\omega$, then with increasing frequency the values of τ_p and τ_m decrease.

Now consider the apparent phase lifetime τ_p as calculated from the measured phase θ of the product of an excited-state reaction: for instance, excited-state dimer formation and dissociation. It will be recalled that $\theta_D = \theta_M + \theta_{D0}$ and $m_D = m_M m_{D0}$, so that τ_p and τ_m are given by the expressions

$$\tan\theta_D = \omega\tau_p = \tan(\theta_M + \theta_{D_0}) \tag{6.176}$$

$$= \frac{\tau_M + \tau_{D0}}{1 - \omega^2 \tau_M \tau_{D0}} \tag{6.177}$$

$$\tau_m = \frac{(1/m^2 - 1)^{-1/2}}{\omega} = \frac{m_M^2 m_{D0}^2}{\tilde{\omega}(1 - m_M^2 m_{D0}^2)} = (\tau_M^2 + \tau_{D_0}^2 + \omega^2 \tau_M^2 \tau_{D0}^2)^{1/2} \tag{6.178}$$

Increasing ω, therefore, yields an increase in τ_m and τ_p, although in practice τ_m is less sensitive to changes in frequency than is τ_p.

A more detailed discussion of the detection of reversible and irreversible reactions via phase fluorimetry complete with model calculations is found in the papers of Lakowicz and Balter (53–55) and Lakowicz (56).

6.2.4.4. *Comparison of Steady-State, Pulse, and Phase Fluorimetry in the Quantitation of Excited-State Rate Parameters*

Steady-State Fluorimetery

Advantage:

- The cost of a steady-state fluorimeter is much smaller than the cost of a single-photon counter or a phase fluorimeter.

Disadvantages

- The determination of excited-state rate parameters by steady-state fluorimetry requires the performance of tedious titrations requiring hours of operator attention.
- Steady-state measurements do not directly yield excited-state rate parameters. $k_a\tau_0$ and $k_b\tau_0'$, not k_a and k_b, are determined from the data generated in a steady-state measurement. Therefore, if the radiative lifetimes of the excited-state reactant and product are not known, one must employ either phase or pulse fluorimetry to determine these rate parameters.

Time Resolved Spectroscopy

Advantages

- One obtains directly a representation of the decay curve, although a distorted one if the lamp and fluorophore have similar decay times.
- The sensitivity is high and when a single-photon counter is employed, signal averaging over a long period (24 hours) is practical.
- The instrumentation is commercially available, modular, compact, extraordinarily stable and reliable and lends itself to computer data processing.

Disadvantage

- For excited-state processes that occur within a time frame comparable to the excitation pulse width, typically several nanoseconds for generally available light sources such as the deuterium and nitrogen flash lamps, numerical deconvolution is an absolute necessity. The available deconvolution procedures permit measurement of 1-ns lifetimes with pulse widths of 2 ns if the shape of the lamp pulse remains constant for a large number of pulses. Although quantitation of subpicosecond processes has been made possible by the introduction of picosecond lasers and streak cameras, this sophisticated equipment is expensive and therefore not widely accessible.

Phase Fluorimetry

Advantages

- Any light source can be used which possesses the required intensity.
- Sensitivity is high with intense sources.
- Measurements of excited-state events occurring on the nanosecond time scale are easily made without the complication of numerical deconvolution necessary in pulse fluorimetry.

Disadvantages

- The quantities measured in phase fluorimetry are related only indirectly to the rate parameters of the monitored emission. Confirmation of the rate law is, therefore, required prior to kinetic analysis.
- A lower limit of approximately 1 nsec presently exists for commercial phase fluorimeters. The time resolution of commercial instruments is limited by the electronic phase shift detector employed and the upper frequency limit of the light modulator. Although the use of lasers, optical phase shift detectors, spectrum analyzers, and so on, has extended the use of phase fluorimetry to the quantitation of picosecond events, these instruments are not generally available.
- A knowledge of RF circuitry and techniques is desirable.

REFERENCES

1. Th. Förster, *Z. Elektrochem.*, **54**, 42 (1950).
2. K. Weber, *Z. Phys. Chem.*, **B15**, 18 (1931).
3. A. Weller, *Z. Elektrochem.*, **56**, 662 (1952).
4. M. Eigen, *Discuss. Faraday Soc.*, **17**, 194 (1954).
5. M. Eigen and L. DeMaeyer, *Z. Elektrochem.*, **59**, 986 (1955).
6. M. Eigen and J. Schonen, *Z. Elektrochem.*, **59**, 483 (1955).
7. M. Eigen and L. DeMaeyer, *Proc. R. Soc. (London) A*, **247**, 505 (1958).
8. M. Eigen and K. Kustin, *J. Am. Chem. Soc.*, **82**, 5952 (1960).
9. A. Weller, *Prog. React. Kinet.*, **1**, 189 (1961).
10. E. Vander Donckt, *Prog. React. Kinet.*, **5**, 274 (1970).
11. S. G. Schulman in E. L. Wehry, Ed., *Modern Fluorescence Spectroscopy*, Vol. 2, Plenum Press, New York, 1976, Chap. 6.
12. J. F. Ireland and P. A. H. Wyatt, *Adv. Phys. Org. Chem.*, **12**, 131 (1976).
13. W. Klöpffer, *Adv. Photochem.*, **10**, 311 (1977).

14. G. J. Woolfe, Ph.D. thesis, University of Melbourne (Australia), 1981.
15. H. Shizuka and K. Tsutsumi, *J. Photochem.*, **9**, 334 (1978).
16. H. Shizuka, K. Tsutsumi, H. Takeuchi, and I. Tanaka, *Chem. Phys. Lett.*, **62**, 408 (1979).
17. H. Shizuka, K. Tsutsumi, H. Takeuchi, and I. Tanaka, *Chem. Phys.*, **59**, 183 (1981).
18. K. Tsutsumi, S. Sekiguchi, and H. Shizuka, *J. Chem. Soc. Faraday Trans. I.*, **78**, 1087 (1982).
19. J. R. Lakowicz and A. Balter, *Chem. Phys. Lett.*, **92**, 117 (1982).
20. S. G. Schulman, W. R. Vincent, and W. J. M. Underberg, *J. Phys. Chem.*, **85**, 4068 (1981).
21. J. N. Brønsted, *Z. Phys. Chem.*, **102**, 169 (1922).
22. J. N. Brønsted, *Z. Phys. Chem.*, **115**, 337 (1925).
23. J. N. Brøted, *Chem. Rev.*, **5**, 231 (1928).
24. A. Weller, *Z. Phys. Chem. (Frankfurt/Main)*, **15**, 438 (1958).
25. A. Weller, *Z. Phys. Chem. (Frankfurt/Main)*, **17**, 224 (1958).
26. S. G. Schulman and B. S. Vogt. *J. Phys. Chem.*, **85**, 2074 (1981).
27. B. S. Vogt and S. G. Schulman, *Chem. Phys. Lett.*, **95**, 159 (1983).
28. B. S. Vogt and S. G. Schulman, *Chem. Phys. Lett.*, **99**, 157 (1983).
29. N. Lasser and J. Feitelson, *J. Phys. Chem.*, **77**, 1011 (1973).
30. S. G. Schulman, B. S. Vogt, and M. W. Lovell, *Chem. Phys. Lett.*, **75**, 224 (1980).
31. B. S. Vogt and S. G. Schulman, *Chem. Phys. Lett.*, **97**, 450 (1983).
32. S. G. Schulman, L. S. Rosenberg, and W. R. Vincent, *J. Am. Chem. Soc.*, **101**, 139 (1979).
33. S. G. Schulman and R. N. Kelly, unpublished results.
34. A. Weller, *Z. Elektrochem.*, 61, 956 (1957).
35. H. Boaz and G. K. Rollefson, *J. Am. Chem. Soc.*, **72**, 3435 (1950).
36. A. C. Capomacchia and S. G. Schulman, *Anal. Chim. Acta*, **59**, 471 (1972).
37. A. Weller, *Z. Phys. Chem.* (Frankfurt/Main), **18**, 163 (1958).
38. H. Leonhardt, Ph.D. thesis, University of Stuttgart, 1958.
39. M. Gurr, Ph.D. thesis, University of Stuttgart, 1959.
40. A. Smoluchowski, *Z. Phys. Chem.*, **4**, 129 (1917).
41. P. Debye, *Trans. Electrochem. Soc.*, **82**, 265 (1942).
42. J. B. Birks, D. J. Dyson, and I. H. Munro, *Proc. R. Soc.*, **275A**, 575 (1963).
43. W. R. Laws and L. Brand, *J. Phys. Chem.*, **83**, 795 (1979).
44. M. R. Loken, J. W. Hayes, J. Gohlke, and L. Brand, *Biochemistry*, **11**, 4779 (1972).
45. W. Ware, L. J. Doemeny, and T. L. Nemzek, *J. Phys. Chem.*, **77**, 2038 (1973).
46. C. M. Harris and B. K. Selinger, *J. Phys. Chem.*, **84**, 891 (1980).
47. F. Gaviola, *Z. Physik*, **42**, 852 (1927).
48. F. Dushinsky, *Z. Physik*, **81**, 7 (1933).
49. G. Weber, in J. B. Birks, Ed., *The Excited States of Biological Molecules*, Wiley, New York, 1976, pp. 363–374.
50. I. V. Veselova, L. A. Limareva, A. S. Cherkasov, and V. I. Shirokov, *Opt. Spectrosc.*, **18**, 202 (1965).

51. I. V. Veselova and V. I. Shirokov, *Izv. Akad. Nauk S.S.S.R. Ser. Fiz.*, **36**, 1024 (1972).
52. R. D. Spencer and G. Weber, *J. Chem. Phys.*, **52**, 1654 (1970).
53. J. R. Lakowicz and A. Balter, *Biophys. Chem.*, **16**, 99 (1982).
54. J. R. Lakowicz and A. Balter, *Biophys. Chem.*, **16**, 117 (1982).
55. J. R. Lakowicz and A. Balter, *Photochem. Photobiol.*, **36**, 125 (1982).
56. J. R. Lakowicz, *Principles of Fluorescence Spectroscopy*, Plenum Press, New York, 1983, Chaps. 3 and 12.

INDEX